Functional Monomers and Polymers

Functional Monomers and Polymers

Procedures, Synthesis, Applications

edited by

Kiichi Takemoto
Yoshiaki Inaki

Faculty of Engineering
Osaka University
Osaka, Japan

Raphael M. Ottenbrite

College of Humanities and Sciences
Virginia Commonwealth University
Richmond, Virginia

Marcel Dekker, Inc. New York and Basel

Library of Congress Cataloging-in-Publication Data

Functional monomers and polymers.

Includes index.
1. Polymers and polymerization. 2. Monomers.
I. Takemoto, Kiichi. II. Inaki, Yoshiaki.
III. Ottenbrite, Raphael M.
QD381.F86 1987 668.9 87-13644
ISBN 0-8247-7373-X

MARCEL DEKKER, INC.
270 Madison Avenue, New York, New York 10016

Current printing (last digit):
10 9 8 7 6 5 4 3 2 1

PRINTED IN THE UNITED STATES OF AMERICA

Preface

It is an appropriate time to publish a book based on the recent discoveries of functional monomers and polymers. For years polymers were known to function basically as materials in the form of films, fibers, and plastic components. However, with the advent of high technology, new needs and demands are being made of all available materials. Polymers have responded in an extraordinary fashion to meet this challenge. Through innovative thinking and extensive research, a myriad of new functions have been created from new monomers, polymeric materials, and processing techniques.

The development of polymers for such applications as optical lenses, conducting and superconducting media, photoresistant and photosensitive components, synthetic reagents and catalysts, biological prosthetic devices, pseudoenzymes, drugs and controlled delivery systems has been phenomenal. This book was written to provide scientists with a basic overview of some of the unique polymer functions that are being achieved as well as the basic procedures for the preparation of these materials.

Functional Monomers and Polymers focuses on the chemistry and technology involved in the functionalization of monomers and the preparation and processing of polymers to serve specific material needs. For example, in Chapter 2, Professor Kondo describes the synthesis of polynucleotides and polysaccharides based on the Merrifield polypeptide synthesis using a polystyrene support. Other synthetic applications of polymeric reagents utilizing one-step chemical processes are also described. Polymeric catalysts and metalloenzyme models and their functions are described in Chapter 3. In Chapter 4,

Professors Takemoto and Inaki summarize the synthesis and characterization of several nucleic acid analogs, focusing on the purine and pyrimide families and their applications to biological systems.

Photosensitive and photoconductive polymers are covered in Chapters 5 and 6, respectively, with emphasis on their preparation and application. The description of analytical applications of crown ether polymers in Chapter 7 is a practical example of polymer functionalization for the selective separation of alkaline cations.

The last three chapters deal with biological applications of polymers. Chapter 8 discusses the utilization of polyanionic polymers as antitumor agents, as well as the use of polymeric materials for artificial human organs and prosthetic devices. The immobilization enzymes for diagnostic procedures and catalysts for low energy chemical processes are explained in Chapters 9 and 10, respectively.

The overall purpose of this book is to provide a comprehensive background to the field of functional polymers as well as some background to basic experimental procedures for practical applications.

Raphael M. Ottenbrite

Contributors

Nobumasa Hojo, Ph.D. President, Shinshu University, Nagano, Japan

Yoshiaki Inaki, Ph.D. Assistant Professor, Department of Applied Fine Chemistry, Faculty of Engineering, Osaka University, Osaka, Japan

Masahiro Irie, Ph.D. Associate Professor, The Institute of Scientific and Industrial Research, Osaka University, Osaka, Japan

Keiichi Kimura, Ph.D. Assistant Professor, Department of Applied Chemistry, Faculty of Engineering, Osaka University, Osaka, Japan

Koichi Kondo, Ph.D. Assistant Professor, Department of Applied Fine Chemistry, Faculty of Engineering, Osaka University, Osaka, Japan

Mikiji Miyata, Ph.D. Department of Applied Fine Chemistry, Faculty of Engineering, Osaka University, Osaka, Japan

Raphael M. Ottenbrite, Ph.D. Professor, Department of Chemistry, College of Humanities and Sciences, Virginia Commonwealth University, Richmond, Virginia

Hirofusa Shirai, Ph.D. Professor, Department of Functional Polymer Science, Faculty of Textile Science and Technology, Shinshu University, Nagano, Japan

Yasuhiko Shirota, Ph.D. Professor, Department of Applied Chemistry, Faculty of Engineering, Osaka University, Osaka, Japan

Toshiyuki Shono, Ph.D. Professor, Department of Applied Chemistry, Faculty of Engineering, Osaka University, Osaka, Japan

Kiichi Takemoto, Ph.D. Professor, Department of Applied Fine Chemistry, Faculty of Engineering, Osaka University, Osaka, Japan

Contents

Functional Monomers and Polymers

1

Introduction

Kiichi Takemoto

Osaka University, Osaka, Japan

The chemistry of the functional monomers and polymers has received considerable attention in recent years, and a variety of such polymers and materials have been produced which now find numerous applications in many fields of modern science and technology. Polymeric materials having high functionalities can be divided into three main categories: (a) functional polymeric materials in a broad sense, (b) functional polymers in a narrow sense, and (c) functional polymeric composites.

The first category, functional polymeric materials in a broad sense, includes a number of engineering plastic materials. These polymers exhibit more particular and more excellent properties if used after processing in microporous form. Microporous polymeric materials as well as microcapsules have been of interest in a variety of industrial fields, including not only the general chemical industry, but also the pharmaceutical, biomedical, and electronics industries. The microporous fluorine-containing resin is important particularly as a safe and durable antithrombogenic biomaterial. The microporous membrane made of poly(vinyl chloride) finds application as a separator or filter in electric cells. Different types of microporous poly(vinyl alcohol) are useful as highly hydrophilic materials not only for the feedwater industry, but also in the cosmetic and hygienic fields.

Recent development in the preparation of microporous polypropylene and polycarbonate resins is also noteworthy. The science and technology of the preparation of microcapsules from different natural and synthetic polymeric materials has made rapid progress. They have also been used in various fields by taking advantage of their ability to solidify liquids; isolate reactive compounds; remove color, odor, and toxicity; and regulate and control the release of the compounds included. Fixation technology of enzymes and the development of polymeric drugs are other examples. In connection with these, highly water- and oil-absorbing resins are of interest and have been developed rapidly in recent years by using grafting and cross-linking techniques of hydrophilic polymers. Knowledge of the inclusion phenomena is also very important.

Transparent polymeric materials with optical functionality have made striking progress. Characterized by their processibility and light weight, a number of resins made of polystyrene, polyacrylates and polymethacrylates, and polycarbonates are now used as plastic lenses. Those that are biocompatible, such as poly(hydroxyethyl methacrylate), serve as material for contact lenses. Plastic optical fibers are also widely used as substitutes for glass and quartz devices in various fields of technology, especially the biomedical and communication sciences.

The so-called functional polymers cover the functional polymeric materials in a narrow sense. This is a very exciting subject which has been developed extensively in the past 10 years.

Some of the most attractive materials in this field are the photosensitive polymers. By using the phenomenon that specially designed polymers undergo a cross-linking reaction to become insoluble, a variety of both negative and positive types of photoresists have been produced. They were first used in the printing and paint and color industries, but are now of great interest also in the electronics and computer industries. The technology of exploiting deep ultraviolet resists having reversible functionality will be one of the most important subjects in the near future. Other subjects of interest in this field are the electron or x-ray-sensitive resists, which are under development, as well as the design of more functional photo-memorizing materials.

Starting with the ion-exchange resins and membranes, the development of functional polymer chemistry and technology has made remarkable progress. Studies on the preparation and design of a number of enzyme and nucleic acid models have advanced dramatically, and the development of the technology of the use of these polymers is now going on. In connection with biomedical polymers, the chemistry of the polymeric drugs is under continuous advancement, and the most effective anticarcinogenic reagents are targeted by the design of specific functional polymers.

The functional polymeric composites are also very attractive. By incorporating functional additives in the usual engineering plastics, a variety of polymeric composites can be prepared. Electron-conducting materials, particularly transparent electron-conducting films and metal- or carbon black-incorporated elastomers, are utilized as new materials. Readily processible, less expensive plastic magnets have a number of advantages over the usual magnetic materials. The technology of metallizing plastic surfaces has recently made considerable progress, and the many metallized plastics that have been developed have found wide application.

This book focuses on the chemistry and technology of the functional monomers and polymers in terms of new aspects with regard to procedures, synthesis, and applicability, particularly with regard to functional polymeric materials. Chapter 2 offers a review of the polymeric reagents with emphasis on the synthetic chemistry through the use of a variety of synthetic macromolecules. In Chapter 3 the polymeric catalysts, particularly metalloenzyme models and their functions, are presented. In Chapter 4, which contains a comprehensive review of nucleic acid analogs, various novel methods of preparing a series of derivatives and their functionalization are summarized. Chapters 5 and 6 deal with new problems regarding the photosensitive and photoresponsive polymers and the electroconducting polymers, respectively. We can expect new developments in these materials in the near future. Functional polymers in analytical chemistry are also very important. Chapter 7 contains a practical review, focused especially on the crown ether-containing polymers. Chapter 8 is concerned with new aspects of both polymeric drugs and biomedical polymers. In this chapter we are shown how many types of human organs could be replaced by well-designed biomaterials. Chapter 9 contains a review of enzyme-like polymers. In connection with the nucleic acid analogs, the chemistry and technology of the immobilization of enzymes are also shown in Chapter 10.

The chapters of this book thus summarize up-to-date work in selected areas. Our aim is to provide a comprehensive background on this wide and promising field.

REFERENCES

1. R. H. Yocum and E. B. Nyquist, Eds., Functional Monomers, Marcel Dekker, Inc., New York (in two volumes) (1973 and 1974).
2. K. Takemoto, Functional Polymers, Asakura-Publishing Co. Ltd., Tokyo (1974).

3. M. Okawara and K. Takemoto, Eds., Functional Polymer Series (in 9 volumes), Kodansha Scientific Ltd., Tokyo (1976-1978).
4. K. Takemoto and I. Tabushi, Biomedical Polymeric Materials, Kodansha Scientific Ltd., Tokyo (1978).
5. K. Takemoto, Molecular Aggregation of Polymers—The Shape of the Molecules, Aggregation and Their Functionalization, Kodansha Scientific Ltd., Tokyo (1981).
6. J. L. Atwood, J. E. D. Davies, and D. D. MacNicol, Inclusion Compounds, Academic Press, Inc., Orlando, Fla. (in three volumes) (1984-1985).

2

Polymeric Reagents for Organic Synthesis

Koichi Kondo

Osaka University, Osaka, Japan

2.1 INTRODUCTION

The first success in the use of polymers was made by Merrifield [1] in polypeptide synthesis using the cross-linked chloromethylated polystyrene support. Following this work, similar techniques have been applied to the syntheses of polynucleotides [2] and polysaccharides [3]. Besides the preparation of biopolymer analogs based on polymer supports, much attention recently has focused on the use of polymers as chemical reagents.

This chapter describes synthetic applications of polymeric reagents associated with one-step processes. The advantage of polymeric reagents over their monomeric counterparts can be summarized as follows:

1. *Ease of processing*. The desired product can be separated from the spent polymer by simple filtration.

2. *Regeneration and reuse of the spent polymer.* The recovered spent polymer can be regenerated by a simple workup and reused for the subsequent reaction.
3. *Ease of handling.* The odor and toxicity associated with several monomeric reagents can be reduced by binding them to polymers.
4. *Stereospecificity effects based on polymer backbone influences.* Regioselective and/or regiospecific features can be expected for substrates adsorbed on the polymers which would restrict a chemical transformation.

These and other uses of polymeric reagents has been discussed in detail in recent reviews [4].

2.2 POLYMERIC SUPPORT MATERIALS

Support materials are classified into organic and inorganic polymers. Most of the work has been done on cross-linked polystyrene resins, which are mechanically stable, easily functionalized, and commercially available in well-defined form. Other promising inorganic supports are based primarily on inorganic oxides such as alumina and silica gel.

2.2.1 Cross-Linked Polystyrene

Cross-linked polystyrene exhibits considerable differences in physical properties based on the cross-linking density. The less cross-linked polystyrenes can be swollen in appropriate organic solvents, whereas the more highly cross-linked polystyrenes (macroreticular or macroporous resins) are rigid and allow diffusion control in the resin pore. Functionalization of the aromatic ring covalent for binding of monomeric reagents can be accomplished by chloromethylation [5], halogenation [6], and lithiation [6]. For example, sulfonates and quaternary ammonium groups allow for ionic binding for use as ion-exchange resins via exchange of the counter ions.

2.2.2 Inorganic Support Materials

A variety of reagents can be adsorbed on or intercalated in inorganic supports such as silica gel, alumina, carbon, graphite, and montmorillonite. The physical properties of the support material itself often reflect the reaction path and the tendency to impart regioselective and/or regiospecific control since the adsorption of organic substrates is varied on different support materials. Reagents supported on inorganic material appear more promising than polystyrene-bound reagents. Subsequently, reagents based on alumina, silica gel, and graphite have become commercially available. The major developments in this field through early 1979 have been discussed by McKillop and Young [7].

2.3 POLYMER-BOUND REAGENTS FOR ORGANIC SYNTHESIS

2.3.1 Polymeric Reducing Reagents

2.3.1.a Polymer-bound tin hydride

Macroreticular polystyrene (Amberlite XE-305)-bound organic tin dihydride (1) can be used for the reduction of aldehydes and ketones to primary and secondary alcohols [8].

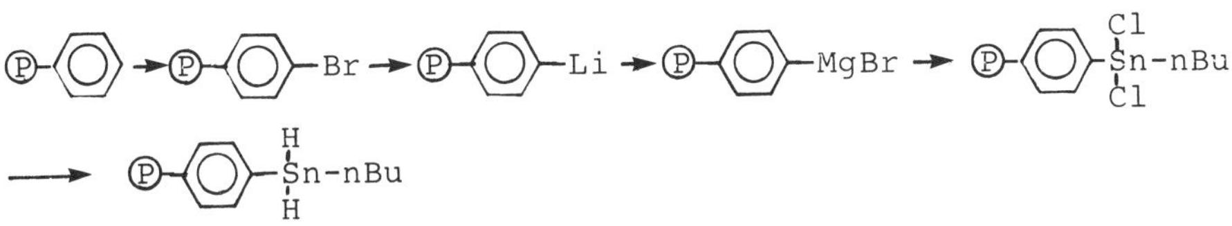

(1)

The selective reduction of terephthaldehyde by 1 gave *p*-hydroxymethylbenzaldehyde in 86% yield. The monoalcohol formation is attributed to the restricted accessibility of the aldehyde group para to the aldehyde group that is bound to the polymer as the tin alkoxide. This reagent also exhibited selectivity in reducing α-bromoacetophenone to acetophenone. The polymeric tin hydride, however, tends to deteriorate slowly on standing and the workup should be carried out in an inert atmosphere. The activity was decreased by 30% for ketone and aldehyde reductions, and 60% for alkylhalide reduction after one generation.

Tributyltin hydride on silica gel reduced aldehydes and ketones to give alcohols in high yields without changing other functional groups, such as sulfoxides, nitro groups, esters, arylnitriles, and alkyl, aryl chlorides. Mixtures of aldehydes and ketones produced the selective reduction of aldehydes. For example, *n*-octanal mixed with cyclohexanone was converted predominately to *n*-octylalcohol in 99% yield [9].

2.3.1.b Polymer-bound borohydride

Sodium borohydride supported on neutral alumina is a quite useful reagent for the reduction of aldehydes, ketones, and acid chlorides under mild conditions [10].

$$PhCHO \longrightarrow PhCH_2OH \ (85\%)$$

$$n\text{-}C_7H_{15}COCl \xrightarrow{\text{Alumina/NaBH}_4} n\text{-}C_7H_{15}CH_2OH \ (90\%)$$

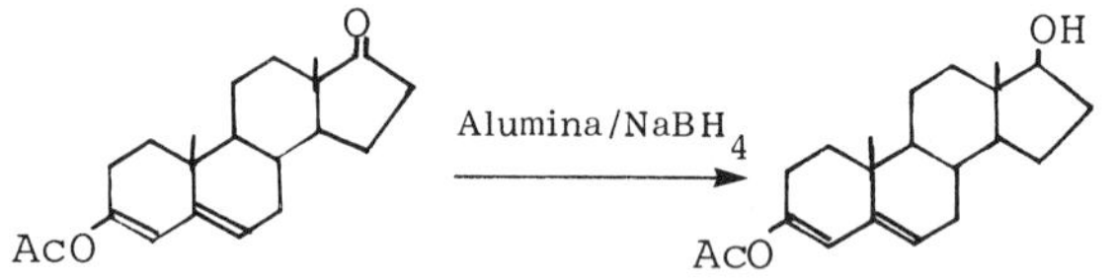

(90 %)

The compounds with highly alkali-sensitive groups can be selectively reduced [11].

$$CH_3C(=O)(CH_2)_2COCl \xrightarrow{\text{Alumina/NaBH}_4} CH_3C(=O)(CH_2)_2CH_2OH \ (80\%)$$

$$PhCH{=}CHCOCl \xrightarrow{\text{Alumina/NaBH}_4} PhCH{=}CHCH_2OH \ (90\%)$$

In contrast, free sodium borohydride gave different products, such as γ-buthyrolactone (40% yield) from 4-methoxy-4-oxobutanoylchloride and 3-phenyl-1-propanol (12%) from cinnamoyl chloride.

A general procedure for the reduction of acid chlorides [11] is carried out as follows. A solution of the acid chloride in ether is added to a mixture of $NaBH_4$-Al_2O_3 in ether under stirring. The reaction is complete within a few hours at ambient temperature; the workup consists of filtration and washing with ether. The filtrate is condensed and the hydroxy compounds are isolated by distillation or recrystallization.

The oxime, which is not reduced by sodium borohydride under normal conditions, can be reduced by silica gel-bound sodium borohydride [12]:

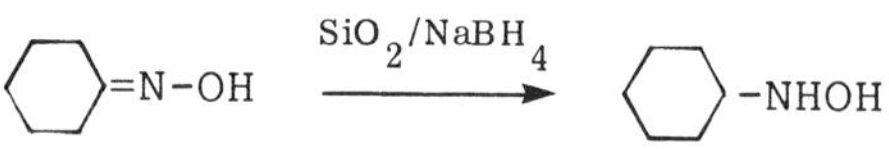

A slightly hydrolyzed silica gel-bound borohydride has been found to be faster for carbonyl reduction than alumina-bound borohydride [13]. Such alumina- or silica gel-bound borohydrides are commercially available (Alfa Division, Ventron Corp.).

Poly(vinyl pyridine) [14] and anion-exchange resin (Amberlite IRA-400) [15] have been used to bind borohydride, and have served for carbonyl reductions.

2.3.1.c Anion-exchange resin-bound cyanoborohydride

Amberlyst 26-bound cyanoborohydride has wide reductive applications similar to sodium cyanoborohydride, such as carbonyl reduction, reductive amination, and dehalogenation. One major advantage of this polymeric reagent is that there is no contamination by cyanide ion since the toxic cyanide is retained on the polymer and not extracted into solvent. Moreover, the spent polymer can be recovered by simple filtration and regenerated quite easily by a column method [16].

2.3.1.d Alumina-impregnated 2-propanol

2-Propanol absorbed on activated neutral alumina smoothly reduces aldehydes to primary alcohols in good yield without any change in other functional groups [17]. The selective reduction of dicarbonyl compounds was achieved by this reagent owing to slow or no reactivity toward the ketone group.

$$CH_3\overset{O}{\overset{\|}{C}}(CH_2)_8CHO \xrightarrow[Al_2O_3]{(CH_3)CHOH} CH_3\overset{O}{\overset{\|}{C}}(CH_2)_8CH_2OH\ (68\%)$$

Such behavior based on this reagent was applied to the selective reduction of the steroidal dione.

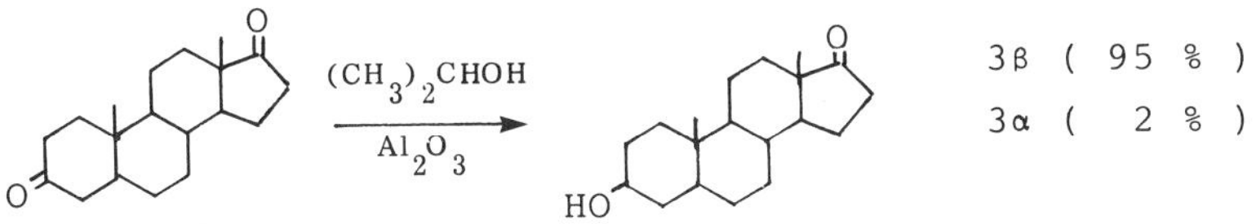

Another useful application of this reagent includes the specific deutration of alcohols as observed in benzyl alcohol-*d* from benzaldehyde treated with 2-propanol-2*d*/alumina [18].

$$PhCHO \xrightarrow[Al_2O_3]{(CH_3)_2CHOD} PhCHDOH\ (80\%)$$

2.3.1.e Potassium in graphite

Potassium on graphite can be prepared by heating a mixture of potassium and graphite to produce various types of reagents, such as C_8K, $C_{24}K$, $C_{36}K$, and $C_{48}K$. Coupling, reduction, and dehydrohalogenation were carried out by C_8K [19]. A variety of ketones have been reduced to secondary alcohols by C_8K and the reaction mechanism is similar to of an electrochemical reaction.

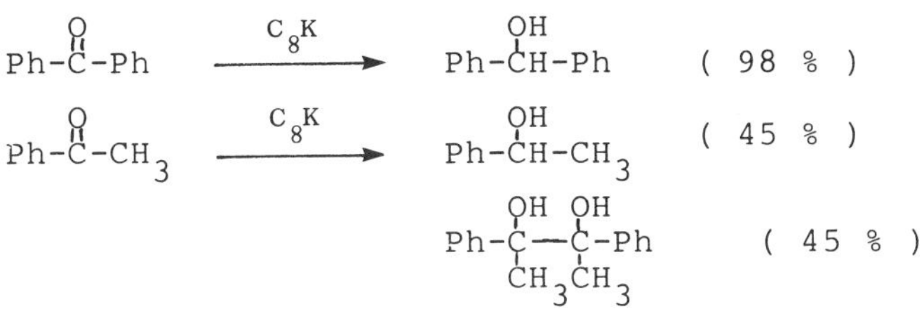

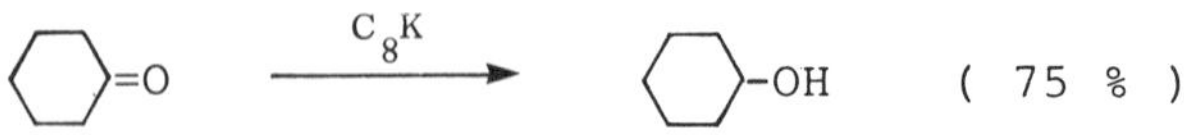

The reduction of camphor by C_8K gives predominantly the exo alcohol, which is different from the usual cases of potassium or sodium in alcohol. The inversion of stereochemistry can be interpreted in terms of the change in adsorption of the ketone on the graphite surface [20]. It is also useful for the reduction of α, β-unsaturated ketones, acids, and esters [21].

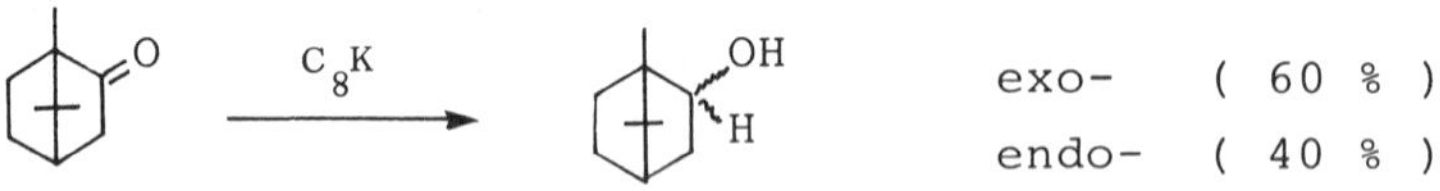

$$\text{Ph-CH=CH-COCH}_3 \xrightarrow{C_8K} \text{Ph-CH}_2\text{CH}_2\text{COCH}_3 \quad (75\ \%)$$

Such a reagent is commercially available (Graphimet 29, Alfa Division, Ventron Corp.).

2.3.2 Polymeric Oxidizing Reagents

2.3.2.a Polymer-bound chromium

Chromium trioxide bound to poly(vinylpyridine) can be used for the oxidation of alcohols to aldehydes and ketones. The advantage of the polymeric reagent is that chromium-containing by-products are retained by the resin and are removed by filtration.

Changes in the preparation of reagents result in different conditions for the oxidation. For example, poly(vinylpyridinium chlorochromate), which was derived from chromium trioxide and poly(vinylpyridine) in concentrated hydrochloric acid, was used in two- to fivefold excess to alcohol to attain good yields [22]. However, an alternative simple treatment of chromium trioxide with the less dense, porous, cross-linked polystyrene in water leads to poly(vinylpyridinium dichromate) (PVPDC, $Cr_2O_7^{2-}$ form) it improved the activity without using an excess of reagent to alcohol and exhibits an excellent yield in aldehyde preparations. Moreover, PVPDC provides no overoxidation products even over a long reaction time. It was used for five reaction cycles without loss of oxidizing capacity [23]. PVPDC is commercially available from Aldrich.

$$\text{PhCH}_2\text{OH} \xrightarrow{\text{PVPDC}} \text{PhCHO (96\%)}$$

$$\text{PhCH=CHCH}_2\text{OH} \xrightarrow{\text{PVPDC}} \text{PhCH=CHCHO (98\%)}$$

$$\text{PhCH(CH}_3\text{)OH} \xrightarrow{\text{PVPDC}} \text{Ph-C(CH}_3\text{)=O (89\%)}$$

Preparation of PVPDC [23]

To a suspension of 200 g of cross-linked poly(vinylpyridine) in 2 l of water is added 200 g of chromium trioxide and stirred for several hours. After filtration, the resin is washed with water until the filtrate is colorless. Generally, the reagent is used without drying.

Oxidation of alcohols with PVPDC [23]. To a suspension of 266 g of PVPDC in 1 l of cyclohexane is added 122 g of 1-phenylethanol and stirred at 65°C. After 48 h the reaction mixture is filtered, and the spent resin is washed with cyclohexane, ether, and dichloromethane. After distillation of the filtrate, the ketone is obtained in 94% yield.

Recycling of PVPDC [23]. The dark spent resin can be recycled by repeated sequential washings with 2 N HCl and 2 N NaOH until the PVP returns to its original white color, and then reacted with CrO_3 to regenerate the reagent.

Amberlyst A26-bound chromium ($HCrO_4^-$ form) is also for the oxidation of alcohols [24] and arylhalides [25] to aldehydes and ketones in excellent yield.

$$\mathrm{Ph{-}CH(OH){-}Ph} \longrightarrow \mathrm{Ph{-}C(=O){-}Ph}\ (77\%)$$

$$\mathrm{PhCH_2X} \longrightarrow \mathrm{PhCHO}\ (95\%)$$

Preparation of Amberlyst A26-bound chromic acid [24]

Amberlyst A26 (Cl^- form) (35 g) is added to a solution of 15 g of CrO_3 in 100 ml of water and stirred for 30 min. The resin is successively rinsed with water, acetone, and ether and then dried in vacuo at 50°C for 5 h.

In contrast, silica gel-supported chromic acid is useful for the oxidation of alcohols with acid labile groups [26].

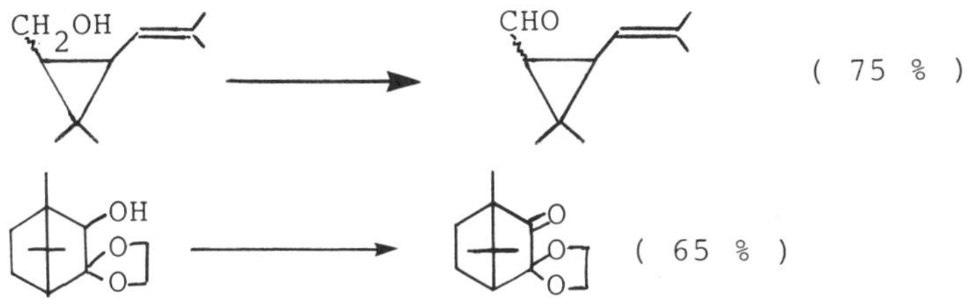

Silica-alumina-bound chromyl chloride is more selective than the homogeneous reagent [27].

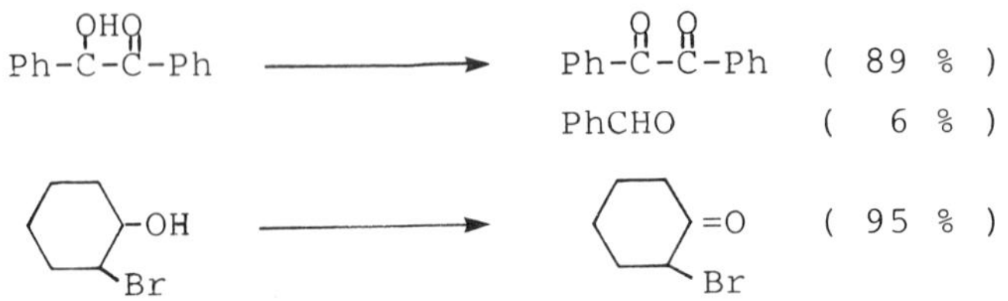

The intercalated chromium trioxide in graphite (commercially available as Seloxcetto from Alfa) is also used for the selective oxidation of primary alcohols to aldehydes, whereas secondary alcohols remain unattacked [20].

Preparation of CrO_3-graphite [20]

Graphite (20g) is heated in a Pyrex tube under reduced pressure at 200°C for 24 h to remove any adsorbed gas. After cooling, 40 g of CrO_3 is added. The tube is then sealed under reduced pressure and heated at 200°C for 48 h. After cooling, the tube is opened and the intercalated product is washed with water, 10% HCl, and acetone and dried at 100°C.

Oxidation of a typical alcohol by CrO_3-graphite[20]. CrO_3-graphite (10 g) and 0.67 g of cinnamyl alcohol are refluxed in 25 ml of toluene for 24 h. After filtration, the filtrate is evaporated to give a quantitative yield of cinnamaldehyde. Pyridinium chromium trioxide on Celite is also an effective oxidizing reagent for the acid-sensitive compounds [28,29].

2.3.2.b Silver carbonate on Celite

Silver carbonate on Celite (Fetizon's reagent), which has been studied extensively, is a remarkably versatile and reproducible oxidizing reagent with wide applications in organic synthesis.

Preparation of silver carbonate on Celite [30]

Celite (30 g) is added to a solution of silver carbonate (34 g) in water (200 ml) and stirred. A solution of sodium carbonate decahydrate (30 g) in water (300 ml) is then added slowly to the homogeneous suspension. After stirring for a further 10 min, the yellow-green precipitate is collected and dried.

In many cases, benzene is a suitable solvent. The residual traces of water are removed beforehand from the reagent under reflux with a Dean-Stark apparatus. The typical synthetic utilities of this reagent are described as follows.

Oxidation of primary and secondary alcohols to aldehydes and ketones

Alcohols having acids and/or basic sensitive groups are smoothly oxidized under mild condition by silver carbonate on Celite to give aldehydes and ketones in excellent yield without overoxidation.

trans-3-Methylhex-2-enal from trans-3-methylhex-2-en-1-ol [31]. Traces of moisture are removed from a suspension of silver carbonate on Celite (8 g) in benzene (100 ml) by azeotropic distillation. To the cooled mixture is added *trans*-3-methylhex-2-en-1-ol (0.8 g) in benzene (10 ml) and then refluxed for 1 h. After cooling and filtration, the filtrate is condensed and distilled to afford the *trans*-aldehyde in 88% yield.

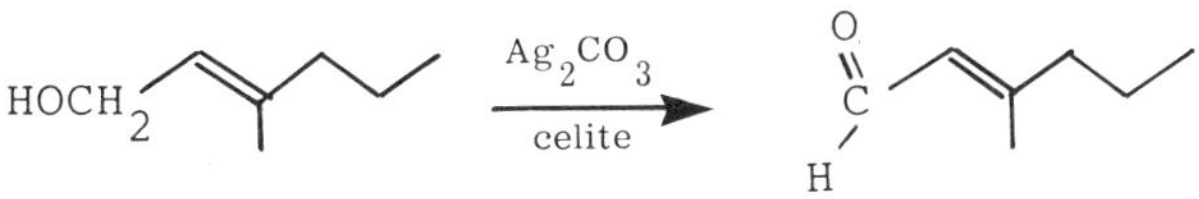

Selective oxidation of diols and triols

Secondary alcohols are more readily oxidized than primary and tertiary alcohols by silver carbonate on Celite [32,33].

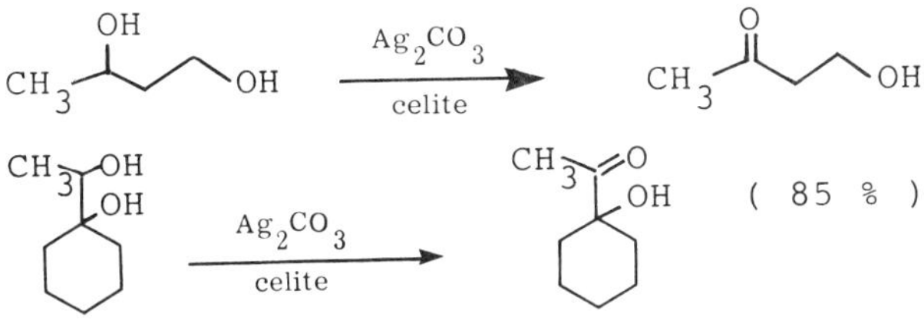

(80 %)

(85 %)

The selective oxidation in a steroid system is quite useful, as shown in 3β-OH oxidation for androstan-3β,5α,6α-triol [34].

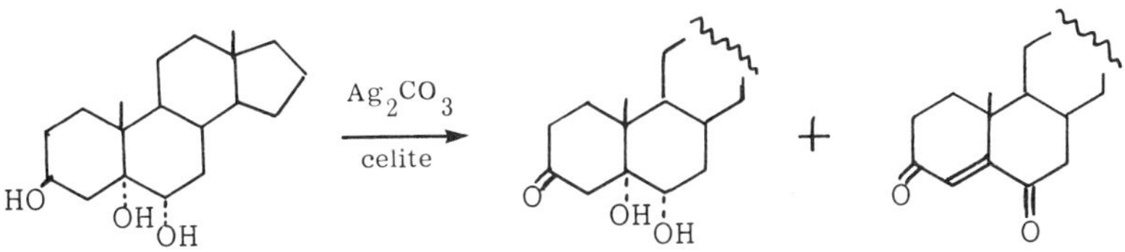

(60 %) (8 %)

Oxidation of α,ω*-diols to lactones* [35]

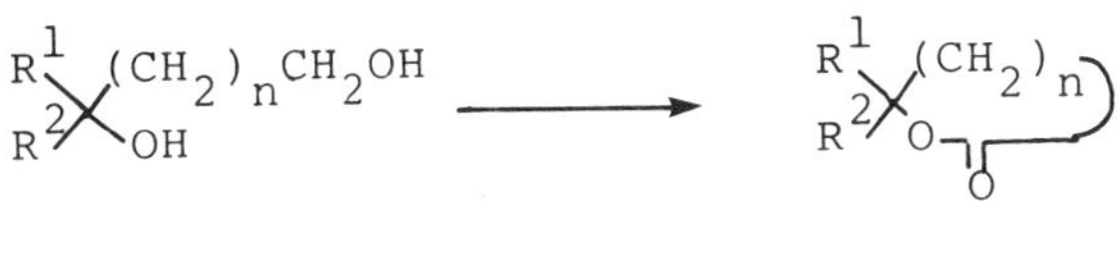

Oxidation of steroid halohydrin to the ring contraction or the epoxide [36]

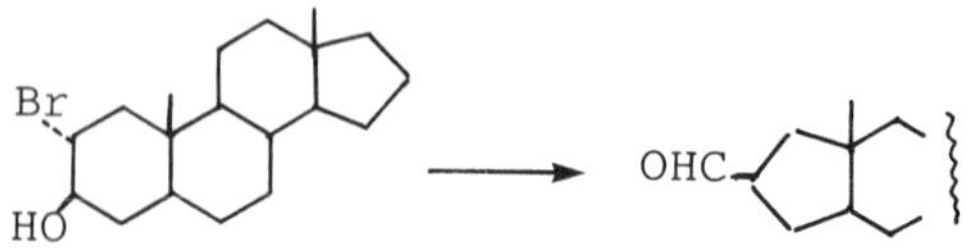

(95 %)

(48%)

Oxidation of hydroquinones [37] *and amine compounds* [38]

The mechanism proposed for the oxidation of alcohols is [39].

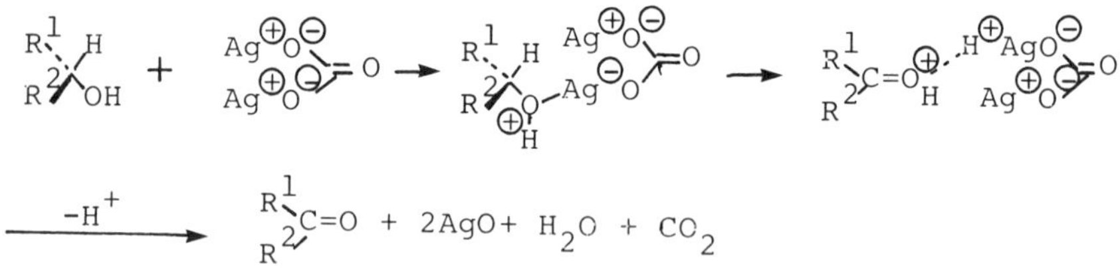

2.3.2.c Other oxidizing reagents based on silica gel

Silver nitrate on silica gel

Silver nitrate on silica gel with peracid is used for the selective epoxidation of alkenes. For example, the sesquiterpene humulene was converted exclusively to the Δ^6-epoxide [40].

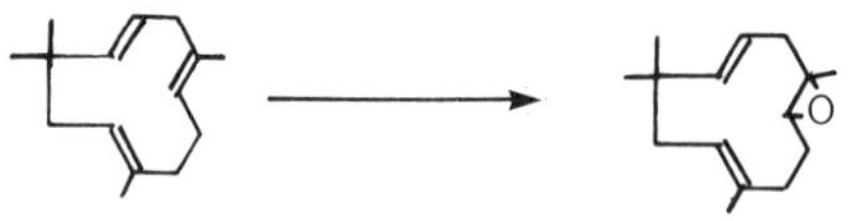

Sulfuryl chloride on wet silica gel

Alkyl, aryl sulfides are quantatively oxidized to the corresponding sulfoxides with sulfuryl chloride on wet silica gel [41].

$$R^1\text{-}S\text{-}R^2 \xrightarrow{AgNO_3/SiO_2} R^1\text{-}\overset{\overset{O}{\uparrow}}{S}\text{-}R^2$$

$R^1=R^2=C_2H_5$ (100%)

$R^1=R^2=Ph$ (100%)

$R^1=Ph, R^2=CH_3$ (95%)

Ozone on silica gel

Ozone is effectively adsorbed on silica gel (ca. 4.5% by weight at -78°C) and the resulting reagent is effective for the direct hydroxylation of hydrocarbons to the tertiary alcohols [42], the oxidation of secondary alcohols to the ketones [42], the direct oxidation of a methylene group to a ketone via initial hydroxylation and subsequent oxidation [43], and the transformation of primary amines to the nitro compounds [44].

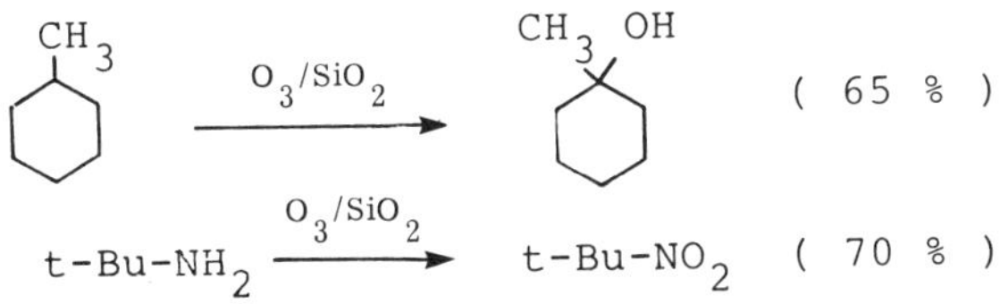

Oxidation of primary amines to nitro compounds [44]. The silica gel (30 g) adsorbed amine (0.1 to 0.2% w/w) is cooled to -78°C and a stream of 3% ozone (in oxygen) is passed into the silica gel. The mixture is allowed to warm to room temperature and the product is separated by filtration.

Sodium methoxide on silica gel

Sodium methoxide can be placed onto silica gel by mixing silica gel with sodium methoxide in methanol followed by drying at 400°C [45]. The resulting reagent is useful for the oxidation of nitro compounds to aldehydes and ketones.

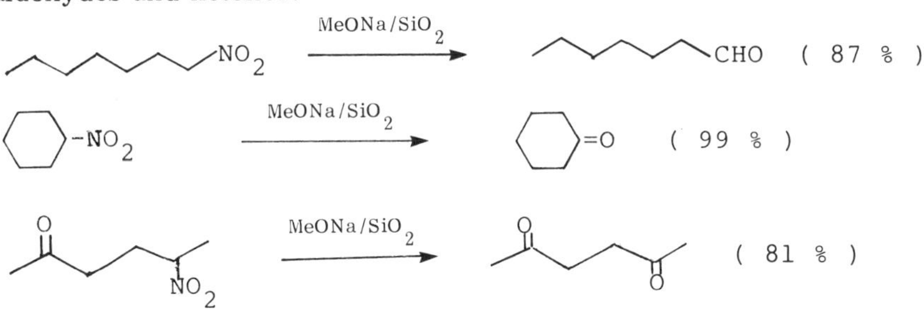

Cyclohexanone from nitrocyclohexane [45]. Nitrocyclohexane (0.5 g) is mixed with 50 g of activated basic silica gel for about 2 min and left in a closed bottle at room temperature for 48 h. Elution of the silica gel with ether gives cyclohexane in 99% yield

The combined silica gel-supported ozone and methoxide was applied to the synthesis of dihydrojasmone from *n*-heptylamine in 68% yield [45].

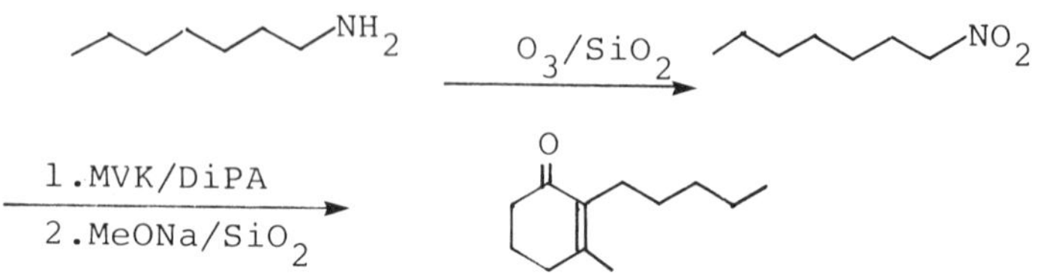

2.3.2.d Miscellaneous oxidizing reagents on inorganic supports

Chloral on alumina

Selective oxidation of diols can be realized with chloral or benzaldehyde supported on alumina. In some cases, the yield is improved over silver carbonate on Celite. In general, secondary alcohols are smoothly oxidized to ketones by chloral on alumina at room temperature [46-48].

$$CH_3CH(OH)\text{-}(CH_2)_8CH_2OH \xrightarrow{PhCHO/Al_2O_3} CH_3C(=O)\text{-}(CH_2)_8CH_2OH \ (65\%)$$

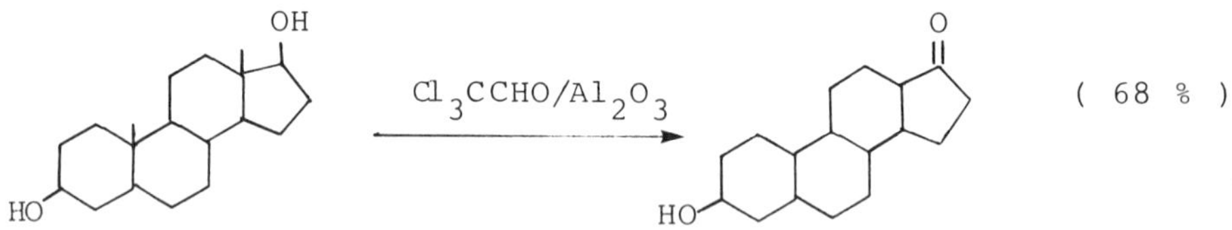

Oxidation of secondary alcohols to ketones [46-48]. Carbon tetrachloride (5 ml) and 5% by weight of the alcohol are added to dehydrated alumina (5 g) under nitrogen. After stirring at room temperature for 10 min, chloral (1 eq) in carbon tetrachloride (1 ml) is added. The mixture is stirred at room temperature for a further 24 h. The solid is removed by filtration and washed with dichloromethane. The filtrate is dried over calcium chloride and the solvent is evaporated to afford the ketone.

Manganese dioxide on carbon

The manganese dioxide can be precipitated from potassium permanganate on activated carbon, and the resulting reagent was found to be highly effective for the oxidation of cinnamyl alcohol to cinnamaldehyde [49]. Other uses of this reagent include the selective oxidation of alcohols and amines without any effect on functional groups such as double and triple bonds [50-52].

$$R\text{-}C{\equiv}C\text{-}CH(OH)\text{-}C{\equiv}C\text{-}R \xrightarrow{MnO_2/C} R\text{-}C{\equiv}C\text{-}C(=O)\text{-}C{\equiv}C\text{-}R \ (85\text{-}90\%)$$

$$Ph\text{-}CH{=}N\text{-}NH_2 \xrightarrow{MnO_2/C} Ph\text{-}CH{=}\overset{+}{N}{=}\overset{-}{N} \ (53\%)$$

Preparation of manganese dioxide on carbon [49]. A solution of 20 g of potassium permanganate in 250 mL of water is heated to reflux and treated portionwise over 5 to 7 min with 6.25 g of activated carbon. After the addition of carbon, the mixture is refluxed until the purple color is completely discharged (2-4 min), allowed to stand at room temperature for 10 to 15 min, and then filtered. The precipitate is washed four times with 50-mL portions of water, and dried in an oven.

Potassium permanganate on molecular sieves, montmorillonite, and silica gel [53,85]

Oxidation of alcohols to aldehydes and ketones are made by potassium permanganate on Linde 13X molecular sieve in modest and excellent yield.

Preparation of potassium permanganate on molecular sieves [53]. Aqueous potassium permanganate (500 mL of 0.06 M) and 20 g of Linde 13X molecular sieve are mixed and the water is then removed by a rotary evaporator. The coated pellets are separated physically from nonadsorbed potassium permanganate by screening (20 mesh).

Thallium on montmorillonite K-10 [54]

Thallium (III) nitrate can be bound to montmorillonite by stirring K-10 with a solution of thallium nitrate in a mixture of methanol and trimethylorthoformate. The resulting reagent is effectively used for the selective oxidation and rearrangement of alkyl, aryl ketones via oxythallation under mild condition.

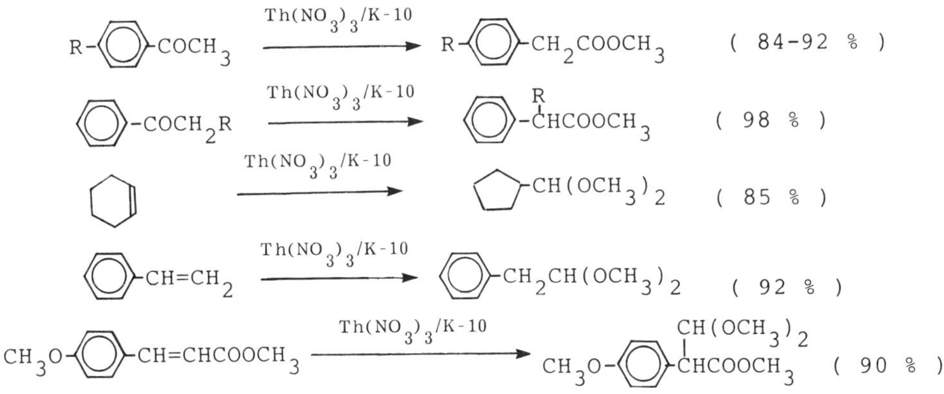

The exact physical nature of this reagent is not clear; however, it is known that both the thallium (III) nitrate and the thallium (I) nitrate generated during the reaction are bound to the K-10 support. Other support materials, such as silica gel, alumina, graphite, and Amberlyst, are less selective.

Preparation of thallium nitrate on K-10 [54]. To a solution of 125 mL of trimethylorthoformate and 100 mL of methanol is added 49 g of thallium nitrate trihydrate and stirred until the thallium salt dissolves. Then 110 g of K-10 clay is introduced and stirring continued for 5 min. The solvent is removed under reduced pressure to a afford pale yellow powder.

Oxidative rearrangement by thallium on K-10 [54]. Thallium nitrate on K-10 reagent is mixed with the substrate in an inert solvent such as heptane, methylene chloride, carbon tetrachloride, toluene, or dioxane. The product is isolated by filtration.

Sodium methaperiodate on alumina [55]

Alumina-supported methaperiodate can be used for the oxidation of sulfides.

2.3.2.e Polymeric peracids

Bifunctional cation exchanger with carboxylic acid and sulfonic acid group can be converted into the percarboxylic acid (2) by aqueous hydrogen peroxide treatment. The reagent (2) oxidized alkenes to α-glycols but failed to give epoxides due to the presence of a sulfonic acid group [56].

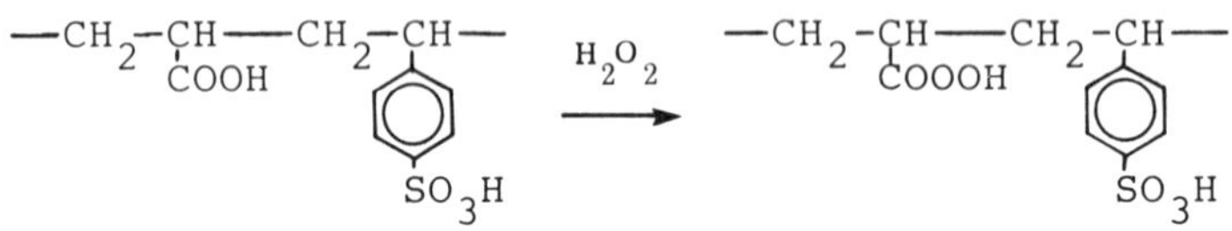

(2)

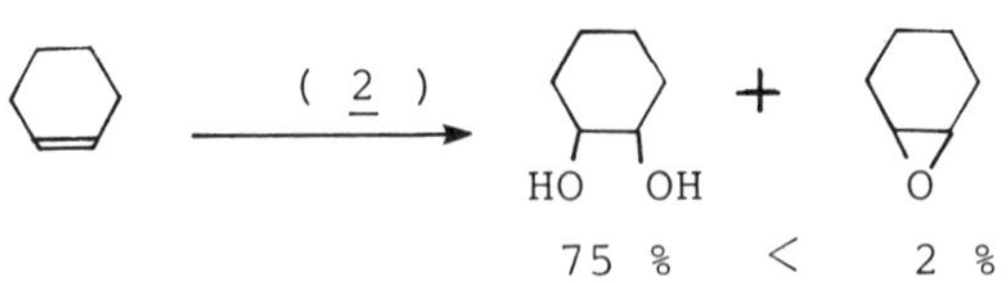

Amberlite XE-89 [cross-linked poly(methacrylic acid)] was treated with methanesulfonic acid-hydrogen peroxide to give the peracid (3), which was used for the oxidation of alkenes to epoxides in good yield [57]. However, this resin was not as stable and could be recycled only a few times:

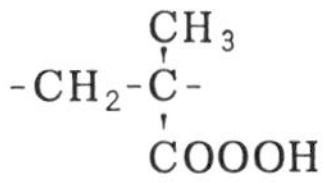

(3)

In contrast, polymeric peracid (4) based on a macroreticular and swellable cross-linked poly(vinylbenzoic acid), which has better stability and recycling capacity, is effective for the oxidation of alkenes to epoxides [58,59].

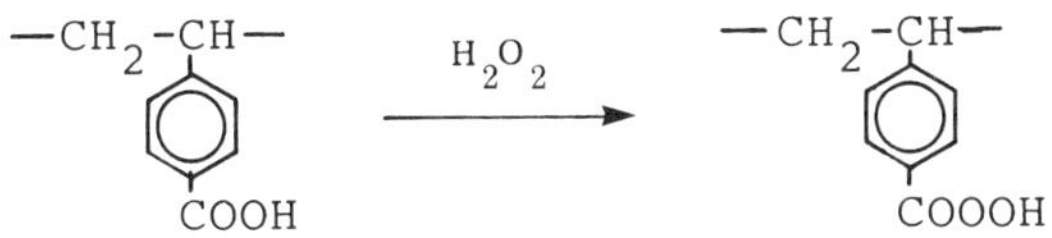

(4)

The swellable support resin gives a better yield of the epoxide formation than does the macroreticular resin.

Preparation of poly(vinylperbenzoic acid) (4) [58]

To a mixture of 1 g of macroreticular poly(vinylbenzoic acid) (1 meq/g) in 4 mL of methanesulfonic acid is added slowly and with stirring 6 mL of 70% hydrogen peroxide while the mixture is kept at 20°C. After 6 h, the mixture is washed with cold methanol (-20°C) until the effluent is free of peroxide.

Epoxidation of cyclohexene by (4) [58]. A suspension of 3 g (2.9 mmol) of a macroreticular poly(vinylperbenzoic acid) in 10 mL of dioxane is stirred with 0.1 mL (2 mmol) of cyclohexene at 40°C for 6 h. After filtration, the epoxide is separated in 65% yield. Similar reaction on a swellable poly(vinylperbenzoic acid) affords the epoxide in 85% yield.

Arsonated polystyrene beads (5) can be changed into an effective catalyst for the Baeyer-Villiger oxidation of ketones since 5 is rapidly converted into the active hydroperoxy form with low or no protonic or Lewis acid to cause hydrolysis [60]. In practice, the Baeyer-Villiger oxidation of cyclic and aliphatic ketones is carried out both in the biphase (in tetrahydrofuran or dioxane) or the triphase (the aqueous H_2O_2-$CHCl_3$) system. The results show that lactones and esters are almost completely nonhydrolyzed and that the peroxyarsonic form (6) could be particularly useful in the triphase system.

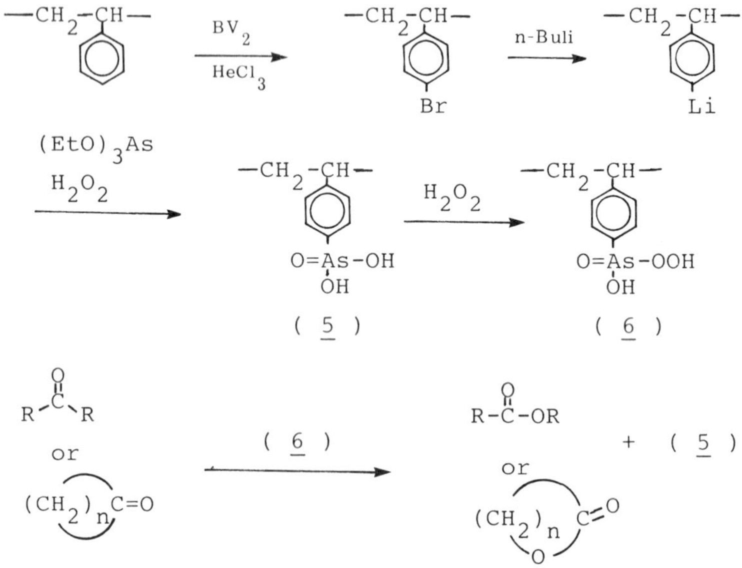

(5)

Preparation of arsonated polystyrene (5) [60]

Cross-linked polystyrene (20 g) and iron trichloride (2 g) are suspended in 100 mL of carbon tetrachloride and cooled to 0°C. To the mixture is slowly a solution of bromine (27 g, 168 mmol) in carbon tetrachloride (100 mL) over a 2-h period. With the exclusion of light the mixture is stirred at room temperature for 72 h. The unreacted bromine and solvent are removed by filtration and the resin is washed with acetone, dioxane-water (1:1), dioxane, tetrahydrofuran, and then ether. The brominated polymer is then suspended in 50 mL of dry toluene under argon. A solution of *n*-butyllithium (145 mmol) in toluene (60 mL) is then slowly added to the suspension at -78°C and allowed to warm to room temperature. After the suspension is heated to 60°C for 3 h, the solvent is removed by filtration. The polymer is washed with toluene and the suspended again in dry toluene (30 mL). Arsenic triethoxide (22 g, 104 mmol) in 30 mL of ether is then slowly added under argon at -78°C. The solution is allowed to warm to room temperature and the suspension is stirred overnight. The beads are collected by filtration, washed with dioxane-water (1:1), water, and dioxane.

This polymeric arsonate (5) can be used for the epoxidation of alkenes by aqueous hydrogen peroxide in excellent yield. In the triphase system the hydrolysis of the epoxide is prevented and the activity still remains after recycling five times.

Preparation of epoxide by 5 [61]. Aqueous 30% hydrogen peroxide (1.7 g, 0.65 mmol) is added to a mixture of 5 (0.25 mmol of As per gram), cyclooctene (0.825, 7.5 mmol), biphenyl (0.10 g, 0.65 mmol) and chloroform (7 g), and heated at 70°C for 45 h. The polymer is recoved by filtration and the organic layer is analyzed by ^{1}H NMR. The spent polymer is thoroughly washed with dioxane and ether and then recycled.

2.3.2.f Oxidizing reagents based on *N*-haloamides

Nylon 6,6 can be choorinated with either aqueous hypochlorous acid or *t*-butylhypochlorite to give *N*-chloronylon (7).

$$\text{Nylon 6,6} \xrightarrow{t\text{-BuOCl}} \text{-}\underset{\text{Cl}}{\text{N}}\text{-}(CH_2)_6\text{-}\underset{\text{Cl}}{\text{N}}\text{-}\underset{\text{O}}{\overset{\|}{\text{C}}}\text{-}(CH_2)_4\underset{\text{O}}{\overset{\|}{\text{C}}}\text{-} \qquad (6)$$

(7)

Preparation of N-chloronylon (7) [62]

To a suspension of nylon 6,6 (10 meq) in 10 mL of tetrachloroethane is added *t*-BuOCl (20 mmol) at 15°C. After stirring for 3 h, the clear solution is poured into 100 mL of ether. The precipitate is filtered and dried in vacuo.

The reactions evoked by 7 include the oxidation of secondary alcohols to ketones, primary alcohols to esters, sulfides to sulfoxides [63,64], and tertiary amines to secondary amines [65].

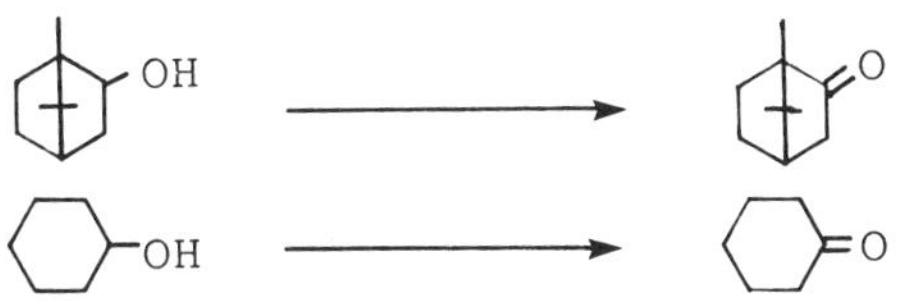

(97%)

(82%)

$PhCH(OH)CH_3$	→	$PhC(=O)CH_3$	(95%)
$PhCH_2OH$	→	PhCHO	(95%)
$CH_3(CH_2)_6CH_2OH$	→	$CH_3(CH_2)_6COO(CH_2)_7CH_3$	(80%)
$PhCH_2SPh$	→	$PhCH_2S(\rightarrow O)Ph$	(70%)
PhSPh	→	$PhS(\rightarrow O)Ph$	(65%)
$PhCH_2N(CH_3)_2$	→	PhCHO	(30%)
		$PhCH_2NHCH_3$	(5%)

The reaction conditions characteristic for this reagent (7) are homogeneous systems since in the beginning of the reaction 7 is soluble in several organic solvents and the subsequent spent polymer (nylon 6,6) precipitates out of solution as the reaction proceeds.

Oxidation of the secondary alcohol by 7 [62]. The secondary alcohol (2 mmol) and pyridine (2 mmol) in 6 mL of benzene are stirred with *N*-chloronylon (2.2 mg atom Cl) at 35°C. After 24 to 48 h, the spent polymer (nylon 6,6) is filtered and washed with ether. The filtrate is condensed, and the residue is dissolved in ethanol and treated with 40 mL of 50% aqueous-ethanol solution containing sodium acetate (0.33 g) and semicarbazide chlorohydrate (0.23 g) at 60°C for 15 min. The mixture is diluted with 100 mL of water, and the semicarbazone is collected.

The quantitative oxidation of sulfides to sulfoxides by 7 is obtained in the presence of water or alcohol, and the ^{18}O incorporation is successfully realized in $H_2{}^{18}O$ [66].

Asymmetric induction to the optically active sulfoxide is made to some extent by using an optically active alcohol [67].

Oxidation of sulfides by 7 [66]. A sulfide (1 mmol) and *N*-chloronylon (1 mg atom Cl) are dissolved in 6 mL of benzene or dioxane, and then alcohol or water (in the case of dioxane) is added quickly. The mole ratio of alcohol or water to the sulfide is 5 to 10:1. The mixture is stirred at 35 to 40°C. When the reaction mixture becomes heavy with precipitated nylon, it is poured into a large amount of ether and the nylon is filtered off. The product is obtained from the filtrate. The use of $H_2{}^{18}O$ in dioxane gives the ^{18}O sulfoxide quantitatively. Other polymeric *N*-haloamides such as poly(*N*-chlorobenzylacrylamide) and poly(*N*-bromosuccimide) can also be used for the oxidation of alcohols [68].

2.3.2.g Polymeric thioanisole

Treatment of a macroreticular poly(lithiostyrene) with methyldisulfide affords poly(methykmercapto styrene) (8), which is similar to thioanisole employed in the Corey oxidation (8). This reagent has several advantages over monomeric thioanisole such as an easy workup free of odor and reuse. Further chlorine treatment of 8 yields the polymeric chlorosulfonium ion (9), which *in situ* can be used for the oxidation of primary and secondary alcohols to aldehydes and ketones in excellent yield [69].

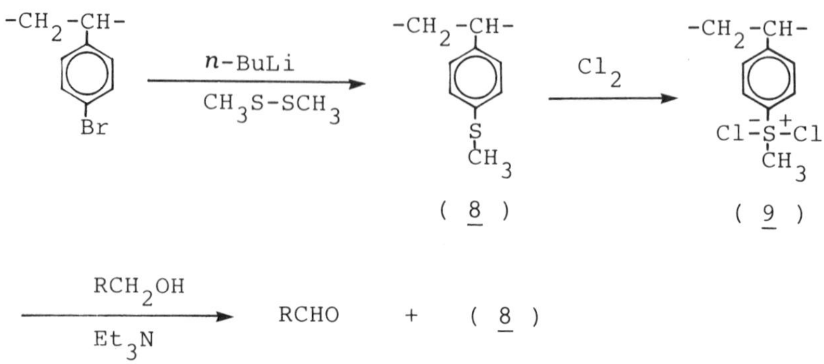

Reagent 9 is effective for the selective oxidation of alcohols that have highly sensitive groups such as a key prostaglandin intermediate.

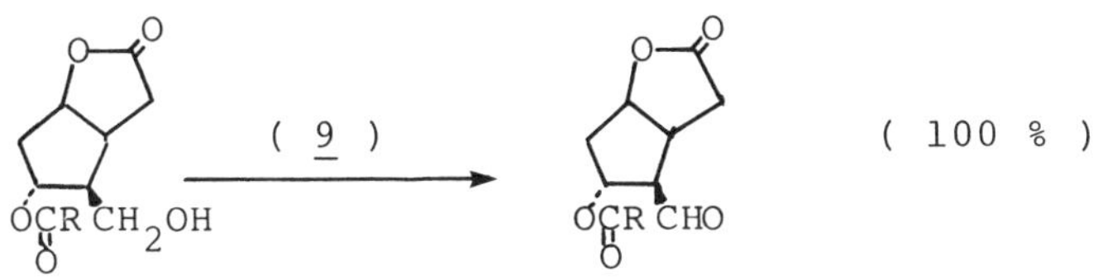

The selective monooxidation of diols is also made by 9. In fact, 1,7-heptanediol is selectively oxidized to the monoaldehyde together with a trace of the dialdehyde. This result suggests that significant site isolation on the polymer surface is due to the ionic repulsion of sulfonium ion groups.

$$HO(CH_2)_7OH \longrightarrow \underset{(50.2\%)}{HO(CH_2)_6CHO} + \underset{(2\%)}{OHC(CH_2)_5CHO}$$

*Preparation of poly (*p*-bromostyrene)* [69]

To 13.0 g (125 mmol) of macroreticular polystyrene (Amberlite XE-305) and 27.6 g (67.5 mmol) of thallic acetate sequihydrate suspended in 200 mL of carbon tetrachloride is added 12.8 g (4.1 mL, 80 mmol) of bromine in 30 mL of carbon tetrachloride and the mixture is then stirred at 50 to 60°C until all bromine color disappears. The beads are filtered and washed with a mixture of 30% aqueous HCl and dioxane, a mixture of water and dioxane, dioxane, and ether and dried in vacuo.

Preparation of poly(p-methylmercaptostyrene) (8) [69]

To a stirred suspension of 14.5 g (51.6 mmol % bromine) of dry poly(*p*-bromostyrene) in 200 mL of tetrahydrofuran cooled to -78°C under an argon atmosphere is added 50 mL of 2.28 M *n*-butyllithium-hexane (114 mmol) solution. The Stirring is continued at -78°C for 0.5 h, then at room temperature for 1 h, followed by removal of the tetrahydrofuran solution from the insoluble polymer by syringe. The beads are recharged at -78°C with 150 mL of tetrahydrofuran and 20 mL of 2.28 M *n*-butyl lithium (68.4 mmol) and stirred at ambient temperature for 1 h. The tetrahydrofuran is removed again and the polymer beads are treated with a mixture of 200 mL of tetrahydrofuran and 20 mL of methyl disulfide with stirring at -78°C for 15 min and at room temperature for 0.5 h and then refluxed under argon for 1 h. The cooled mixture is diluted with 100 mL of water and filtered. The beads are washed with water, a 3:1 mixture of dioxane-water, and dried. Finally, the air-dried beads are refluxed in a 2:1 mixture of benzene and dioxane in a Dean-Stark trap and dried overnight in vacuo at 70°C.

Oxidation of prostaglandin alcohol [69]. To a stirred suspension of 2.0 g of poly(*p*-methylmercaptostyrene) (2.66 mmol of sulfur per gram of resin) in 20 mL of dichloromethane is added 395 mg (5.0 mmol) of chlorine in 11 mL of dichloromethane at -10°C. Then 352 mg (1.0 mmol) of prostaglandin intermediate alcohol in 10 mL of dichloromethane is added and the mixture stirred at -40°C. After 4 h, 871 mg (8.6 mmol) of triethylamine in 2 mL of dichloromethane is added and the reaction mixture stirred at -40°C for 5 min and then removed from the cooling bath. After 15 min, the mixture is filtered and the beads are washed with three 25-mL portions of dichloromethane. The combined filtrates are washed with three 120-mL portions of ice-cold 1% hydrochloric acid, water, and brine and dried over magnesium sulfate while keeping the solution at 0°C. The dried solution is concentrated to 20 mL under a rotary vacuum at room temperature. The concentrate is then chilled to 0°C and petroleum ether or *n*-hexane is added to precipitate the product.

2.3.3 Group Transfer Reagents

2.3.3.a Polymer-bound nucleophile

A variety of nucleophiles can be bound to polymers by either ionic or covalent bonding. The ionic bonding is simple in practice. For example, anion-exchange resins are commonly used for binding nucleophiles in which the counteranion on the starting resins can be replaced by nucleophiles through a simple mixing method in protic solvents.

The advantages of polymeric nucleophiles in organic synthesis include (a) simple workup, (b) easy product isolation, (c) mild reaction conditions, and (d) easy regeneration of reagent. Typical polymer-bound nucleophiles are given in Table 1 together with the synthetic applications.

Preparation of Amberlyst A26–Br_3^- [76]

Amberlyst A26-Cl^- form (35 g) is washed (on column) with 1 N aqueous sodium hydroxide, then with distilled water until neutral, and a negative test for the chloride ($AgNO_3$) is observed. The polymer is then suspended in 1 N aqueous hydrogen bromide (200 mL), stirred overnight at room temperature and filtered. The product is washed with water, acetone, ether, and dried at 50°C overnight under vacuum to give Amberlyst A26-Br^- form.

The Amberlyst A26-Br^- form (14.9 g) is suspended in tetrachloromethane (160 mL) and stirred for 0.5 h, then bromine (2.8 mL) in tetrachloromethane (28 mL) is added slowly. After 7 h, the polymer is filtered, thoroughly washed with tetrachloromethane, extracted in a Soxhlet with tetrahydrofuran and ether until a colorless organic layer results, and the solid is dried under vacuum at room temperature for 15 h to give a stable yellow polymer.

Bromination of cholestan-3-one with Amberlyst A26-Br_3^- [76]. To a stirred solution of cholestan-3-one (0.5 g, 1.29 mmol) in tetrahydrofuran (5 mL) is added Amberlyst A26-Br_3^- form (1.0 g, 1.55 meq) at room temperature in one portion. After 0.5 h, the decolored resin is recovered by filtration and washed with a saturated aqueous sodium hydrogen carbonate solution and then water. The filtrate is concentrated under reduced pressure. The white residue is chromatographed on silica gel with 7:3 *n*-hexane-ethyl acetate as eluent to recover pure 2α-bromocholestan-3-one; yield 0.4 g (78%), mp 164 to 166°C.

Preparation of the fluoride anion-exchange resin [80]

Oven-dried anion-exchange resin (Bio-Rad AG1-X10 Cl^- form, 40 g) is suspended in water and transferred to a burette. The resin is eluted with 5.0 M aqueous potassium fluoride at a rate of 150 mL/h. Approximately 3 L of 5.0 M KF solution is needed to elute the chloride from the resin. The fluoride form is washed with three 100-mL portions of distilled water, dried in an oven at 115°C for 2 h, and stored in a desiccator. This preparation yields 38.7 g of dried resin.

Synthesis of sulfonyl fluorides [80]. The oven-dried fluoride resin (5 mL) is transfered to a 10 B-D disposable syringe and washed with 20 mL of acetonitrile. The appropriate sulfonyl chloride (1.0 g) is dissolved in 10 mL of acetonitrile and passed through a resin at a rate of approximately 1 mL/min. The resin is then washed with 5-mL portions of acetonitrile and the combined eluate and washings are evaporated to near dryness on a rotary evaporator.

Preparation of Amberlyst A26-benzenesulfinate [73]

A 0.1 M aqueous solution of sodium benzensulfinate is slowly percolated through a column filled with Amberlyst A26-Cl^- form until a negative test for chloride ion in the eluate is obtained. This reagent can be more quickly prepared using a system of two columns in series, the upper one filled with Amberlyst H15 resin (16.6 g, H^+ form, average capacity 4 meq/g) and the lower one with Amberlyst A26 resin (20 g, OH^- form, average capacity 3.5 meq/g). A slight excess of sodium benzenesulfinate (11.8 g, 72 mmol) dissolved in methanol (50 mL) is eluted through the system until an acidic reaction of the eluate is observed. The resins prepared by either of the methods are successively washed with water, acetone, and ether, and dried in vacuo at 50°C for 5 h.

Synthesis of n-*octyl phenyl sulfone* [73]. Amberlyst A26-benzenesulfinate form (11 g, 39 meq) and 1-bromooctane (5.8 g, 30 mmol) are vigorously stirred in refluxing benzene (30 mL) for 3 h.

Table 1. Polymer-Bound Nucleophiles

Type of resin	Anion bound to resin	Reaction	References
Amberlite IRA-400(Cl^-) Dowex 21K(Cl^-)	CN^-	$RBr \longrightarrow RCN$	70
Amberlite IRA-904(OH^-)	CH_3COO^-	$RX \longrightarrow CH_3COOR$	71
Amberlyst A-26(Cl^-)	CN^-	$RX \longrightarrow RCN$	72
	SCN^-	$RX \longrightarrow RSCN$	72
	$PhSO_2^-$	$RX \longrightarrow PhSO_2R$	73
	OCN^-	$RX \longrightarrow ROCN \longrightarrow RNHC(=O)NR$	74
	SCN^-	$RX \longrightarrow RSCN, RNCS$	74
Amberlite IRA-900(OH^-)	PhO^-	$RX \longrightarrow ROPh$	75
	NO_2^-	$RCH(X)COOR \longrightarrow RCH(NO_2)COOR$	75
	β-Diketonate	$RX \longrightarrow RC(=O)CH(R)C(=O)R$	75
Amberlyst A-26(Cl^-)	Br_3^-	$PhC(=O)CH_3 \longrightarrow PhC(=O)CH_2Br$	76
		$RCH_2CHO \longrightarrow RCH(Br)CHO$	77
		$RCH_2C(=O)CH_3 \longrightarrow RCH(Br)C(=O)CH_3$	77
			77

(Table 1 continued)

Type of resin	Anion bound to resin	Reaction	References
Amberlyst A-26(Cl^-)	Br_3^-	RC≡CH → R-C(Br)=CH(Br)	77
	$BrCl_2^-$	RC≡CH → R-C(Cl)=CH(Br)	77
Amberlyst A-26(OH^-)	CH_3COS^-	RX → CH_3COSR	78
	F^-	RX → RF	79
		$PhC(O)CH_3$ → $PhC(O)CH_2F$	79
Bio-Rad AG1(Cl^-)	F^-		80
		CH_3SO_2Cl → CH_3SO_2F	80
Amberlyst A-26(OH^-)	$(RO)_2P(O)CH-X$ X:CN COOMe	$R(R')C=O$ → $R(R')C=CHX$	81
Dowex 1-4X	$^{18}F^-$	$BrCH_2COOEt$ → $^{18}FCH_2COOEt$	82
Al_2O_3	CN^-	Ph-Br → Ph-CN	83
	I^-	R-Br → R-I	84,85
	Cl^-	R-Br → R-Cl	
	CH_3COO^-	R-Br → CH_3COOR	
SiO_2	SCN^-		86
	CN^-		87

The resin is then filtered off, washed with dichloromethane (30 mL), and the solvent is removed *in vacuo*. Distillation of the crude product gives the sulfone; yield 7.16 g (92%).

Preparation of Amberlyst A26-phosphonate [81]

Amberlyst A26-OH^- form (5 g) and diethylphosphonoacetonitrile (1.78 g, 10 mmol) in methanolic solution are shaken mechanically for 2 h. The resin is then filtered off and thoroughly washed with tetrahydrofuran.

Synthesis of p-*chlorocinnamonitrile* [81]. *p*-Chlorobenzaldehyde (0.49 g, 3.5 mmol) and Amberlyst A26-phosphonate (5 g, 5.6 mmol) in wet form are stirred at room temperature in tetrahydrofuran (10 mL). The reaction, followed by gas-liquid chromatography is complete in 1 h.

The resin is recovered by filtration and the solvent is removed by gentle distillation to give p-chlorocinnamonitrile in 83% yield.

Preparation of Amberlite IRA 900-phenoxide [75]

Amberlite IRA 900-Cl^- form (1 eq) packed in a column is washed with 5% aqueous sodium phenoxide until complete removal of chloride ion. The resin is then washed with water, ethanol, and benzene, and is finally dried in vacuo at 40°C for 4 h.

O-alkylation of phenol [75]. Amberlite IRA 900-phenoxide (15 mmol) in benzene (10 mL) is stirred with *n*-butylbromide (5 mmol) at 50°C for 6 h. The resin is filtered off and washed with benzene, the filtrate is dried over sodium sulfate, and the solvent is removed to give the product in 100% yield.

Preparation of Amberlite IRA 400-cyanate [70]

Amberlite IRA 400 or Dowex 21K (chloride ion form) is washed several times with distilled water and then three times with three times its volume of 20% aqueous sodium cyanide. The mixture of resin and cyanide solution is stirred for 5 min, diluted with distilled water (0.5 volume), and stirred for an additional 5 min. The cyanide solution is decanted each time and the resin is finally washed with distilled water until the supernatant liquid gives a negative cyanide test (silver nitrate).

Synthesis of nitriles by Amberlite IRA 400-cyanate [70]. A mixture of 8.6 g (0.05 mol) of benzyl bromide in 100 mL of 95% ethanol and 50 mL of Amberlite IRA 400-cyanate (0.05 eq) is stirred for 1.5 h. The temperature of the reaction mixture is maintained at 65°C. The resin is removed by filtration and washed with five 10-mL portions of ethanol. The combined filtrates are concentrated under reduced pressure and the residue distilled to give 3.1 g of phenylacetonitrile (53%).

2.3.3.b Halogenating reagents

Poly(*N*-bromomaleimide) is similar to monomeric *N*-bromosuccimide (NBS) on the bromination and oxidation of olefins and ketones. The parent polymaleimide can be obtained by a copolymerization of maleimide with divinylbenzene (2.5 to 5%). The treatment of the polymer with bromine in aqueous alkaline solution affords the *N*-bromomaleimide from 10 [88].

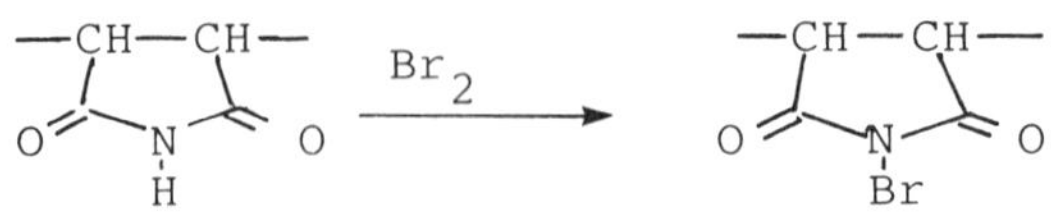

(10)

Alkenes and alkyl benzene are brominated by this reagent (10). However, the unusual bromination of cummen is observed to give predominately α, β, β-tribromocummene together with α,β-dibromocummene. The change in mechanism, which involves the dehydrobromination followed by subsequent hydrobromination, is attributed to the polar environment of the neighboring succimide group in the polymer. In contrast, poly(*N*-chloromaleimide) produces aromatic substitution as shown by the conversion of ethylbenzene to ethylchlorobenzene [89].

Poly(vinylpyridinium hydrobromide perbromide) (11) derived from the hydrogen bromide-bromine treatment of cross-linked poly(4-vinylpyridine-co-styrene) is used for the bromination of alkenes and ketones to give the dibromo and α-bromo compounds in excellent yield. The regeneration of the spent polymer is made without loss of activity [90].

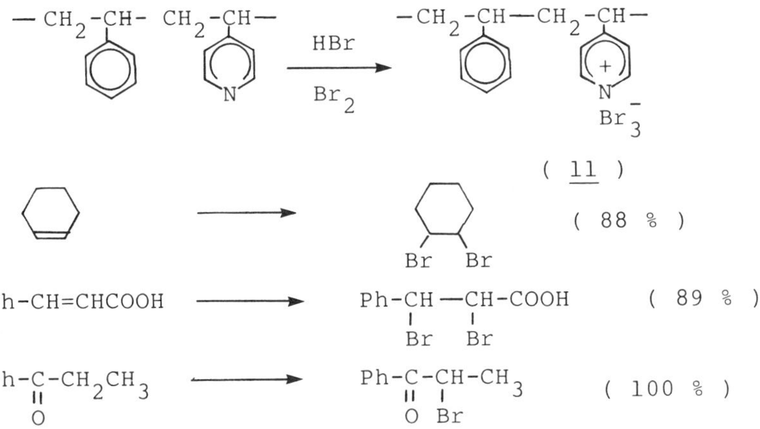

Preparation of poly(vinylpyridinium hydrobromide perbromide) (11) [90]

Cross-linked 4-vinylpyridine-styrene copolymer (1.49 g) is swollen in dioxane (25 mL) and the suspension is stirred for 1 h with 2.5 mL of 48% hydrobromic acid. After cooling in an ice bath, 2 mL of bromine is added, and the mixture is stirred for an additional 4 h. After filtration, the bright orange resin is dried under reduced pressure to yield 3.1 g of a product containing 5.6 meq Br per gram.

Bromination of ketones with 11 [90]. To a solution of 1.4 g of propiophenone in 30 mL of methanol is added 5.71 g of 11. The mixture is stirred slowly and heated to reflux until the bright orange color of the resin becomes lighter (40 min). The resin is then filtered and washed three times with dichloromethane. The combined filtrates are evaporated under reduced pressure to yield phenyl-1-bromoethyl ketone.

Regeneration of the spent polymer [90]. After each reaction, the resin is washed with aqueous sodium hydroxide, then with water. The regeneration is accomplished by reaction with excess 48% hydrogen bromide followed by addition of bromine to give 11. The polymeric trisubstituted phosphines (12a and 12b) can be transformed to the phosphine dichloride forms via phosgene treatment of the phosphine oxide form (13a and 13b).

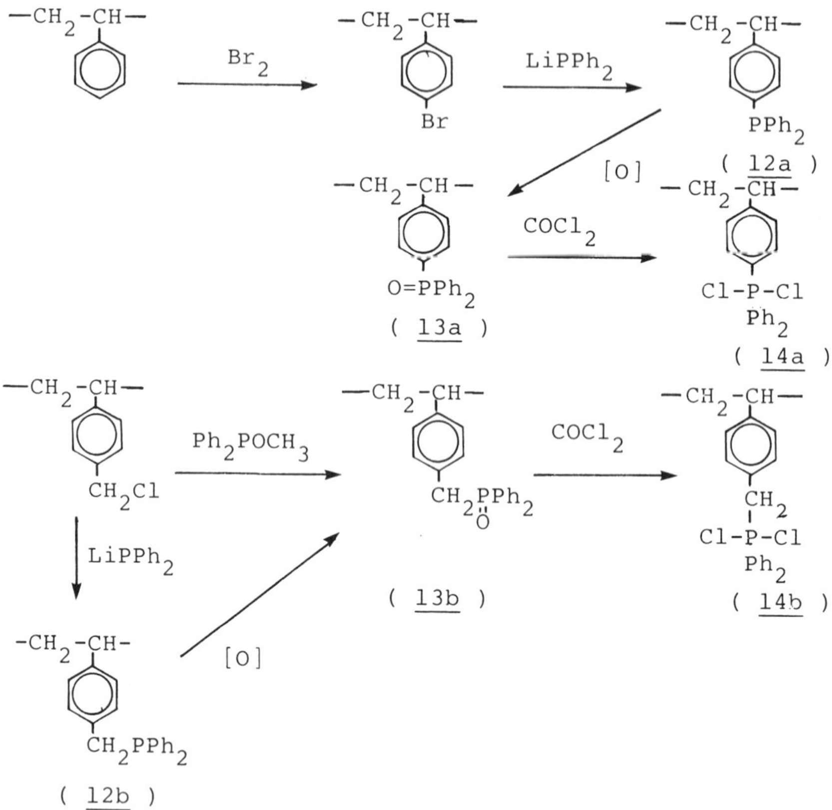

14a and 14b are used for the preparation of acid chlorides, amidoyl chlorides, nitriles, alkyl chlorides, and chloroolefins in excellent yield. The spent polymer (12a and 12b) can be regenerated [91].

$$PhCH_2COOH \xrightarrow{14\ (a\ or\ b)} PhCH_2COCl \quad (100\%)$$

$$PhCONHPh \xrightarrow{14\ (a\ or\ b)} PhC(Cl){=}NPh \quad (93\%)$$

$$PhCONH_2 \xrightarrow{14\ (a\ or\ b)} PhCN \quad (78\%)$$

$$PhCH_2OH \xrightarrow{14\ (a\ or\ b)} PhCH_2Cl \quad (88\%)$$

$$PhCOCH_3 \xrightarrow{14\ (a\ or\ b)} PhC(Cl){=}CH_2 \quad (75\%)$$

Similarly, the polymeric trisubstituted phosphine in tetrachloromethane is effective for the synthesis of alkyl chlorides and acid chlorides. The resulting phosphine oxide formed can be reconverted to the initial phosphine form with chlorosilane treatment. This behavior is quite similar to that of triphenyl phosphine in tetrachloromethane [92-94].

Preparation of the polymeric trisubstituted phosphine (*12a*) [91]

A suspension of 2% cross-linked polystyrene (83.68 g) in 500 mL of nitromethane containing BF_3 (56 g, 0.8 mol) is stirred at 25°C in the dark, while 256 g (1.60 mol) of Br_2 is added dropwise during 0.5 h. The mixture is stirred for an additional 18 h at 25°C, and the beads are filtered and washed with dichloromethane-methanol and then dichloromethane. The beads are dried in vacuo at 100°C. The poly(bromostyrene) thus obtained (36.92 g) is stirred for 1 h under nitrogen with 900 mL of tetrahydrofuran. A solution of 88 g (0.40 mol) of chlorodiphenylphosphine in 300 mL of tetrahydrofuran is added, followed by 6.4 g (0.93 mol) of lithium metal. The mixture is stirred at 25°C for 18 h after which time the excess pieces of lithium are carefully removed under nitrogen. The solution is then heated under nitrogen at reflux for 4.5 h. Methanol is then added; the beads are removed by filtration and washed with dichloromethane-methanol and then dichloromethane. The beads are dried in vacuo at 100°C overnight.

Oxidation of the polymeric trisubstituted phosphine: Preparation of 13a [91]. A suspension of 53.45 g (0.200 mol) of 12a in 50 mL of dichloromethane and 50 mL of acetic acid is stirred at 25°C while 70 mL of 40% peracetic acid in acetic acid is slowly added. After stirring at 25°C for 3 h, the beads are filtered and washed with dichloromethane-methanol and then dichloromethane. The beads are dried in vacuo overnight at 100°C.

Preparation of (*14a*) [91]. 13a (14.40 g) is stirred in 100 mL of dichloromethane as excess phosgene is added to the solution. The mixture is stirred at 25°C for 1 h. Excess phosgene is removed by filtering the beads and then washing them five times with 50-mL portions of dry dichloromethane. Poly(*p*-vinylbenzoyl chloride) is used for the preparation of acid chloride from carboxylic acid, as well [95].

The polymeric aryl iodine difluoride (15) is effective for the preparation of the *gem*-difluoride from alkenes in excellent yield [96].

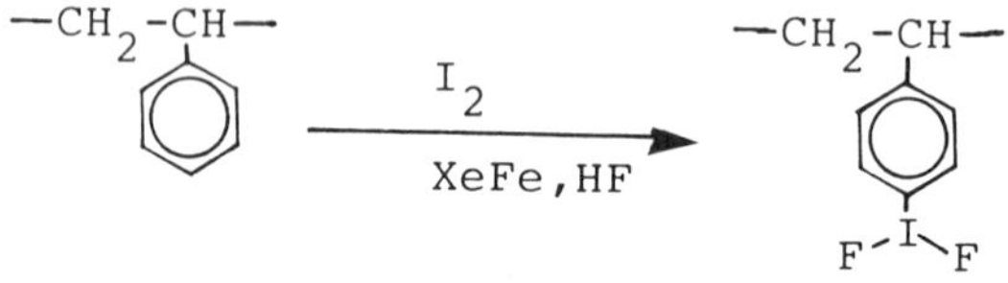

(15)

In contrast, bromine intercalated in graphite produces regioselective aromatic bromination as observed in the case of the bromination of 1,1-binaphthyl [97]. The intercalate $C_{24}SbCl_5$, which is prepared from antimonyl pentachloride and graphite, is more stable to moisture than antimonyl pentachloride and can be used for the halogen exchange of secondary alkyl and cycloalkyl bromides (or tosylate). The product distribution is different from antimony pentachloride [98]. Similarly, antimony pentafluoride and xenon hexafluoride can be intercalated into graphite using fluorinating reagents more improved in stability [99,100]. Bromine adsorbed on molecular sieve-5A or carbon undergoes selective bromination of styrene mixed with cyclohexene to α,β-dibromostyrene [101] together with the slight ring bromination [102].

2.3.3.c Polymeric Wittig and ylide reagents

A major advantage of a polymeric Wittig reagent over a monomeric reagent is to suppress the by-product formation. Particularly, excess lithium ions, which lead to the formation of *trans*-alkanes, are washed away after the polymeric betaine formation [103-106].

(12a)

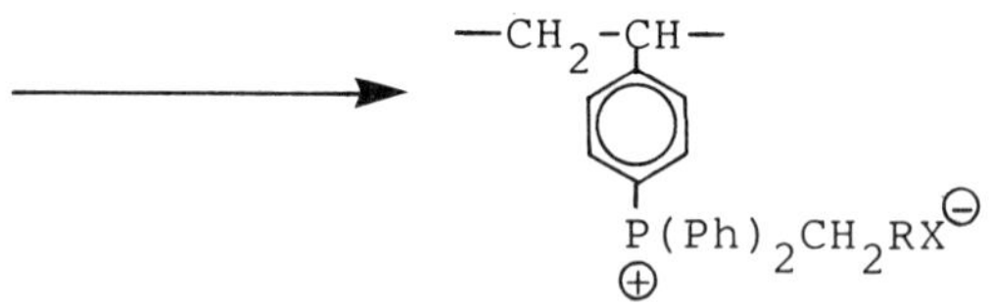

(16)

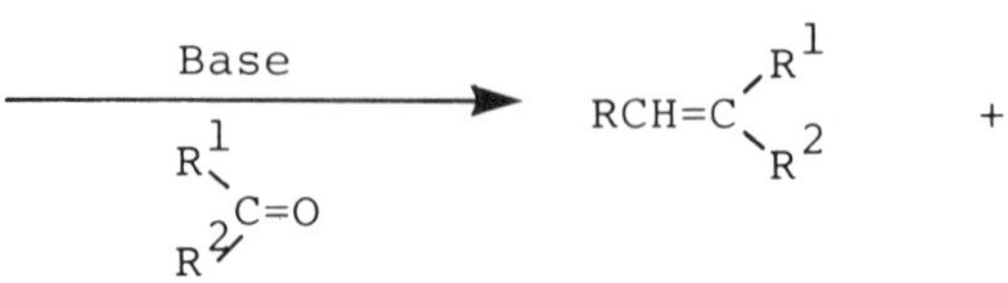

(13a)

Preparation of polymeric alkyl triphenylphosphonium bromide [105]

Polymeric trisubstituted phosphine (12a) [7.3% of P, 0.0705 mol of triphenylphosphane group (30 g)] is stirred with *N*-butyl bromide (11.6 g, 0.0841 mol) in tetrahydrofuran (80 mL) under reflux for

24 h. The polymer is then filtered and extracted with tetrahydrofuran for 24 h in a Soxhlet for 8 h and then dried for 24 h at 40°C in vacuo.

Preparation of polymeric Wittig reagent [105]

Polymeric *n*-butyl triphenylphosphonium bromide (19 g), dried and saturated with nitrogen, is slurried with anhydrous dioxane (40 mL). Butyllithium (20% solution) (13 mL) is added at 20°C. The supernatant solvent is removed and anhydrous dioxane (30 mL) is added. The mixture is stirred and filtered again. The washing procedure is repeated with dioxane until the Gilman test is negative.

Synthesis of olefins with polymeric Wittig reagent [105]. The polymeric Wittig reagent obtained above is suspended in dioxane (40 mL) at 20°C and vigorously stirred while a solution of acetophenone (2.95 g) in dioxane (10 mL) is added. After 8 h of stirring at 60°C, the polymer is extracted with aqueous tetrahydrofuran in a Soxhlet for 12 h. Removal of the solvent gives 4 g of extract, which is distilled in vacuum to afford 2-phenyl-2-hexene in 97% yield.

The phase-transfer-catalyzed method is also applied to soluble and insoluble polymeric betaines (16) [107].

$$(\underline{16}) \xrightarrow[\substack{CH_2Cl_2/50\%\ NaOH \\ PTC(CTAB)\ \text{or}\ Bu_4N^+Cl^-}]{R^1\text{-}CHO} RCH{=}CHR^1 + (\underline{13a})$$

Polymeric sulfur ylides (17) can be used for the epoxide formation. The advantage is the nonodor and easy handling over the monomeric sulfur ylide. An early attempt to epoxide formation based on 17 resulted in an inactive regeneration accompanied by the loss of some sulfur [108]. To avoid such a rapid loss of reactivity, the use of PTC under milder conditions showed no appreciable loss of sulfur and/or decrease in reactivity after five successive reaction recycles for obtaining epoxides in excellent yield [109].

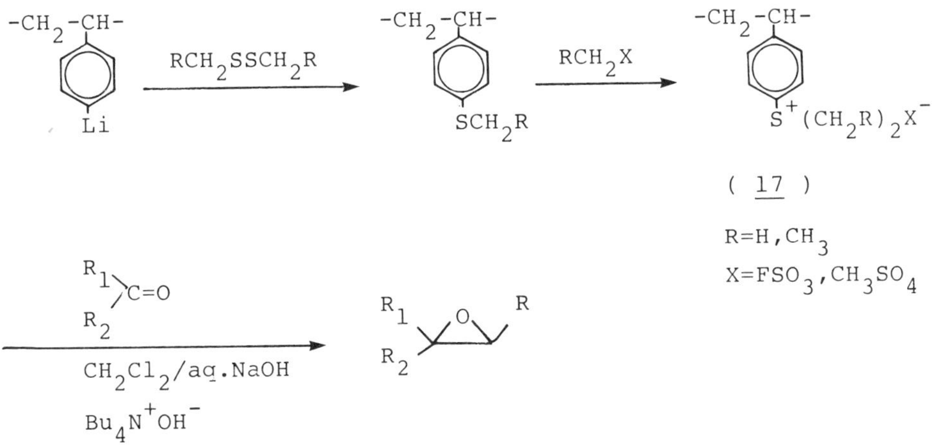

2.3.3.d Polymeric alkylating reagents

The polymeric phenylthiomethyllithium reagent (18) can be used with alkyliodides to afford the one-carbon-lengthened alkyliodide in good yield. The use of α,ω-diiodoalkane leads to the selective two-carbon-lengthened α,ω-diiodoalkane [110]. This reaction involves a two-step site-site interaction of a polymer ion cluster. This reaction is in contrast to the monooxidation of α,ω-diols by the chlorinated thioanisole polymer (9) [69].

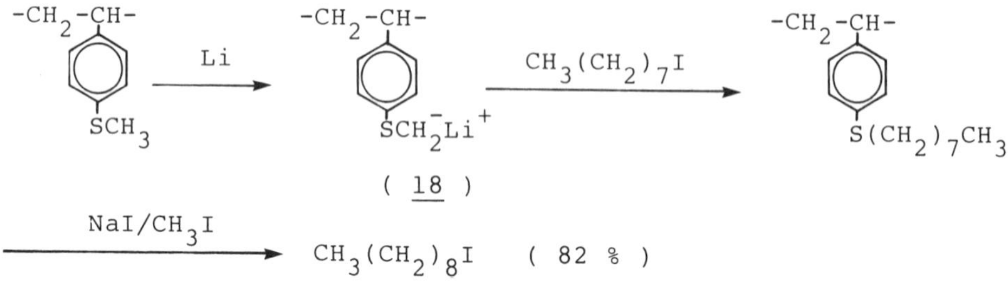

$$(\underline{18}) + I(CH_2)_4I \longrightarrow I(CH_2)_6I + I(CH_2)_5I$$

Anionic activation leading to C-C bond formation is realized in the reagents based on inorganic supports such as sodium methoxide on alumina or silica gel and sodium on charcoal (graphite) or alumina.

Alumina (or silica gel)-bound sodium methoxide undergoes the selective monoalkylation of malonic ester with α,ω-dibromoalkane. An increase in methoxide on alumina gives only the cyclized product [111].

$$CH_2(COOMe)_2 + Br(CH_2)_5Br \xrightarrow{MeONa/Al_2O_3} Br(CH_2)_5CH(COOMe)_2$$

Sodium on the surface of amorphous carbon, graphite, and alumina is quite effective for the monoalkylation of ketones in good to excellent yields. The process is interpreted in the terms of the metal enolate adsorbed on the surface [112].

Tetrabutylammonium fluoride can be changed into the more stable, nonhygroscopic reagent by being bound to silica gel, which is conveniently used for Michael addition, C-alkylation, and oxidation [113].

$$MeNO_2 + PhCH{=}CHCOPh \longrightarrow PhCH(CH_2NO_2)CH_2COPh \quad (86\%)$$

$$EtI + PhCOCH_2COPh \longrightarrow PhCOCH(Et)COPh \quad (55\%)$$

Preparation of silica gel-bound tetrabutylammonium fluoride [113]

A slurry of 20 g of 20% aqueous tetrabutylammonium fluoride and 10 g of 60 to 120-mesh silica gel purified by treatment with hydrochloric acid, washed with water, and dried at 150°C is made. Water is removed under reduced pressure. After most of the water is removed, 100 mL of methanol is added and evaporated. The resulting solid is then washed with ether and dried in air at 100°C.

Synthesis of 4-nitro-1, 3-diphenylbutanone [113]. Silica gel-bound tetrabutylammonium fluoride (1 g with 0.1 mmol eq of fluoride) is suspended in 20 mL of tetrahydrofuran containing 6.1 g (0.1 mol) of nitromethane and 1.04 g (5 mmol) of chalcone. The resulting mixture is stirred at 20°C for 6 h. Separation and recrystallization of the product from ethanol gives 1.08 g (75%) of 4-nitro-1, 3-diphenylbutanone.

Similarly, Celite-bound potassium fluoride is an effective reagent for C-, N-, O-, and S-alkylation [114].

$$(CH_3CO)_2CH_2 + CH_3I \longrightarrow (CH_3CO)_2CHCH_3 \quad (100\%)$$

$$PhOH + CH_3I \longrightarrow PhOCH_3 \quad (89\%)$$

$$PhSH + CH_3I \longrightarrow PhSCH_3 \quad (100\%)$$

$$PhNH_2 + CH_3I \longrightarrow PhNHCH_3 \quad (100\%)$$

Preparation of Celite-bound potassium fluoride [114]

Celite (20 g) is mixed with an equal weight of potassium fluoride in 500 mL of water, and the water is removed at 60°C with a rotary evaporator. The Celite obtained is then shaken in 100 mL of acetonitrile, filtered, washed with two 50-mL portions of acetonitrile, and dried in a desiccator at room temperature.

2.3.3.e Polymeric acylating reagents

N-Trifluoroacetyl-nylon 6,6 prepared from nylon 6,6 and trifluoroacetic anhydride is a mild trifluoroacetylating reagent (19) for amines and alcohols. The trifluoroacetylation can be carried out under mild conditions in excellent yield. The product is easily isolated from the spent polymer [115].

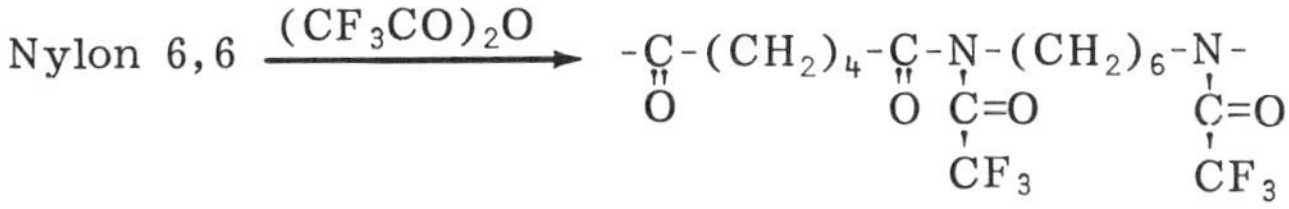

(19)

Preparation of N*-trifluoroacetyl-nylon 6,6 (19)* [115]

Nylon 6,6 (1.13 g) is suspended in 10 mL of dried ether and cooled to 0°C. To the mixture is added 2.1 mL (15 mmol) of trifluoroacetic anhydride. The mixture is stirred at room temperature for 15 h.

The polymer is recovered by filtration, washed with ether, and dried *in vacuo*.

Mixed carboxylic anhydrides attached to polystyrene (20 and 21) can also be used for the acylation of amines and alcohols [116-118].

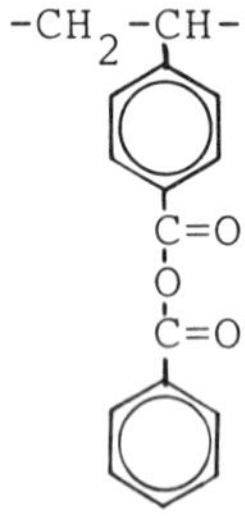

(20)

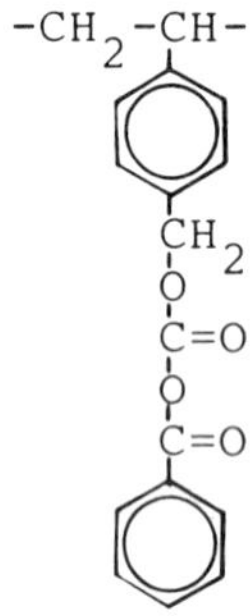

(21)

Poly(3-acyl-2-oxazolone) (22) can be used for the highly selective acylation of amines, amino alcohols, and polyols. Preferential acylation of the less hindered amino group and exclusive primary alkyl ester formation occurs with diols in the presence of Lewis acid [119].

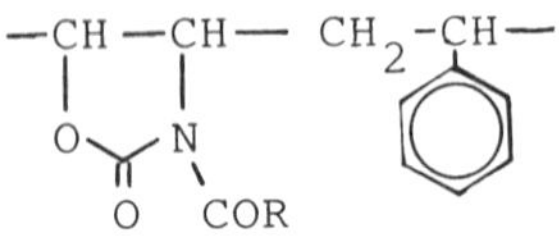

(22)

Polymeric yneamine (23) is an efficient acylating reagent for amides, carboxylic anhydrides, and esters synthesis. This reagent is presumed to react with carboxylic acid to give the intermediate, similar to the case of carbodiimide [120].

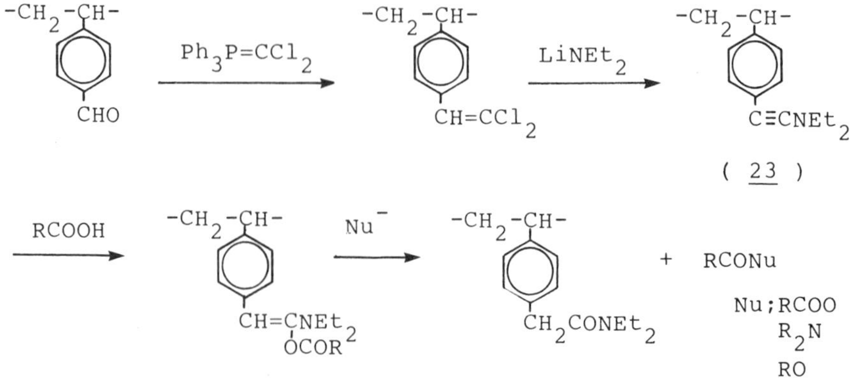

(23)

2.3.3.f Polymer-bound tosyl azide

Treatment of Amberlite XE-305 with chlorosulfonic acid followed by sodium azide affords the diazo transfer reagent (24), which is stable at room temperature and does not detonate on shock [121].

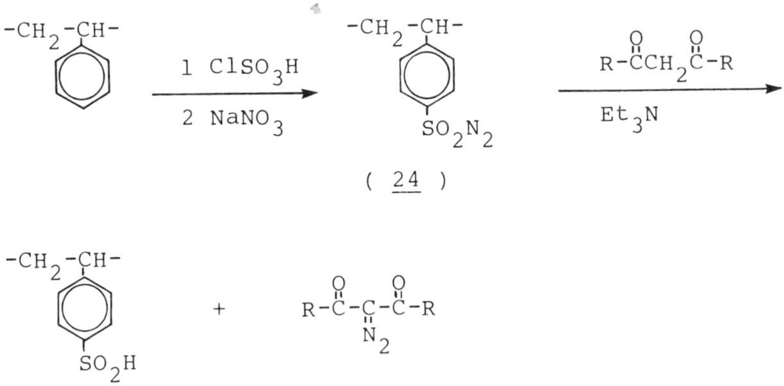

2.3.4 Polymeric Coupling Reagents

Chloromethylated polystyrene can be converted into the polymeric carbodiimide (25 to 27) by a method similar to conventional carbodiimide synthesis.

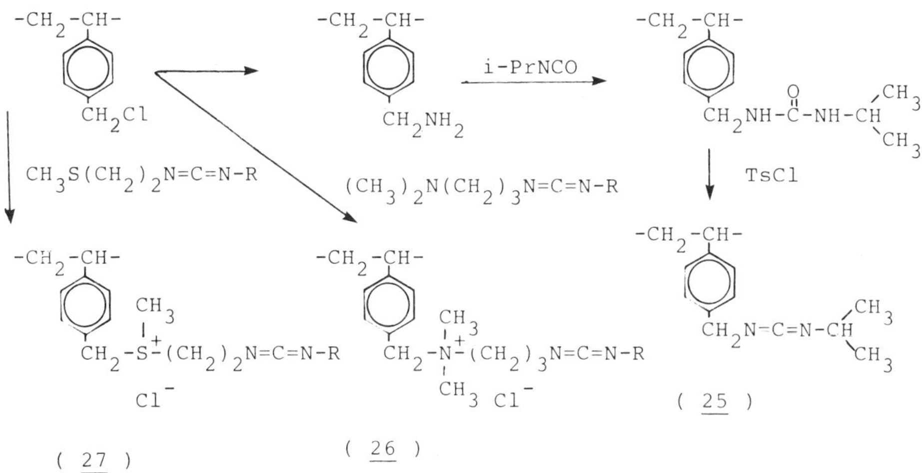

Other types of polymeric carbodiimides (28 and 29) are derived from the dehydroxylation of main-chain or side-chain urea groups.

Table 2. Miscellaneous Reactions Based on Polymeric Reagents

Polymeric reagent	Type of reaction	Yield (%)	References
SiO_2/H_2O		92 - 95	132
$SO_2Cl_2/SiO_2/H_2O$		100	41, 133
Montmorillonite K-10/ $CH(OCH_3)_3$	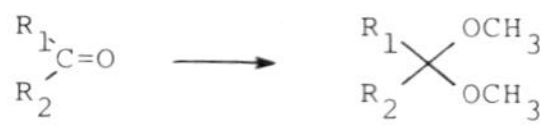	100	134
Al_2O_3/ROH, RSH, PhSeH, RNH_2, CH_3COOH	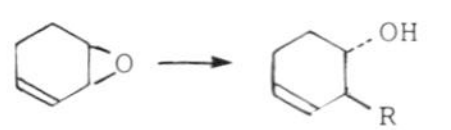 $R=CH_2CH\text{-}CH_2O$ $n\text{-}C_4H_9NH$ CH_3COO	67 75 33	135
	 $R=CH_3O$ $n\text{-}C_4H_9NH$	97 95	
	 $RY=PhS$ PhSe CH_3COO	64 72 52	
Montmorillonite K-10/ H_2O	$2\ RCH=CH_2 \longrightarrow R(CH_3)CH\text{-}O\text{-}CH(R)CH_3$		136

Al_2O_3/t-BuOOH	$R_1CH_2-CH(SeCH_3)-R_2$ →	$R_1HC=CHR_2$		80 - 96	137
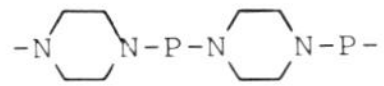	R-S-S-R →	R-S-R		80	138
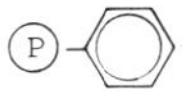 -SeCl/H_2O_2	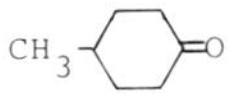 →	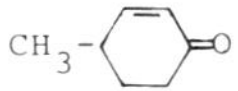		91	139
-SeNa/H_2O_2	$CH_3CHBrCOOEt$ →	$CH_2=CHCOOEt$		98	
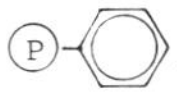 /BF_3OEt_3	$HO(CH_2)_nCOOH$ →	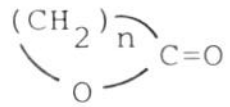	n=11	41	140
			17	78	
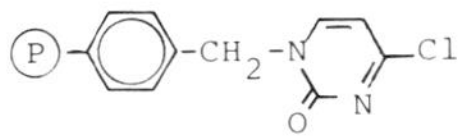	RNHC(=S)NHR →	RN=C=NR		60	141
	RNHC(=S)-S^- →	RNCS		38	
	RCOOH/R_1NH_2 →	$RCONHR_1$		90	
	RCOOH/R_1OH →	$RCOOR_1$		26	
	RCH=NOH →	RCN		80	
	$HO(CH_2)_nCOOH$ →	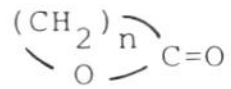		30	
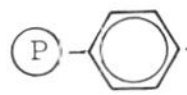 $-CH_2N^+(CH_3)_3$ $NaCO_3^-$ (Amberlyst A-26/ $NaCO_3$)	 →			90	142

$-(CH_2)_6-N=C=N-$ $-CH_2-CH(-N=C=N-R)-$

(28) (29)

The reactions using these polymeric carbodiimides include peptide synthesis [122,123], carboxylic anhydride synthesis, and Moffatt oxidation [124].

Preparation of polymeric carbodiimide (25) [124]

Chloromethylated polystyrene (3.1 mmol of Cl per gram) is reacted with potassium phthalimide (2.5 eq/Cl) in DMF at 100°C for 5 h followed by hydrazinolysis (hydrazine hydrate, 1 mL per gram of polymer in EtOH, reflux 6 h). Treatment with isopropylisocyanate (2 eq THF, 25°C, 15 h) and dehydration of the resulting urea (2 eq TsCl, Et_3N, CH_2Cl_2, reflux 15 h) affords the polymeric carbodiimide (25).

The polymeric analog of mesitylene sulfonylchloride (30) is quite suitable as a coupling reagent for oligonucleotide synthesis since the spent polymer can easily separate from the product [125,126].

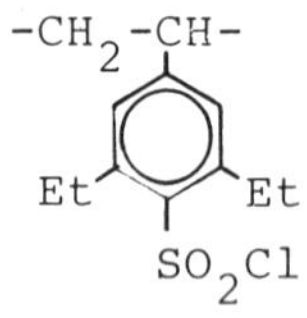

(30)

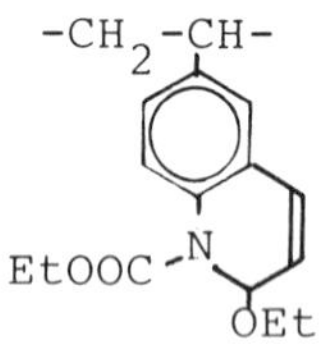

(31)

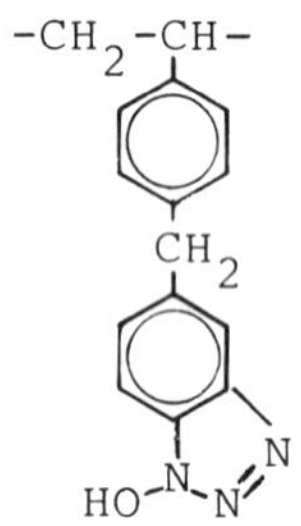

(32)

Other polymeric coupling reagents for peptide synthesis include polymer-bound 2-ethoxy-1-ethoxycarbonyl-1,2-dihydroquinoline (31) [127], *N*-hydroxybenzotriazole (32) [128,129], and triphenyl phosphine [130,131].

2.3.5 Miscellaneous Polymeric Reagents

A variety of reactions using other polymeric reagents are summarized in Table 2.

Dethioacetalization with silica gel-bound sulfuryl chloride [41]

A solution of sulfuryl chloride (1.2 mmol) in dichloromethane (3 mL) is added dropwise at room temperature (or 0°C) to a stirred mixture of wet silica gel (0.2 to 0.5 g of silicic acid, Mallinckrodt, 100 mesh, and 0.2 to 0.5 g of water) and thioacetal (1.0 mmol) in dichloromethane (6 mL). After stirring for 2 to 3 h at room temperature (or 0°C), finely powdered anhydrous potassium carbonate (0.3 to 0.6 g) is added to the reaction mixture and stirring is continued for another 20 min. Filtration and evaporation give the corresponding carbonyl compound.

Selective opening of epoxides by alumina-bound alcohols [135]

Cyclohexene oxide (1.0439 g, 10.6 mmol) is reacted in ether with 80.4 g of alumina (W-200-N) doped with allyl alcohol (3.22 g, 55.5 mmol) for 10 min at room temperature. The mixture is then filtered, and the filtrate is evaporated *in vacuo* to give *trans*-2-allyloxycyclohexanol in 57% yield.

Macrolide synthesis by polystyrene-boron trifluoride system [140]

A solution of 15-hydroxypentadecanoic acid (0.5 g, 1.94 mmol) in 1,2-dichloroethane (250 mL) containing polystyrene beads (Rohm and Haas XE-305, 0.5 g) is heated to reflux and boron trifluoride etherate (3.0 mL) is added. After 7 days at reflux, the mixture is cooled, filtered, and washed with chloroform. The combined filtrate is evaporated in vacuo to give macrolide in 76.6% yield.

REFERENCES

1. R. B. Merrifield, J. Am. Chem. Soc., 85, 2149 (1963).
2. R. L. Letsinger and V. Mahadevan, J. Am. Chem. Soc. 88, 5319 (1966).
3. J. M. J. Fréchet and C. Schuerch, J. Am. Chem. Soc. 93, 492 (1971).
4. a. A. Akelah and D. C. Sherrington, Chem. Rev. 81, 557 (1981).
   b. W. H. Daly, Makromol. Chem. Suppl. 2, 3 (1979).
   c. J. M. J. Fréchet, Tetrahedron 37, 663 (1981).
   d. P. Hodge and D. C. Sherrington, Eds., Polymer Supported Reactions in Organic Synthesis, John Wiley & Sons Ltd., Chichester, West Sussex, England.
   e. M. A. Kraus and A. Patchornik, J. Polym. Sci. Macromol. Rev. 15, 55 (1980).
   f. N. K. Mathur, C. K. Narang, and R. E. Williams, Polymers Aids in Organic Chemistry, Academic Press, Inc., New York (1980).
5. a. F. S. Feinberg and R. B. Merrifield, Tetrahedron 30, 3209 (1974).

b. G. A. Olah, D. A. Beal, S. H. Yu, and J. A. Olah, Synthesis 560 (1974).
c. K. W. Pepper, H. M. Paisley, and M. A. Young, J. Chem. Soc. 4097 (1953).
d. J. T. Sparrow, Tetrahedron Lett. 4637 (1975).
6. M. J. Farrall and J. M. J. Fréchet, J. Org. Chem. 41, 3877 (1976).
7. A. McKillop and D. W. Young, Synthesis 401, 481 (1979).
8. N. M. Weinschenker, G. A. Grosby, and J. Y. Wong, J. Org. Chem. 40, 1966 (1975).
9. N. H. Fung, P. deMayo, J. H. Schauble, and A. C. Weedon, J. Org. Chem. 43, 3977 (1978).
10. F. Hodosan and N. Serban, Rev. Roum. Chim. 14, 121 (1969).
11. a. E. Santaniello, C. Farachi, and A. Manzocchi, Synthesis 912 (1979).
b. E. Santaniello, F. Ponti, and A. Mazocchi, Synthesis 891 (1978).
12. V. Ciurdaru and F. Hodosan, Rev. Roum. Chim. 22, 1027 (1977).
13. G. Bram, E. Dincan, and A. Loupy, J. Chem. Soc. Chem. Commun. 1066 (1981).
14. M. L. Hallensleben, J. Polym. Sci. Polym. Symp. 47, 1 (1974).
15. H. W. Gibson and F. C. Bailey, J. Chem. Soc. Chem. Commun. 815 (1977).
16. R. O. Hutchins, N. R. Natale, and I. M. Taffer, J. Chem. Soc. Chem. Commun. 1088 (1978).
17. G. H. Posner and A. W. Runquist, Tetrahedron Lett. 3601 (1975).
18. G. H. Posner, A. W. Runquist, and M. J. Chapdelaine, J. Org. Chem. 42, 1201 (1977).
19. F. Glockling and D. Kingston, Chem. Ind. 1038 (1961).
20. J.-M. Lalancette, G. Rollin, and P. Dumas, Can. J. Chem. 50, 3058 (1972).
21. M. Contento, D. Savoia, C. Trombini, and A. U-Ronchi, Synthesis 30 (1979).
22. J. M. J. Fréchet, J. Warnock, and M. J. Farrall, J. Org. Chem. 43, 2618 (1978).
23. J. M. J. Fréchet, P. Darling, and M. J. Farrall, J. Org. Chem. 46, 1728 (1981).
24. G. Cainelli, G. Cardillo, M. Orena, and S. Sandri, J. Am. Chem. Soc. 98, 6737 (1976).
25. G. Cardillo, M. Orena, and S. Sandri, Tetrahedron Lett. 3985 (1976).
26. R. P. Singh, H. N. Subbarao, and S. Dev, Tetrahedron 35, 1789 (1979).
27. J. S. Filippo, Jr., and C. I. Chern, J. Org. Chem. 42, 2182 (1977).
28. N. H. Andersen and H.-S. Uh, Synth. Commun. 3, 115 (1973).

29. N. H. Andersen and H.-S. Uh, Tetrahedron Lett. 2079 (1973).
30. V. Balogh, M. Fetizon, and M. Golfier, J. Org. Chem. 36, 1339 (1971).
31. N. Khan, D. E. Loeber, T. P. Toube, and B. C. Weedun, J. Chem. Soc. Perkin Trans. 1, 1457 (1975).
32. M. Fetizon, H. Golfier, and J.-M. Louis, J. Chem. Soc. Chem. Commun. 1102 (1969).
33. J. Bastard, M. Fetizon, and J. C. Gramain, Tetrahedron 29, 2867 (1973).
34. M. Fetizon and P. Mourgnew, Tetrahedron 30, 327 (1974).
35. M. Fetizon, M. Golfier, and J.-M. Louis, J. Chem. Soc. Commun. 1118 (1969).
36. M. Fetizon, M. Golfier, and J.-M. Louis, Tetrahedron Lett. 1931 (1973).
37. V. Balogh, M. Fetizon, and M. Golfier, Angew. Chem. Int. Ed. Engl. 8, 444 (1969).
38. M. Fetizon, M. Golfier, and I. Papadakis, Tetrahedron 31, 165 (1975).
39. F. S. Kakis, M. Fetizon, N. Dorchkine, M. Golfier, P. Hourqres, and T. Prange, J. Org. Chem. 39, 523 (1974).
40. N. P. Damodaran and S. Dev, Tetrahedron 24, 4123 (1968).
41. M. Hojo and R. Masuda, Tetrahedron Lett. 613 (1976).
42. Z. Cohen, E. Keinan, Y. Mazur, and T. H. Varkony, J. Org. Chem. 40, 2141 (1975).
43. E. Proksh and A. deMeijer, Tetrahedron Lett. 4851 (1976).
44. E. Keinan and Y. Mazur, J. Org. Chem. 42, 844 (1977).
45. E. Keinan and Y. Mazur, J. Am. Chem. Soc. 89, 3861 (1977).
46. G. H. Posner, R. B. Perfetti, and A. W. Runquist, Tetrahedron Lett. 3499 (1976).
47. G. H. Posner and M. J. Chapdelain, Tetrahedron Lett. 3227 (1977).
48. G. H. Posner and M. J. Chapdelain, Synthesis 555 (1977).
49. L. A. Carpino, J. Org. Chem. 35, 3971 (1970).
50. H. Hauptmann, Tetrahedron 32, 1293 (1976).
51. L. A. Carpino and J. P. Masaracchia, J. Org. Chem. 37, 1851 (1972).
52. L. A. Carpino, L. V. McAdams III, R. H. Pynbnandt, and J. W. Spiewak, J. Am. Chem. Soc. 95, 476 (1971).
53. S. L. Regen and C. Koteel, J. Am. Chem. Soc. 99, 3837 (1977).
54. E. C. Taylor, C.-S. Chiang, A. McKillop, and J. F. White, J. Am. Chem. Soc. 98, 6750 (1976).
55. K. T. Liu and Y.-C. Tang, J. Org. Chem. 43, 2717 (1978).
56. F. Helfferich and D. B. Luten, Jr., J. Appl. Polym. Sci. 8, 2899 (1964).
57. T. Takagi, J. Polym. Sci. Polym. Lett. Ed., B5, 1031 (1967).

58. J. M. J. Frechet and K. E. Haque, Macromolecules 8, 130 (1975).
59. C. R. Harrison and P. Hodges, J. Chem. Soc. Chem. Commun. 1009 (1974).
60. S. E. Jacobson, F. Mares, and P. H. Zambri, J. Am. Chem. Soc. 101, 6938 (1979).
61. S. E. Jacobson, F. Mares, and P. H. Zambri, J. Am. Chem. Soc. 101, 6946 (1979).
62. H. Schuttenberg and R. C. Schulz, Makromol. Chem. 143, 153 (1971).
63. H. Schuttenberg and R. C. Schulz, Angew. Makromol. Chem. 18, 175 (1971).
64. H. Schuttenberg, G. Klump, U. Kaczmar, S. R. Turner, and R. C. Schulz, J. Macromol. Sci. Chem. A7(5), 1085 (1973).
65. T. Sato and R. C. Schulz, Makromol. Chem. 180, 299 (1979).
66. Y. Sato, N. Kunieda, and M. Kinoshita, Chem. Lett. 1023 (1972).
67. Y. Sato, N. Kunieda, and M. Kinoshita, Makromol. Chem. 178, 683 (1977).
68. K. Hahn and R. C. Schulz, Angew. Makromol. Chem. 50, 53 (1976).
69. G. A. Crosby, N. M. Weinshenker, and H. S. Uh, J. Am. Chem. Soc. 97, 2232 (1975).
70. M. Gordon, M. L. Depamphilis, and C. E. Griffin, J. Org. Chem. 28, 698 (1963).
71. G. Cainelli and F. Manescalchi, Synthesis 723 (1975).
72. C. R. Harrison and P. Hodge, Synthesis 299 (1980).
73. F. Manescalchi, M. Orena, and D. Savoia, Synthesis 445 (1979).
74. G. Cainelli, F. Manescalchi, M. Panuzro, and O. Eilia, Synthesis 141 (1979).
75. G. Gelbard and S. Colonna, Synthesis 113 (1977).
76. S. Cacchi and L. Caglioti, Synthesis 64 (1979).
77. A. Bongini, G. Gainelli, M. Contento, and F. Manescalchi, Synthesis 142 (1980).
78. G. Cainelli, M. Contento, F. Manescalchi, and M. C. Mussatto, Synthesis 302 (1981).
79. G. Gainelli, F. Manescalchi, and M. Panunzio, Synthesis 472 (1976).
80. C. L. Borders, Jr., D. L. MacDonell, and J. L. Chambers, Jr., J. Org. Chem. 37, 3549 (1972).
81. G. Cainelli, M. Contento, F. Manescalchi, and R. Regnoli, J. Chem. Soc. Perkin Trans. 1, 2516 (1980).
82. G. D. Robinson, Jr., Radiopharm. Label. Compounds Pro. Symp. 1, 423 (1973); Chem. Abst. 81, 91003q (1974).
83. J. R. Dalton and S. L. Regen, J. Org. Chem. 44, 4443 (1979).
84. S. L. Regen, S. Quici, and M. D. Ryan, J. Am. Chem. Soc. 101, 7629 (1979).

85. S. Quici and S. L. Regen, J. Org. Chem. 44, 3436 (1979).
86. M. O. Brimeyer, A. Mehrota, S. Quici, A. Nigam, and S. L. Regen, J. Org. Chem. 45, 4254 (1980).
87. K. Takahashi, T. Nishizuka, and H. Iida, Tetrahedron Lett. 2389 (1981).
88. C. Yaroslavsky, A. Patchornik, and E. Katchalski, Tetrahedron Lett. 3629 (1970).
89. C. Yaroslavsky and E. Katchalski, Tetrahedron Lett. 5 (1972).
90. J. M. J. Fréchet, M. J. Parrall, and L. J. Nuyens, J. Macromol. Sci. Chem., All(13), 507 (1977).
91. H. M. Relles and R. W. Schluenz, J. Am. Chem. Soc. 96, 6469 (1974).
92. S. L. Regen and D. P. Lee, J. Org. Chem. 40, 1669 (1975).
93. D. C. Sherrington, D. J. Craig, J. Dagleish, G. Domin, J. Taylor, and G. V. Meehan, Eur. Polym. J. 13, 73 (1977).
94. P. Hodge and G. Richardson, J. Chem. Soc. Chem. Commun. 622 (1975).
95. M. L. Hallesleben, Angew. Makromol. Chem. 31, 143 (1973).
96. M. Zupan and A. Pollak, J. Chem. Soc. Chem. Commun. 715 (1975).
97. H. B. Kagan, Chem. Tech. 6, 510 (1976).
98. J.-L. Luche, J. Bertin, and H. B. Kagan, Tetrahedron Lett. 759, 763 (1974).
99. J.-M. Lalancette and J. Lafontaine, J. Chem. Soc. Chem. Commun. 815 (1973).
100. H. Selig, M. Rabinovitz, I. Agrant, C.-H. Lin, and L. Ebert, J. Am. Chem. Soc. 98, 1601 (1976).
101. P. A. Risbood and D. M. Ruthven, J. Am. Chem. Soc. 100, 4919 (1978).
102. S. H. Stoldt and A. Turk, J. Org. Chem. 34, 2370 (1969).
103. F. Camps, J. Castells, J. Font, and F. Vela, Tetrahedron Lett. 1715 (1971).
104. S. V. McKinly and J. W. Rakshys, Jr., J. Chem. Soc. Chem. Commun. 134 (1972).
105. W. Heitz and R. Michels, Angew. Chem. Int. Ed. Engl. 11, 298 (1972).
106. W. Heitz and R. Michels, Ann. 767, 277 (1973).
107. S. D. Clarke, C. R. Harrison, and P. Hodge, Tetrahedron Lett. 1375 (1980).
108. S. Tanimoto, J. Horikawa, and R. Oda, J. Chem. Soc. Jpn. Ind. Chem. Sect. 70, 1269 (1967).
109. M. J. Farrall, T. Durst, and J. M. J. Fréchet, Tetrahedron Lett. 203 (1979).
110. G. A. Crosby and M. Kato, J. Am. Chem. Soc. 99, 278 (1977).

111. G. Bram and T. F. Khan, J. Chem. Soc. Chem. Commun. 522 (1979).
112. H. Hart, B.-L. Chen, and C.-T. Peng, Tetrahedron Lett. 3121 (1977).
113. J. H. Clark, J. Chem. Soc. Chem. Commun. 780 (1978).
114. T. Ando and J. Yamawaki, Chem. Lett. 45 (1979).
115. E. J. Gunster and R. C. Schulz, Makromol. Chem. 180, 1891 (1979).
116. M. B. Shambhu and G. A. Digenis, Tetrahedron Lett. 1627 (1973).
117. M. B. Shambhu and G. A. Digenis, J. Chem. Soc. Chem. Commun. 619 (1974).
118. G. E. Martin, M. B. Shambhu, S. R. Shakhshir, and G. A. Digenis, J. Org. Chem. 43, 4571 (1978).
119. T. Kunieda, T. Higuchi, Y. Abe, and M. Hirobe, Tetrahedron Lett. 23, 1159 (1982).
120. J. A. Moore and J. J. Kennedy, J. Chem. Soc. Chem. Commun. 1079 (1978).
121. W. R. Roush, D. Feitler, and J. Rebek, Tetrahedron Lett. 1391 (1974).
122. Y. Wolman, S. Kivity, and M. Frankel, J. Chem. Soc. Chem. Commun. 629 (1967).
123. H. Ito, N. Takamatsu, and I. Ichikizaki, Chem. Lett. 577 (1975).
124. N. M. Weinshenker and C.-M. Shen, Tetrahedron Lett. 3281, 3285 (1972).
125. M. Rubenstein and A. Patchornik, Tetrahedron Lett. 2881 (1972).
126. M. Rubenstein and A. Patchornik, Tetrahedron Lett. 31, 1517 (1975).
127. J. Brown and R. E. Williams, Can. J. Chem. 49, 3765 (1971).
128. R. Kalir, M. Fridkin, and A. Patchornik, Eur. J. Biochem. 42, 151 (1974).
129. R. Kalir, A. Warshawsky, M. Fridkin, and A. Patchornik, Eur. J. Biochem. 59, 55 (1975).
130. R. Appel, W. Struver, and L. Willms, Tetrahedron Lett. 905 (1976).
131. K. Horiki, Tetrahedron Lett. 4103 (1976).
132. F. Huet, M. Pellet, and J. M. Conia, Tetrahedron Lett. 3505 (1977).
133. M. Hojo and R. Masuda, Tetrahedron Lett. 3883 (1977).
134. E. C. Taylor and C.-S. Chiang, Synthesis 467 (1977).
135. G. H. Posner and D. Z. Rogers, J. Am. Chem. Soc. 99, 8208, 8214 (1977).
136. J. M. Adams, Angew. Chem. Int. Ed. Engl. 17, 282 (1978).
137. D. Laber, L. Hevesi, W. Dumont, and A. Krief, Tetrahedron Lett. 1141, 1145 (1978).

138. D. N. Harpp, J. Adams, J. G. Gleason, D. Mullins, and K. Stelion, Tetrahedron Lett. 3989 (1978).
139. R. Michels, M. Kato, and W. Heitz, Makromol. Chem. 177, 2311 (1976).
140. L. T. Scott and J. O. Naples, Synthesis 738 (1976).
141. K. Kondo, K. Hyodo, M. Murakami, and K. Takemoto, Makromol. Chem. 182, 3411 (1981).
142. G. Cardillo, M. Orena, G. Porzi, and S. Sandri, Synthesis 793 (1981).

3

Polymeric Catalysts

Hirofusa Shirai and Nobumasa Hojo
Shinshu University, Nagano, Japan

3.1 INTRODUCTION

A metal complex with polymeric ligands has interesting and specific characteristics that are different from those of a low-molecular-weight analog due to the polymer matrix. In recent years metal-polymer complexes have been of interest to many chemists, as these polymers have highly functional characteristics. A variety of studies of these polymers, such as electrical conductivity, thermal stability, separation, photofunctions, carrier gas, catalytic activity, pharmaceutical activity, and biomedical effects have been carried out.

In this chapter the synthesis of metal-polymer complexes is reviewed and experimental methods are described. Most metalloenzymes are metal complexes of protein with a metal ion. Therefore, metal-polymer complexes are important as models for metalloenzymes. Specific catalytic actions of the polymer complexes have received particular attention as models for metalloenzymes, because researchers have come to understand that these are not only good approximate models, but also lead to the development of highly efficient catalysts.

Other nonenzyme polymeric metal complexes have been found to be useful as insoluble catalysts. The polymeric-metal complexes as models of metalloenzyme and immobilized catalyst for practical use are also described in this chapter. Many reviews (1-5) and books (6-10) concerning metal-polymer complexes have appeared in recent years and are appropriately referenced.

3.2 METAL POLYMER COMPLEXES AND THEIR SYNTHESIS

3.2.1 Direct Complexation of Polymeric Ligand with Metal Ion

3.2.1.a Homogeneous complex formation

This type of metal-polymer complex is synthesized by the reaction of synthetic or natural polymers containing donating groups such as hydroxyl, carboxylic acid, ketone, phosphonic acid, amine, heterocyclic nitrogen, or thiol with metal ions. The typical polymeric ligands are shown in Table 1. In the case of complex soluble polymeric

Table 1. Typical Polymeric Ligands

Polyamine

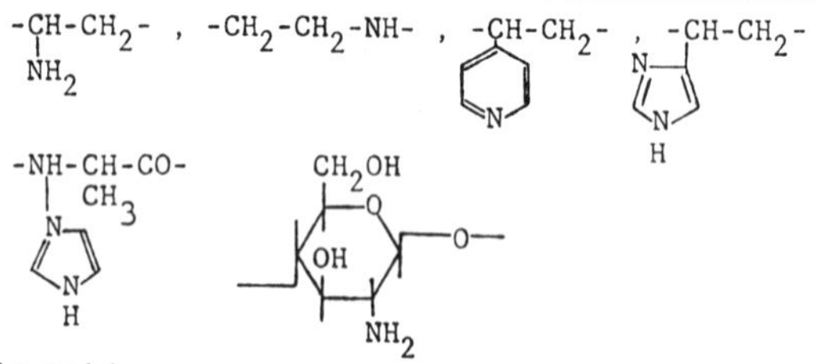

Polycarboxylic acid

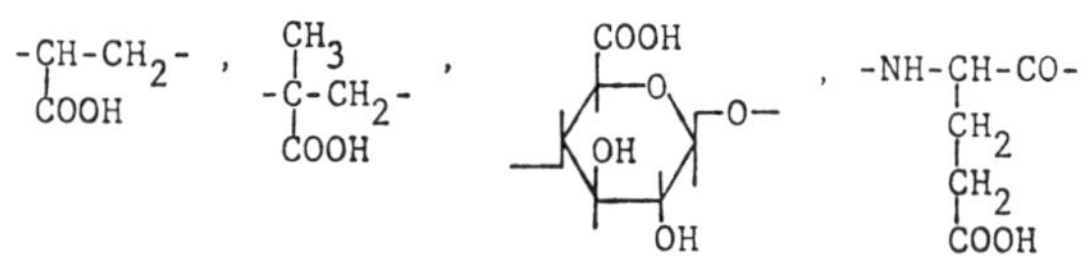

Polyalcohol

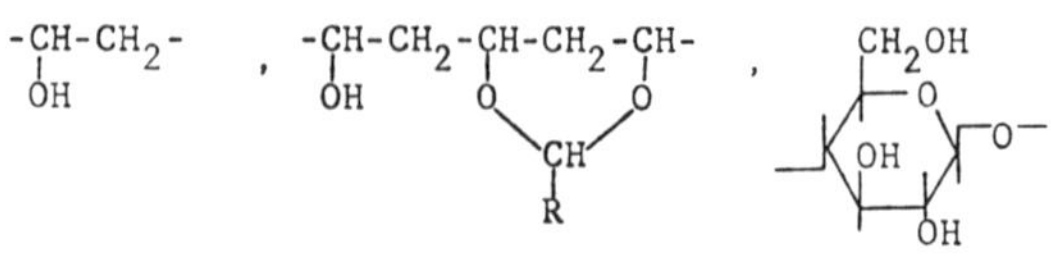

Polyphosphoric acid

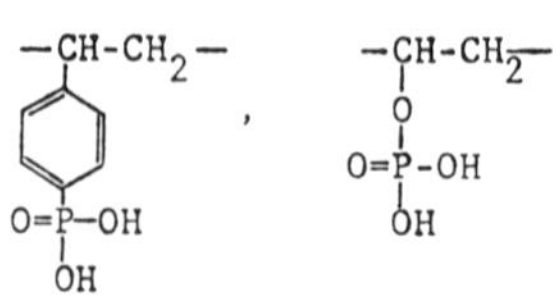

(Table 1 continued)

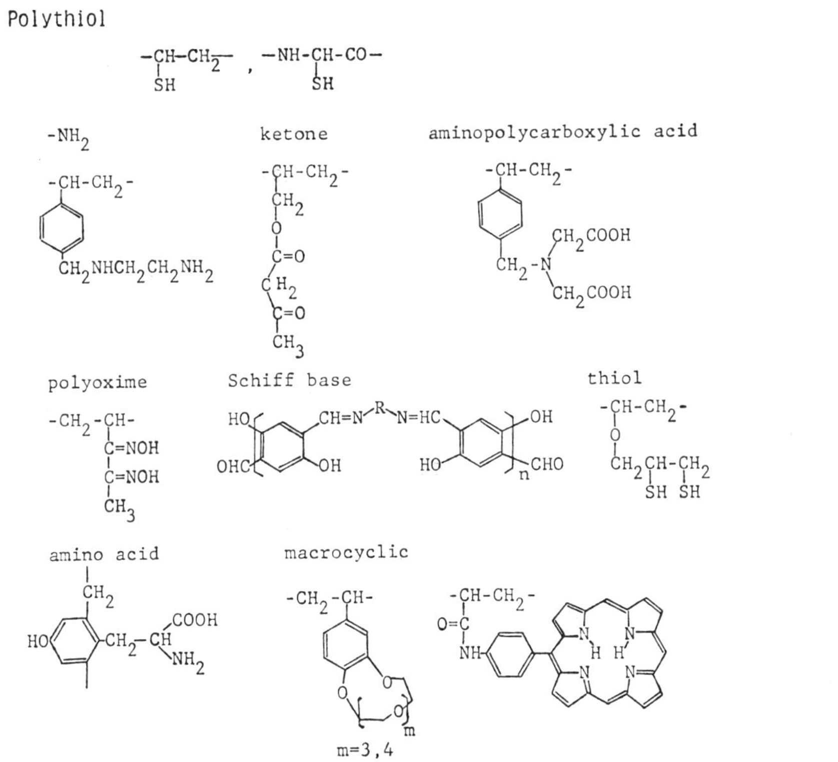

ligands with metal ion, the complex formation is often achieved simply by mixing the two solutions. If necessary, the pH of the solution is adjusted to the desired value with an alkaline solution of KOH, NaOH, or NH_4OH.

Complexations of this type often result in inter- or intramolecular chelate bridging, as shown in Reactions 1 and 2, respectively. In the case of complex formation by intermolecular chelate bridging, the resulting metal-polymer complex is often insoluble in water and in organic solvents. Furthermore, the coordination structure is not clear. Intramolecular metal-polymer complexes are, however, often soluble either in water or in organic solvents. The metal-polymer complexes with uniform structures are formed when a metal ion such as a cupric and a ferric ion with a labile ligand exchanges for a polymeric ligand. The composition of the complex is almost constant due

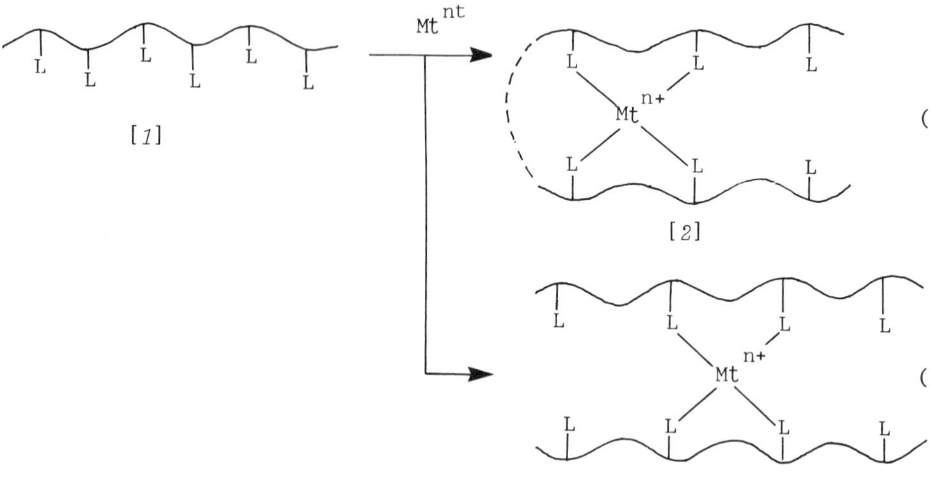

[*1*] [*2*] [*3*]

(1)

(2)

to the large stability of intrapolymeric chelate, as illustrated in Reaction 1. With low T_{Mt}/T_L, the ratio of the molar concentration of metal ion to the molar concentration of monomeric unit of the polymer, the resulting metal-polymer complex is usually soluble.

Preparation of Cu(II)-poly(vinyl alcohol) complex (11-13)

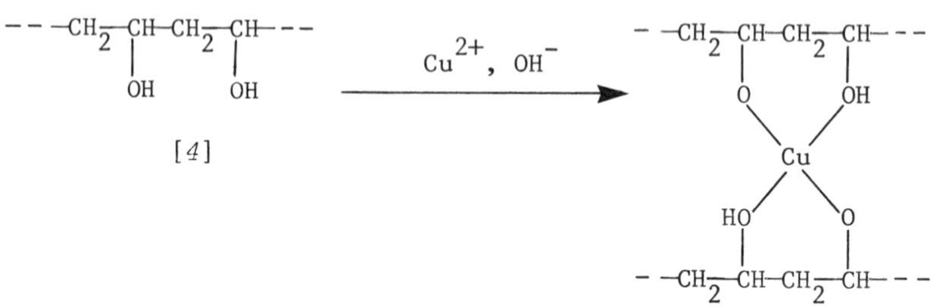

[*4*] [*5*]

(3)

The poly(vinyl alcohol) (PVA) is separated into several fractions of different average molecular weight according to the nonsolvent addition method, and each fraction is completely hydrolyzed by alkali in methanol. An aqueous Cu(II) ion solution, 5 mL (0.1 M), is added to 20 mL of an aqueous solution of PVA ($\bar{P}_n$ = Ca. 500, 0.2 M, molar concentration of monomeric units of PVA) containing a neutral salt. The mixture is stirred while an aqueous KOH solution (0.5 N) is added dropwise to the aqueous solution. When the pH of the mixed solution is adjusted to above 7.3 the color of the solution changes from light blue to green. The volume is adjusted to 100 mL with water, and the system is kept at 25°C for 24 h. Since the $T_{Cu^{2+}}/T_{HL}$ [the ratio of molar concentration of Cu(II) ion to the molar concentration of vinyl alcohol units] is 0.125, the resulting Cu(II)-PVA complex is soluble in water.

The Cu(II)-PVA complex obtained is dialyzed against distilled water for 3 days to remove surplus reagents. A film of Cu(II)-complex is made by casting this solution onto a chromium, polyethylene, or poly(methyl methacrylate) plate and drying for about 3 days at room temperature.

A powder of Cu(II)-PVA complex is produced by slowly adding to the dialyzed Cu(II) complex solution a nonsolvent such as methanol with vigorous agitation. The greenish gel-like precipitate is filtered, washed with acetone, and then dried at 30°C in vacuum for 3 days over phosphorus pentoxide.

A pH titration of Cu(II)-PVA complex indicates that four hydroxyl groups combine with one Cu(II) ion above pH 7.3 (14). The Cu(II)-PVA complex in aqueous solution has absorption bands at 640 nm (log ε = 1.5), 350 nm (log ε = 3.0), and at 260 nm (log ε = 3.5). The band at 640 nm is assigned to the d-d band of the Cu-O complex in the square planar ligand field. The viscosity of the Cu(II)-PVA aqueous solution diminishes sharply with complex formation. The polymer metal complex probably assumes a tightly packed conformation. A Cu-O stretching band in the polymer complex film appears at 605 cm^{-1} (15). Other metal complexes, such as Fe(III)-PVA, were prepared in a similar way (16).

Preparation of Cu(II)-polyvinylimidazone complex (17)

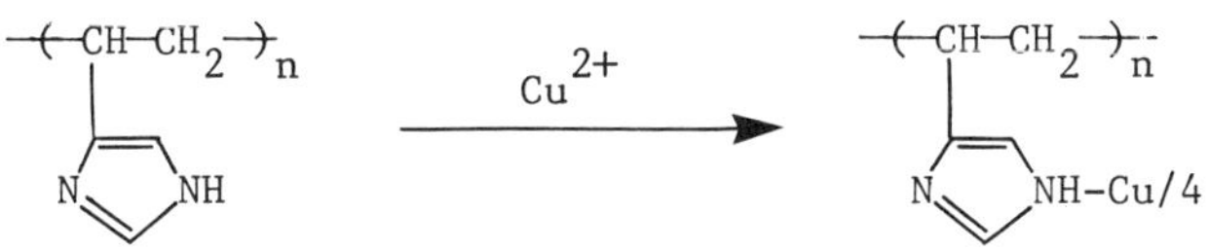

[*6*] [*7*]

(4)

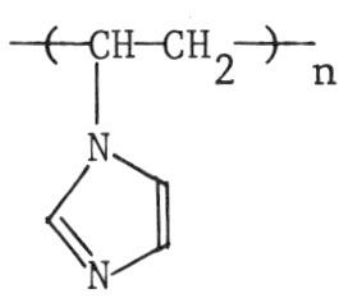

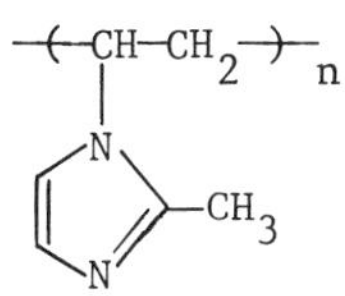

[*8*] [*9*]

Poly(1-vinylimidazole) (P-1-Vim) [*8*], poly(2-methyl-1-vinylimidazole) (PVMIm) [*9*], and poly(4- or 5-vinylimidazole) (P-4- or 5-VIm) [*7*] are obtained by the polymerization of the corresponding monomers. The Cu(II)-PVIm complexes are prepared by adding aqueous Cu(II) neutral salt solution to an aqueous solution of PVIm containing small amounts of perchloric acid to dissolve the polymeric ligand. The Cu(II)-PVIm complexes have characteristics that the low-molecular-weight complexes do not. Depression of the molecular motion of the polymeric ligand by these polymer complexes can easily be effected by changing the pH or ionic strength of the aqueous solution. In addition, copper-copper interactions are observed (18).

3.2.1.b Heterogeneous complex formation

The metal-polymer complexes can also be prepared effectively by the interfacial procedure. The extent of chelation to polymer is higher in the interfacial reaction method than in the homogeneous reaction method. Cu(II)-polyhydroxamato complexes were produced by this method (19,20).

Preparation of Cu(II)-polyhydroxamato complex (20)

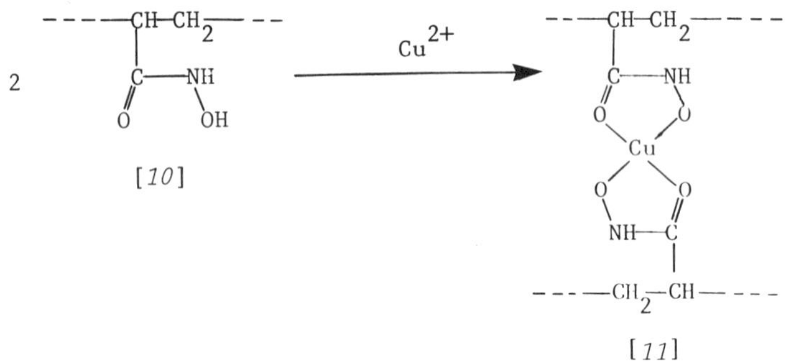

(5)

Poly(ε-chloroacrylhydroxamic acid) (0.1 mol of the monomer unit) was dissolved in 100 mL of cyclohexanone and was added to an aqueous solution of cupric acetate (0.5 mol) at a constant temperature with stirring. Chelate formation occurs at the interface between the cyclohexanone and aqueous layers.

The formation of films, fibers, or other moldings containing donating groups with metal ion can easily be achieved. The heterogeneous complexation of water-insoluble PVA film and Cu(II)-amine complex aqueous solution was reported by Shirai et al. (21,22). The coordination bond between Cu(II) ion and hydroxyl groups of PVA molecules was confirmed by electronic and infrared spectra. The heterogeneous complexation was effected by (a) the ratio of molar concentration of ammonia, $T_{Cu^{2+}}/T_{NH_4Cl}$, (b) the anion of $[Cu(NH_3)_n]X_2$ ($X = ClO_4$, NO_3, Cl, Br^-, and SO_4), (c) the degree of crystallinity in PVA film, and (d) the degree of orientation of crystallites in PVA film.

Preparation of Cu(II)-poly(vinyl alcohol) complex film (12)

The water-insoluble PVA films were prepared by spreading a 2% aqueous solution of PVA on a polyethylene plate and heating to dryness for about 2 h at about 80°C. The film strips were peeled off the plate are heated further at 110°C for 1.5 to 2 h. The dried film (about 0.03 40 40 mm) swelled isotropically when suspended in water. When the film was suspended in an aqueous solution containing 2×10^{-2} M Cu(II) ion, 0.2 M NH_4OH, and 0.1 M KNO_3, a greenish film was obtained. The film was washed with 0.2 M NH_4OH solution and then with methanol and was dried at 30°C in vacuum for 3 days over phosphorus pentoxide.

The amount of Cu(II) absorbed by the PVA was nearly proportional to the absorbance of the film at 640 nm (21,23). An increase in the amounts of Cu(II) ion bonded by PVA resulted in a decrease in crystallinity, melting point, swelling index, and solubility in water of the film. As the annealing temperature of these PVA films was raised up to 190°C, the amounts of Cu(II) ion bound by PVA decreased. On the other hand, when the films were drawn by monoaxial stretching, the amounts of Cu(II) ion bound by PVA were influenced by the draw ratio. A maximum amount of bound metal was observed at about a 2.0 draw ratio. The microstructure of the ligand groups in these films is destroyed by the intramolecular complex formation as shown in Reaction 1. The complex formation between PVA films and Cu(II) ion is also affected by the microstructure of PVA films. In this reaction the progress of the reaction into polymer ligand substrate was reported by the cylindrical film roll method (23).

3.2.2 Synthesis of Pendant-Type Metal-Polymer Complexes

This type of metal-polymer complex is obtained by a substitution reaction between a polymer ligand and an inert metal complex, as presented in Reaction 6. A polymer ligand is coordinate to the vacant

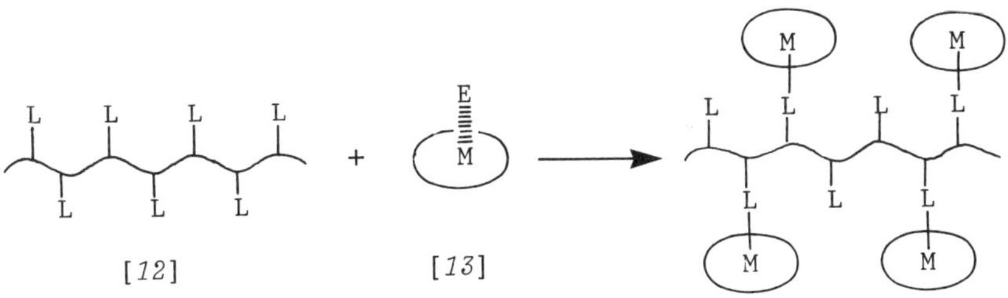

(6)

[12] [13] [14]

site of a previously prepared, stable, low-molecular-weight inert metal complex, such as a cobalt complex. The complex formed has a simple monodentate-type structure [*14*]. The important characteristics of this type of metal-polymer complex are: (a) The coordination structure is very clear; (b) the effects of polymer chains are exhibited clearly; and (c) the polymer complex is often soluble in either water or in organic solvents, since it contains few, if any, bridged structures which would reduce the solubility. Well-known examples of polymer ligands and low-molecular-weight metal complexes are listed in Table 2.

3.2.2.a Preparation of Co(III)-poly(4-vinylpyridine) complex (24)

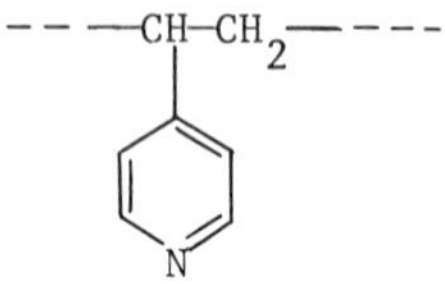

[*15*]

+ Cis-$[Co(en)_2Cl_2]Cl$ ⟶ EtOH/H_2O

[*16*]

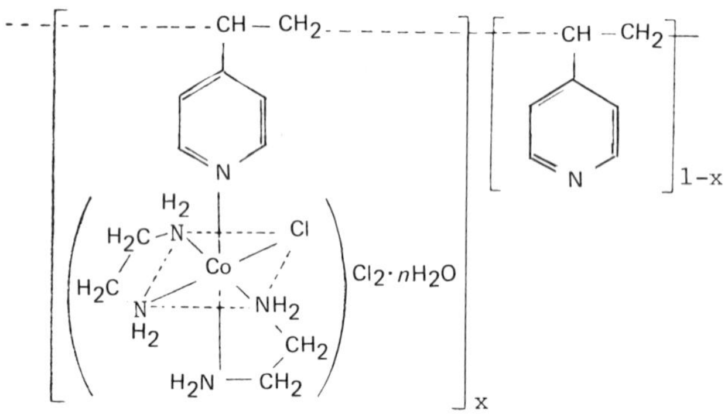

(7)

[*17*]

X^-: Cl^-, Br^-

N-N: ethylenediamine

An aqueous solution of cis-$[Co(en)_2Cl_2]Cl$ (en: ethylenediamine) was added to an ethanolic solution of poly(4-vinylpyridine) (PVP, $\bar{P}_n = 98$), and the mixture warmed at 80°C for 2 to 6 h in a water bath. During this time, the color of the solution changed from violet to reddish violet. The solution was condensed to 20 to 30 mL at 70°C by a rotary evaporator, and filtered. The filtrate was dialyzed in cold

Table 2. Examples of Polymer Ligands and Low-Molecular-Weight Cobalt Complexes

PL		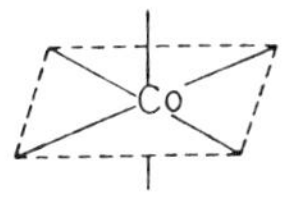	
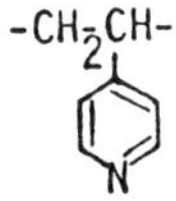	PVP	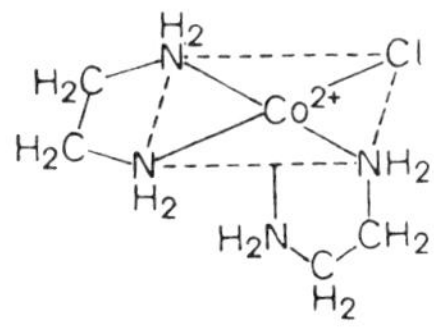	$2\,Cl^-$
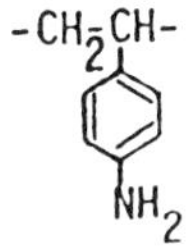	PAS	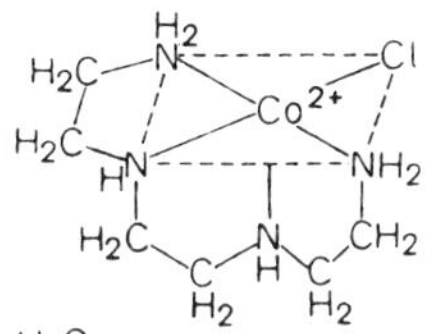	$2\,Cl^-$
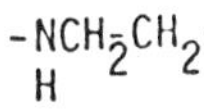	LPEI		
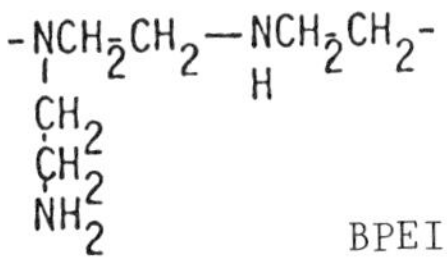	BPEI	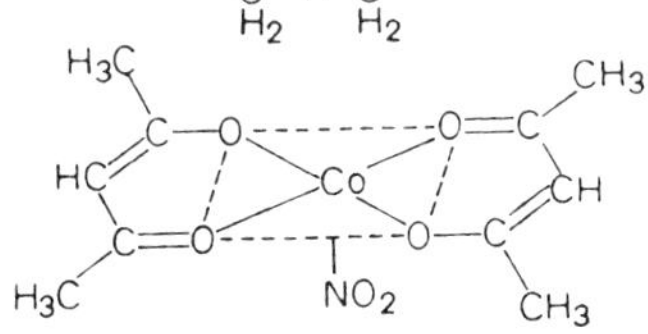	
		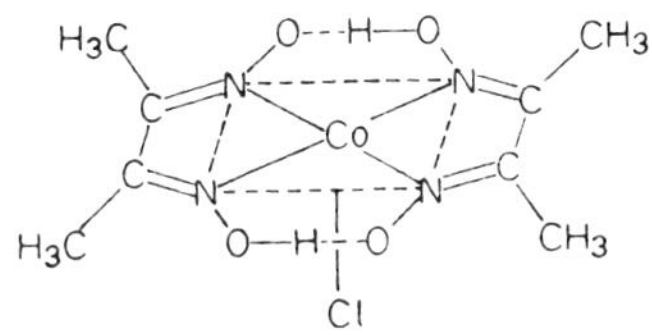	
		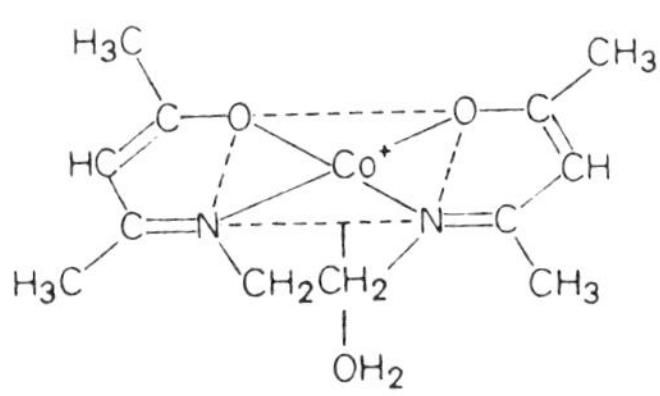	ClO_4^-

Source: Ref. 1.

Table 3. Infrared Spectrum of the Co(III)-PVP Complex

Ligands and complex	Assignment (cm^{-1})				
	$\nu_{C=C,C=N}$	$\nu_{C=C}$	δ_{CH} (in plane)	δ_{CH} (out of plane)	$\Delta\nu_{C=C}$; $\nu_{C=N}$
Pyridine	1580	1485	–	–	–
PVP[a]	1600	1492	992	822	–
cis-$[Co(en)_2Py \cdot Cl]Cl_2$	1610	1496	–	–	30
cis-$[Co(en)_2PVP \cdot Cl]Cl_2 \cdot 2H_2O$	1600, 1620	1502	998	824	20

[a]$\bar{P}_n$ of PVP = 108, KBr tablet method.

water. Thin films of a transparent reddish-violet PVP complex were obtained after water evaporation. These were pulverized and dried at 50°C. The coordination of PVP to Co(III)-complex was confirmed by an infrared (IR) spectrum. The absorption peaks of the cis-$[Co(en)_2]$, the corresponding monomer complex cis-$[Co(en)PyCl]Cl_2$ (Py: pyridine), and the free PVP are shown in Table 3. The absorption by Py at 1580 cm^{-1}, which is assigned to $\nu_{C=C}$ or $\nu_{C=N}$, is shifted to a higher wave number by about 30 cm^{-1} in cis-$[Co(en)_2PyCl]Cl_2$, where Py is coordinated to Co(III) ion.

Tsuchida et al. also prepared a series of Co(III)-polymer complexes by the reaction of poly(4-vinylpyridine) with several Co(III) chelates as presented in Reaction 1 and reported the factors that influence the formation and structure of the Co(III)-polymer complexes (25,26). Steric and electrostatic factors caused by the high density of the Co(III)-chelate along the polymeric ligand chain are important to the characteristic formation and structure of Co(III)-polymer complexes. The degree of coordination (x) or saturation of $[Co(acac)_2PVP\text{-}(NO_2)]$ (acac: acetylacetonato) complex was calculated statistically. The conformational profile of $[Co(acac)_2(PVP)\ (NO_2)]$ is illustrated schematically in Figure 1. The Co(III)-acac chelate is a bulky group with a diameter of about 10 Å. When the Co(III) chelates coordinate along a chain of PVP in a close packing arrangement one of four Py units must remain uncoordinated to avoid steric collision between the Co(III)-chelates. The calculated x-value of saturation was based on the following three assumptions: (a) the Co(III) chelate is allowed to coordinate continuously with PVP up to three successive Py units; (b) the coordination is as complete as possible; and (c) the degree of polymerization of PVP is infinite. Based on these assumptions, the permutation of the coordinated Py units and the uncoordinated Py

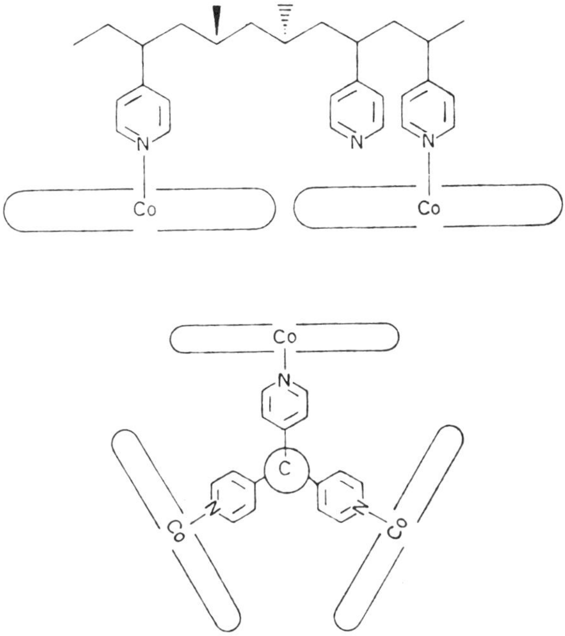

Figure 1. Conformational profile of $[Co(acac)_2(PVP)(NO_2)]$.

units in PVP complex is considered to be the combination of four sequences from C_1 to C_4, as shown in Table 4. In Table 4, P shows the facile formation of each sequence related to C_1. For example, the sequence C_2, which is composed of two coordinated units, is twice as difficult to form than the sequence C_1, which is composed of only one coordinated unit; thus the P value of C_2 is 0.5. The sequence C_1 cannot connect to C_1 itself and C_4; similarly, C_2 cannot connect to C_4, because the uncoordinated units remain below assumption a; therefore, the probability of junction of each sequence to the other sequences is represented as Q. The R value, which is the fraction of the product of P and Q, shows the proportion of each sequence in the PVP complex. Consequently, the statistical value of x was calculated according to Equation 8, in which ℓ and m are the numbers of the coordinated and uncoordinated units, respectively. The x-value obtained, 0.63, is nearly equal to the experimental value of 0.65.

$$x = \frac{\sum_{i=1}^{4} \ell_i R_i}{\sum_{i=1}^{4} (\ell_i + m_i) R_i} \tag{8}$$

Table 4. Statistical Calculation of the X-Value[a]

Sequence		P	Q	R
C_1	●○	1	2/4	0.325
C_2	●●○	1/2	3/4	0.243
C_3	●●●○	1/3	4/4	0.216
C_4	●●●○○	1/3	4/4	0.216

[a]●, coordinated pyridine unit in PVP; ○, uncoordinated pyridine unit in PVP; P, ease in the formation of each sequence related to C_1 (= 1/the number of the coordinated unit), Q, probability of junction; $R = P_iQ_i/\sum_{i=1}^{4} P_iQ_i$.

The polymer metalloporphyrin complexes have been studied from the polymeric models of hemoglobin (27). The polymer-metalloporphyrin complexes are also formed by Reaction 6. Metal-porphyrins coordinate to various donating groups in the polymer at the axial position. Many other reports have been made with poly(L-lysine) (28, 29) poly(L-histidine) (30), polyvinylimidazole (31), and polyvinylpyridine as the polymer ligand. Many investigations on the redox properties (34) and the reversible binding of molecular oxygen (27) have been carried out. Tsuchida et al. reported the coordination number (n), which is the molar ratio of axial ligand to the ferri- and ferroprotoporphyrin IX [*18*]. The formation constants (K) of these complexes were deter-

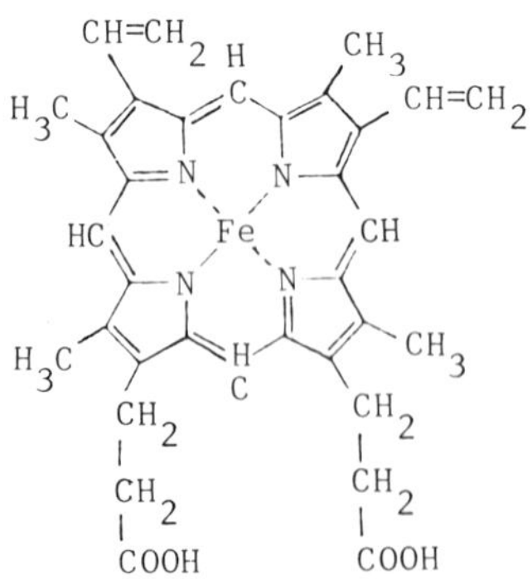

[*18*]

mined by spectrometric titration using Miller-Drough's method (33). These values are listed in Table 5 (32).

Table 5. Axial Coordination Number ($\bar{n}$) and Formation Constant (K) for Polymer-Ferriheme and Ferroheme Complexes

		$\bar{n}$		K(l/mol for n = 1; l^2/mol^2 for n = 2)	
Ligand	Solvent	Ferri-heme	Ferro-heme	Ferri-heme	Ferro-heme
PVP	DMF-MeOH(7/3)	0.95	1.3	3.8×10^2	1.1×10^3
Pyridine	DMF-MeOH(7/3)	1.0	1.1	6.0	5.1×10
PVMI	DMF-MeOH(1/3)	1.1	1.2	1.8×10^2	2.6×10^3
Imidazole	DMF-MeOH(1/3)	1.1	1.2	9.5	4.6×10
QPVP	H_2O-DMF(9/1)	1.3	1.3	2.6×10^2	2.8×10^4
PLL	H_2O(pH 12)	2.0	2.2	4.2×10^4	3.5×10^9
BPEI	H_2O(pH 12)	1.0		10^2	
Butylamine	H_2O(pH 12)	1-2		—	
PBLG-im	DMF-MeOH(7/3)	1.0		3.9×10^2	

3.2.2.b Preparation of copolymer of 1-vinylimidazole with styrene-coordinated iron prophyrin (35)

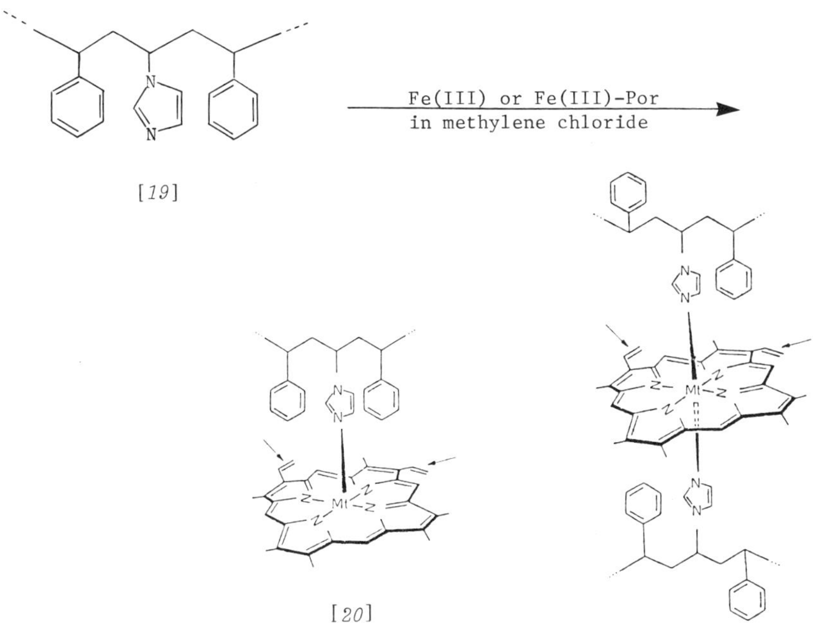

[19] [20] [21] (9)

Styrene is copolymerized with 1-vinylimidazole in the presence of 2,2'-azoisobutyronitrile (AIBN). The copolymer is redissolved in benzene and reprecipitated with methanol or acetone-water mixtures. The Fe(III) or Fe(II)-protoporphyrin is added to a benzene solution of the copolymer and the resulting solution is then evaporated to dryness or the complex polymer-iron porphyrin is precipitated with an acetone-water solution.

3.2.3 Polymerization of Metal Complex Monomer

Monomers containing metal complexes can be polymerized to produce metal-polymer complexes. Three fundamentally different polymerization techniques that lead to metal-polymer complexes have been reported in the literature:

1. Vinyl polymerization or copolymerization of vinyl compounds containing metal complexes:

(10)

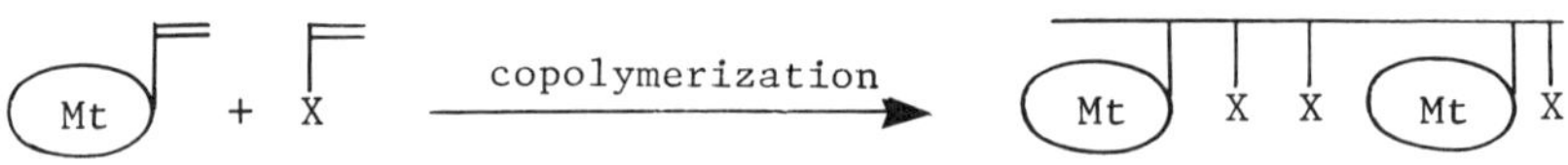

(11)

2. Polycondensation or copolycondensation of metal complex containing free functional groups:

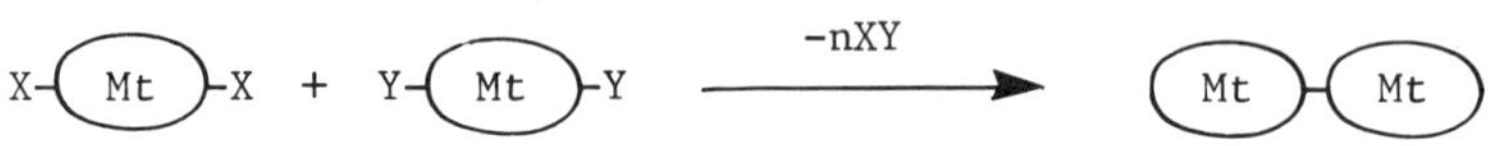

(12)

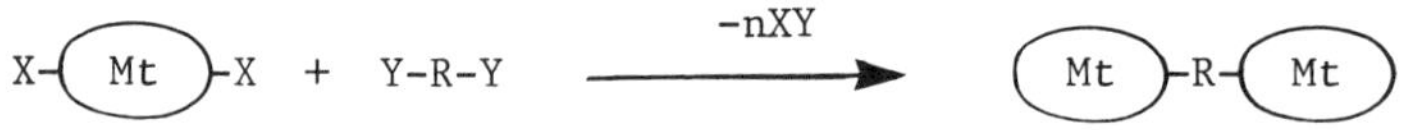

(13)

3. Other types of polymerization or copolymerization, such as

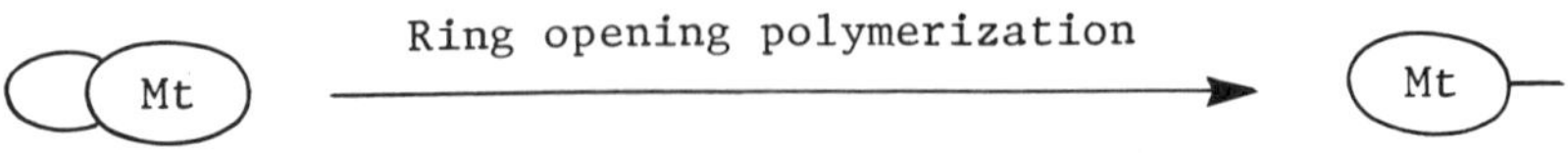

(14)

Tazuke has reported the radical polymerization of 4-vinylpyridine in the presence of Cu(II) or Fe(III) (36) which is strongly inhibited. However, Zn(II)-vinylpyridine complex has been polymerized to form Zn(II)-poly(vinylpyridine) (37). Methacrylate has been coordinated with amine cobalt(III) complexes to form methacrylatopentaammine cobalt(III) perchlorate [*22*] and *cis*-dimethacrylatotetraammine cobalt(III) perchlorate [*23*]. Homo- and copolymerization with methacrylic acid has been carried out in an aqueous medium at pH 3 or pH 7

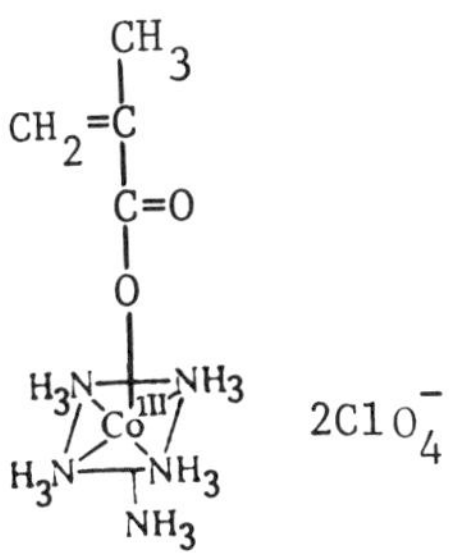

[*22*]

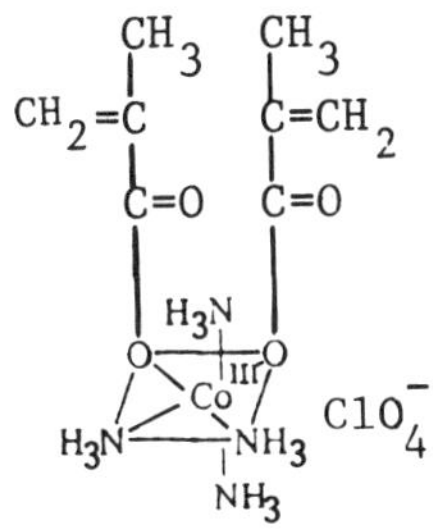

[*23*]

using a redox initiator system at 30°C. Complex [*23*] can be homopolymerized at pH 7, although the product contains several amounts of methacrylic acid.

3.2.3.a Polymerization of methacrylic acid coordinated to Co(III)-complex methacrylic acid (38, 39)

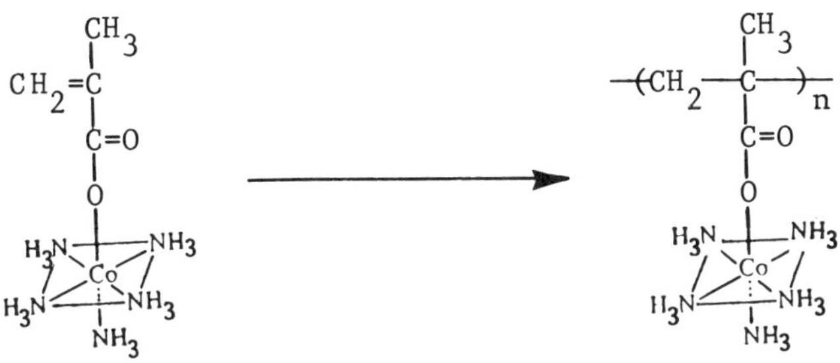

[*22*] [*24*] (15)

Polymerization of [*22*] is carried out at 30°C in deionized distilled water at pH 7 using a $K_2S_2O_8$ initiator. The pH of the polymerization medium is established by $HClO_4$ or NaOH. The polymerization product is allowed to dialyze for 2 days and is precipitated in ethyl acetate-acetone from water ($\bar{P}_n$ = 40).

The ferrocene-containing polymers have attracted special interest as transition-metal-containing polymers. Ferrocene has low toxicity and high ultraviolet and γ-radiation absorption. Ferrocene in polymeric form is easily oxidized to the stable ferricenium ion (41). A series of these mixed-valence Fe(II) and Fe(III) polymers has been demonstrated to be semiconducting (42). Vinyl ferrocene and its derivatives can be polymerized by radical, cationic, or anionic initiators to form high-molecular-weight polymers. Many investigations of the radical-initiated homo- and copolymerization of vinylferrocene (43-45), acrylates, and methacrylates containing ferrocene (47) have been carried out. Acryloylferrocene was homopolymerized by $LiAlH_4$ and by butyl lithium initiation (48). Poly(ferrocenylmethyl methacrylate) is commercially available (47).

3.2.3.b Polymerization of ferrocenyl methacrylate (FMMA) [*25*] (47)

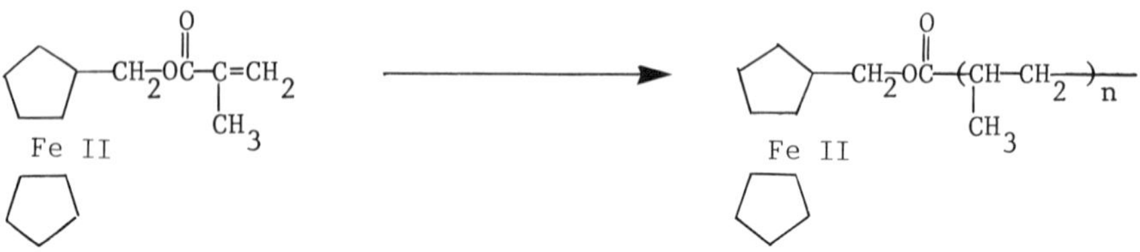

(16)

[*25*] [*26*]

Monomer [*25*] is prepared from *N*,*N*'-dimethylaminomethylferrocene. Polymerization is carried out under high vacuum in sealed glass reactors. The initiators used are sodium naphthalide, potassium naphthalide, butyl lithium, EtMgBr, PhMgBr, and $LiAlH_4$. The initiator and monomer [*25*] solutions are placed in separate break-seal glass vessels on a vacuum line. Each solution is transferred into the reactor by breaking the seals (the initiator solution is added first). The reaction solution is stirred by a magnetic stirrer and is allowed to stand for 20 h at 30 or -78°C. The polymer is precipitated by the addition of an excess of petroleum ether, filtered, redissolved in THF, and reprecipitated in petroleum ether two times. The polymer is soluble in methanol-water systems.

The preparation of metal-polymer complex by condensation has been reported by Hartwell and Bailar (49). Polyesters and polyamides of bis(tyrosinato)metal(II), (Cu, Ni) and bis(lysinato)metal(II), and dichloride (Cu, Zn) complexes have been synthesized by interfacial polycondensation with aryl diacid chlorides.

3.2.3.c Preparation of tyrosine Cu(II)-chelate polyester (49)

Preparation of bis(L-tyrosinato)copper(II) dihydrate

A solution of cupric acetate in a minimum amount of methanol is added to a suspension of a stoichiometric amount of DL-tyrosine in methanol. After stirring for 12 to 18 h, the solution is filtered. The violet precipitate is washed with methanol and dried at 110°C.

Esterification

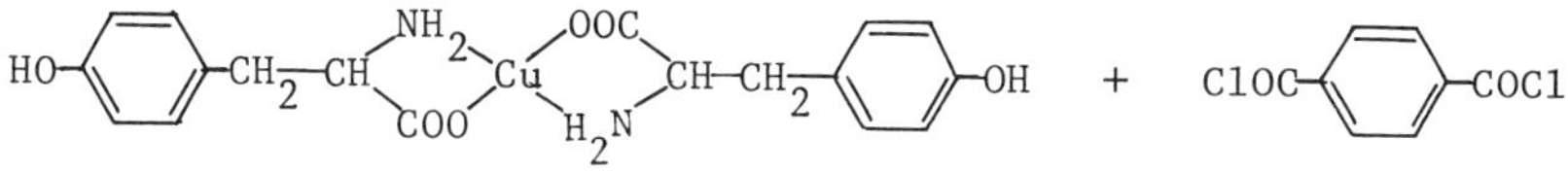

[27]

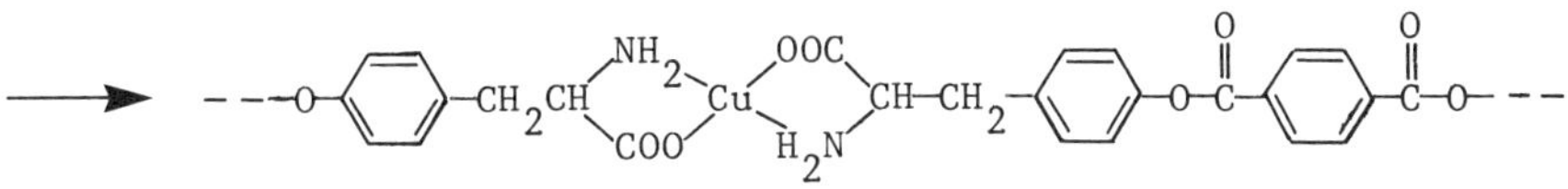 (17)

[28]

A solution of sodium hydroxide (0.005 mol) in 10 mL of water is added to a suspension of Cu(II)-bis(tyrosinato) complex (0.0025 mol) in 10 mL of water. A few crystals of tetramethylammonium chloride are added as a catalyst. Terephthaloyl chloride (0.0025 mol) dissolved in 10 to 15 mL of chloroform, is added in one portion to the rapidly stirred basic solution. A precipitate forms almost immediately and stirring is continued for 15 to 25 min. The polymer is collected, washed in succession with water, acetone, and ether, and then extracted overnight with 97% ethanol in a Soxhlet extractor. The polymer is dried to constant weight at 110°C. The degree of polymerization is 4 to 5.

The metal-phthalocyanine polymers have been extensively studied for their catalytic and electrical interest. A novel class of polyimides containing Mt(II)-phthalocyanine rings [Mt(II) = Co(II), Ni(II)] were synthesized by the solution condensation in *N*-methyl-2-pyrolidone (NMP) of Mt(II)-[2,9 or 10 (or 2 or 17)-bis(3,4-dicarboxybenzoyl)]-phthalocyanine dianhydride with aromatic diamines such as 4,4'-methylene-bis(aniline), followed by thermal imidization in vacuum at 100°C. These polyimides containing Mt-phthalocyanine rings show room-temperature direct-current conductivity in the range of 10^{-11} to 10^{-6} s/cm (50,51).

3.2.3.d Synthesis of polyimides containing Co(II)-phthalocyanine (50,51)

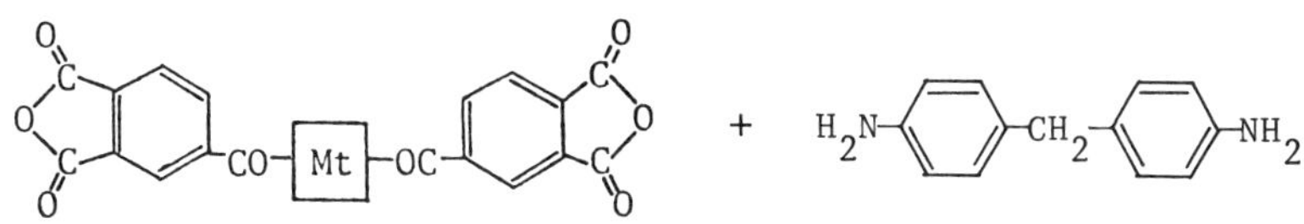

[29] [30]

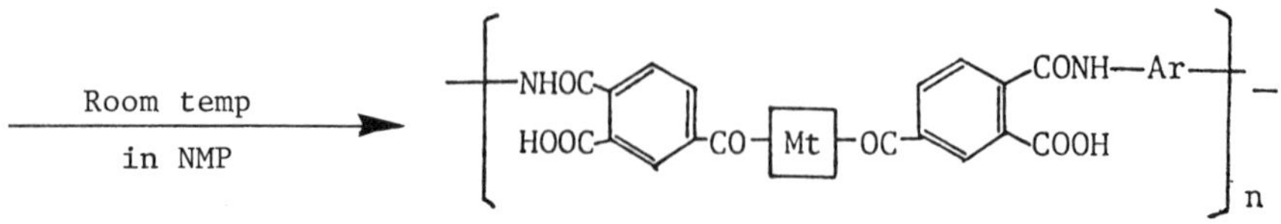

[31]

(18)

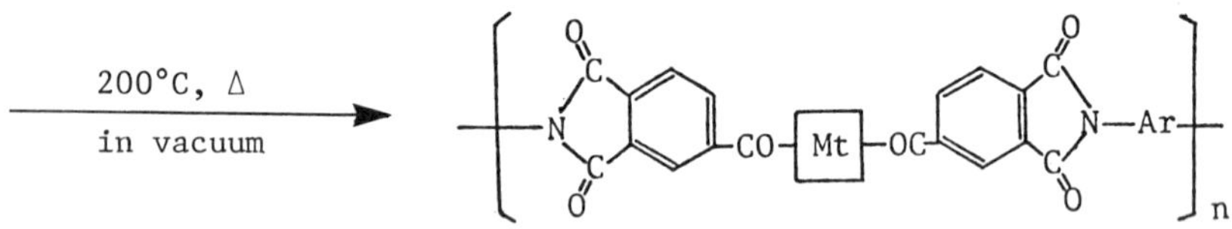

[32]

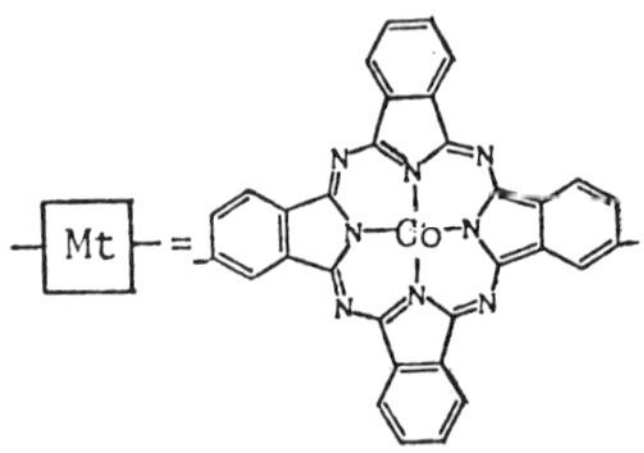

Preparation of Mt(II)-[2,9 or 10 (or 2,16 or 17)-bis(3,4-dicarboxybenzoyl)]phthalocyanine dianhydride [Mt(II)-dahPc] [29]

The monomer, 2 g (2.2×10^{-3} mol) of Mt(II)-[2,9 or 10 (or 2,16 or 17)-bis (3,4-dicarboxybenzoyl)]phthalocyanine, is refluxed for 10 h with 10 mL (0.12 mol) of acetyl chloride in 30 mL of benzene. The product is filtrated, washed with benzene, and dried at 50°C in vacuum. The product is then refluxed for 4 h with 20 mL (0.18 mol) of acetic anhydride, filtered, washed with benzene, and dried at 50°C in vacuum (100% yield).

Polymerization

To a solution of Mt(II)-dahPc (5×10^{-4} mol) in 40 mL of NMP, 4,4'-methylene-bis(aniline) (5×10^{-4} mol) is added at room temperature. The mixture is vigorously stirred for 1 h under a stream of nitrogen. The resulting solution is poured into 200 mL of ether to yield a dark-colored solid. The polymer is washed with methanol and diethyl ether, and dried at room temperature in vacuum.

The crude polymer is purified by using NMP-diethyl ether as solvent-nonsolvent. The polyamic acid so prepared is cured at 200°C for 1 h in vacuum to bring about ring closure, and the polyimide-containing metal phthalocyanine rings are obtained. [*n*] = 0.71 in NMP. The simplest structures of polymetallocene are shown in [33] to [*36*].

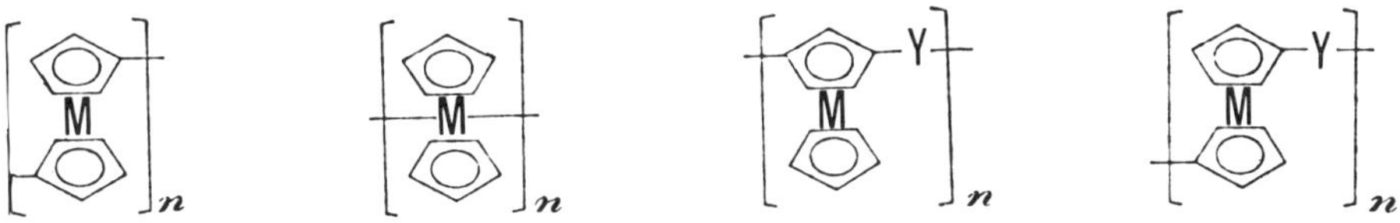

[*33*] [*34*] [*35*] [*36*]

Polyferrocenylenes were synthesized by a step-growth poly-coupling reaction involving 1,1'-dilithioferrocene (chelated with *N*,*N*,*N*',*N*'-tetramethylethylenediamine) and 1,1'-diiodoferrocene as monomers (52).

3.2.3.e Preparation of 1,1'-ferrocenylenes (52)

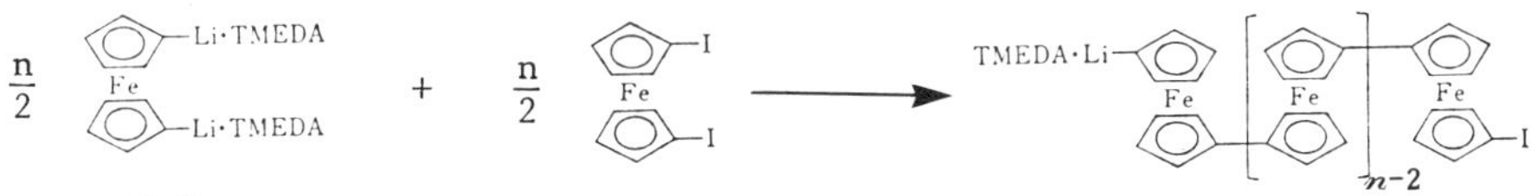

[*37*] [*38*] [*39*]

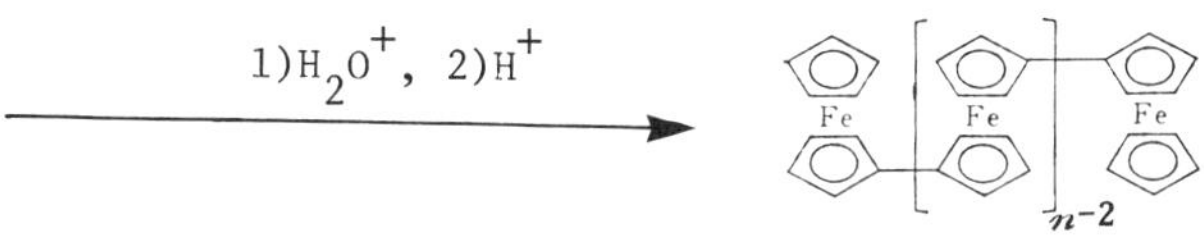

(19)

[*40*]

THF (27 mL) is added under argon to the solid chelate [*37*] contained in a reaction flask. To the stirred suspension, cooled to 0°C, finely powdered copper(II) chloride is added from a side tube preattached to the flask. The mixture is allowed to warm up to 25°C, with continued stirring at this temperature for 2 h to ensure complete reduction of Cu(II) to Cu(I) and organocuprate formation. Precooling this mixture to 0°C is followed by the addition of [*38*] dissolved in 27 mL of 1,2-dimethoxyethane with continual stirring in the dark.

3.2.3.f Synthesis of macrocylic beryllium polymers (53)

Macrocyclic beryllium β-dicarbonyl chelates [*42*] have been prepared by vacuum sublimation of the polymer obtained from the interaction of a bis-β-carbonyl compound [*41*] with beryllium salts in a neutral medium. On heating above their melting points these macrocycles were converted to high-molecular-weight polymers (53).

$$R'COCH_2CO\text{-}R\text{-}COCH_2COR' \xrightarrow[\text{2. 150-200° (0.05 mm)}]{\text{1. } Be^{2+}\text{, pH 7}}$$

[41]

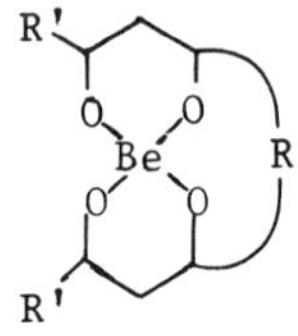

[42]

$\xrightarrow{200°C}$

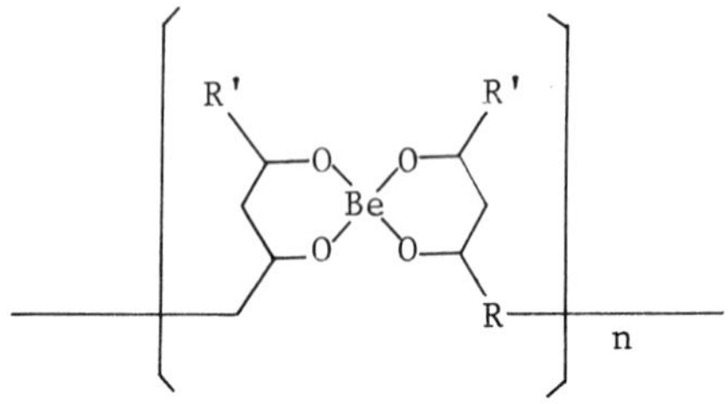

[43]

(20)

$R = -(CH_2)_{6,12}-$, $-O-(CH_2)_{6,12}-O-$,

$-NH(CH_2)_{6,10}-$

$R' = -CH_3$

Preparation of cyclic monomer [42]

2,4,13,15-Tetraketohexadecane 3.000 g (0.0106 mol) and 1.881 g (0.0106 mol) of beryllium sulfate tetrahydrate are stirred in a 125 mL flask containing 50 mL of distilled water and 10 mL of purified dioxane. The reaction mixture is adjusted to pH 8 with sodium hydroxide, stirred for an additional 16 h, and the insoluble product removed by filtration, and solvent extracted. This material contains low-polymer starting tetraketone and beryllium salts. The reduced viscosity of the benzene-soluble portion is 0.05 (0.2 g per 100 mL of benzene at 25°C). Sublimation of the crude product at 200°C (0.2 mm) gives 2.3 g (74%) of pure cyclic monomer, m.p. 138 to 140°C.

Polymerization of macrocycles

Upon heating above their melting points these crystalline macrocycles spontaneously polymerize, producing amorphous materials that dissolve slowly in benzene. These polymers have intrinsic viscosities in the range 0.1 to 2.7, characteristic of high-molecular-weight polymers; $R' = CH_3$, $R = -(CH_2)_8-$; intrinsic viscosity 0.9. They are flexible at room temperature (glass transition temperature 35°C) and have a tensile modulus at 25°C of 122,000 psi and a tensile strength (100% elongation (minute) of 2710 psi.

3.2.4 Polymer Reaction of Peripheral Functional Groups in Metal Complex with Polymer

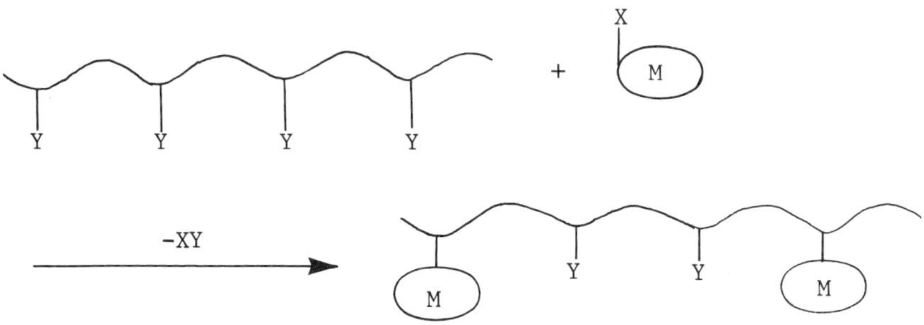

(21)

This type of metal-polymer complex is synthesized by the reaction of neutral polymers and copolymers with peripheral functional groups that form metal complexes. The coordination structure of the metal-polymer complex of this type is clear. These polymer complexes are often soluble in either water or in organic solvents because of the presence of unreacted soluble parts in the polymer.

Almost all industrial catalysts are heterogeneous, with the reaction taking place on a solid surface. The polymeric catalysts are formed by chemically bonding a transition-metal complex to an insoluble polymer; they are then useful as heterogeneous catalysts. Cross-linked polystyrene-divinylbenzene gel, phosphonated polystyrene-divinylbenzene, ion-exchange resin, and silica are well-known insoluble polymers that have been used. Examples of polymer reactions are as follows.

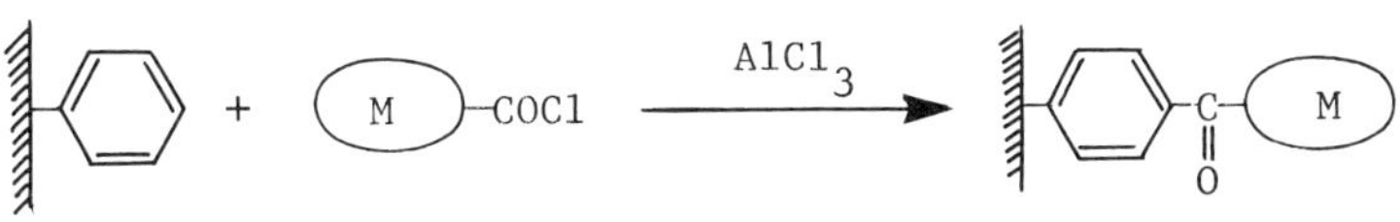

(22)

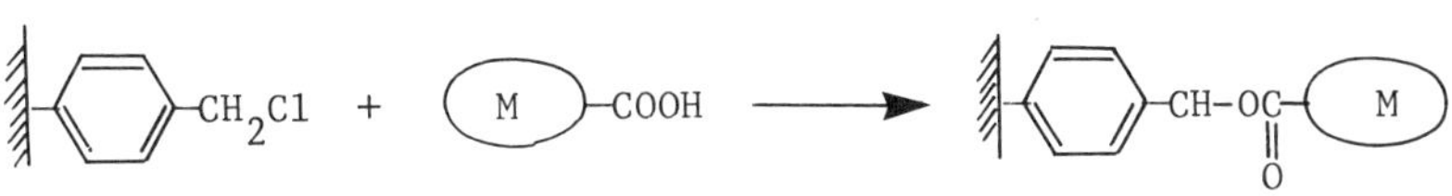

(23)

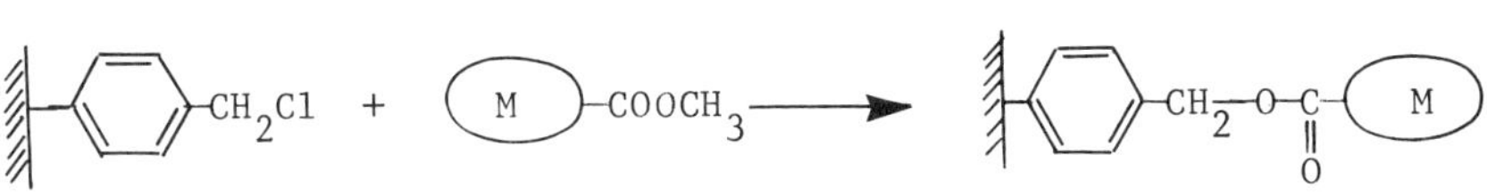

(24)

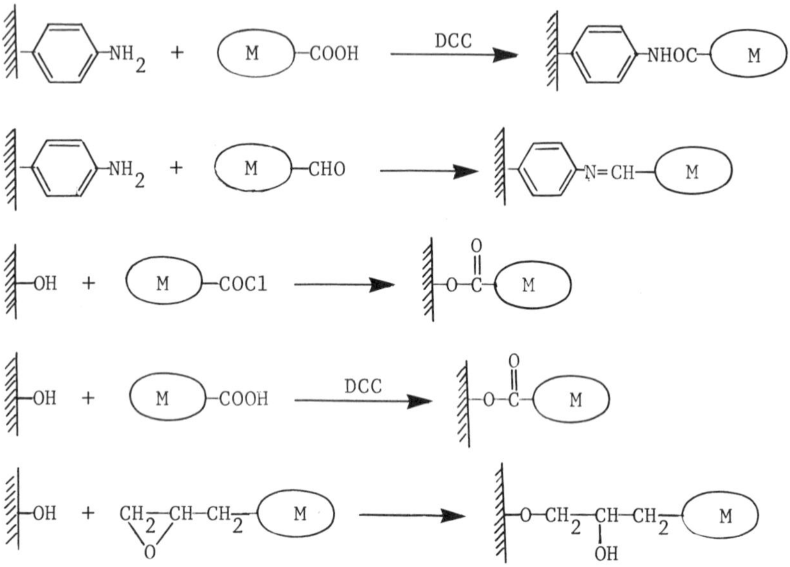

(25)

(26)

(27)

(28)

(29)

Polymer-attached metal porphyrins have attracted wide interest as organic semiconductors, catalysts, and enzyme models. The fixation of metal-porphyrins into a polymer matrix has been achieved by the reaction of peripheral functional groups in the natural porphyrins with reactive polymers such as chloromethylated polystyrene. Lautsh and co-workers prepared various porphyrin polymers and investigated their functions as enzyme models (54). Kamogawa has prepared chlorophyll polymers to obtain vinyl polymers capable of catalyzing photoredox reactions (55). He reported the incorporation of several porphyrin functions, such as protoporphyrin IX and chlorine, as well as metalloporphyrin functions such as Mg(II)- and Cu(II)-chlorophyllin and Fe(III)- and Co(II)-protoporphyrin IX into vinyl polymers (55). Rollmann used macroreticular resins to synthesize polymer-bonded metalloporphyrins in which the porphyrin moieties were monomolecularly dispersed (57). Yamakita et al. have tried to graft copolymerized chloromethystyrene onto porous polyolefins such as polyethylene and polypropylene and to make an ester linkage between the chloromethyl groups and carboxyl groups of metalloporphyrin (58). They have also tried to use polymer-bonded cobalt protoporphyrins as the catalysts in some oxidation reactions.

3.2.4.a Fixation of Mg(II)-chlorophyll into preformed polymer by chemical modification (56)

Ester-forming reaction

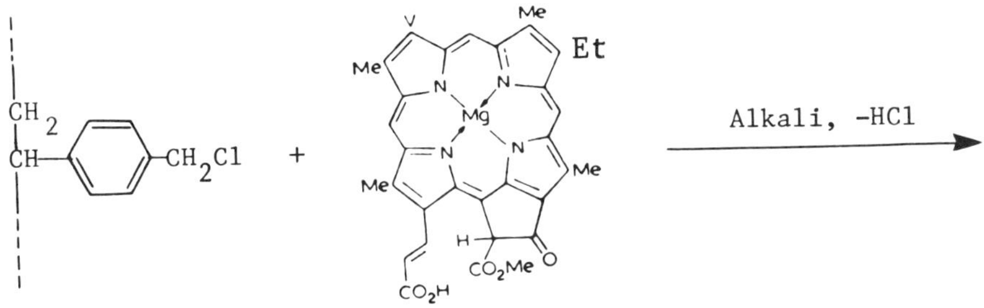

[*44*]

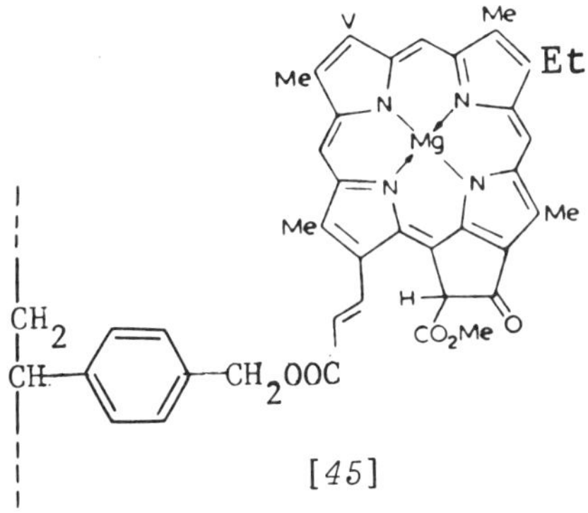

[*45*]

Freshly distilled *p*-chloromethylstyrene (0.2 mL), 5 mL of *N*-vinylpyrrolidone, and 0.05 g of α,α'-azobisisobuthylonitrile are dissolved in 10 mL of tetrahydrofuran. The solution is put into a glass ampul, vacuum sealed, and heated for 72 h at 80°C to provide a yellowish viscous solution. To this solution are added 0.4 g of the metal chloride complex prepared from Mg(II)-chlorophyll A by hydrochloric acid treatment, and 30 mL of dimethylformamide to afford a clear dark green solution. To this solution, 0.5 mL of triethylamine is added and the solution is heated at 90°C for 2 h. Upon cooling, the reaction product is dialyzed against water through visking cellulose tubing for 1 week to remove water-soluble impurities. The solution is freeze dried, and the product redissolved in water, filtered, and freeze dried to afford 2.2 g of purified green powder. The visible absorption spectrum in water revealed typical porphyrin absorptions at 400 nm together with a 660-nm band characteristic of chlorophyll derivatives. The formation of the ester linkage by this reaction is confirmed by infrared spectrum (1730-cm^{-1} band).

Schiff base-forming reaction

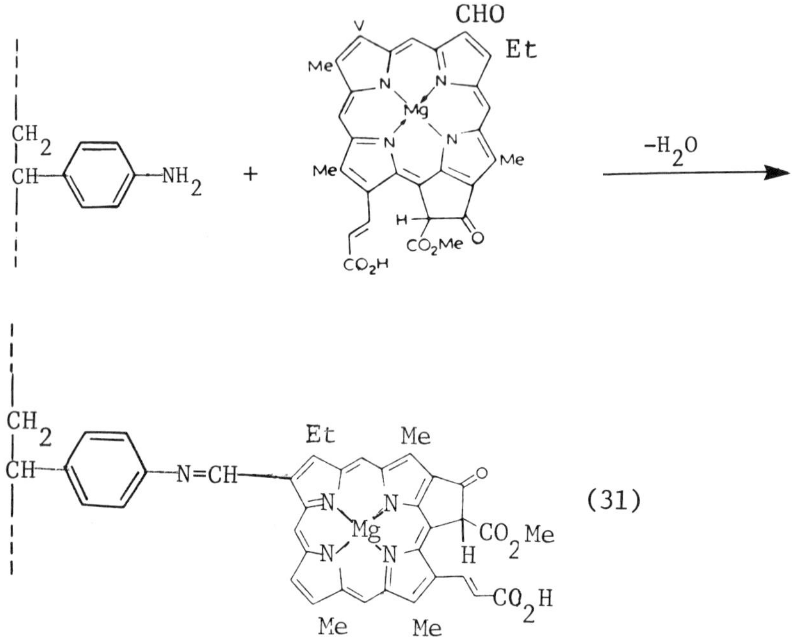

(31)

[47]

Freshly distilled *p*-aminostyrene (0.02 mL), 1 mL of *N*-vinylpyrrolidone, and 0.01 g of α,α'-azobisisobutyronitrile in 2 mL of ethanol are put into a Pyrex glass ampul, which is sealed under vacuum and then heated at 80°C for 72 h. The viscous, pale yellow polymer solution produced is the preformed polymer, which is then mixed with 0.1 g of Mg(II)-chlorophyll B extracted from fresh spinach in 20 mL of ethanol. The resulting green solution is stirred at 50°C for 10 h and then diluted with water to double the volume, followed by centrifugation. The supernatant green solution has typical chlorophyll absorptions at 420 and 660 nm. Despite the insolubility of chlorophyll B in water, the resulting green polymer is soluble in water.

3.2.4.b Preparation of porous polymer-bonded Co(III)-porphyrin (57)

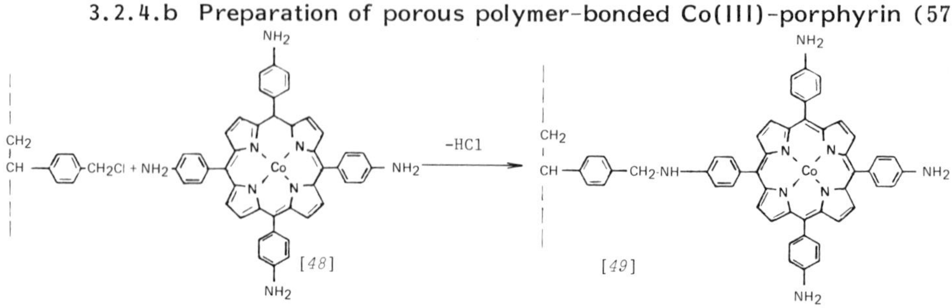

[48] [49]

Porous, macroreticular, styrene-based resin cross-linked with divinylbenzene or with ethylene glycol dimethacrylate is used. Chloromethylated resin (10 g, 17% Cl, 0.05 mol of Cl) is added to a solution of 1.0 g of Co(III)-*meso*-tetra(*p*-aminophenylporphyrine) Co(III)-TPP-NH [*48*] (1.5×10^{-3} mol) in 200 mL of DMF. The reaction mixture is heated at 100°C for 2 to 3 h with stirring. After cooling, the resin is filtered and washed with DMF and 1 MHCl until a colorless filtrate is obtained. The green polymer product, obtained after washing with 0.1 M NaOH and water, contains 1.1% nitrogen or about 6% porphyrin. The polymer product also retains 5.2% Cl.

3.2.4.c Synthesis of polystyrene-bonded Fe(III)-4,4',4'',4'''-tetracarboxyphthalocyanine [Fe(III)-taPc] by Friedel-Crafts reaction (58,59)

Preparation of Fe(III)-4, 4', 4'', 4'''-tetracarboxamidephthalocyanine [51]

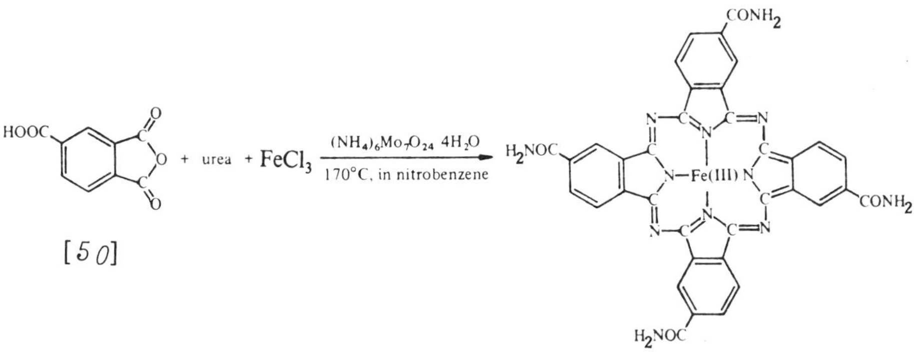

[*50*]

[*51*]

Trimellitic anhydride [*50*] (10 g, 0.05 mol) of urea, 0.03 mol of metal salt $FeCl_3$, 1 g (0.001 mol) of ammonium molybdate, and 150 mL of nitrobenzene are mixed in a mortar and then placed into a 200-mL round-bottomed flask equipped with a reflux condenser. The reaction mixture is heated at 150 to 170°C for 3 h, after which the blue-green black solid obtained is washed from the flask with 200 mL of methanol and collected by filtration. The solid is washed with methanol until no nitrobenzene can be detected, and dried for 24 h at 60°C in vacuum; yield 85%, IR(KBr) 1660 cm^{-1} (carboxamide C=O).

Preparation of Fe(III)-4,4′,4″,4‴-tetracarboxyphthalocyanine [52]

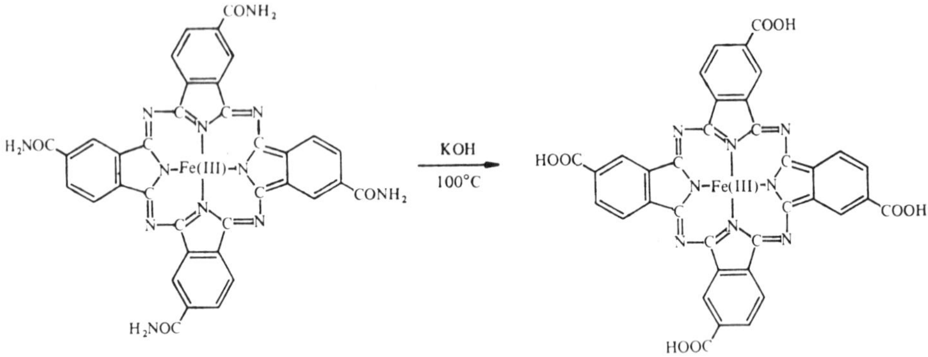

[51] [52]

A mixture of 20 g, 2.5×10^{-2} mol of Fe(III)-4,4′,4″,4‴-tetracarboxamidephthalocyanine and 500 g of KOH in 500 mL of water is heated with stirring for 24 h at 100°C. The reaction mixture is then diluted with 100 mL of water, filtered through a sintered-glass funnel, and slowly acidified to pH 2 with 6 M HCl. At this point, the product completely precipitates as a blue-green flocculent solid. This is collected on a sintered-glass funnel, washed in succession with 0.1 M HCl, water, acetone, and ether, and dried in air. The reaction product is a blue-green-black solid. This is soluble in dilute base, DMF, DMA, and DMSO. The product is added to 2 L of 0.5 M KOH aqueous solution and then filtered through a sintered-glass funnel. The filtrate is acidified to pH 2 with 0.2 M HCl solution. This procedure is repeated three times. After standing for several hours, the mixture is centrifuged, the supernatant liquid decanted, and the precipitate collected on a Büchner funnel, washed with water until the washings are nearly neutral, and then washed with ethanol and ether. The product is dried at reduced pressure over phosphoric anhydride for 24 h; yield 80%, IR(KBr) 1705 cm^{-1} (carboxy C=O stretching), UV (water) 675 (log ε 4.33), 638 (4.62), and 330 nm (4.62).

Preparation of Fe(III)-4, 4', 4'', 4'''-tetracarboxyphthalocyanine-tetraacid chloride [53]

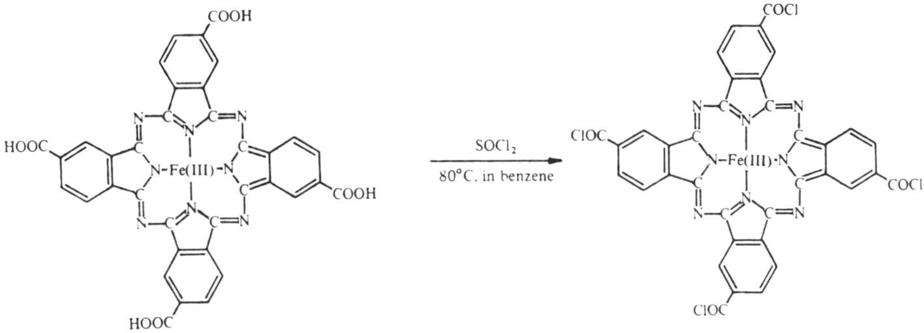

[52] [53]

A mixture of 1 g (0.0013 mol) of Fe(III)-4,4',4'',4'''-tetracarboxyphthalocyanine, 3 mL (0.003 mol) of thionyl chloride, and a few drops of pyridine in 10 mL of benzene is refluxed for 10 h. The product is recovered by filtration, washed with dry benzene, and dried at 70°C in vacuum; yield 100%; IR(KBr) 1775 cm^{-1} (acid chloride C=O stretching).

Covalently bonding of Fe(III)-phthalocyanine ring into polystyrene backbone [54]

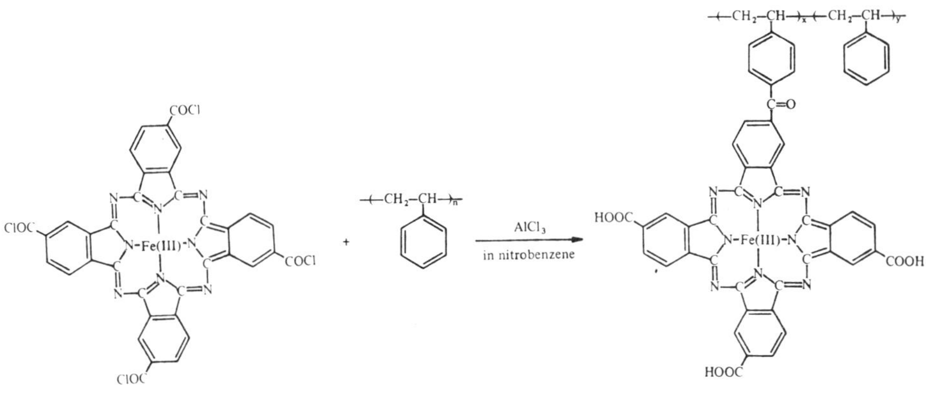

[53] [54]

To a solution of 3 g (0.03 mol) of polystyrene ($\overline{P}_n$ = 1700) and 0.5 g (0.61 mmol) of Fe(III)-4,4',4'',4'''-tetracarboxyphthalocyaninetetraacid chloride in nitrobenzene, 1 g (0.0075 mol) of aluminium chloride is added. After stirring for several hours at room temperature, the reaction mixture is then steam-distilled until no nitrobenzene can be detected in a sample of the aqueous distillate. The product is recovered by filtration, washed with water and methanol, and dried at room temperature. The light blue solid is ground to a powder and washed with an alkaline aqueous-methanol solution in order to remove the unreacted Fe(III)-taPc. The infrared spectrum of Fe(III)-taPc bonded to polystyrene exhibits the carboxyl carbonyl stretching band as well as ketone carbonyl at 1700 and 1680 cm^{-1}.

The electronic spectrum in DMF solution has absorption bands at 634 and 660 nm. The polymer contains 4.0 mol % Fe(III)-Pc rings which are covalently bonded to polystyrene. If the amount of Fe(III)-Pc attached to the polymer is less than 4 mol %, the polymer is soluble in DMF or benzene. A Fe(III)-Pc-containing film is obtained by casting from a benzene solution. Co(II), Ni(II), and Cu(II)-phthalocyanine can be bonded covalently to polystyrene in a similar way.

3.2.5 Synthesis of Coordination Polymers (2,60)

A low-molecular-weight compound having multidentate ligand groups on both ends of the molecule give a linear sheetlike or network polymer complex, as shown in Table 6. The polymer complex is formed through metal ion bridging. The ligand refers to bridging ligand. The coordination polymers were synthesized in 1927, and since 1950 they have been investigated for the purpose of obtaining thermally stable polymers, catalysts, and organic semiconductors. The polymerization occurs by the complex formation of the following ligands with metal ion; (a) compounds having more than two donating groups, and (b) compounds not forming stable intramolecular bridging. The typical coordination polymers are shown in Table 7. When the complex formation of ligands containing four coordinating groups reacts with a metal ion, the resulting polymer complex usually forms network structures. The polymer formation is achieved by mixing solutions of metal ions and polymers. If necessary, the pH of the mixed solution is adjusted to the desired value with an alkaline solution such as KOH, NH_4OH, or various amines. Water, dioxane, alcohol, tetrahydrofuran, and aprotic polar solvents such as *N,N'*-dimethylformamide (DMF) are used as solvents for the complex formation. Most polymer-metal complexes of this type are insoluble in water and in organic solvents, and are infusible. The degree of polymerization of coordination polymer is low, because of insolubility, side reactions with end groups, and unstable coordinate bonds in the polymer.

There has been considerable interest in polymeric transition metal complexes in which the metal ions are bridged by heterocyclic aromatic

Table 6. Various Types of Coordination Polymers

Coordination numbers of metal	Ligand	Types of polymers
2	2 A – A	----A–A–M–A–A–M----
	3 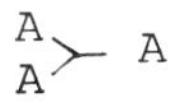	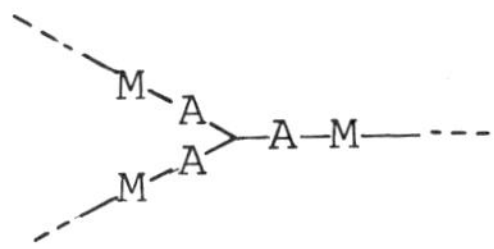
4	2 A – A	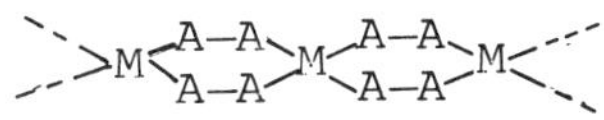
	4 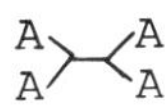	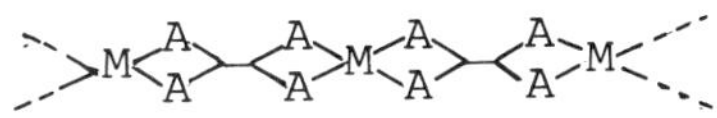
6	2 A – A	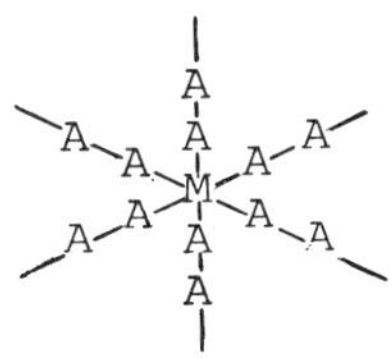
	4 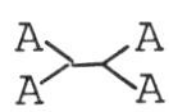	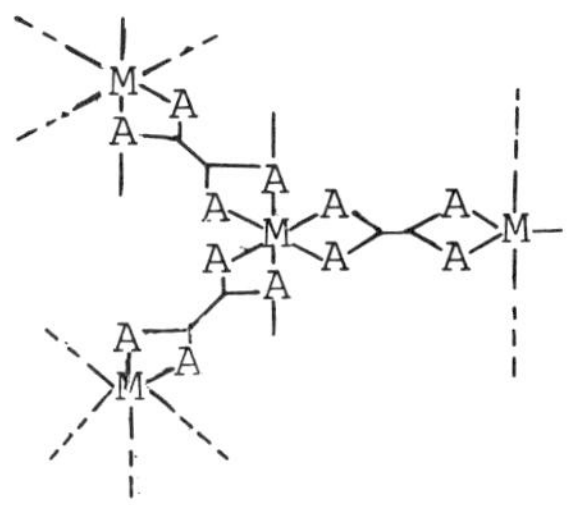

diamines, especially pyrazine (Pyr) [*56*] (61-63). Magnetic susceptibility measurements, made at low temperature on a single crystal (64, 65) of the pyrazine-bridged linear polymer, revealed antiferromagnetic chainlike behavior which could be described by the Heisenberg linear chain model (66). In addition to the pyrazine bridge in [Cu(Pyr)$(NO_3)_2$], there are also weakly bonded nitrate bridges between Cu(II) ions of the chains.

Table 7. Typical Bridging Ligands

Bridging ligands

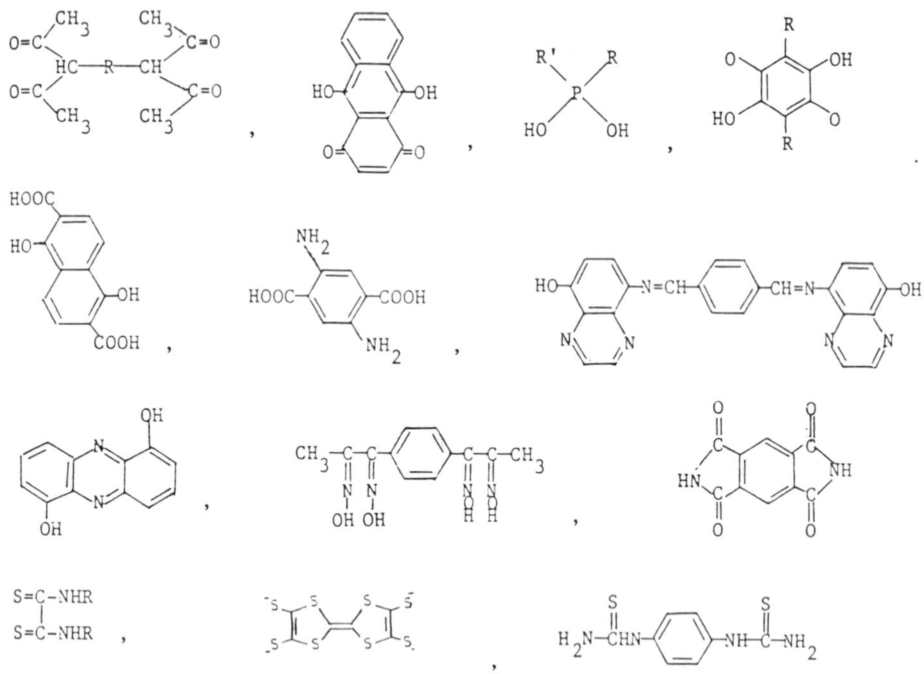

Poly(metal phosphinates) were prepared from phosphinic acid and metal ion such as Zn(II) or Cr(III) (67,68). These polymers contain tetrahedral Zn(II) centers and symmetrical bridging O,O'-phosphinate groups in polymeric structure. The zinc phosphinate [55] with alkyl chains on the phosphorus is melted.

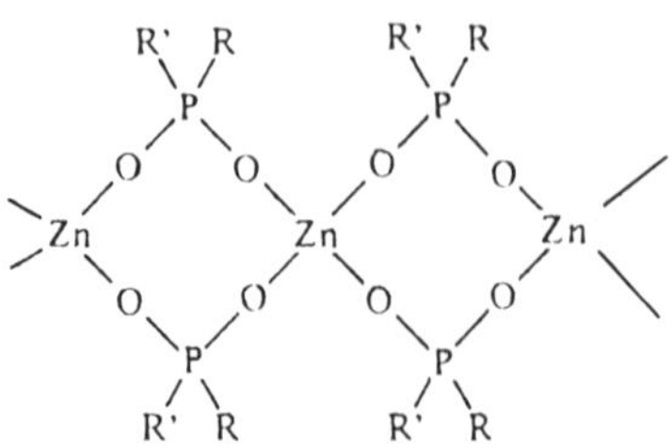

[55]

3.2.5.a Preparation of polymeric pyrazine-bridged copper(II) complexes (63)

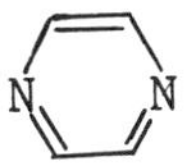

+ $Cu(NO_3)_2 \cdot H_2O$ ⟶

[*56*]

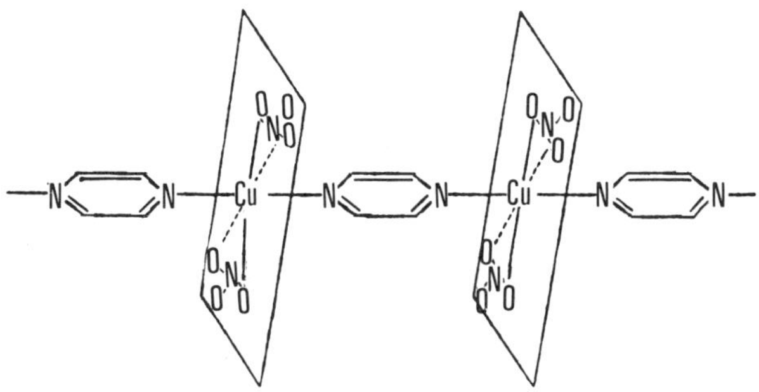

(37)

[*57*]

An ethanol solution of Cu(II) nitrate trihydrate is treated with 2,2'-dimethoxypropane to remove the water of hydration. The resulting solution is added to an ethanol solution of [*58*]. The blue precipitate is immediately collected, washed with ethanol and then ether, and dried in air.

2,5-Dihyroxy-1,4-benzoquinones [*58a*] are capable of bridging transition-metal ions to form coordination polymers (59). Polymers with 1:1 coordination of the dianion of chloranilic acid (R = Cl) [*58b*] with Co(II), Ni(II), Cu(II), Sn(IV), Pt(II), and Pb(II) were prepared and their optical and magnetic properties measured. Wrobleski et al. reported synthesis and characterization of polymeric Fe(II) complexes of dianions of [*58a*] and [*58b*] (59).

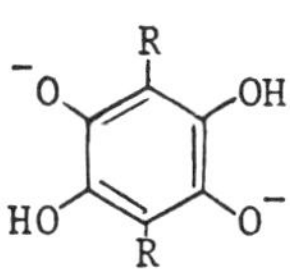

R = H [*58a*], Cl [*58b*]

3.2.5.b Preparation of iron(II)-complex polymers containing chloranilic acid (CA) [*59b*] and 2,5-dihydroxy-1,4-benzoquinone (DHBQ) [*59a*] (59)

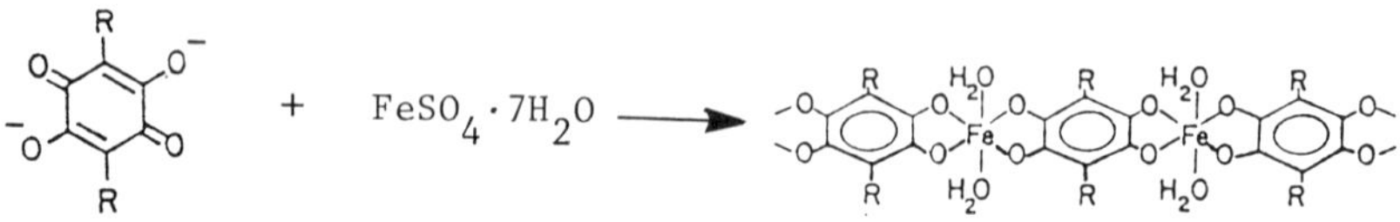

(38)

[*59-a*] R = H

[*59-b*] R = Cl

[*60*]

A degassed aqueous solution of $FeSO_4 \cdot 7H_2O$ (1.14 g, 4.05 mmol) is slowly added to a nonaqueous solution of chloranilic acid (0.85 g, 4.05 mmol). The resulting solution is stirred at reflux for 15 min and then quickly cooled to room temperature and filtered. The dark green precipitate is washed with water and then ethanol, and dried in vacuum at 100°C. Polymeric (diaquo) dihydroxybenzoquinonato iron(II) is prepared and isolated in a manner similar to that of $[Fe(Ca)(H_2O)_2]_n$ or $[Fe(DHBQ)(Pyr)]_n$.

3.2.5.c Preparation of iron(II)-mixed complex polymer containing chloranilic acid or 2,5-dihydroxy-1,4-benzoquinone and pyrazine (56)

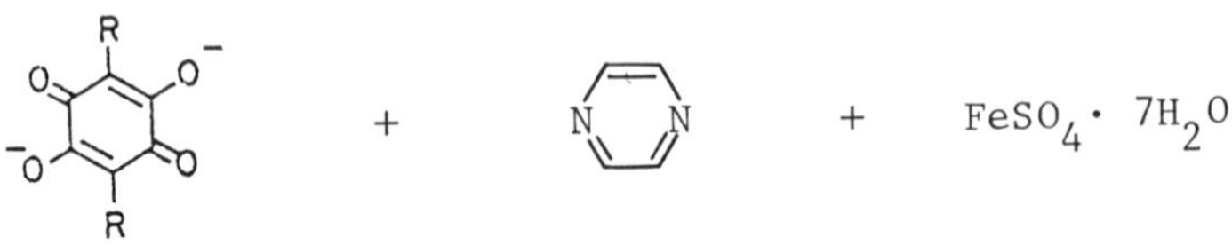

[*59a*] R = H

[*59b*] R = Cl

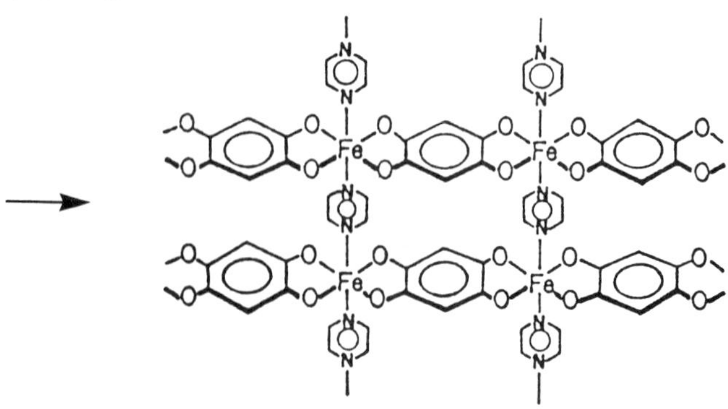

(39)

[*61*]

Solid $FeSO_4 \cdot 7H_2O$ (5.64 g, 20.3 mmol) is added to a refluxing methanol solution (250 mL) containing pyrazine (1.63 g, 203 mmol). The red-orange iron(II)-pyrazine complex which precipitates is redissolved by adding 25 mL of water to this mixture. A concentrated aqueous solution of chloranilic acid (4.24 g, 20.3 mmol) is added, which produces a brown-black precipitate. The precipitate is washed with 500 mL of hot water and then with 250 mL of ethanol, and air dried at room temperature over P_2O_5. Infrared spectral results are consistent with the presence of bridging bidentate Pyr in these complexes and corresponding polymer. The thermally stable coordination polymers are prepared by the complexation of tetracarboxylic diimidates such as pyromellitic diimide (PMDI) with the salts of cobalt, nickel, copper, zinc, and mercury in DMF solution (70-73). The linear coordination polymer of tetracarboxylic diimidate with zinc or mercury, is thermally stable above 550°C. Hojo et al. also reported on the preparation and thermal stability of a series of aliphatic amine complexes of bivalent transition metals such as cobalt, nickel, copper, zinc, and mercury with tetracarboxylic acid diimidates (72).

3.2.5.d Preparation of mixed complexes of copper(II)-pyromellitic diimidate (72)

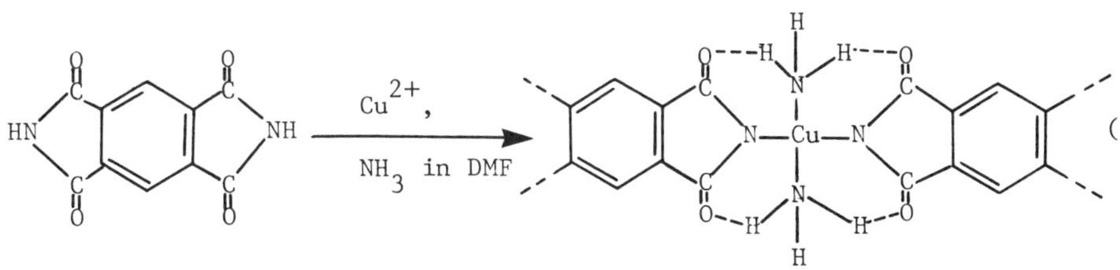

(40)

[62] [63]

A 25 mL flask, fitted with a stirrer and flushed with nitrogen, is charged with 2×10^{-3} mol (0.432 g) of PMDI in DMF. With stirring under N_2, 2×10^{-3} mol NH_3 are added to the solution. A purple-red gel is formed, which is washed with hot DMF and then ethanol and dried at 100°C in vacuum; yield 0.5 g (99.0%), IR (Nujol mull) 3300 (ν_{NH}), 1670 ($\nu_{C=O}$), 408 cm^{-1} (ν_{Cu-N}), MW(Rast's mcthod) 2000.

Thermal gravimetric analysis (TGA) curves of mixed amine complexes of metal-PMDI are shown in Figure 2. These polymers undergo extensive decomposition at 500 to 570°C.

By reacting tetrasodium tetrathiafulvalenetetrathiolate with transition-metal salts, tetrathiafulvalene-metal bisdithiolene oligomers were synthesized by Rivera and Engler (74). Treatment of [64] with trialkyl phosphites provided the dimer dithiopendione [65], which was subsequently treated to give the thermally sensitive and air-sensitive tetrathiafulvalenetetrathiolate intermediate [66]. Reaction of [66] with transition-metal salts such as Cu, Fe, Pt, and Pd led to the

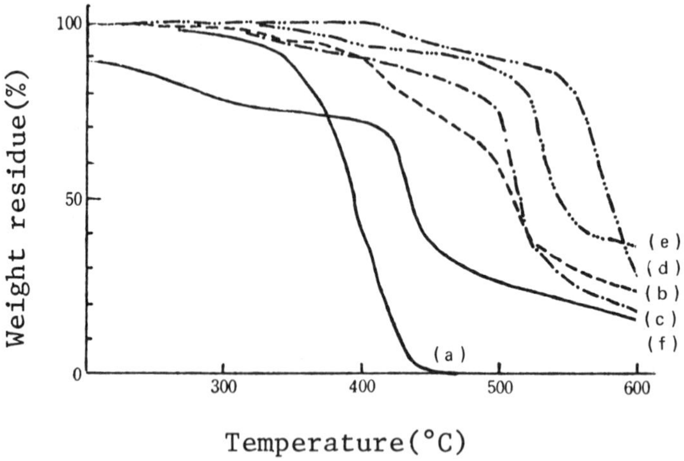

Figure 2. TGA curves for $[M(PMDI)(NH_3)_2]_n$ polymers. M = (a) PMDI, (b) Cu, (c) Ni, (d) Zn, (e) Cd, (f) Co.

precipitation of an insoluble, amorphous powder [67]. These dithiolate metal complexes are believed to have a one-dimensional structure. The nickel derivative possesses high conductivity (ca. 30 S/cm).

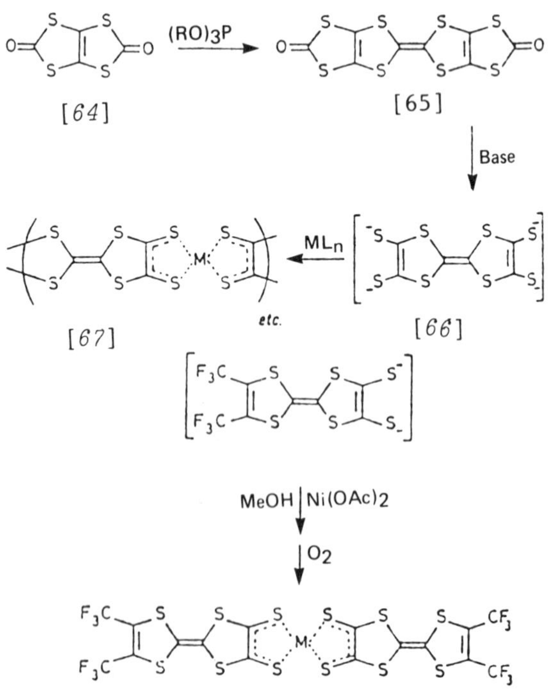

(41)

A series N,N-disubstituted dithiooxamide Cu-(R-DTOA) [*71*]-Cu(II) complexes have been studied as organic and metal-organic semiconductors (75). Kanda et al. prepared the coordination polymer by the "bulk method" (77). From chemical composition and properties, a two-dimensional polymeric structure [*68*] has been proposed for these compounds; however, the preparation of Cu-(R-DTOA) by the bulk method produces isometric mixtures. Synthesis of the stereospecific coordination polymers was attempted by an interfacial technique [*69*] (75).

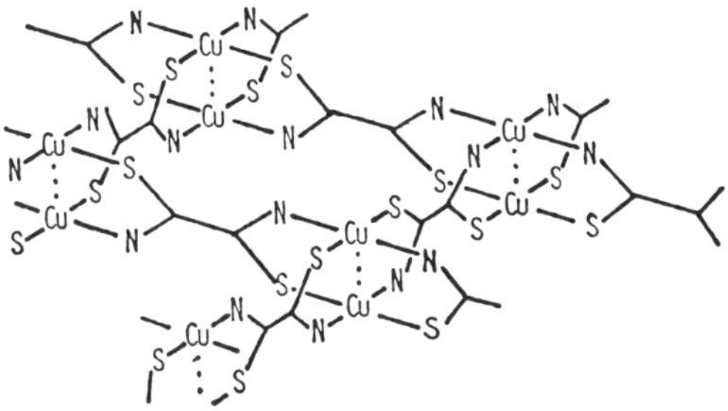

[*68*]

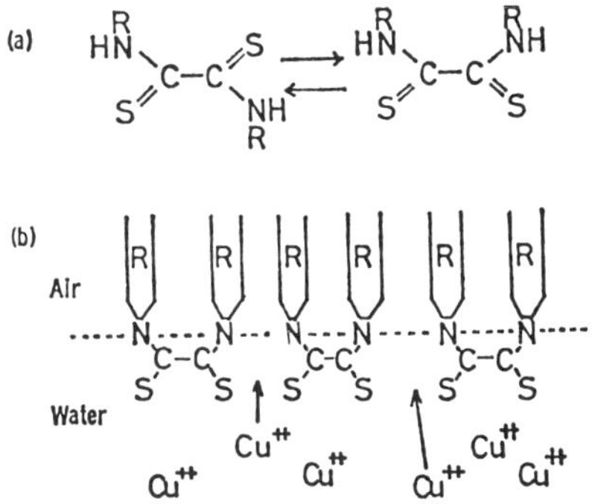

[*69*]

3.2.5.e Preparation of stereospecific N,N'-disubstituted dithiooxamide-Cu(II) coordination compound

[*70*]

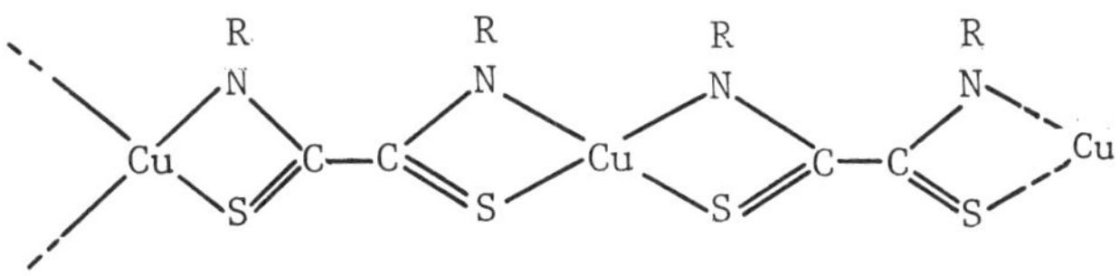

[*71*]

(42)

R: $C_6H_5CH_2$-, C_6H_{11}-, $C_{12}H_{25}$-, $C_{12}H_{25}OCOCH_2$-

Cupric sulfate is heated for 2 h at 350°C in a furnace to diminish the surface-active contaminants. About 2 mg of the ligand is dissolved in 10 mL of binary mixed solvents composed of ethanol, propanol, benzene, or hexane. The ligand solution is spread onto an aqueous cupric solution and is allowed to stand for 30 to 60 min for the complete coordination reaction. The complex monolayer obtained is transferred to the water surface of a second section to wash away unreacted Cu(II) ions absorbed beneath the complex monolayer. It is then transferred to the third section, filled with pure water, where the build-up procedure is carried out.

Takahashi et al. synthesized a series of interesting metal-poly-acetylene polymers which have the transition metals linked through bonds in the polymer backbone. A condensation reaction was carried out between a metal chloride and an acetylene or a metal acetylido complex in the presence of cuprous iodide as a catalyst. The metal-poly-acetylene polymers are soluble in several organic solvents, such as *n*-decanacyclohexane, benzene, and tetrahydrofuran. The poly-acetylene polymers have rodlike structures, as indicated by the viscosities of the solution independent of the solvent, and by a high value (1.7) in the Mark-Houwink-Sakurada equation ($[\eta] = KM$). The concentrated solutions of these polymers produce liquid crystalline structures (9,78-80).

3.2.5.f Preparation of poly[*trans*-bis(tri-*n*-butylphosphine)platinum-1,4-butadiyenediyl] [*74*] (80)

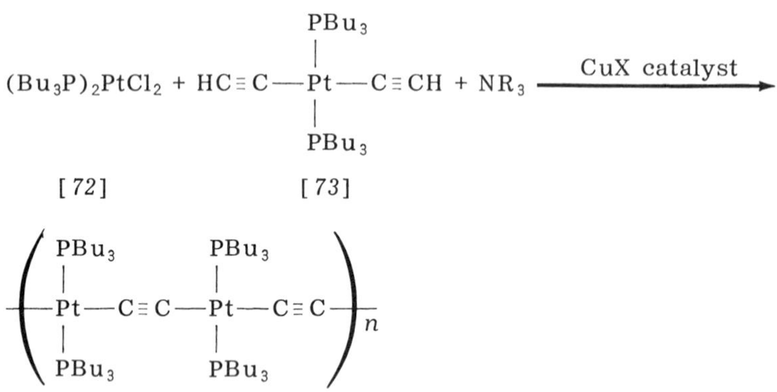

[*72*] [*73*] [*74*] (43)

Cuprous iodide (7 mg, 0.037 mmol) is added to a deoxygenated solution of *trans*-bis(tri-*n*-butylphosphine)bis(1,4-butadiynyl)platinum (1.3968 g, 2.0 mmol) and dichlorobis(tri-*n*-butylphosphine)platinum (1.3404 g, 2.0 mmol) in 50 mL of diethylamine under a nitrogen atmosphere. The reaction mixture is heated to a gentle reflux with stirring for 24 h and then evaporated to dryness under reduced pressure. To remove the cuprous compound, the residue is dissolved in methylene chloride and the solution is passed through an alumina column. After evaporation of the eluent, a pale yellow product is collected and purified by repeated precipitation from methylene chloride into methanol. Finally, a benzene solution (100 mL) of the product is freeze dried to afford a pale yellow polymer of [*74*], having [η] of 1.20 dL/g (in THF, at 25°C); yield 2.5 g (96%).

3.2.5.g Preparation of poly(copper phthalocyanine) (81)

Poly(metal phthalocyanine) is formed by the reaction of tetracyanobenzene with metal salts, and urea. This type of polymer metal complex makes use of a template reaction. Generally the reaction of ligands containing four functional groups with a metal ion results in a two-dimensional-network chelated structure.

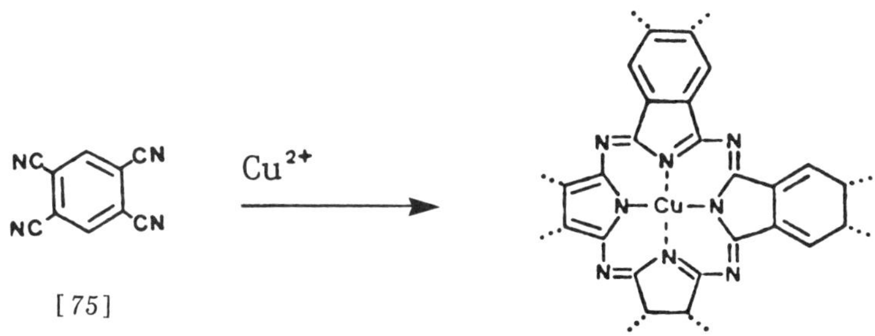

[*75*] [*76*] (44)

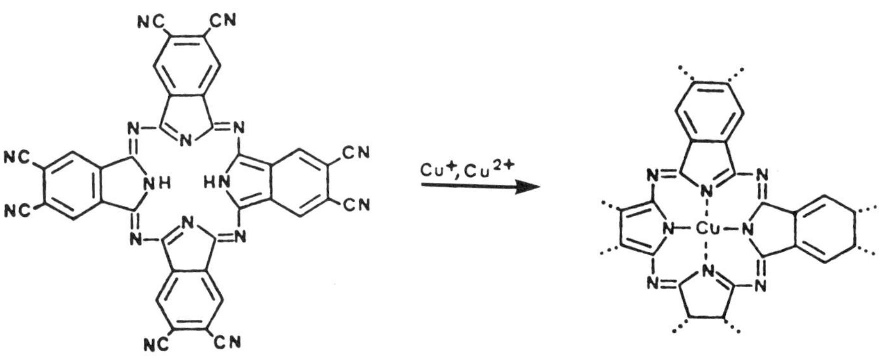

[*77*] [*78*] (45)

Reaction of 1, 2, 4, 5, -tetracyanobenzene with a copper plate

A polished copper plate (1 cm^2) is heated in a pressure vessel with 2 mg of 1,2,4,5,-tetracyanobenzene [75] at 400°C for 24 h. The small amount of [76] vaporizes and reacts at the surface of the metal plate, forming a homogeneous layer of blue-colored polymer. The polymer [76] is obtained by dipping the plate into concentrated sulfuric acid.

Reaction of octacyanophthalocyanine with copper salts

Octacyanophthalocyanine [77] 0.75 g (1 mmol) and 2 mmol of CuCl are heated for 6 h at 300°C or in a high-boiling solvent (quinoline, ethylene glycol) under reflux for 24 h. The dark brown or black polymer is extracted with acetone and DMF in a Soxhlet apparatus. The yield of DMF insoluble dark polymer is 0.46 to 0.78 g (53 to 89%).

3.3 MODELS FOR METALLOENZYMES

Many trace metals are found in living organisms. A classification scheme for some of the types of metallobiomolecules is presented in Figure 3. Here the principal distinction is made between enzymes, catalysts of bond-breaking and bond-making transformations, and other types of metallobiomolecules which themselves are noncatalytic in this sense. This classification identifies metals involved in some fundamental biological processes, such as electron storage and transfer (Fe, Cu), dioxygen binding, storage, and activation (Fe, Cu), and substrate activation and catalysis (Mg, Mn, Fe, Co, Cu, Zn, Mo). Most metallobiomolecules are metal complexes of biopolymers such as protein. Therefore, metal-polymer complexes are important as models for metallobiomolecules. Specific catalytic actions of metal-polymer complexes have received particular attention as models for metalloenzymes. The characteristic properties of metalloenzymes are due to the polar and/or nonpolar interactions which may be induced by the residual groups of their amino acids. Investigations of the catalytic actions of metal-polymer complexes have been made in order to clarify the structural effects of a protein, and the coordination structure. Many metal-polymer complexes have been found to exhibit catalytic efficiencies that are high compared with those of low-molecular-weight analogs, and the catalytic behavior of the complexes has been compared with those of metalloenzymes.

3.3.1 Models for Hydrolytic Metalloenzymes

Metal ions play a functional role in a number of hydrolytic enzymes, such as carboxypeptidase, alkaline phosphatase, and thermolysin.

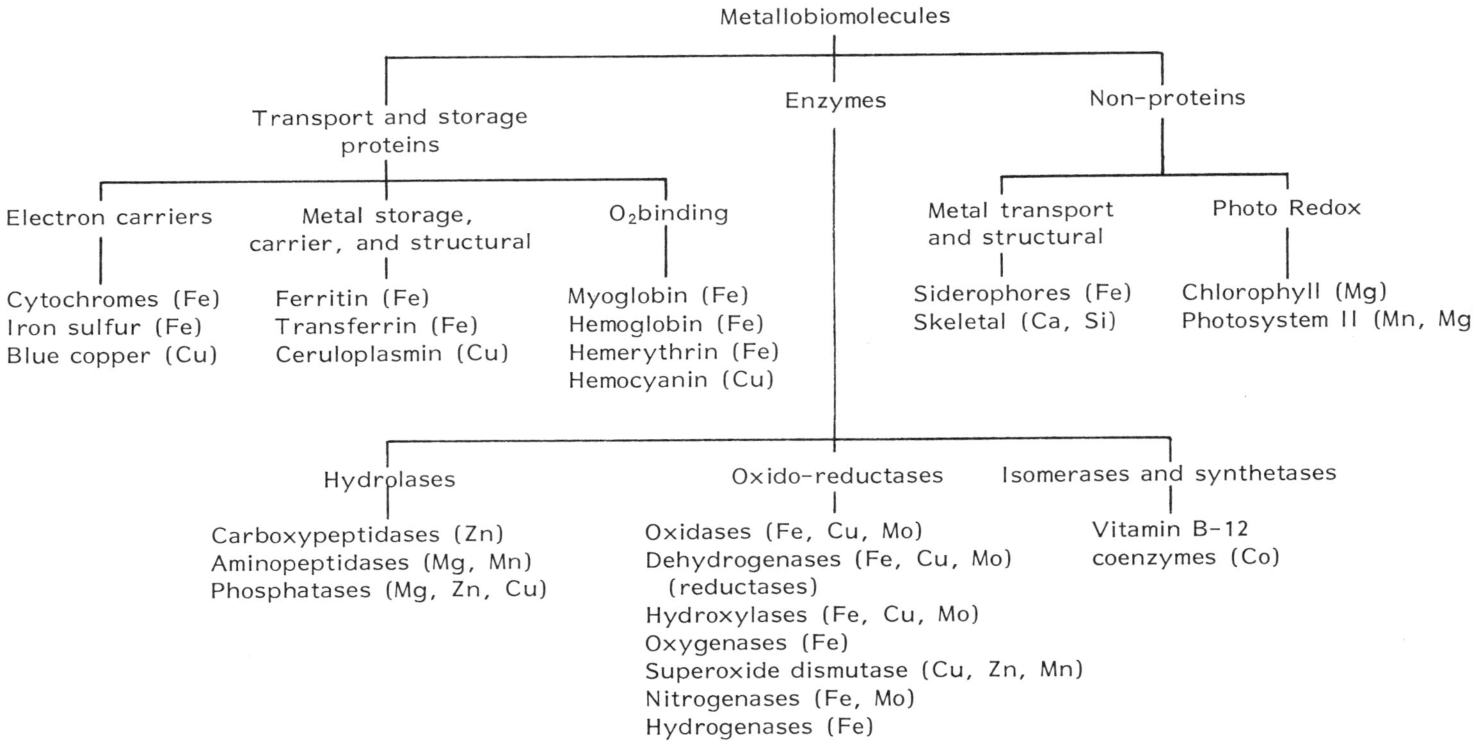

Figure 3. Partial classification of metallobiomolecules in terms of biological function. Examples rather than complete tubulations for each class are given. (From Ref. 82.)

The effects of metal ions in the synthetic polymer-catalyzed ester solvolysis were investigated as simplified kinetic models for hydrolytic metalloenzymes. Aflrossman and Murray (83) have reported that the proton-catalyzed hydrolysis of an olefinic ester is accelerated characteristically by a sulfonic resin partially exchanged with silver ion (84). Takeishi and Hayama investigated the silver ion effects for ester solvolyses catalyzed by nucleophile groups located on synthetic polymer poly(acrylic acid) (84,85). Breslow et al. have reported the reactions of coordinated substrates as models for a metalloenzymes carboxypeptidase A (86). This is an excellent example of an "artificial metalloenzyme," where a substrate is bound by single recognition, as shown by [*80*] in reaction (46), as far as binding characteristics are

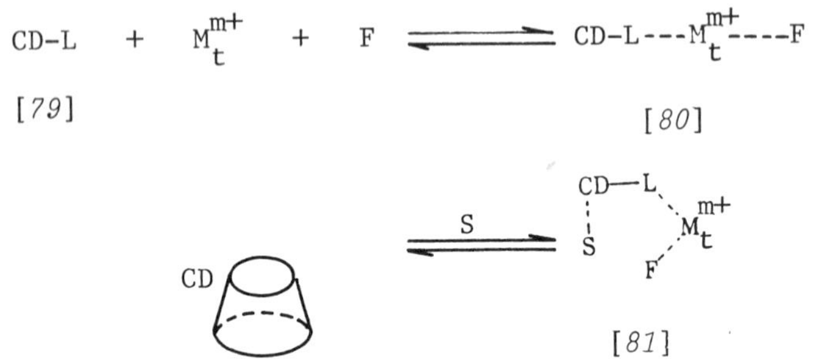

(46)

concerned (CD = cyclodextrin, M^{m+} = metal ion, F = catalyst, S = solvent, and binding interactions are shown by dotted lines). Tabushi et al. found several interesting catalysts by functionalizing cyclodextrins (87). Cyclodextrins functionalized with polyamines [*83a*] and [*83b*] strongly interact with metal ions such as Cu(II), Zn(II), or Mg(II) ion. The resulting cyclodexatrins are capped by metal ions [*84c*] to [*84e*] that can bind several anions with hydrophobic moieties much more strongly than parent or functionalized cyclodexatrin [*82*] without metal coordination, as shown in Table 8. The host [*83*], with Mt = Zn^{2+}, bound adamantan-2-one-L-carboxylate 330 times stronger than CD. For substrate binding, metal-ligand recognition is cooperative with the hydrophobic recognition by CD, and the energetics of the additivity of these two processes are shown schematically in Figure 4. Total binding energy of double recognition (ΔG) is 7.4 kcal.

An asymmetrically selective hydrolysis by asymmetric metal-polymer complexes has also been investigated. Nozawa et al. reported that Cu(II)-poly(L-lysine) complexes exhibit an asymmetrically selective catalysis on the hydrolysis of phenylalanine esters at pH 7 (88). They proposed that the enantioselective effects were due to different stabilities of diasterisomeric complexes formed by the aminoester enantiomers and the catalyst. Yamskov et al. have studied the

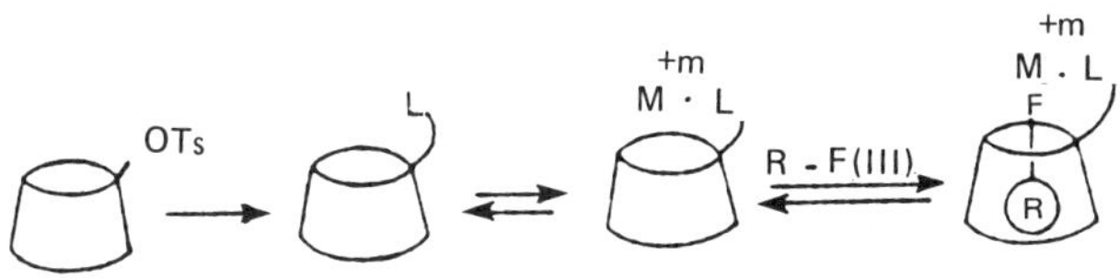

(47)

[82] [83] [84] [85]

(a) L : $-NHCH_2CH_2NHCH_2CH_2NH_2$

(b) L : $-NHCH_2CH_2NHCH_2CH_2NHCH_2CH_2NH_2$

(c) M : Cu^{++}

(d) M : Zn^{++}

(e) M : Mg^{++}

(f) 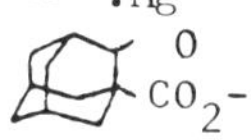

(g) 

(h) 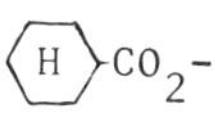

(i) 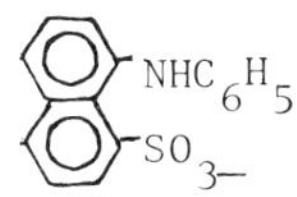

(j) 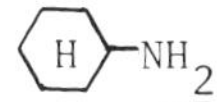

(k) 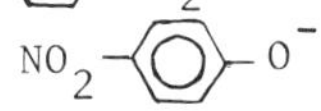

(l) 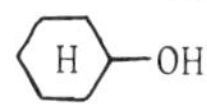

(m) 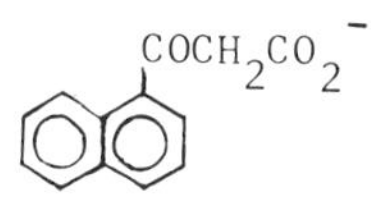

hydrolysis of a number of amino acid methyl esters in the presence of polystyrene-type asymmetric catalysts containing groups of *S*-(2-aminoethyl)-L-cycteine, *S*-(carboxymethyl)-L-cysteine, and *N*-carboxymethyl-L-valine, coordinated with copper or nickel ions (89). Takagishi and Klotz reported that the effects of metal ions on the hydrolysis of p-nitrophenyl caproate by modified polyethylenimines containing imidazole moieties were enhanced markedly in the presence of divalent metal ions (90). The order of effectiveness is Cu(II) > Co(II) > Zn(II) > Mn(II) and Cu(II), which increase the rate about 20-fold.

The imidazole groups in the polymer play a crucial role in creating sites with a high affinity for Cu(II) ions and may also be influential in the kinetic pathway. The mechanism involves an imidazole-linked Cu(II) ion to act as an electrophile, while an adjacent free ligand imidazole functions as a nucleophile, [*86*] and [*87*].

Table 8. Dissociation Constants of Holohost-Substrate Complexes

	Guest	Host	K_d,M	$K_d(\beta\text{-CD}) / K_d(\text{holohost})$	
(f)	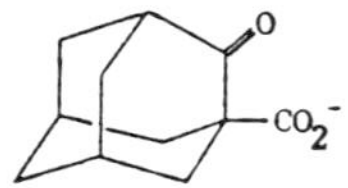	β-CD	1.2×10^{-3}		a
		Ia–Zn^{2+}	3.6×10^{-6}	330	
(g)	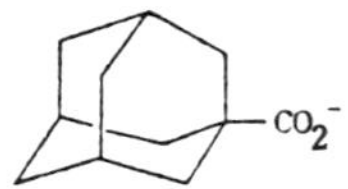	β-CD	4.3×10^{-3}		a, c
		Ia–Zn^{2+}	1.9×10^{-4}	22.6	
(h)	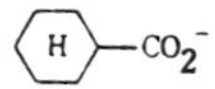	β-CD	7.1×10^{-3}		a
		Ia–Zn^{2+}	5.3×10^{-4}	13.4	
(i)	1,8-ANS	β-CD	1.3×10^{-2}		b
		Ib	7.8×10^{-3}		
		Ib–Zn^{2+}	1.9×10^{-3}	6.8	
		Ia–Zn^{2+}	4.5×10^{-3}	2.9	
(m)	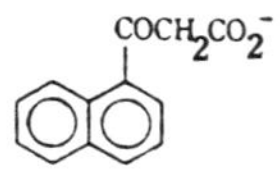	β-CD	1.4×10^{-3}		a
		Ia–Zn^{2+}	2.4×10^{-4}	5.8	
(j)	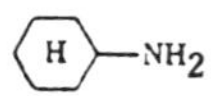	β-CD	2.2×10^{-2}		b
		Ia–Zn^{2+}	4.7×10^{-3}	4.7	
(k)	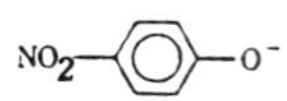	β-CD	2.1×10^{-3}		a
		Ia	1.8×10^{-3}		
		Ia–Zn^{2+}	8.3×10^{-4}	2.5	
(l)	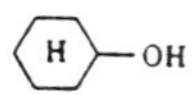	β-CD	2.0×10^{-3}		a
		Ia–Zn^{2+}	2.4×10^{-3}	0.83	

[a] In 0.05 M borate buffer (pH 10.0) at 25°C.
[b] Reported value, 1.60×10^{-3} M, in a buffer of pH 9, $I = 0.2$, 0.5-0.75% (v/v) CH_3CN involved and 0.7×10^{-3} M.
[c] In 0.05 M borate buffer (pH 11.0) at 25°C.

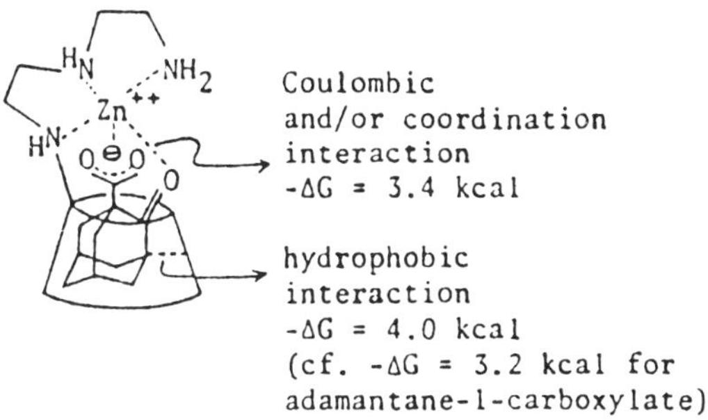

Figure 4. Schematic energetics of "additivity" for Zn^{2+}-β-CD.

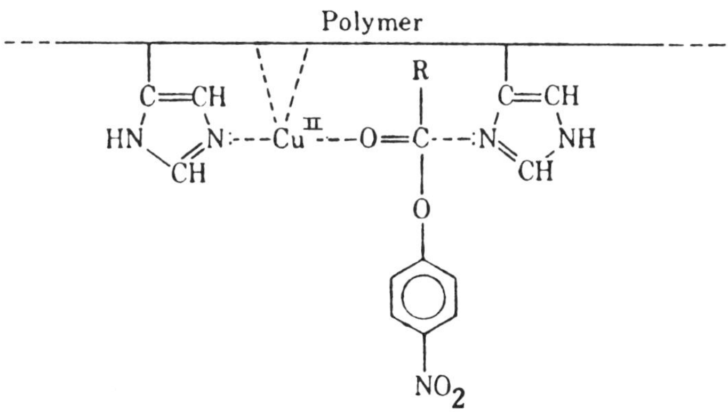

[86]

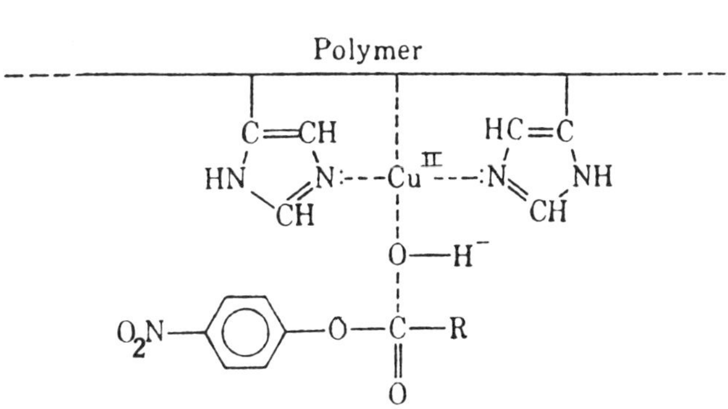

[87]

3.3.1.a Hydrolysis of *p*-nitrophenyl caproate by laurylated polyethyleneimine with imidazole groups in the presence of Cu(II) ion

Preparation of laurylated polyethyleneimine

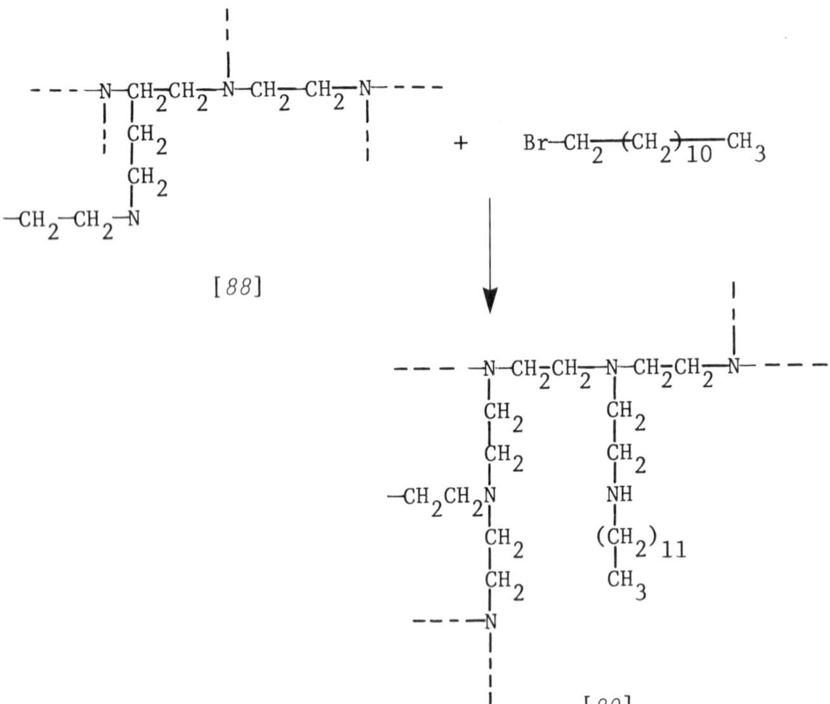

[88] [89] (48)

A 2.21 g portion of polyethylenimine 600 (PEI 600, Dow Chemical Co.), MW 60,000, is filtered through a PM-30 membrane with 12 L of water, lyophilized, and then dissolved in 100 mL of absolute ethanol. To this solution are added 1.28 g of 1-bromodecane and 13.39 g of diisopropylethylamine. The reaction mixture is refluxed overnight under nitrogen. At the conclusion of the reaction, all volatile materials are removed on a rotatory evaporator. The residue is dissolved in water and the solution is ultrafiltered with 12 L of water and then lyophilized. The lauryl group content of the modified polymer is determined by NMR spectroscopy and elemental analysis; 13% of the ethylenimine residues are alkylated. The stoichiometric composition is represented as $\text{-}\!\!\!+\!(C_2H_4N)_{0.87m}(C_2H_4NH_2)_{0.31m}(C_{12}H_{25})_{0.13}\!+\!\!\text{-}$, where m is about 1400.

Preparation of laurylated polyethyleneimine with imidazole groups

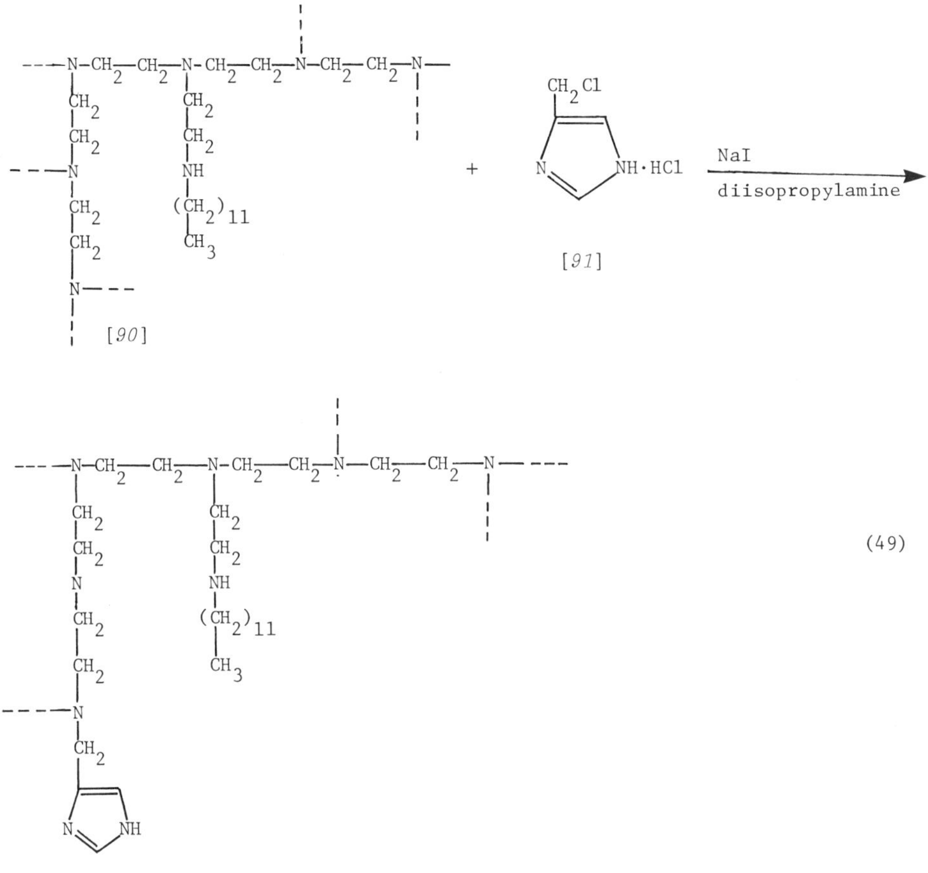

[90] [91] [92]

(49)

To a solution of 1.50 g of the laurylated polymer in 50 mL of absolute ethanol is added 0.80 g of 4-(chloromethyl)imidazole hydrochloride, which was prepared from 4-(hydroxymethyl)imidazole hydrochloride according to the procedure of Turner et al. (91). This is followed by the addition of sodium iodide and 18 mL of diisopropylethylamine. The reaction mixture is refluxed under nitrogen for 7 days. The mixture is ultrafiltered with 12 L of water and then lyophilized. The imidazole groups were attached to 11% of the ethylenimine residue.

Hydrolysis of p-nitrophenyl caproate

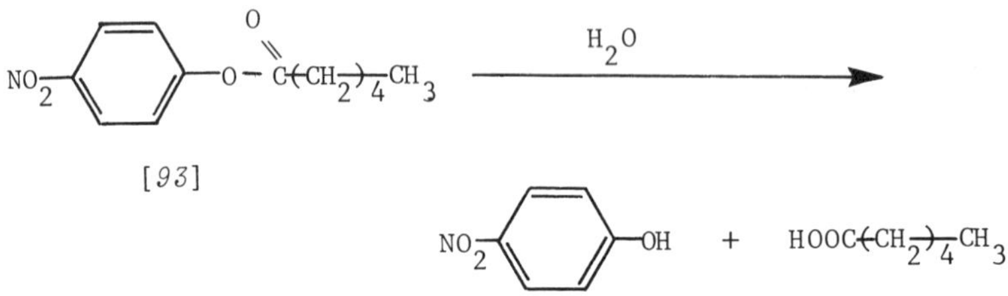

[93]

[94]

(50)

The rate of hydrolysis of *p*-nitrophenyl caproate is measured by following the appearance of *p*-nitrophenolate ion spectrophotometrically at 400 nm with a spectrophotometer. The cell compartments are maintained at 25°C by a circulating water bath. Pseudo-first-order rate constants are calculated from plots of $\log[(OD) - (OD_0)]/[(OD)_t]$ versus time, where (OD_0), (OD_t), and (OD) represent the optical densities (or absorbances) at time 0, t, and infinity, respectively. The reactions are run in 0.01 M bis-tris buffer containing 0.5% acetonitrile, 1.5×10^{-6} to 1.50×10^{-3} mol of Cu(II) ion, 1.21×10^{-3} mol of polymer, and 5×10^{-5} mol of *p*-nitrophenyl caproate at pH 7.3.

3.3.1.b Asymmetrically selective hydrolysis phenylalanine esters catalyzed by Cu(II)-poly(L-lysine) complex (88)

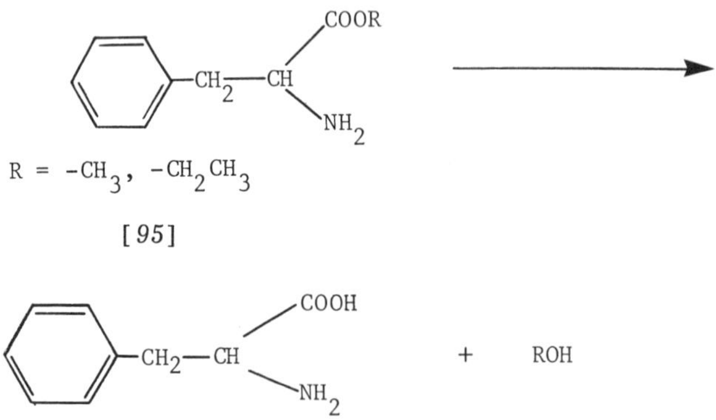

R = $-CH_3$, $-CH_2CH_3$

[95]

[96]

(51)

The hydrolysis of phenylalanine ester is followed by the pH-stat technique. As the ester is hydrolyzed, phenylalanine and alcohol are produced. The acid proton of the phenylalanine is titrated to maintain a constant pH with 0.1 M sodium hydroxide solution.

Standardized potassium nitrate (20 mL of 0.1 M) solution is added to the thermostated beaker, which is equipped with a magnetic stirrer followed by a weighed amount of poly(L-lysine). The ratio of the molar concentration of Cu(II) ion to the residue concentration of PLL (T_{Cu}/T_{HL}) of 0.13 to 2.0 is achieved by adding 25 mL of standard Cu(II) nitrate solution (0.05 M). The pH of the resultant solution is 5, which is adjusted to the desired pH value with 1 M sodium hydroxide solution. The rate data at 25°C are obtained by use of the recording pH stat under a flow of nitrogen to eliminate the effect of CO_2. The reaction vessel is a jacketed 100-mL beaker which is temperature controlled by means of water from a constant-temperature bath.

3.3.2 Models for Oxidoreductases

Biological oxidation is brought about by one of the following processes (S = oxidized substrate, SH_2 = reduced substrate): (a) electron transfer with oxygen (or another species) as the terminal electron acceptor, (b) hydrogen atom abstraction by Equation 52 or 53:

$$2SH_2 + O_2 \longrightarrow 2S + 2H_2O \quad (52)$$

$$SH_2 + O_2 \longrightarrow S + H_2O_2 \quad (53)$$

(c) hydride group transfer, (d) oxygen atom incorporation, or (e) hydroxyl group incorporation.

Equations 52 and 53 represent an overall reaction. In practice, the reaction is represented by Equation 54, where A^n and A^nH_2 represent the oxidized and reduced states of a series of acceptors, respectively.

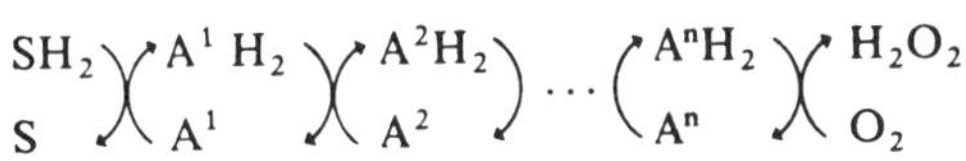

(54)

Enzymes of this type are classified as follows: (a) heme iron enzymes, (b) non-heme-iron enzymes, (c) copper enzymes, or (d) molybdenum enzymes. The catalytic effects of metal-polymer complexes have received attention as models for these metalloenzymes (92).

Pecht et al. reported the copper-catalyzed oxidation by molecular oxygen of ascorbate and homogentisate ions, hydroquinone as well as tetramethyl-*p*-phenylenediammonium and 1-(2,5-dihydroxyphenyl)-isopropylammonium cations, in the presence of poly(*L*-histidine) (PLH) (93). The rate of oxidation of ascorbic and homogentisic acids and of hydroquinone by Cu(II)-PLH was followed as a function of the substrate concentration (Figure 5).

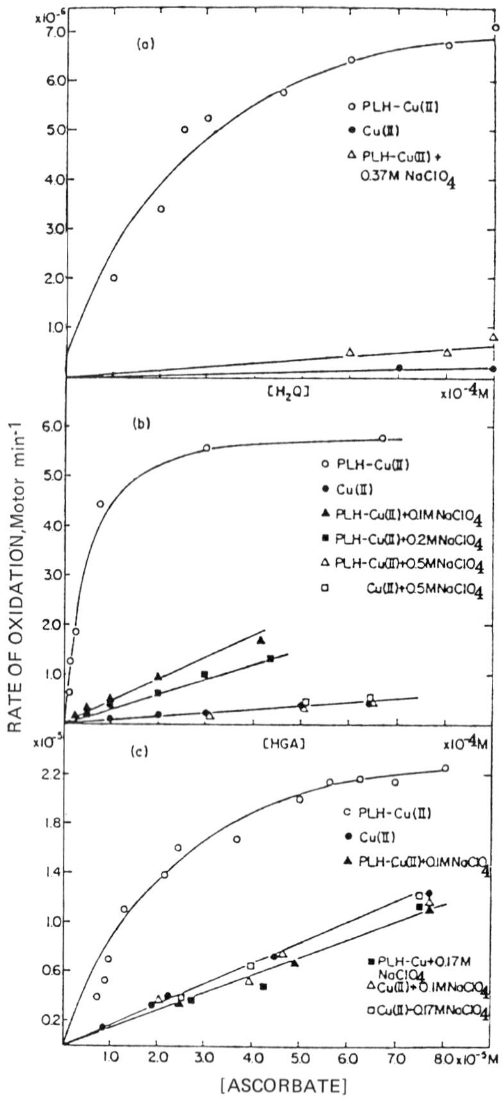

Figure 5. Rate of oxidation as a function of the substrate concentration: (a) H_2O oxidation in sodium acetate buffer, 0.02 M, pH 4.8, Cu(II)8 × 10^{-4} M, PLH 4 μg/mL (4 × 10^{-7} M in polymer or 2.65 × 10^{-5} in histidyl residues). (b) HGA oxidation in sodium acetate buffer, 0.02 M, pH 5.0, Cu(II) and PLH concentration same as in part (a). (c) Ascorbate oxidation in sodium acetate buffer, 0.02 M, pH 4.3, Cu(II) 6 × 10^{-4} M, PLH 6 μg/mL (4 × 10^{-4} M in histidyl residues).

The typical Michaelis-Menten kinetic behavior, which indicates the existence of a catalyst-substrate complex, is not exhibited by the aquoacetato complex in the same range of substrate concentrations. The mechanism of the copper-catalyzed reactions is described by the following steps:

$$Cu^{2+} + H_2A \underset{k_2}{\overset{k_1}{\rightleftharpoons}} CuAH^+ + H^+ \quad (55)$$

$$CuAH^+ \xrightarrow{k_3} Cu^+ + AH \quad (56)$$

$$Cu^+ + O_2 \xrightarrow{\text{fast}} Cu^{2+} + O_2^- \quad (57)$$

$$Cu^+ + O_2^- \ 2H^+ \longrightarrow Cu(II) + H_2O_2 \quad (58)$$

where H_2A represents the substrate and AH the radical produced on single-equivalent oxidation. Reaction 56 is the rate-determining step, whereas Reaction 55 is a preequilibrium. Hydrogen peroxide is obtained by Reaction 58. The rate of reaction is expressed by the experimental conditions $(k_2[H^+] + k_3)k_1[H_2A]$; the observed specific

$$-\frac{d[H_2A]}{dt} = \frac{k_1 \cdot k_3[Cu^{2+}][H_2A]}{k_2[H^+] + k_1[H_2A] + k_3} \quad (59)$$

rate constant is equal to $k_1 \cdot k_3/k_2[H^+]$. Therefore, Equation 59 is simplified to Equation 60.

$$-\frac{d[H_2A]}{dt} = \frac{k_1 \cdot k_3[Cu^{2+}][AH_2]}{k_2[H^+] + k_3} \quad (60)$$

In the case of the Cu(II)-PLH-catalyzed reaction,

$$Cu(II)\text{-}PLH + H_2A \underset{k_{2a}}{\overset{k_{1a}}{\rightleftharpoons}} Cu(II)\text{-}PLH\text{-}AH + H^+ \quad (61)$$

$$AH\text{-}Cu(II)\text{-}PLH \xrightarrow{k_{3a}} Cu(II)\text{-}PLH + AH \quad (62)$$

the rate equation is given by

$$-\frac{d[H_2A]}{dt} = \frac{k_{3a}[Cu(II)\text{-}PLH][H_2A]}{(k_{2a}[H^+] + k_{3a})/k_{1a} + [H_2A]} \quad (63)$$

The Cu(II)-PLH-catalyzed reaction follows Michaelis-Menten kinetics because the terms $k_{2a}[H^+] + k_{3a}$ and $k_{1a}[H_2A]$ are of the same order of magnitude. The addition of PLH enhances the catalytic efficiency

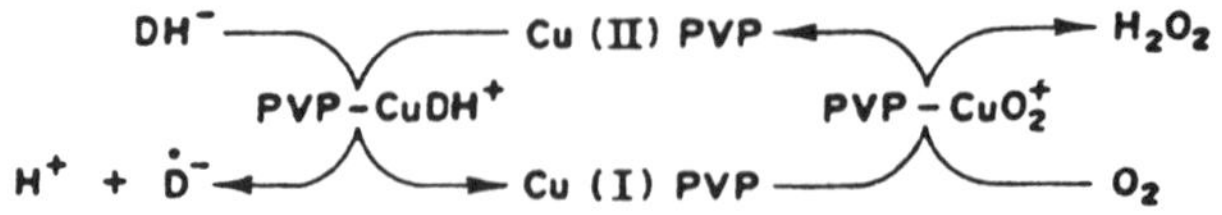

Figure 6. Cyclic mechanism of oxidation of ascorbic acid catalyzed by Cu(III)-PVP complex.

of Cu(II) toward negatively charged substrates such as ascorbate, homogentisate anion, and hydroquinone and inhibits catalysis toward positively charged substrates such as tetramethylenediamine ammonium and 1-(2,5-dihydroxyphenyl)isopropylammonium cations.

The specific effect of PLH on the Cu(II)-catalyzed reactions, which is abolished in the presence of moderate concentrations of inert electrolytes, is interpreted as an electrostatic effect of the polyelectrolyte reactants involved in the close vicinity of the bound copper.

Skurlatov et al. reported the mechanism of ascorbic acid (AH_2) oxidation with molecular oxygen catalyzed by Cu(II)-poly(4-vinylpyridine) (PVP) [97], partially quarternized by dimethyl sulfate (94).

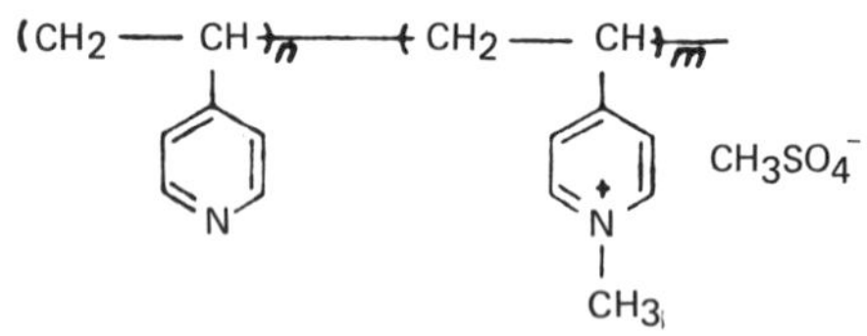

[97]

The half-conversion time of AH_2 with Cu(II)-PVP under anaerobic conditions is independent of [Cu(II)]. At pH 3.5, to $t_{0.5}$ (s) = $0.8 + 5 \times 10^{-4}[AH_2]$. The rate constant of the Cu(II)-PVP · AH-ternary complex dissociation is $0.4\ s^{-1}$. The reaction of Cu(II)-PVP with O_2 is not accompanied by the formation of O_2^- outside the macromolecule bulk. The rate constant of this reaction is $1.3 \times 10^2\ M^{-1}\ s^{-1}$. A proposed cyclic mechanism of the catalytic reaction is shown in Figure 6. A similar mechanism has been suggested for the effect of ascorbate oxidase. Kinetic parameters for the reaction are listed in Table 9. Many Cu-polymer complexes have been reported to catalyze the oxidation of ascorbic acid, such as Cu(II)-poly(4-vinylpyridine (partially alkylated by bromoacetic acid) (153).

Oxidation of hydroquinone catalyzed by Cu(II)-poly(1-vinylimidazole) was studied and showed catalytic activity 20 times greater than cupric ions. The oxidation mechanism is shown in Reaction 64 (95).

Table 9. Kinetic Parameters of Copper Complexes for Oxidation of Ascorbic Acid

Ligand	$\phi_0(Cu^{2+}/Cu^{-})(V)$	k_{bi}[a] $(M^{-1}\ s^{-1})$	K_t (M^{-1})	$k_3(Cu^{-} + O_2)$[b]
H_2O	0.15	1	2×10^2	5×10^4
bipy	0.12	0.5	2.8×10^2	1.5×10^4
Ascorbate oxidase	–	1.1×10^7	10^4 [c]	4.8×10^6 [d]
PVP	0.42	4×10^3	10^4	1.3×10^2

[a] $k_{bi} = k_1K_e$.
[b] Effective bimolecular constant of Cu^{-} oxidation rate $M^{-1}\ s^{-1}$.
[c] Estimated from nmL DH^{-}) [11] by extrapolating oxygen partial pressure to zero.
[d] At t = 1.7.
bipy: 2,2'-bipyridine.
PVP: poly(4-vinylpyridine).

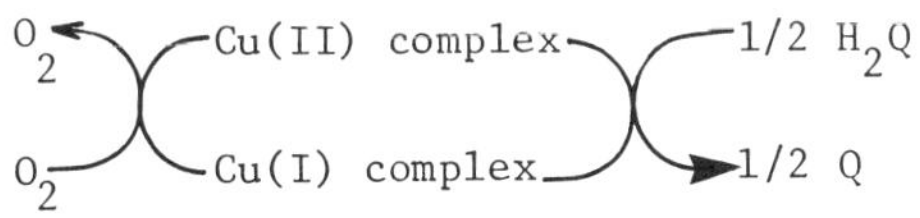

(64)

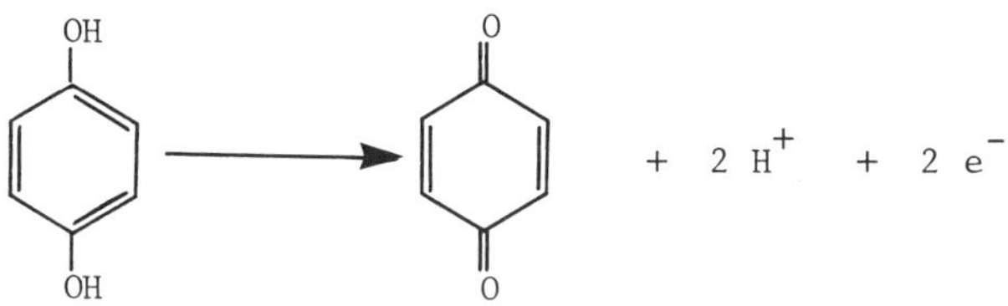

(65)

[98] [99]

3.3.2.a Catalytic effects of Cu(II)-poly(l-histidine) for the oxidation of ascorbate by molecular oxygen (93)

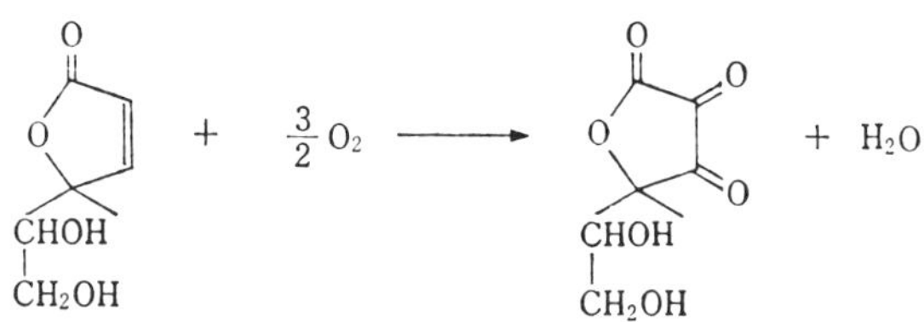

(66)

[100] [101]

Table 10. Molar Extinction Coefficient of Ascorbic Acid

pH	$\varepsilon(M^{-1}\ cm^{-1})$ at 265 nm
3.40	6,500
3.90	10,000
4.40	13,000
4.80	13,600
5.45	13,600
5.45	14,300
5.85	14,300

All oxidations of the substrate are conducted in quartz cuvettes in a final volume of 3.0 mL containing the components specified below. The solution is saturated with pure oxygen prior to addition of the substrate. In the kinetic experiments, an initial rate is determined for each substrate concentration. The rate of ascorbate oxidation is followed spectrometrically at 265 nm at 25°C. The molar extinction coefficient of ascorbic acid is determined within the pH range 4.0 to 5.9 using 0.02 M sodium acetate buffer (Table 10). The rate of oxidation of ascorbic acid by Cu(II)-PLH is followed as a function of the substrate concentration [Figure 5(a)]. Assuming the existence of a catalyst-substrate complex as an intermediate (which obeys the steady-state approximation d[substrate-Cu(II)-PLH]/dt = 0), Equation 68 is obtained.

$$H_2A + Cu(II)\text{-}PLH \underset{k_2}{\overset{k_1}{\rightleftharpoons}} (\text{substrate-PLH-}Cu(II)) \xrightarrow{k_3} AH \tag{67}$$

$$V = \frac{d[H_2A]}{dt} = \frac{k_3[Cu(II)\text{-}PLH][AH_2]}{((k_2 + k_3)/k_1) + [AH_2]} \tag{68}$$

$$\frac{1}{V} = \frac{1}{k_3[Cu(II)\text{-}PLH]} + \frac{k_1 + k_3}{k_1 \cdot k_3[Cu(II)\text{-}PLH][AH_2]} \tag{69}$$

A plot of the reciprocal of the rate as a function of the reciprocal of the substrate concentration gives a straight line (Figure 7). Values of K_m, V_{max}, and k_3 (known as the "turnover number," that is, the moles of substrate transformed per mole of catalyst per minute) are summarized in Table 11.

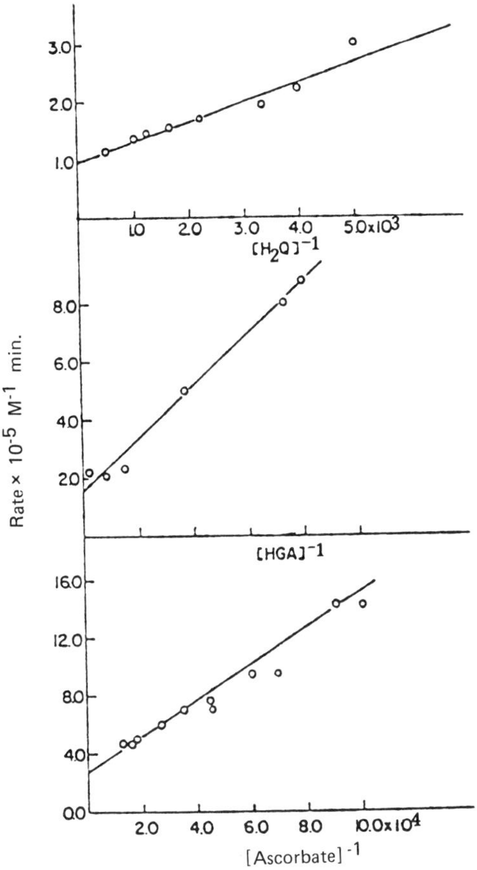

Figure 7. Lineweaver-Burk plot of the PLH-Cu(II)-catalyzed oxidations.

Table 11. Kinetic Parameters for the PLH-Cu(II)-Catalyzed Oxidation

Substrate	pH of assay	K_m (M)	V_{max} ($M \times min^{-1}$)	k_3[a] (min^{-1})	k_{obsd} (min^{-1}): PLH-Cu(II)	k_{obsd} (min^{-1}): Cu(II)
Ascorbic acid	4.3	4.3×10^{-5}	3.6×10^{-5}	6.0	1.0	0.16
Homogentisic acid	5.0	6.0×10^{-6}	0.7×10^{-5}	0.88	0.1	0.00058
p-Hydroquinone	4.8	4.0×10^{-4}	1.1×10^{-5}	1.37	0.19	0.002
TMPD	4.9	—	—	—	0.025	0.25

[a]k_s (turnover number) = moles of substrate/mole of copper × min.
[b]Calculated graphically from the slope of the linear parts of Figure 7.

3.3.2.b Oxidation of hydroquinone catalyzed by Cu(II)-poly(4,5-vinylimidazole) VIm) (95)

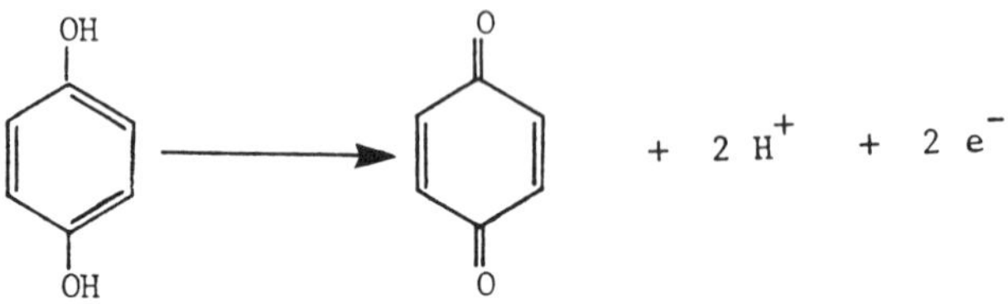

(70)

[102] [103]

The aqueous solution containing Cu(II) ion and complexing agents is prepared in acetic buffer solution and is saturated with pure oxygen prior to adding hydroquinone to the system. The initial rate of hydroquinone oxidation is determined spectrophotometrically at 245 nm.

3.3.2.c Oxidation of 3,4-dihydroxyphenylalanine by Cu(II)-poly(L-lysine) complex

In the oxidation of 3,4-dihydroxyphenylalanine (dopa) the effect of the asymmetrical structure of Cu(II)-poly(*L*-lysine) Cu(II)-PLL on the asymmetrical selectivity ratio was investigated by Nozawa and Hatano (96,97). From a comparison of the asymmetric selectivity with the structure of the catalyst, the combination of two or more asymmetric cupric complex units was considered as the cause for the asymetrically selective catalysis, where α-helical structure may serve to hold the two or more cupric complex units in a special steric configuration. Moreover, a plausible mechanism for the asymmetrically selective oxidation was discussed based on the evidence from the addition effects of catechol or phenylalanine and from the asymmetrically selective interaction between Cu(II)-PLL and dopa.

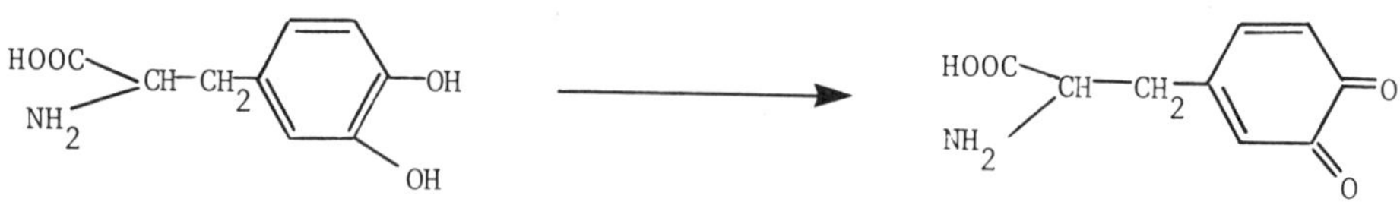

[104] [105]

(71)

An aqueous solution of *L*- or *DL*-dopa is introduced by a microsyringe into a quartz cell containing a solution of Cu(II)-PLL at pH 10.5 under a nitrogen atmosphere. The rate of the oxidation is followed by the increase in absorbance at 340 nm, where dopa has no absorption. The oxidation product has a molar extinction coefficient of 3000, which is determined by the method described below. The oxidation rate of dopa at pH 6.9 is measured by the same method as

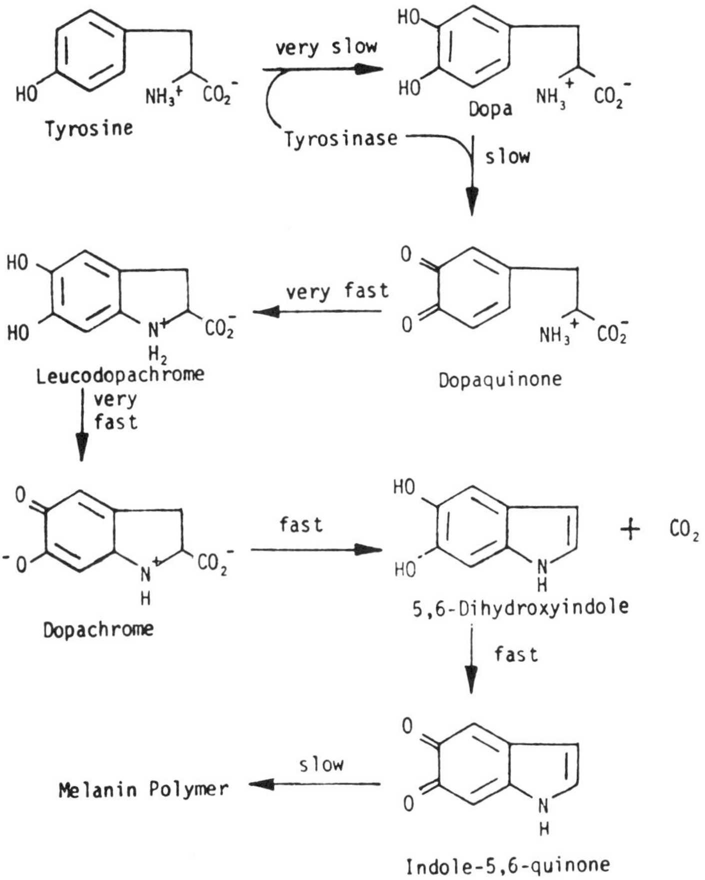

Figure 8. Pathway from tyrosine to melanin.

at pH 10.5. The pH is adjusted by using phosphate buffer at pH 6.9 and *o*-cresol-NaOH buffer solution at 10.5. The initial concentrations of dopa at pH 10.5 and 6.9 are determined from the absorbances at 300 and 270 nm, respectively, on the basis of their molar extinction coefficients, which are 3200 and 1540 at the given wavelengths, respectively.

The melanin pigments that appear on the surface of the human skin are quinoid polymer formed by the oxidation of 3,4-dihydroxyphenylalanine (dopa). The dopa is formed by the hydroxylation of tyrosine (Tyr) catalyzed by tyrosinase, which is a copper complex. The pathway from tyrosine to melanine is described as Figure 8. The formation of melanin by oxidative polymerization of dopa catalyzed by polymer-Cu complexes was studied as a model for tyrosinase by Tsuchida and Nishide (99). The melanin formation catalyzed by Cu(II)-polymer complexes are shown

Table 12. Melanin Formation Catalyzed by Metal Complexes

Catalyst	Melanin formation rate constant[a] (1/mol-s) pH 7	pH 4	Rate-determining step
Cu(II)	0	0	Dopa → (pH 4)
Im-Cu	23	0	
PVIm-Cu	23	0.17	
$PIPo_{20}$-Cu[b]	40	0.16	Dopa → (pH 7)
$PIPo_{35}$-Cu[b]	38	0.16	Dopachrome → (pH 7)
PAA-Cu	0	0	
$NaIO_4$-PVIm-Cu	—	1.2	Dopachrome →
$NaIO_4$-PAA-Cu	—	0.4	
PVIm-Ni	—	0.04	
PVIm-Fe	—	0.04	Dopa →
Autoxid.	—	0	
Tyrosinase	900		Dopa →, Dopachrome →

Im, imidazole; PVIm, poly(vinylimidazole); PIPo, poly(1-vinylimidazole-co-1-vinyl pyrrolidone); PAA, poly(acrylic acid).

in Table 12. The Cu complexes with imidazole-containing ligands exhibited much higher catalytic activity among other complexes, such as Cu(II)-polyamine, poly(aminocarboxylic acid), or pyridine. The formation rate of dopachrome formation is considered to involve two molecules of dopa to one O_2, through the transfer of four electrons from dopa to O_2 to produce two molecules of dopaquinone and of water, respectively. The active intermediate was presumed to be the binuclear α-dioxo complex [*106*]. Nishide and Tsuchida speculate that the active site of tyrosinase can be described by [*107*].

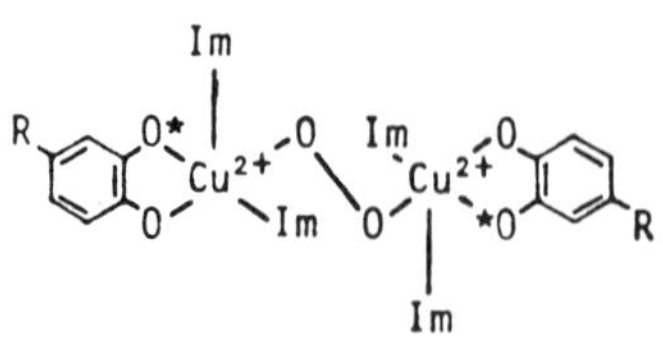

[*106*]

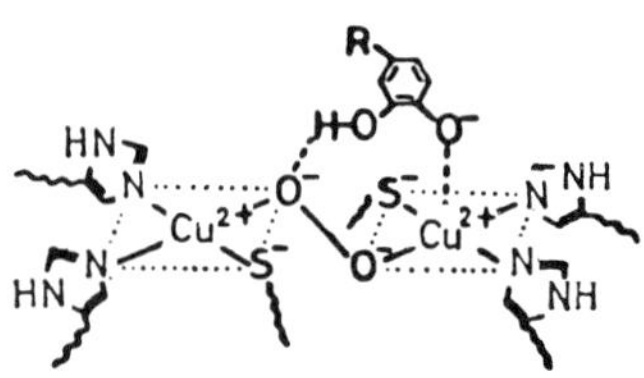

[*107*]

3.3.2.d Formation of melanin from 4-(3,4-dihydroxyphenyl)-L-alanine catalyzed by Cu(II)-polymeric amine complex (107)

The oxidation reaction of dopa is followed by spectral change recorded with a highly sensitive ultraviolet (UV) spectrophotometer. The ligand-containing buffered solutions are mixed in a 1-cm quartz cell. A spectrum of the 480-nm chromophore which is characteristic of dopachrome that is produced during the oxidation of dopa to dopaquinone, is observed at 1-min intervals. This intermediate is the rate-determining step of dopachrome formation. The dopachrome formation rate per Cu^{2+} ion is designated as the rate constant, k_D. The general increase in absorbance in the visible region indicates melanin formation; however, the formation rate is difficult to determine because the molar extinction coefficient of melanin varies with the degree of coupling and polymerization. A method of integration was developed to calculate the rate constant of melanin formation, k_M, in which the amount of dopachrome converted into melanin is plotted against the integral of dopachrome in accordance with

$$\text{dopa} \xrightarrow[k_D]{\text{Cu(II)},\ O_2} \text{dopachrome} \xrightarrow[k_M]{\text{Cu(II)},\ O_2} \text{melanin} \qquad (72)$$

$$\frac{-d[\text{dopa}]}{dt} = k_D[\text{Cu(II)}] \qquad (73)$$

$$\frac{d[\text{melanin}]}{dt} = k_M[\text{Cu(II)}][\text{dopachrome}] \qquad (74)$$

$$\text{nelanin}] = k_D[\text{Cu(II)}]_t - \frac{A^t_{480} - 1.5A^t_{650}}{\varepsilon_{480}(\text{dopachrome})} = k_M[\text{Cu(II)}] \int_0^t [\text{dopachrome}]\, dt \qquad (75)$$

where

$[\cdot]$ = concentration in mol/L
t = reaction time (s)
A^t = absorbance at $\times$ (nm) at time t.

Catechol is oxidatively cleaved in the presence of the Cu(II)-pyridine complex with a small amount of methanol to yield the monomethyl ester under oxygen atmosphere; this is similar to the model dioxygenase reaction. The ring cleavage is an intermolecular reaction which competes with polymerization caused by the intermolecular coupling of oxidized catechol, as represented by Reaction 76.

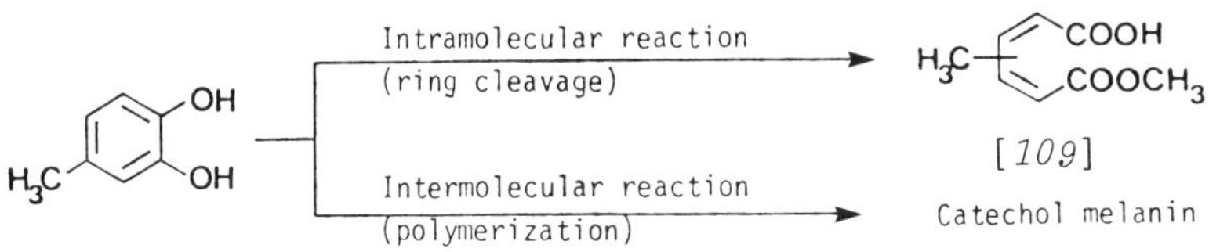

[109] (76)

[108] [110]

Nishide et al. reported that a Cu-complex bound to a benzene-soluble poly(styrene-co-4-vinylpyridine) was also effective for the cleavage to give the muconate in 80% yield (99).

3.3.2.e Oxidative cleavage of 4-methylcatechol with Cu(II)-poly(styrene-co-4-vinylstyrene) (99)

Cuprous chloride [0.5 g (5 mmol)] is dissolved in 1 L of methanol. A 500-mL benzene solution of poly(styrene-co-4-vinylpyridine) (PSP_{38}) 6.9 g, n = 25 mmol, is added slowly to the copper solution. The mixture is stirred under oxygen for 5 h. The initial yellow-green solution turns to a deep green. The solution is poured into petroleum ether and the precipitate is collected and dried under vacuum; yield 8.2 g, Cu content 6.2%, N/[Cu] = 4.8. The PSP-Cu complex (1.5 g) is dissolved in 30 mL of benzene containing 1 mL of methanol. A 20-mL benzene solution of 4-methylcatechol (0.124 g, 1 mmol) is added slowly to the PSP-Cu solution over a 10-h period. The addition is monitored by the rate of oxygen uptake due to the oxidation of 4-methylcatechol. The reaction mixture is treated with 100 mL of 1 N NaOH solution and the muconate is extracted with 100 mL of dichloromethane. The dichloromethane solution is evaporated to give a white powder. The monoethyl ester of *cis*, *cis*-muconic acid is obtained as colorless needles by recrystallization from *n*-hexane; mp 116 to 119°C; IR(KBr) 1660, 1710 cm^{-1} ($\nu_{C=O}$); NMR (CCl_4) 2.18 (s, 2H), 3.9 (m, 3H), 6.2 to 6.5 (m, 2H), 7.55 (m, 1H), 8.8 to 9.0 (m, 1H).

Nitrogenase contains Mo, eight Fe, and six labile inorganic S atoms in a polypeptide structure. EXAFS investigation of the Fe-Mo cofactor indicates the presence of a mixed Mo-Fe sulfur cluster at the molybdenum site. This cofactor has catalytic activity in the selective reduction of acetylene to ethylene with $NaBH_4$ as the reducing agent. Low-molecular-weight models for nitrogenase have been reported by Schrauzer and others. Dioxomolybdenum complex containing 1-cysteine and other compounds bearing thiolato groups as ligands have catalytic activity for the reduction of dinitrogen as well as compounds such as acetylene. Various other modifications of these catalyst systems have also been achieved. In native nitrogenase, the "Fe-Mo-Co" is surrounded by polypeptide containing probably 1-cysteine residues which stabilize the active sites of nitrogenase. Oguni et al. reported the nitrogenase model employing the di-μ-sulfido-bis[thioxanthato(GXO)molybdenum] complex supported on co-polypeptide consisting of L-cysteine and α-benzyl-*L*-glutamate (100). The catalysis for the reduction of phenylacetyene and azobenzene with $NaBH_4$ was evaluated. A binary catalytic system consisting of polypeptide-molybdenum complexes and ferredoxin model complex was found to be superior to that of the corresponding unsupported molybdenum complex. Reduction of phenylacetyene with $NaBH_4$ by the catalysis of a mixture of copolypeptide-molybdenum complexes and $[Fe_4S_4(SPh)_4]^{2-}$ are shown in Table 13. The mixed system of the copolypeptide-molybdenum complex and ferredoxin model complex have the highest catalytic activity among the mixed systems. Oguni et al. have proposed the mechanism shown in Figure 9. Oguni et al. (100) have also reported the electron-transfer reduction of acetylene catalyzed by partially *p*-mercaptomethyl-substituted polystyrene-molybdenum (IV) complexes. The system had excellent catalytic activity for the reduction of acetylene, compared to the corresponding unsupported molybdenum complexes.

3.3.2.f Electron-transfer reduction of phenylacetyene by catalysis of copolypeptide molybdenum complex (100)

Preparation of Mo-poly(α-benzyl-L-glutamete-co-S-benzylthiomethyl-L-cysteine)

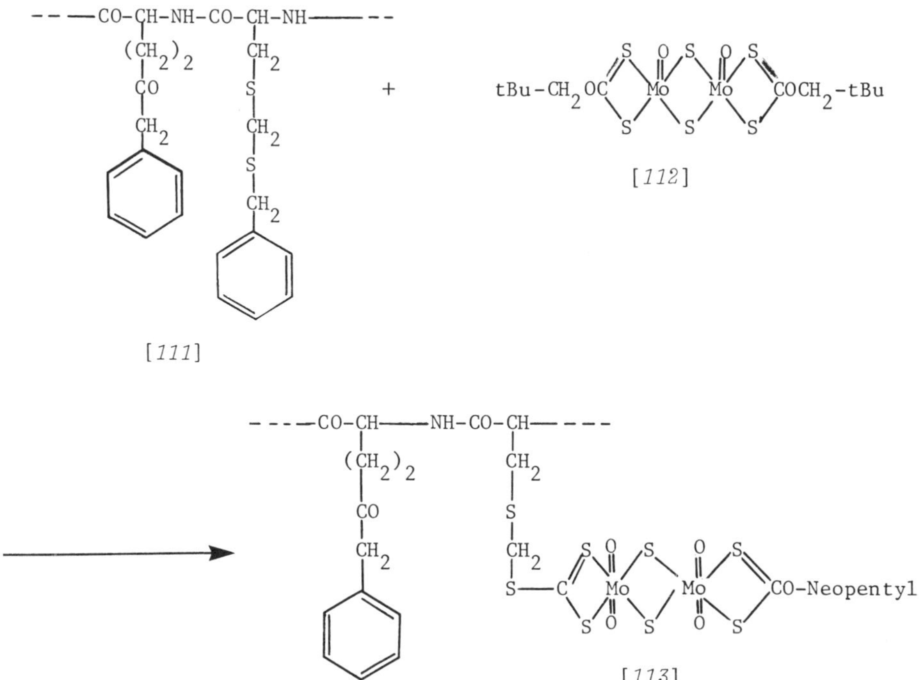

[111] [112] [113] (77)

The molybdenum complex [112] (2 of cysteine residues) is added to a THF solution (90 mL) of copolypeptide (1.0 g) in an ampul under a dry argon atmosphere. The mixture is allowed to stand at 50°C for 62 h. The reaction mixture is concentrated to one-third of its

Table 13. Reduction of Phenylacetylene and Azobenzene with $NaBH_4$ by the Catalysis of a Mixture of Copolypeptide-Molybdenum Complexes and $(Fe_4S_4(SPh)_4]^{2-}$ [a]

Catalyst system	Products from PhC≡CH (mmol)			Products from PhN=NPh (mmol)	
	$PhCH{=}CH_2$	$PhCH_2CH_3$	Reduced dimer and polymer	PhNHNHPh	$PhNH_2$
6A	1.43	0	Trace	0	0
$(Fe_4S_4(SPh)_4]^{2-}$	0.11	0	0.02	0.24	0.21
$Mo_2O_2S_2(S_2CO\text{-neopentyl})_2$ + $[Fe_4S_4(SPh)_4]^{2-}$	3.61	0.34	0.73	0.51	0.39
6A + $[Fe_4S_4(SPh)_4]^{2-}$	24.0	0.66	4.83	12.1	5.24
6B + $[Fe_4S_4(SPh)_4]^{2-}$	21.9	0.34	1.78	—	—

[a]The value of substrate consumption is based on 0.1 mmol of Mo. Conditions: [substrate], 4.9 mmol; $[NaBH_4]$, 24.5 mmol; time, 7 h; temperature, 30°C; solvent, EtOH-THF-MeOH (3.5:6:7 for PhC≡CH, 3:8:6 for PhN-NPh); ratio of Mo-Fe, 1:3; amount of catalyst (mmol of Mo), 6A, and 6A + $[Fe_4S_4(SPh)_4]^{2}$ - 0.78×10^{-2}, $[Fe_4S_4(SPh)_4]^{2-}$ 0.14, $Mo_2O_2S_2(S_2CO\text{-neopentyl})_2$ + $[Fe_4S_4(SPh)_4]^{2-}$ 0.49×10^{-1}, 6B + $Fe_4S_4(SPh)_4]^{2-}$ 0.7×10^{-2}.

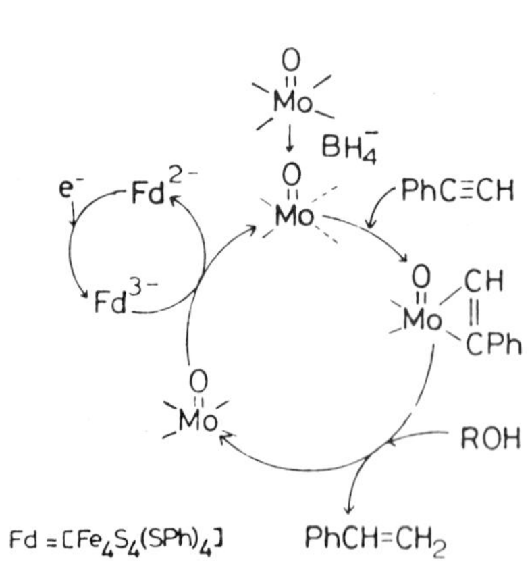

Figure 9. Proposed mechanism for the reduction of phenylacetylene with the molybdenum complex.

original volume by the removal of solvent under reduced pressure. It is then added dropwise to a large amount of methanol with stirring. The precipitated copolypeptide-molybdenum complexes are separated by filtration and washed with methanol several times to remove any of the unreacted molybdenum complexes. The molybdenum content in the copolypeptide-molybdenum complexes is analyzed by an atomic absorption spectrometry in aqueous $HClO_4$. In THF, the product is separated into two fractions, a soluble and an insoluble portion.

Reduction of phenylacetylene with copolypeptide-molybdenum complex

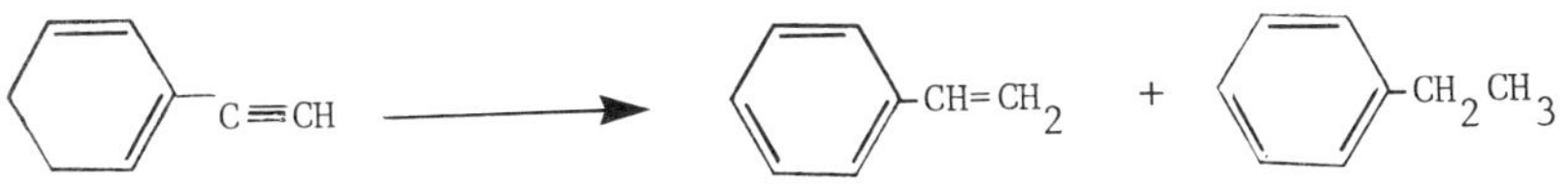

(78)

[*114*] [*115*] [*116*]

The copolypeptide-molybdenum complex (0.5 mmol of molybdenum) and ferrodoxin model compound, $[Fe_4S_4(SPh)_4]^{2-}$ $[Nn\text{-}Bu_4]_2^{2+}$ (0.037 mmol), are prepared by the method of Holm (101) and mixed in a solution of ethanol (3.5 mL), THF (6.0 mL), and methanol (7.0 mL) under a nitrogen atmosphere. Added to this catalyst solution is sodium borohydride (25 mmol) as a reducing agent. After the evolution of hydrogen gas ceases, phenylacetylene (4.9 mmol) is added to the mixture and the reaction vessel is shaken in a water bath kept at 30°C for 24 h. Hexane is then added to precipitate the catalyst. The clear upper layer is evaporated under reduced pressure and the distillate collected in a cold trap at -78°C is analyzed by gas-liquid chromatography (GLC) using toluene as a reference.

3.3.3 Models for Hydroperoxidase

Peroxidases are heme proteins that catalyze oxidations by hydrogen peroxide, while catalase catalyzes the disproportionation of hydrogen peroxide (92).

$$H_2O_2 + SH_2 \xrightarrow{\text{peroxidase}} 2H_2O + S \tag{79}$$

$$H_2O_2 + H_2O_2 \xrightarrow{\text{catalase}} 2H_2 + O_2 \tag{80}$$

Catalase activity may be regarded as a special example of peroxidase activity in which the substrate oxidized by hydrogen peroxide is another molecule of hydrogen peroxide. Catalases are heme proteins with a molecular weight of 240,000, containing four heme groups.

The catalase molecule can decompose 6×10^6 molecules of peroxide per second. The exact mechanism has been the subject of controversy. It appears that the catalase operates through a Fe(III)-Fe(II) cycle. A reasonable alternative is that it serves to bring together substrate and oxidant. It is known that peroxide coordinates readily with catalase (replacing carboxylate). It is assumed that the second peroxide molecule will displace a ligand at the fifth coordination position and that the ferric ion mediates the transfer of an electron between the two peroxide molecules.

For the decomposition of H_2O_2, metal-polymer complexes very often have much higher catalytic activity than those of the corresponding monomeric analogs. The reasons for this are that the polymer (a) contains incomplete coordination sites due to the steric hindrance of polymer complex formation; (b) contains relatively weak coordination bond between metal and ligand, allowing the substrate to coordinate more frequently; and (c) the concentration of substrate is locally enhanced in metal-polymer complex systems. Many metal-polymer complexes have been reported as catalase models: Cu(II)-poly(-diketone) (102), Cu(II)-poly(methacryl acetone) (103), Cu(II)-poly(acrylhydroxamic acid) (104), Fe(III)-poly(ethylenimine-co-acrylic acid), Fe(II)-poly(methacrylic acid), Fe(III)-poly(4-vinylpyridine) (105), Cu(II)-poly(amino acid) (106,107), Cu(II)-DNA or RNA (109), Fe(III) complexes supported on poly(L-glutamate) (110), Cu(II)-alginate (108), and Cu(II)-poly(vinyl alcohol).

Hojo et al. examined the decomposition of hydrogen peroxide (H_2O_2) catalyzed by Cu(II)-PVA complexes (111). This complex exhibited a large catalytic activity after forming a stable planar complex structure. The mechanism in the first step of reaction for H_2O_2 decomposition was considered, since an amine substituent has a critical influence on the rate of H_2O_2 decomposition. The relation between the initial rate and the initial concentration of H_2O_2 varied in accordance with the Michaelis-Menten type rate expressions.

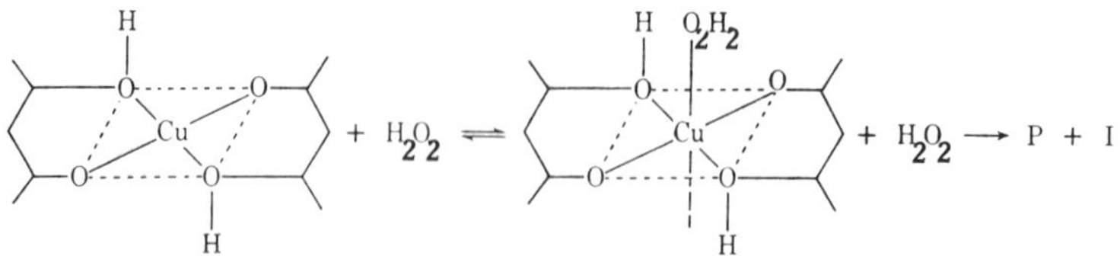

(81)

[117] [118]

At pH >8.2:

$$H_2O_2 \rightleftharpoons OOH^- + H^+ \tag{82}$$

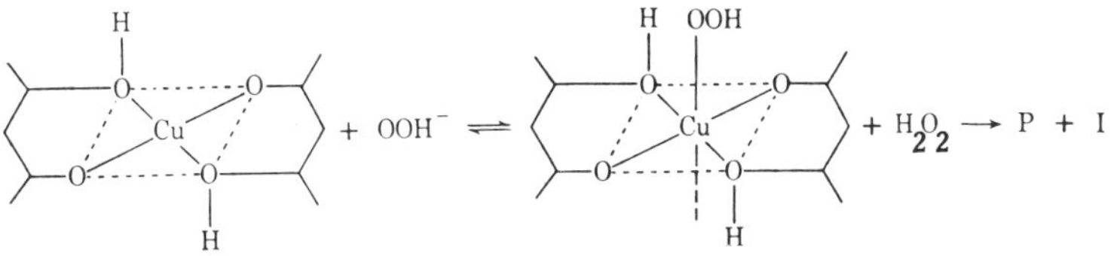

(83)

[*117*] [*119*]

3.3.3.a. Method 1: Decomposition of hydrogen peroxide by Cu(II)-poly(vinyl alcohol) (PVA) complex

Preparation of Cu(II)-PVA complex solution

An aqueous solution of Cu(II) nitrate is added to the PVA solution, and the pH of the mixed solution was adjusted to the desired value with KOH. The ionic strength of Cu(II)-PVA complex solution is maintained at 0.2 M with KNO_3. The Cu(II)-PVA complex solution is maintained at room temperature for 24 h before each experiment.

Measurement of the catalytic activity

To each reaction vessel was added 50 mL of H_2O_2 of known concentration, Cu(II)-PVA complex solution, KOH to adjust the pH of the reacting solution, KNO_3 stock solution to keep the ionic strength of the reacting medium constant at 0.2 M, and water to adjust the volume to 100 mL. The decomposition of H_2O_2 is done in the pH range 5.3 to 12.0, and in the complex concentration range 10^{-4} to 10^{-8} M. A 5-mL sample of experimental solution of the reaction vessel is pipetted into 10 mL of 2 N H_2SO_4 every 10 min. The H_2O_2 decomposed is measured by titration with potassium permanganate solution. The initial rate of the decomposition reaction of H_2O_2 is obtained from the slope of the tangent to the curve that represented the concentration of H_2O_2 reaction time relationship. Thus the initial rate V_o is expressed as follows:

$$V_o = \left(\frac{dC}{dt}\right)_{t=0,C=0}$$

The V_o-S_o relationship follows the Michaelis-Menten equation. A plot of $1/S_o$ against $1/V_o$ according to the Lineweaver-Burk procedure gives a straight line.

An incomplete coordinated structure of a metal-polymer complex is effective as a catalyst in H_2O_2 decomposition; therefore, incomplete coordination structures of metal-polymer complex, Cu(II) and Fe(III) mixed PVA complexes, were prepared (112). The complex formation of PVA with the Fe(III) ion proceeds smoothly in an acid solution, while the corresponding reaction with the Cu(II) ion takes place in the neutral or alkaline region. A Cu(II) and Fe(III) mixed complex is formed by increasing the pH of the solution of the Fe(III)-PVA complex in the presence of Cu(II) ion. Under a limited metal concentration ($T_{Fe^{3+}}/T_{HL} < 0.014$, $T_{Cu^{2+}}/T_{HL} < 0.08$), the mixed metal complexes were obtained. On the other hand, the Fe(III)-PVA complexes containing a larger amount of Fe(III) ion (Fe^{3+}/T_{HL} over 0.014) have a compact structure, and therefore additional complex formation with Cu(II) ion is not expected. The Cu(II) complexes are unstable. For the complex formation of Cu(II) ion with residual ligand in Fe(III)-PVA complex, the formation constants and thermodynamic parameters have been determined by applying the modified Bjirrum's and Ringbom's methods. The second successive stability constants (log k_2), $-\Delta H°$, and $-\Delta S°$ values for the complex formation decrease with increasing $T_{Fe^{3+}}/T_{HL}$ in the Fe(III)-PVA complexes. Schematic representations of the reaction are given in (84) and (85). The effect of $T_{Fe^{3+}}/T_{HL}$ for initial Fe(III)-PVA complex on the H_2O_2 by Cu(II)- and Fe(III)-PVA mixed complex are shown in Figure 10, compared with Fe(III)-

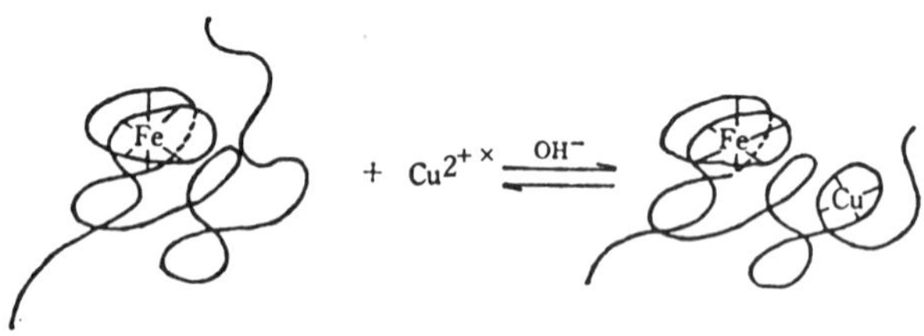

(84)

[*120*] [*121*]

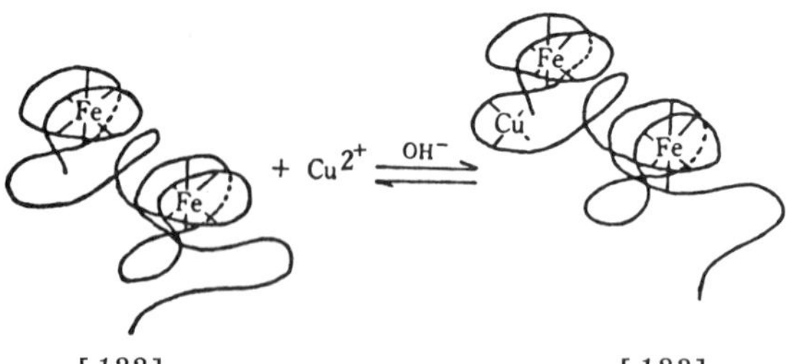

(85)

[*122*] [*123*]

PVA only. The rate of H_2O_2 decomposition increases with increasing $T_{Fe^{3+}}/T_{HL}$ values, because of an increase in the complete four-coordinated planar structures in the mixed metal-polymer complexes. Branca et al. have reported the binding of pseudo-octahedral *trans*-

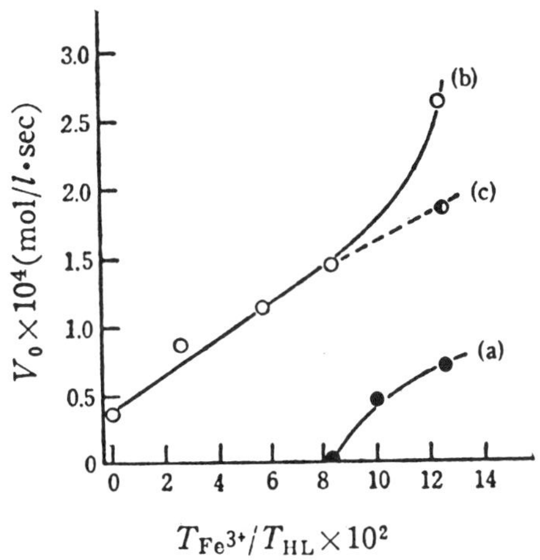

Figure 10. Effect of $T_{Fe^{3+}}/T_{HL}$ on the decomposition of hydrogen peroxide by the Cu(II) and Fe(III)-PVA mixed complexes and by the Fe(III)-PVA complexes. [Cu(II), Fe(III)-PVA](Cu(II) ion conc. constant) = 2×10^{-5} mol L^{-1}; [H_2O_2] = 4×10^{-2} mol L^{-1}, μ = 0.1(KNO_3), pH = 10.5, 25°C; [Fe(III)] = 2×10^{-4} to 1.25×10^{-3} mol L^{-1}, $T_{Cu^{2+}}/T_{HL}$ = 0.05. (a) Fe(III)-PVA complex. (b) Cu(II), Fe(III)-PVA mixed complexes. (c) (b)-(a).

and *cis*-Fe(III) complex, [*124*] and [*125*], to sodium poly(L-glutamate) (PLG) (113,114).

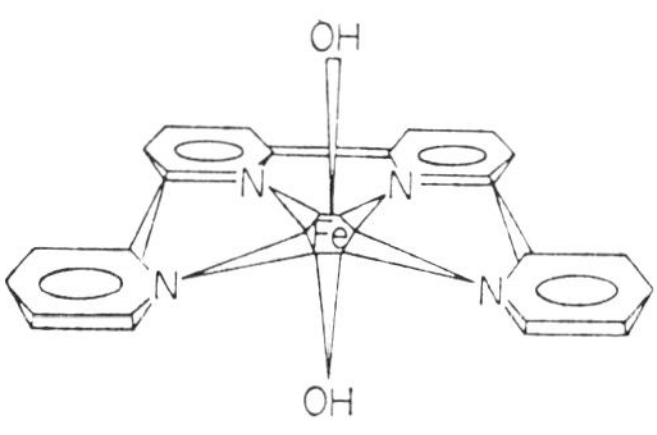

[*124*]

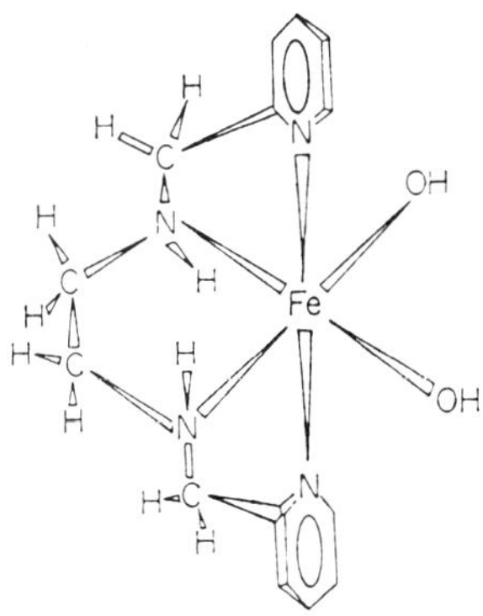

[*125*]

The association of the cis complex was found to be governed predominantly by electrostatic forces. The trans compound is also involved in coordination bonding between an apical site of the complex and a carboxylate group of the side chains of the polypeptide. To determine whether the diverse topology of the active site influences H_2O_2 decomposition, Barteri et al. studied the rate-determining step of the reaction, which involves the breakup of the O-O bond of the coordinated substrate species. At 25°C the rate constant for the irreversible decomposition of Michaelis complex was k_3 = 5.3 and d2.7 s^{-1} for $[Fe(Pken)(OH_2)]^+$ and $[Fe(tetpy)(OH_2)]^+(Glu)_n$, respectively. The activation energy of the reaction was 9.5 and 11.0 kcal/mol, respectively. The catalytic systems used differ from each other chiefly because the bound cis derivative has two vicinal sites available for the entering H_2O_2 molecule, while the bound trans complex ions have only one site (110).

3.3.3.b Method 2: Decomposition of hydrogen peroxide by Fe(III) complex supported on poly(L-glutamate) (110)

Poly(L-glutamic acid) (MW = 30,000) is converted to the sodium salt in 0.1 M NaOH. This stock solution is then exhaustively dialyzed against water to eliminate excess sodium ions and the material recovered by freeze drying. A 0.01 M (hydroxymethyl)aminomethane is used as a buffer at pH 7 to 8. In the acid region, the pH of polymers and complex solution is adjusted each time by adding the appropriate amounts of standard acid or base. Kinetic experiments are carried out with freshly prepared solutions using doubly distilled water with a conductively of less than $2 \times 10^{-6}\ \Omega^{-1}\ cm^{-1}$ (20°C). Equal volumes of buffered complex-(Glu)n and H_2O_2 solutions are kept at constant temperature before mixing. The reaction mixtures have a total volume of 10 mL and are maintained under vigorous stirring and constant temperature. The reaction is stopped after a short time by adding 3 N H_2SO_4 containing 0.05 M KF into the mixture. The unreacted substrate is then directly titrated, employing standard ceric sulfate with *o*-phenenthroline iron(II) as indicator. Blank experiments rule out any influence of the polymeric matrix on the rate of reaction or on the method for analysis. The initial reaction velocity, $V_o = -d[H_2O_2]/dt$ (mol L^{-1} s^{-1}) is evaluated from the slope of H_2O_2 concentration-time curves at t = 0, the extent of decomposition generally being limited to about 20% of the original concentration. In the steady state, the rate of reaction is found in both cases to obey a first-order equation with respect to both H_2O_2 and the complex.

A model of such a mechanism, the decomposition of H_2O_2 by Fe^{3+} complex of poly(acrylic acid), which is partly amidated by diethylenetriamine (dien), was studied by Pshezhetski et al. From the kinetic studies the diene groups were considered to be a cofactor in the reaction as shown in [86] (115).

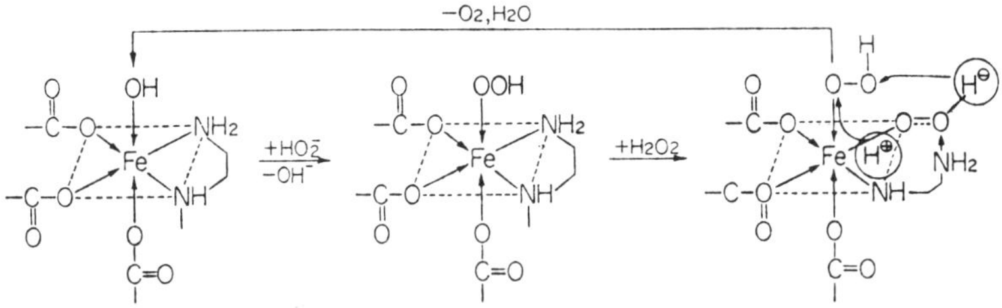

(86)

[*126*] [*127*] [*128*]

Fe(III)-phthalocyanine (Pc) compounds have structures similar to heme proteins. Fe(III)-Pc catalyze the decomposition of hydrogen peroxide. Sigel et al. have studied the mechanism of the H_2O_2 decomposition reaction Fe(III)- and Co(II)-tetrasulfo-Pc (116,117). Shirai et al. reported the decomposition of H_2O_2 by using Fe(III)-3,4,3',4',3'',4'',3''',4'''-octacarboxyphthalocyanine [Fe(III)-oaPc] [*129*] as a catalyst (118). Fe(III)-oaPc is a remarkably effective catalyst for the reaction and is comparable to hemin and other metal complexes. From the kinetic studies, the mechanism for the reaction was proposed as follows. At pH 7.0:

$$H_2O_2 \underset{}{\overset{K^H_{H_2O_2}}{\rightleftharpoons}} H^+ + OOH^- \tag{87}$$

$$\text{Fe(III)-oaPc} + OH^- \underset{k_2}{\overset{k_1}{\rightleftharpoons}} \text{Fe(III)-oaPc} \cdot OOH^- \tag{88}$$

$$\text{Fe(III)-oaPc} \cdot OOH^- \xrightarrow{k_3} X \tag{89}$$

$$X + H_2O_2 \xrightarrow{\text{fast}} \text{Fe(III)-oaPc} + H_2O + OH^- + O_2 \tag{90}$$

$$\text{Fe(III)-oaPc} + H_2O_2 \underset{k_2'}{\overset{k_1'}{\rightleftharpoons}} \text{Fe(III)-oaPc} \cdot H_2O_2 \tag{91}$$

$$\text{Fe(III)-oaPc} \cdot H_2O_2 \xrightarrow{k_3'} X \tag{92}$$

$$X + H_2O_2 + \xrightarrow{\text{fast}} \text{Fe(III)-oaPc} + 2H_2O + O_2 \tag{93}$$

3.3.3.c Method 3: Decomposition of hydrogen peroxide by Fe(III)-oaPc

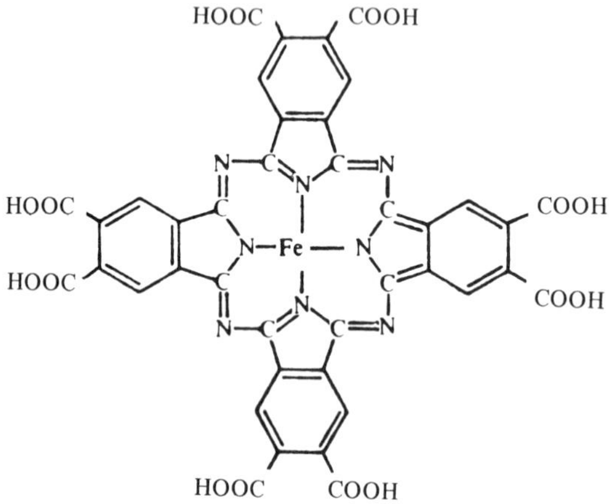

Fe (III)–oaPc

[*129*]

The decomposition rates of hydrogen peroxide in the presence of metal complexes are measured by the determination of the concentration of oxygen originated in the aqueous reaction media using a Warburg respirometer. Standard reactions (5 mL) contained $[\text{complex}]_{tot}$ = 5×10^{-5} M. Starting with this solution, the concentrations of complex and H_2O_2 are varied. Before adding H_2O_2 the pH is adjusted with 0.1 M HCl. After the addition of H_2O_2, the pH is measured. The ionic strength of the reaction solution is usually kept constant at 0.1 M (KCl). The exact concentration of the H_2O_2 stock solution is determined by titration with $KMnO_4$ solution. Shown in Figure 11 is an example of these measurements. The initial reaction velocity, $V_0 = -d[H_2O_2]/dt$ (in $M \cdot s^{-1}$), is taken from the slope of the tangent of the concentration of H_2O_2 decomposed (which is calculated from the concentration of O_2 originated in the reaction solution) versus time curve at the t = 0.

Shirai et al. also reported synthesis and catalase-like activity of Fe(III)-taPc bound to polystyrene. The activation energies of the polymer-bound Fe(III)-Pc are about half of those with free Fe(III)-taPc. From continuous-flow experiments in a column, the polymer-attached catalyst was found to be very stable compared with free Fe(III)-taPc (119,120).

The H_2O_2 decomposition proceeds smoothly in the higher-pH region because of the formation of the peroxo complex of OOH^- with catalyst (118). The catalysts possess active sites and basic sites in cooperative

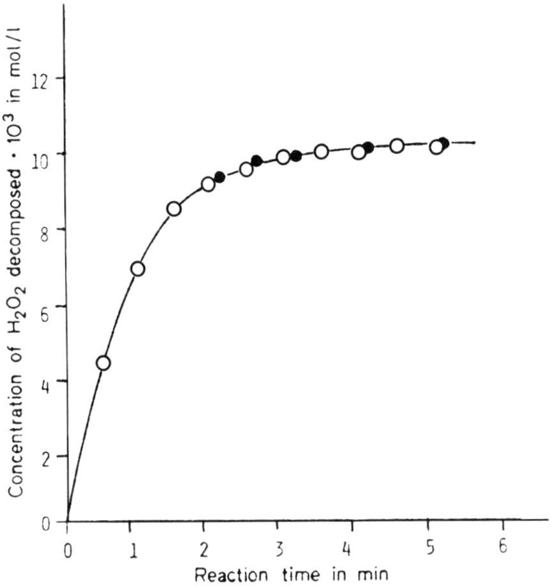

Figure 11. Decomposition of hydrogen peroxide by Fe(III)-oaPc. [Fe(III)-oaPc] = 5×10^{-5} mol L^{-1}, [H_2O_2] = 10^{-2} mol L^{-1}, pH = 7.0, 25°C. ○, values calculated from the O_2 evolution; •, titration with $KMnO_4$.

interaction. Shirai et al. synthesized poly(2- and 4-vinylpyridine-co-styrene) with covalently bonded Fe(III)-Pc [*130*] (121,122).

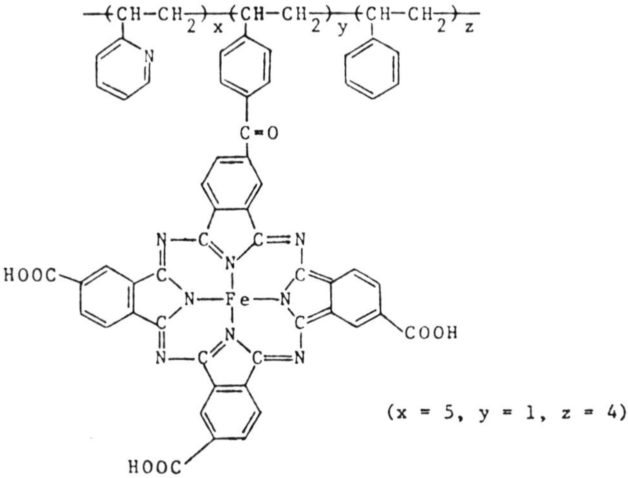

[*130*]

The copolymer-bounded Fe(III)-Pc dispersed in water is the five-coordinate high-spin type similar to catalase and peroxidaes as evidenced by the ESR signals with g = 4.27 (Figure 12). One coordination site at the axial position of Fe(III)-Pc ring is occupied by the nitrogen of the pyridine groups. High-spin Fe(III)-porphyrin with coordination number five is useful in the decomposition of H_2O_2 (122). The kinetic data for H_2O_2 decomposition reaction by copolymer-bounded Fe(III)-Pc is comparable to free Fe(III)-Pc compounds and catalase (Table 14). V_{max} for the copolymer-attached Fe(III)-Pc is 20 times greater than the free Fe(III)-Pc, and the turnover number is about 10 times greater. Fe(III)-Pc supported on the copolymer is a remarkably effective catalyst suspended in aqueous solution, a typical competitive inhibition was observed and the ESR signal is typical for a low-spin iron(III) (g = 1.997) octahedral ligand field [Figure 3(b)]. The six-coordination position of Fe(III)-Pc in the copolymer matrix is occupied by CN^-. The catalytic function appears to operate by exchange of OOH^- in this position.

Barteri et al. reported the stereoselective peroxidase-like action of $[Fe(tetpy)(OH\ )\ ]^+$-poly(L-glutamate) [*125*] that occurs according to reaction [*94*] (123):

$$AH_2 + H_2O_2 \xrightarrow{\text{cat.}} A + 2H_2O \qquad (94)$$

Kinetic experiments on the oxidation of L(+)-ascorbic acid by H_2O_2 in the presence of $[Fe(tetrapy)(OH)_2]$-$(L\text{-}Glu)_n$ or of the enantiometric system $[Fe(tetrapy)(OH_2)_2{}^+$-$(D\text{-}Glu)_n$ showed that stereo-specific catalysis occurred which was strongly dependent on the conformational dissymmetry of the active site.

3.3.3.d Peroxidatic asymmetric oxidation of L(+)-ascorbic acid by $[Fe(tetrapy)(OH)_2]^+$-$(L\text{-}Glu)_n$ or -$(D\text{-}Glu)_m$ (123)

Kinetic experiments are performed by measuring the disappearance of ascorbic acid at 265 nm, under pseudo-first-order conditions. A typical run consisted of adding H_2O_2 by a microsyringe into the 1-cm optical cell containing 2 mL of catalyst and ascorbic acid mixture, both systems being thermostated at 25°C. The experimental conditions normally used are: ascorbic acid concentration $[AH^-] = 1 \times 10^{-4}$ M, $[H_2O_2] = 1 \times 10^{-2}$ M, complex concentration $[c] = 0.5 \times 10^{-5} - 6 \times 10^{-5}$ M, and polymer concentration $[P] = 1 \times 10^{-4}$ and 5×10^{-4} M (referred to the monomeric unit). Plots of $\log(A_t - A_\infty)$ versus time are linear over several half-lives and the observed rate constants k (s^{-1}) obtained are consistent. A slight oxidation of ascorbic acid by H_2O_2 in the presence of only the polypeptide matrix also occurs. This reaction amounted to about 20% of the total rate. All pseudo-first-order rate

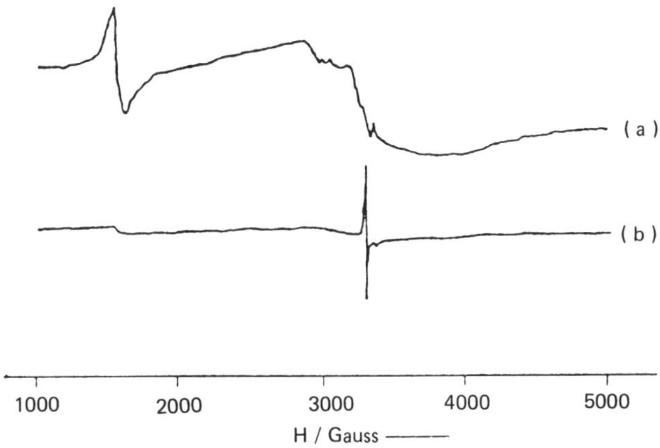

Figure 12. ESR spectra of Fe(III)-taPc covalently bonded on P2VP-St dispersed in water at 213°K. (a) Fe(III)-taPc-P2VP-St. (b) System adding KCN to Fe(III)-taPc-P2VP-St.

Table 14. Kinetic Data for the H_2O_2 Decomposition Reaction by Fe(III)-taPc Covalently Bonded to Poly(2-vinylpyridine-co-styrene)

Catalysts	V_{max} (mol L^{-1} min^{-1})	k_m (mol L^{-1})	k_3 (min^{-1})
Fe(III)taPc-P2VP-St[a,b]	1.2×10^{-2}	2.2×10^{-2}	200
Fe(III)taPc-PSt[c]	1.0×10^{-3}	2.8×10^{-2}	20
Fe(III)-taPc[d]	5.2×10^{-4}	6.6×10^{-3}	19
Hemin	—	—	35
Catalase	—	—	5×10^{6}

[a] [Fe(III)-Pc] = 7.44×10^{-4} g/5 mL, [H_2O_2] = 3.56×10^{-2} mol L^{-1}, pH 7.0, 25°C.
[b] Fe(III)-taPc covalently bonded to poly(2-vinylpyridine-co-styrene).
[c] Fe(III)-taPc covalently bonded to polystyrene.
[d] [Fe(III)-taPc] = 5×10^{-5} mol L^{-1}.

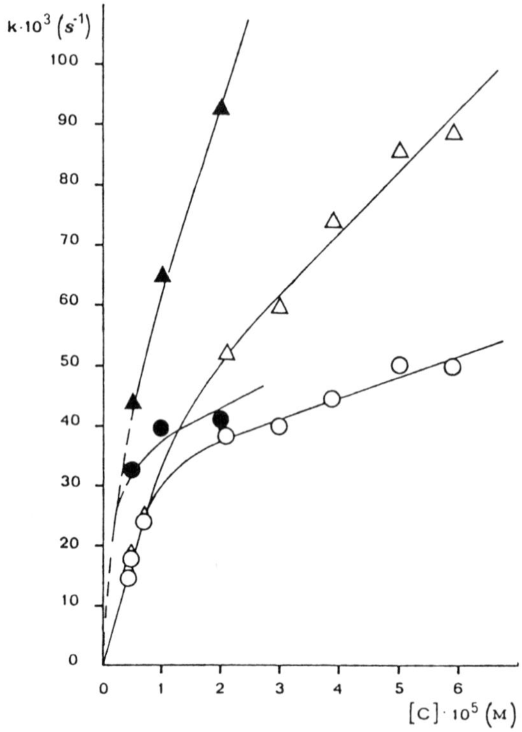

Figure 13. Stereoselective peroxidase-like activity of $[Fe(tetpy)(OH)_2]^+$ ions bound to $(L\text{-}Glu)_n$ (circles) or to $(D\text{-}Glu)_n$ (triangles) in the oxidation of L(+)-ascorbic acid at pH 7.0 (t = 25.9°C). k is the difference between pseudo-first-order rate constants in the presence and absence of bound counterions. Polypeptide concentration [P] = 1×10^{-4} M (shaded symbols) and 5×10^{-4} M (open symbols), referred to the monomeric unit. [C] is the concentration of complex ions.

constants are properly corrected for this effect. The value of the oxidation reaction of L(+)-ascorbic acid on complex ion concentration [P] = 1×10^{-4} and 5×10^{-4} M is presented in Figure 13.

3.3.3.e Peroxidative oxidation of guaiacol by Fe(III)-oaPc/H_2O_2 or Fe(III)-oaPc/H_2O_2 and polyelectrolyte (124)

Shirai et al. reported the peroxidative oxidation of guaiacol by Fe(III)-oaPc/H_2O_2 and Fe(III)-oaPc-polyelectrolyte systems. The maximum rate of oxidation was about 1000 times greater than in the absence of

H_2O_2. The kinetic investigation indicated that the reaction can be regarded as "an ordered two-substrate reaction" and that the oxidation proceeds according to the peroxidase-like mechanism. The reaction is inhibited by adding cyanide (124).

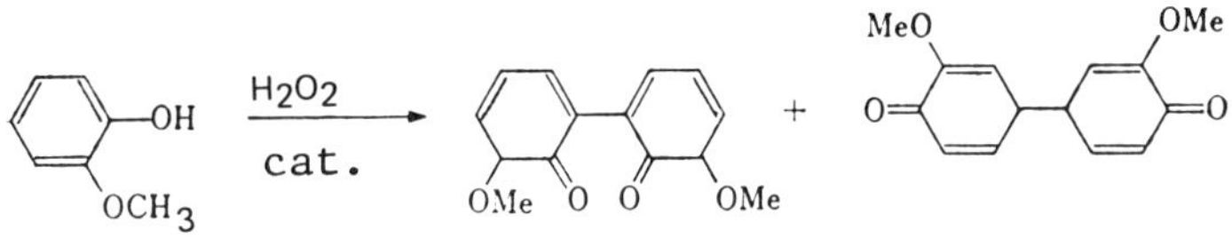

(95)

[*131*] [*132*] [*133*]

The rates of oxidation of guaiacol by Fe(III)-oaPc-polyelectrolyte-H_2O_2 systems are measured by rapidly mixing an aqueous solution containing H_2O_2 and guaiacol with the complex system in a stopped-flow spectrophotometer with a 1-cm cell at 25°C. The reaction can easily be followed by directly monitoring the increase in absorbance at 436 nm, where guaiacol has no absorption. A "standard" reaction solution contains $[\text{Fe(III)-oaPc}]_{tot} = 5 \times 10^{-5}$ M, $[\text{guaiacol}]_{tot} = 2.4 \times 10^{-3}$ M, and $[H_2O_2]_{tot} = 8.8 \times 10^{-3}$ M. Starting from this solution, the concentrations of Fe(III)-oaPc, guaiacol, or H_2O_2 are varied. Before mixing, the pH of both solutions is adjusted as desired with 0.1 M KOH or KCl. The pH of the solutions used for the kinetic runs ranged from 4 to 11. The initial rate, V_o, is calculated according to the equation

$$V_o = \left(\frac{dA_{436}}{dt}\right)_{t=0,A=0} \qquad (96)$$

where A_{436} is the change in absorbance at 436 nm. From the kinetic results, the scheme shown in Figure 14 was proposed.

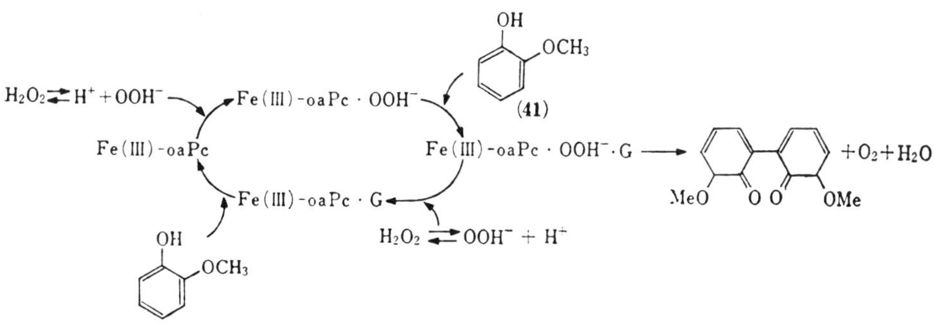

Figure 14. Reaction scheme for the oxidation of guaiacol catalyzed by Fe(III)-oaPc. G: guaiacol.

3.4 NONENZYMIC CATALYTIC ACTIVITIES OF METAL-POLYMER COMPLEXES

The metal-polymer complexes, which are made by attaching the active metal complex to an insoluble polymer support, are markedly useful for practical use. Such catalytic systems are of particular interest for:

1. Simplicity of separation of the products from the reaction mixture and easy catalyst recovery and reuse
2. Stabilization of active site
3. High selectivity
4. Prevention of reaction vessel corrosion

3.4.1 Oxidative Reaction

The oxidation of thiosalts by molecular oxygen catalyzed by Cu(II)-poly(4-vinylpyridine) was studied as a catalyst for the removal of polythionate contaminants in effluent wastes (125).

$$3[S_2O_3]^{2-} + 2O_2 + H_2O \longrightarrow 2[S_3O_6]^{2-} + 2OH^- \tag{97}$$

$$[S_2O_3]^{2-} + 2O_2 + 2OH^- \longrightarrow 2[SO_4]^{2-} + H_2O \tag{98}$$

The rate of thiosulfate oxidation on polymer-attached Cu(II)-bipyridine complex was examined in terms of the Langmuir-Hinshelwood model, which is derived from considering the reaction of two different absorbants on the surface, and the Rideal model, which is applicable for the surface reaction of an absorbed molecule with the other substrate species in solution. Assuming that (a) the equilibrium concentration of $S_2O_3^{2-}$ and O_2, represented by A and B, respectively, are established on the catalyst surface; (b) the reaction between the two absorbed species is rate controlling; and (c) the reaction is first order with respect to both A and B, a Langmuir-Hinshelwood type of rate law can be derived as

$$r = \frac{kK_AK_BC_AC_B}{(1 + K_AC_A + K_BC_B)^2} \tag{99}$$

where

k = overall rate constant for oxidation of thiosulfate by molecular oxygen

K_A and K_B = absorption equilibrium constants for thiosulfate and dissolved oxygen, respectively

C_A and C_B = concentrations of S_2O_3

At a constant pressure of O_2 (i.e., C_B = constant), Equation 99 can be rewritten as

$$\left(\frac{C_A}{r}\right)^{0.5} = \frac{1 + K_B C_B}{(kK_A K_B C_B)^{0.5}} + \left(\frac{K_A}{kK_B C_B}\right)^{0.5} C_A \tag{100}$$

and at a constant concentration of thiosulfate (i.e., C_A = constant) as

$$\left(\frac{C_B}{r}\right)^{0.5} = \frac{1 + K_A C_A}{(kK_A K_B C_A)^{0.5}} + \left(\frac{K_B}{kK_A C_A}\right)^{0.5} C_B \tag{101}$$

Equations 100 and 101 were tested by plotting $(C_A/r)^{0.5}$ versus C_A and $(C_B/r)^{0.5}$ versus C_B, based on experimental data, as shown in Figures 15 and 16.

The linearity existing in these plots shows the validity of the Langmuir-Hinshelwood model. To obtain the values for the constants, Equation 3 was rearranged to the form

$$\left(\frac{C_A C_B}{r}\right)^{0.5} = \alpha + \beta C_A + \gamma C_B \tag{102}$$

where

$$\alpha = 1/K_A K_B$$
$$\beta = K_A/kK_B$$
$$\gamma = K_B/kK_A$$

The data for the initial rate of O_2 uptake versus thiosulfate concentration at constant O_2 partial pressure (Figure 17) and initial rate versus equilibrium of O_2 at constant thiosulfate concentration (Figure 18) were fitted to Equation 102 to obtain α, β, and γ, which then yielded $k = 8.48 \times 10^{-6}$ mol^{-1} (g cat.)$^{-1}$, $K_A = 11.41$ mol^{-1}, and $K_B = 113.71$ mol^{-1}. The values of the initial rate estimated from these constants agree well with the experimental values, the advantage absolute deviation being 4.5%.

The dependence of the initial rate on the concentration of thiosulfate C_A should have a maximum in a Langmuir-Hinshelwood type of equation. Thus the concentration $(C_A)_{max}$ at the maximum rate can be determined by differentiation of Equation 99 with respect to C_A:

$$K_A(C_A)_{max} = 1 + K_B C_B \tag{103}$$

Hence one obtains the following expression for the maximum initial rate of O_2 consumption for thiosulfate oxidation:

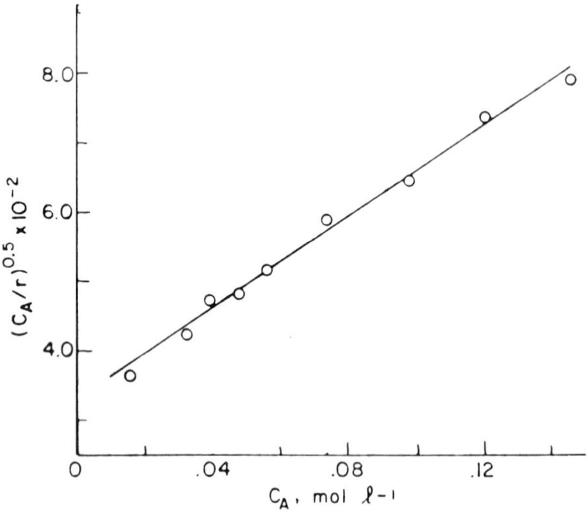

Figure 15. Applicability of Langmuir-Hinshelwood model (Equation 100).

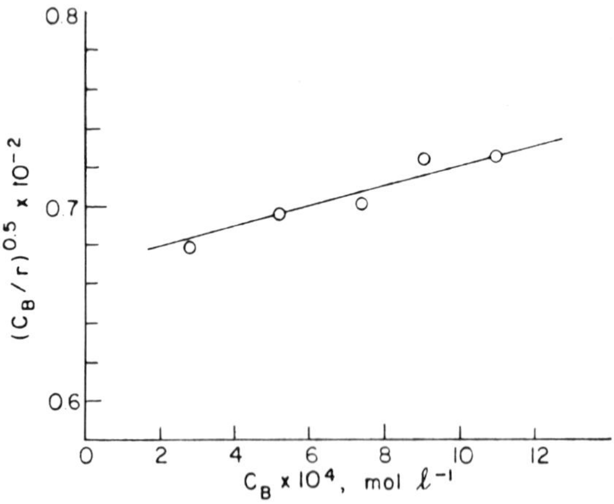

Figure 16. Applicability of Langmuir-Hinshelwood model (Equation 101).

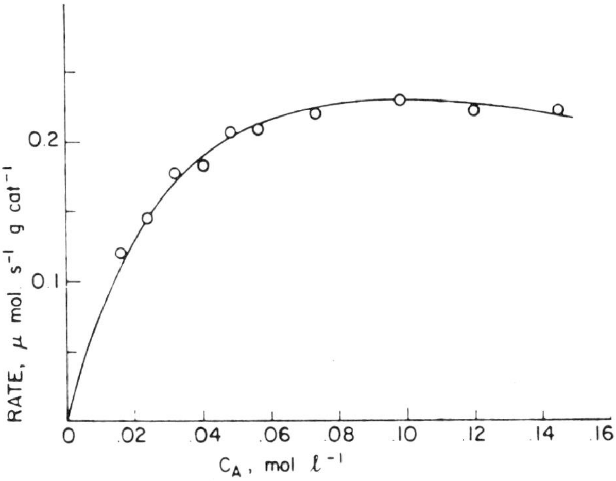

Figure 17. Effect of concentration of thiosulfate (C_A) on initial rate of O_2 consumption at constant pressure of O_2 (93.0 kPa) at 30°C.

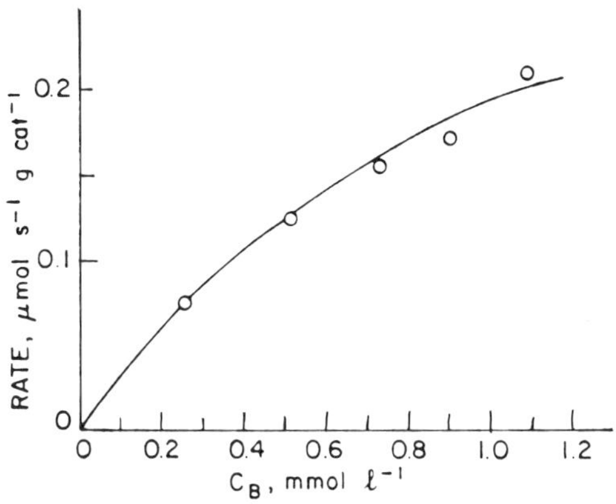

Figure 18. Effect of concentration of dissolved oxygen on initial rate of oxygen consumption at constant concentration of thiosulfate (48.4 mmol L^{-1}) at 30°C.

$$r_{max} = \frac{kK_B C_B}{4(1 + K_B C_B)} \tag{104}$$

The value of $(C_A)_{max}$ calculated from Equation 103 at 93.0 kPa O_2 using the foregoing values of K_A and K_B is 0.098 M, which is in good agreement with the experimental data plotted in Figure 16.

The kinetics of oxidation of thiosulfate by molecular oxygen in the presence of copper(II) complex of this polymer-bound ligand can be explained by a Langmuir-Hinshelwood type of rate law, while a Rideal type of rate applies only at low concentrations of thiosulfate (< 0.05 M).

3.4.1.a Oxidation of thiosalts by molecular oxygen by Copper(II) complex of polystyrene-bound 2,2,-bipyridine (125,126)

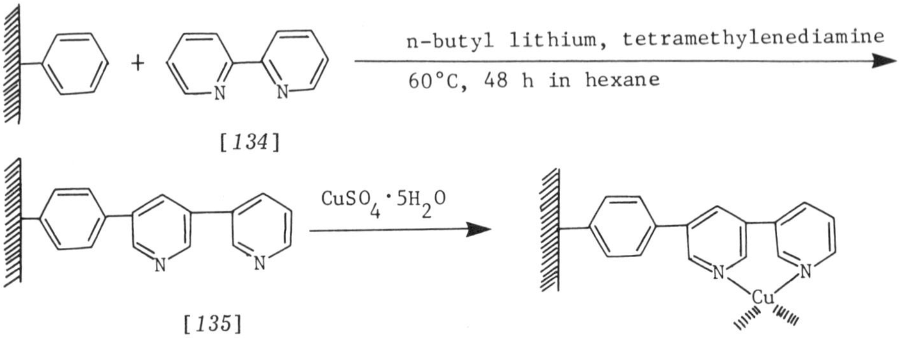

[134] [135] [136] (105)

Preparation of copper(II) complex of polystyrene-bound 2,2'-bipyridine

A 5-g sample of porous polystyrene-4% divinylbenzene copolymer beads less than 38 m in size is suspended in 50 mL of freshly distilled cyclohexane in a flask purged with argon. *n*-Butyl lithium (35 mL 1.6 M in hexane) and 10 mL of tetramethylenediamine (TMEDA) are added to this with a syringe and the mixture is heated to 60°C for 48 h under a continuous stream of argon. After cooling, the polymer is washed twice with cyclohexane and three times with dry, oxygen-free tetrahydrofuran (THF). Solid bipyridine (5 g) is added to a suspension of the lithiated polymer in 50 mL of THF, and the resulting solution is brought to reflux. After stirring for 16 h under a continuous flow of argon, the suspension is cooled to room temperature and air is bubbled through the solution until a color change from dark purple to golden yellow is observed. The polymer-containing attached bipyridine is separated by filtration and washed thoroughly with THF, methanol, and water in succession.

The golden yellow polymer, after drying, is stirred in a solution of $CuSO_4 \cdot 5H_2O$ in a methanol-ethyl acetate (1:1 by volume) mixture for 12 h. The greenish-blue complex formed is separated from the unreacted polymer by repeated washing with water. The copper content of the polymer-attached bipyridine complex of copper sulfate was determined, by leaching with mineral acid followed by volumetric analysis, to be 1.6 mmol/g. These data suggest that there are three bipyridine units per 10 styrene units in the complex.

Determination of reaction rates

The rate of oxygen consumption for thiosalt oxidation is measured using the constant-pressure gas uptake apparatus and procedure described by James and Rempel (127). The partial pressure of oxygen is varied from 23.6 to 93.0 kPa and the corresponding concentration in solution is estimated from the solubility data of Seidell (128). Except where specified, the reaction conditions are standardized as follows: 5 mL of an aqueous suspension containing 48.4 mmol of thiosalt and an amount of complex containing 19 mg of complexed copper is shaken at 220 rpm through a stroke of 5 cm at 30°C under 93.0 kPa O_2. The reproducibility of measurements is to within 10%.

Determination of reaction products

Thiosulfate, trithionate, and tetrathionate are determined using the photomeric procedure and sulfate is determined gravimetrically as $BaSO_4$. The catalytic autooxidation of mercaptans (RSH) is appreciably accelerated by addition of a base, the generated thiolate anion (RS^-) being the species susceptible to reaction with oxygen.

$$2RSH + 2OH^- \longrightarrow 2RS^- + 2H_2O \qquad (106)$$

$$2RS^- + 2H_2O + O_2 \xrightarrow{\text{cat.}} RSSR + H_2O_2 \qquad (107)$$

Metal ions and metal complexes can be used as catalysts in this oxidation. A class of highly active catalysts is formed by transition-metal phthalocyanine compounds. These metal complexes have industrial applications as catalysts in many processes. The reaction requires base (alkaline hydroxides), which is usually added separately to the reaction medium. Schutten et al. showed that a particularly active catalytic system could be obtained using a polymeric base instead of alkaline hydroxide with cobalt(III)-4,4',4'',4'''-tetrasulfophthalocyanine as the oxidation site (129,130). Such a bifunctional catalyst acts both as a supplier of basic sites and as a carrier for the oxidation catalyst cobalt(II)-4,4',4'',4'''-tetracarboxyphthalocyanine. Polymer attachment of the cobalt-phthalocyanine to poly(vinyl amine) was achieved by complexation through amine groups [*137*] as well as by formation of a peptide linkage between the phthalocyanine ring system and the amines on the polymeric carrier [*138*]. Both methods of coupling provided bifunctional catalysis with extremely high activities without requiring additional base. The amount of basic groups incorporated in the polymer necessary to get high activity is less than the amount of NaOH used in the corresponding NaOH/Co-Pc system by about a factor of 100 to 1000. The polymeric catalysts present here can be applied in the sweetening process of petroleum fractions, which involves the autoxidation of mercaptans to disulfides. These processes are usually carried out in a two-phase system, an oil phase containing

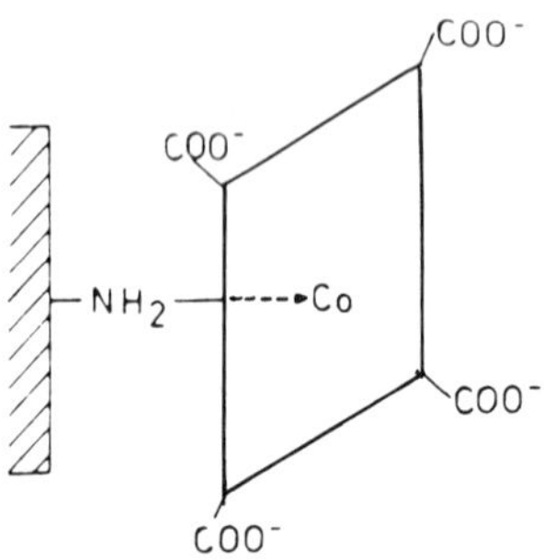

[*138*]

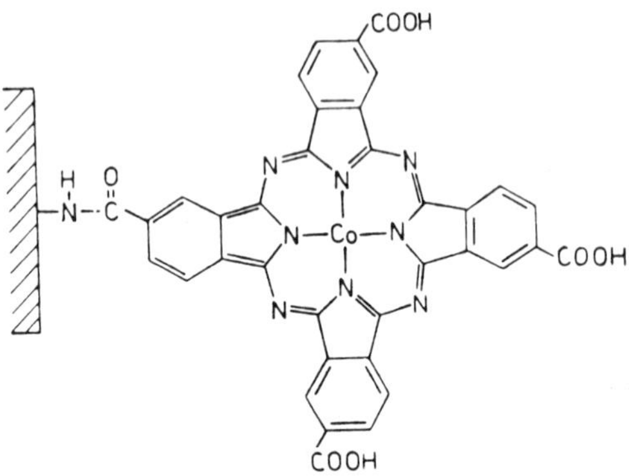

[*139*]

the mercaptans and an aqueous phase containing the required base. The polymeric catalysts do not require additional base as a promoter, so these catalysts provide the possibility of avoiding the aqueous phase. No activity of the functional catalyst is observed in pure toluene as a reaction medium. When a slight amount of water is added (0.1 mL), a dramatic change in activity occurs. The addition of more water gives rise to further increases in catalytic activity, up to a maximal value (Figure 19).

3.4.1.b Autooxidation of mercaptans promoted by a polymer attachment of cobalt-phthalocyanine

Activity measurements are carried out in an all-glass, thermostated double-walled Warburg apparatus with a glass stirrer. The rate of oxidation is determined by measuring the rate of consumption of the first 20 mL of oxygen at constant oxygen pressure $P(O_2)$ = 1 atm and a constant stirring speed (v_s = 3000 rpm). Because of this high stirring speed, no oxygen-transport limitation is observed. Unless otherwise stated, water is used as the solvent; the total reaction

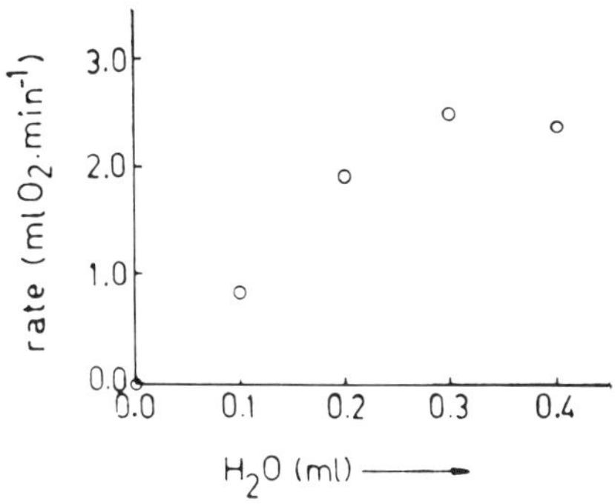

Figure 19. Relationship between the catalytic activity of catalyst II (4.9 mg) in toluene (70 mL) and the amount of water present (substrate: 1 mL of 2-mercaptoethanol).

volume is 75 mL. The amount of catalyst is chosen in such a way that an appropriate rate of oxygen consumption results (0.5 to 10.0 mL/min). The substrate, 2-mercaptoethanol, is distilled before use and kept under nitrogen. Accumulated H_2O_2 is analyzed spectrophotometrically using $TiCl_3$-H_2O_2 as reagent.

The oxidation reaction of ethylbenzene in the liquid phase in the presence of complexes of cobalt sorption copolymer of diethyl ester of vinylphosphonic acid cross-linked by methylenediacrylamide (131) was studied in a gasometric unit (Equation 108). The catalyst can be used repeatedly without losing its catalytic activity.

3.4.1.c Liquid-phase oxidation of ethylbenzene over cobalt complex supported by polymeric materials (142)

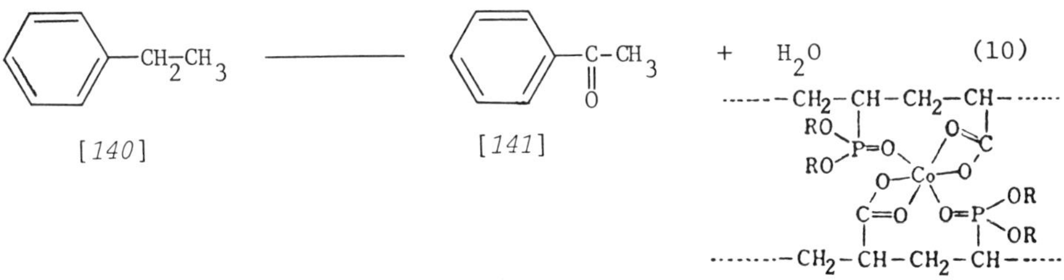

[*140*] [*141*] [*142*]

Bulk copolymerization of diethyl ester of vinylphosphonic acid with acrylic acid is carried out by photochemical initiation with 1 wt % cumene hydroperoxide in a sealed quartz glass tube which is evacuated to 10^{-5} torr. UV light is supplied from a high-pressure 300-W mercury lamp. The resulting copolymer is dissolved in ethanol and then precipitated in diethyl ether. After reprecipitation the copolymer is dried under vacuum in a desiccator at only a slightly elevated temperature. A fairly dilute ethanol solution of copolymer (0.8 g/100 mL) is mixed with a double amount of 0.075 M solution of cobalt chloride salt adjusted to pH 1.1 with HCl. The resulting mixture is then slowly treated with NH_4OH until a pH of 5.20 is achieved. The rate of titration is such that the pH increase is not greater than 0.2 h^{-1}. The precipitated metal-polymer complex is filtered, repeatedly washed with distilled water until the washings gave a negative test for cobalt, and dried in vacuum at 35 to 40°C. The dry metal-polymer complex is mixed with methylene diacrylamide as the cross-linking agent and ground in a vibrating ball mill having a ball diameter of 0.8 cm for 1 min. The resulting mixture has a particle size in the range 150 to 200 mesh. A small amount of the smaller particles is set aside. Tablets with a thickness of 0.025 cm and a diameter of 0.8 cm are prepared, evacuated at 10^{-4} torr, and heated in sealed tubes at 150°C for 5 h. The cobalt in the cross-linked copolymer is fully desorbed into 1 M HCl. Cross-linked tablets of specially prepared cobalt ion polymers are treated with $CoCl_2$ solution to obtain the complexes. The complexes, containing 4.3 meq of Co per gram of dry sorbant, are used to study the catalytic activity. Tablets are repeatedly washed with distilled water to remove any excess of cobalt, dried at 105°C, and ground to a powder. The liquid-phase oxidation of ethylbenzene is carried out in the gasometric unit under conditions such that oxidation of ethylbenzene would not take place in the absence of a catalyst. Gaseous oxygen is used as the oxidizer at atmospheric pressure in the absence of solvent. The catalytic activity of the sample is evaluated according to the oxygen absorption rate and the maximum amount of O_2 consumed in the reaction.

3.4.2 Reduction

A wide range of hydrocarbon reactions have been catalyzed by non-Werner-type metal complexes for hydrogenation (Reaction 109), hydrosilynation (Reaction 110), hydroformylation (Reaction 111), acetoxylation (Reaction 112), polymerization, and oligomerization (2,8). The effects of polymer ligands on the catalysis of non-Werner-type complexes are as follows: (a) Catalysts can easily be separated from the reaction mixture, regenerated, and recycled; (b) catalysts are not susceptible to poisoning by impurities because the catalytic sites are protected by the polymer matrix; (c) activities are sometimes higher than the monomer complex of unsaturated coordination structure; and (d) prevention of dimerization of metal complex due to steric effect of polymer matrix. Styrene-divinylbenzene copolymer is most often used as the polymer matrix.

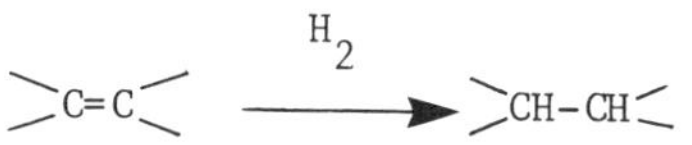

(109)

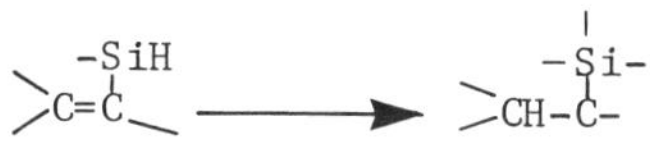

(110)

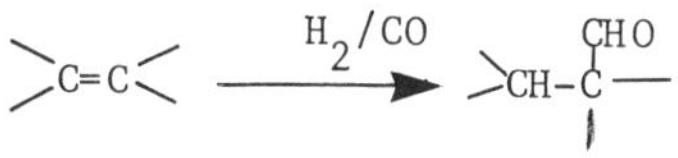

(111)

$$RX + CH_3COOK \longrightarrow CH_3COOR + KX \qquad (112)$$

Titanocene and its hydrides are particularly useful catalysts in the hydrogenation of various unsaturated organic compounds and molecular nitrogen. However, the complexes readily polymerize (132), according to the following reaction, to form catalytically inactive materials.

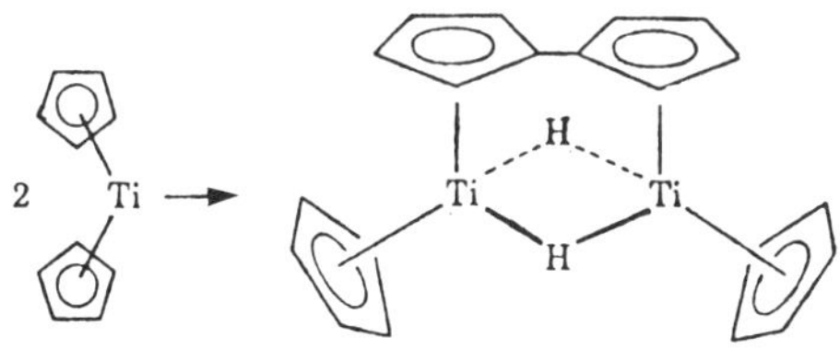

(113)

[*142*] [*143*]

Bonds et al. reported the preparation of polystyrenedivinylbenzene copolymer resin supporting titanocene, $TiCp_2Cl_2$. The catalyst efficiency is greater than a corresponding nonattached species (133).

3.4.2.a Polystyrene-attached titanocene species preparation and reactions (133)

Preparation of cyclopentadienyl-substituted copolymer (134)

An oven-dried 2-L flask is charged with 200.0 g of copolymer beads which were previously washed free of styrene monomer with benzene and 1 L of freshly distilled chloromethyl ethyl ether. A cautiously prepared solution of 35 mL of $SnCl_4$ in 125 mL of ice-chilled chloromethyl ethyl ether is introduced. After vigorous stirring of the reaction mixture at room temperature for 3 days, the ether is removed and the residue is washed with 2-L portions of 50% aqueous dioxane containing 10% HCl (v/v), and finally dry dioxane. Chloride analysis of the chloromethylated copolymer, after drying for 2 days in vacuum, indicated 1.20 meq of Cl^- per gram or a 13.8% chloromethylation of the styrene rings. The product obtained is added to 300 mL of dry, oxygen-free THF and 316 mL of 1.58 M sodium cyclopentadienide in THF. After the mixture is stirred for 5 days at room temperature, excess sodium cyclopentadienide is removed by filtration and the product is washed with 1:1 dry, oxygen-free ethanol:THF until it is free of chloride ion. The product is then washed with three 1-L portions of THF and dried in vacuum for several days to yield a substance containing 0.17 meq of Cl^- per gram.

Preparation of copolymer-attached $TiCp_2Cl_2$

To a 500-mL Schlenk flask maintained under argon are added 75 g of cyclopentadiene-substituted copolymer, 150 mL of THF, and 100 mL of 2.25 M methyl lithium in diethyl ether. The reaction mixture is stirred for 2 days, filtered, and washed with three 300-mL portions of benzene. A solution containing 35.89 g (0.16 mol) of $TiCpCl_3$ dissolved in 350 mL of benzene is introduced and the mixture stirred for 3 days. Extraction of excess $TiCpCl_3$ with hot benzene followed by drying gives a salmon-red $TiCpCl_2$-substituted copolymer.

Use of the reduced copolymer complexes as catalyst

Four types of catalysts are used, all with cyclohexene as the olefin. In each case, the catalyst is weighed into the reaction flask, placed under an atmosphere of hydrogen, suspended in 8 mL of hexane, and treated with 1 mL of 2.1 M BuLi in hexane for 1 h. The appropriate olefin is then added by means of a syringe. The rate of hydrogen uptake is determined by the use of an electronic monitoring device. The rates of reduction of other olefins (Table 15) have been determined in a similar fashion using noncrushed copolymer beads.

Colloidal transition metals have been prepared by reduction of the corresponding metal salts and by pulverization of the metallic mass. The colloidal metals prepared by the former procedure are generally superior to those prepared by the latter procedure with respect to size and the narrow size distribution of particles. In the preparation of colloidal metals in aqueous solution, natural polymers, such as gelatin and dextrin, were added as a protective colloid to prevent coagulation and precipitation of metallic particles. Hirai et al. reported a novel method for the preparation of colloidal dispersion of rhodium by refluxing a solution of rhodium(III) chloride and poly(vinyl alcohol) in methanol-water (135,136). The particle size of rhodium in the colloidal dispersion was sharply distributed in the range 30 to 70 Å, which comprised 94% of the particles. The average particle diameter was estimated to be 40 Å. The colloidal dispersion was effective for hydrogenation of cyclohexene at 30°C under 1 atm hydrogen pressure.

Table 15. Hydrogenation Rates with Polymer-Attached Titanocene Species

Olefin (0.5 M in hexane)	Hydrogenation rate (mL of H_2/min/meq of Ti)
1-Hexene	213.0
Styrene	243.0
Cyclohexene	90.3
1-Methylcyclohexene	1.0
1,2-Dimethylcyclohexene	0.0
1,3-Cyclooctadiene	216.0
1,5-Cyclooctadiene	183.0
3-Hexyne	149.0
Diphenylacetylene	40.6
Cholestenone	0.0
Vinylacetate	0.0

3.4.2.b Preparation of colloidal rhodium in poly(vinyl alcohol) by reduction with methanol (136)

Poly(vinyl alcohol) (147 mg, 3.33 mmol) and rhodium(III) chloride (818 mg, 0.033 mmol) are dissolved in methanol (25 mL) to form a rose-pink solution. Refluxing the solution under argon or air on a steam bath for 0.2 to 16 h gave a homogeneous light-brown solution with a colloidal disperson of rhodium. Shown in Figure 20 is the size distribution of the large particles as they appear in an electron micrograph of the colloidal dispersion of rhodium.

3.4.2.c Hydrogenation of cyclohexene

The hydrogenation of cyclohexene is carried out in methanol at 30.0°C under 1 atm of hydrogen. Into a flask flushed with hydrogen, methanol (18.7 mL) and the colloidal dispersion of rhodium (0.30 mL; Rh 2.0×10^{-7} mol) are added in that order. The solution is stirred at 30.0°C under atmospheric hydrogen pressure for 1 h. Then 1.0 mL of solution of cyclohexene (0.5 mmol) in methanol is added to the flask, with the pressure being kept at 1 atm. The initial rate of hydrogen uptake is determined by the initial slope of hydrogen uptake-time curves. The

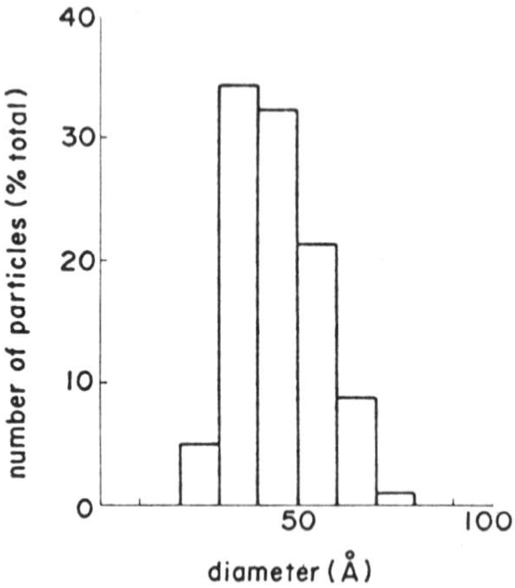

Figure 20. Particle size distribution of colloidal rhodium in poly(vinyl alcohol) prepared by refluxing for 4 h.

product is identified as cyclohexane by GLPC with a 2-m column of β,β'-thiodipropionitrile at 50°C.

In the earliest stage (within 1 min) of reaction, the colloidal dispersion exhibited a remarkably high rate of hydrogenation catalyst activity, as shown in Figure 21.

3.4.3 Polymerization

Organic compounds having labile hydrogen, such as phenols, phenylenediamine, disulfides, and acetylene, are oxidatively coupled by metal complex to give polymeric materials as shown in Reactions 114 and 115 (1,2). These reactions are called oxidative polymerizations. Tsuchida et al. studied the oxidative polymerization of 2,6-dimethylphenol (XOH) as a redox reaction catalyzed by Cu(II)-polyvinylpyridine complexes, and have proposed a detailed mechanism (137-139).

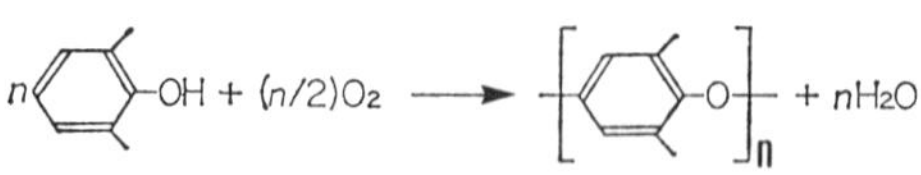

(114)

(115)

Since the oxidative polymerization of phenols is an industrial process used to manufacture poly(phenylene oxide)s, the application of polymer

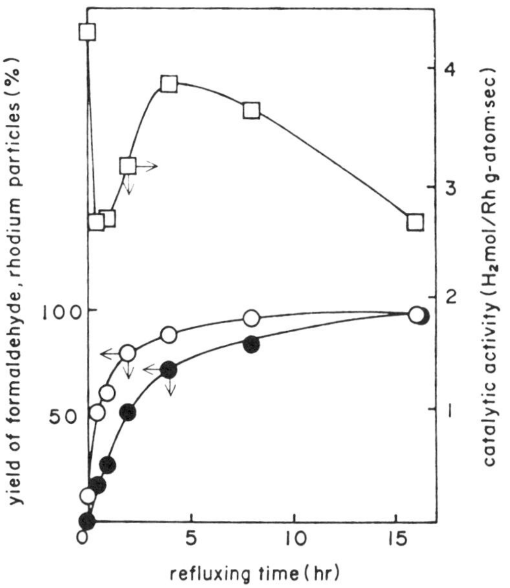

Figure 21. Dependence on reflux time of yields of (○) formaldehyde and (●) rhodium particle and (□) of catalytic activity for hydrogenation of cyclohexene.

catalysts for this process may well be of interest. The mechanism for the oxidative polymerization of XOH by metal-polymer complex is given by Reaction 116.

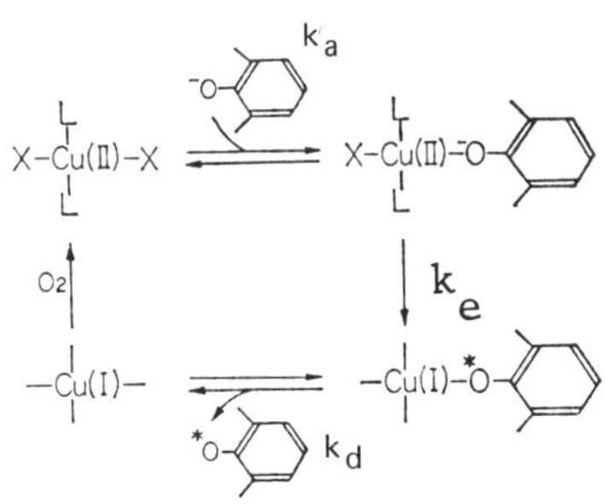

(116)

Tsuchida et al. also kinetically studied the effect of polymer ligand for process 116 and determined the rate constants k_a, k_e, and k_d (139). The rate constants thus determined and the ratio of catalyst species are given in Table 16. The value of k_e is the lowest. The

Table 16. Rate Constants of Elementary Reactions in Cu-Catalyzed Oxidation of 2,6-Dimethylphenol[a]

Catalyst	Solvent	System	k_a (mol min^{-1})	k_e (min^{-1})	k_o (min^{-1})	[Cu-XOH] / Σ [Cu]	[Cu(II)] / Σ [Cu]
QPVP-Cu	DMSO	Homogeneous	19	0.027	0.95	–	0.24
Pyridine-Cu	DMSO	Homogeneous	71	0.011	11	–	0.28
PVP-Cu	Benzene	Heterogeneous	2.5	0.0015	0.075	0.81	0.62

[a]k_a = rate constant of coordination step of substrate, k_e = rate constant of electron-transfer step, k_o = rate constant of reoxidation step of Cu catalyst, [Cu XOH] or [Cu(II)]/Σ|Cu] = fraction of the substrate-coordinated Cu catalyst or of Cu(II) catalyst in the steady state of the reaction.

smaller rate constant k_a for the polymer complex catalyst compared to the monomeric analog is explained by steric hindrance bought about by the bulky polymer ligand.

3.4.3.a Oxidative polymerization of 2,6-dimethylphenol catalyzed by the Cu(II)-polyvinylpyridine (PVP-Py) (137,138)

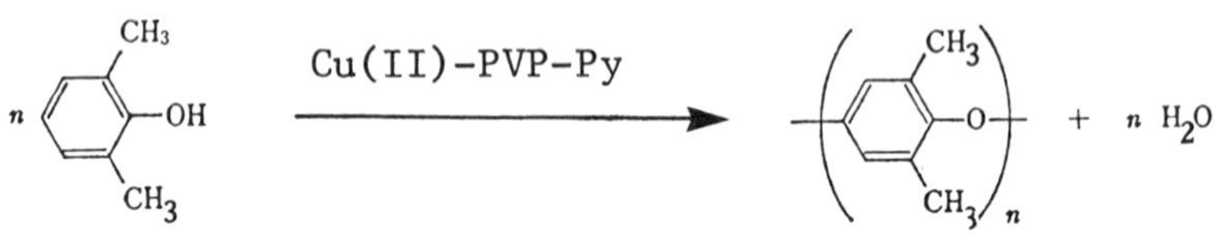

(117)

Homogeneous catalysis

A homogeneous catalyst solution of Cu(II)-PVP-Py complex is prepared by adding polyvinylpyridine to a Cu(II)-Py complex solution. The phenol is added to the catalyst solution and the polymerization is carried out under an oxygen atmosphere at 30°C in a reaction vessel that was designed to keep the pressure constant. The polymerization rate is determined by measuring the oxygen uptake. The completion of the process is confirmed by the cessation of oxygen uptake. The solution is poured into a hydrochloric acid-methanol solution to precipitate the polymer as a white powder. The polymer is filtered, purified by the reprecipitation, and dried under vacuum. The polymer is identified as poly(2,6-dimethylphenol) by IR spectroscopy and elementary analysis. The molecular weight of the polymer is measured by vapor preosmometer (MW = 28,000 to 71,000).

Heterogeneous catalysis

The Cu-PVP complexes, which are insoluble but catalytically active, are prepared by the treatment of the $CuCl_2$-PVP complex with water or alkaline water. The $CuCl_2$-Py complex is catalytically inactive; however, the $CuCl_2$-PVP complex washed with water is catalytically active. During the washing, a chloride ion is exchanged for a hydroxyl ion yielding a $CuCl_{0-1}(OH)_2$-PVP complex. The water-insoluble PVP ligand fixes the catalytically active CuCl(OH) complex and the chloride ions are washed away with water. The active Cu-PVP complex may also be obtained by treatment of the $CuCl_2$-PVP complex with potassium hydroxide (118). In an alkyl-coated vessel the Cu-PVP complex is

$$\text{PVP-CuCl}_2 + \text{OH}^-(\text{H}_2\text{O}) \longrightarrow \text{PVP-CuCl(OH)} + \text{Cl}^-(\text{HCl}) \quad (118)$$

suspended in a benzene solution of XOH, and the polymerization is promoted by vigorous stirring. After polymerization the Cu-PVP catalyst is removed by filtration and the reaction mixture is poured into a hydrochloric acid-methanol solution and the polymer is precipitated. The filtered catalyst is washed with nitrobenzene and the recovered amount of the catalyst is determined.

Cu(II)-polymer complexes initiate radical polymerization and often show higher activity than the corresponding monomeric analogs. The systems of Cu(II)-ion with nylon (141), α-aminocaproic acid (140), nylon oligomers (140), polyethylene-polyamine (142,143), poly(vinylamine-co-vinyl alcohol) (144), and cellulose have been made by Imoto and Takemoto as initiation systems for free-radical polymerization of vinyl monomers such as methyl methacrylate (MMA). Inaki et al. reported that Cu(II)-polyvinylamine complexes in the presence of CCl_4 in an aqueous solution showed higher activity as an initiator than Cu(II)-diaminopropane complex for the radical polymerization of methyl methacrylate, acrylonitrile, and styrene (145). The pH dependence for activity indicates that the free amine groups on the poly(vinylamine) chain are involved in the catalysis. The initiation mechanism is proposed as follows:

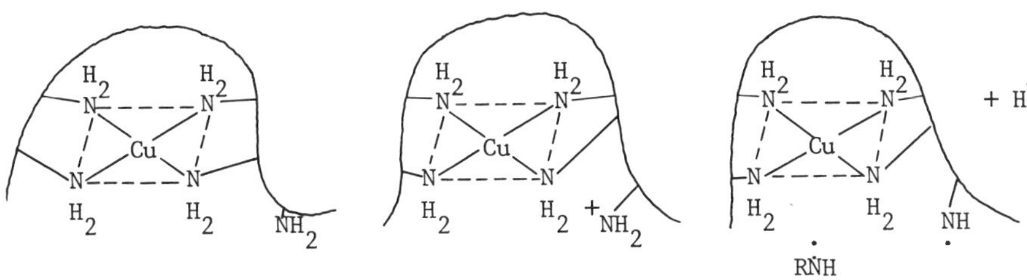

(119)

[*144*] [*145*] [*146*]

$$R\dot{N}H + CCl_4 \longrightarrow RNHCl + .CCl_3 \quad (120)$$

$$R\dot{N}H \text{ or } .CCl_3 + M \longrightarrow M\cdot \quad (M\cdot\text{:monomer}) \quad (121)$$

3.4.3.b Vinyl polymerization by Cu(II)-polyvinylamine-CCl_4 system

The initiation systems of the ligands with copper(II) salt in aqueous solution are adjusted to various pH values with sodium hydroxide, while keeping the ionic strength of the reaction system constant by adding sodium chloride. Polymerization is carried out in sealed tubes in the usual manner. Vinyl monomer (2.0 mL) and carbon tetrachloride (1.0 mL) are added to 4.0 mL of aqueous Cu(II)-chelate solution. The tubes are degassed and thermostated at 60°C for 3 h. The polymer is precipitated by pouring the solution into a large excess of methanol, and dried in vacuum.

3.4.4 Photoreaction

Production of H_2 and O_2 by the photodecomposition of water solves the problem of conversion and storage of solar energy. The tris(bipyridine)ruthenium(II) dichloride complex $[Ru(bpy)_3]Cl_2$ can theoretical photoreduce water molecules to give H_2 because it has a small reduction potential at the existing ^3CT state . An example of model reactions for the photodecomposition of water is shown in Reaction 122.

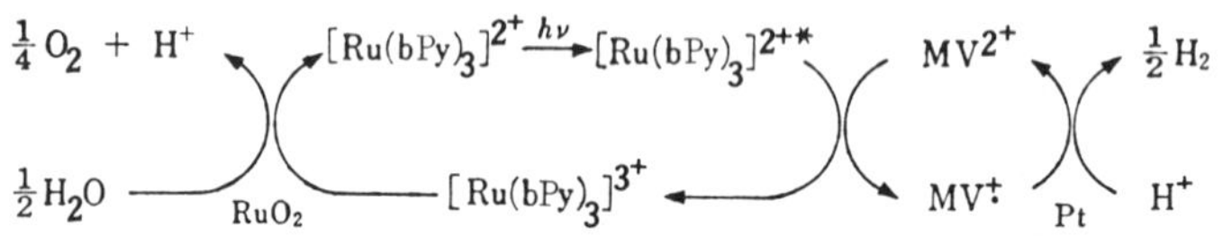

(122)

Kaneko et al. reported the use of $[Ru(II)(bPy)_3]$ immobilized on a synthetic polymer support (146) as a solid-phase photocatalyst. Polystyrene pendant $[Ru(II)(bPy)_2]$ was also synthesized by Kaneko et al. by the reaction of bipyridylated polystyrene(PSt-bPy) with dichlorobis(2,2'-bipyridyl)ruthenium(II) (147). When a mixture of PSt-$(bPy)_3$Ru and EDTA was put into a methanol solution of MV^{2+} and the solid-liquid interface was irradiated with visible light, photo-induced charge separation at the interface occurred, resulting in the electron transfer from EDTA in the solid phase to MV^{2+} in liquid phase. Hydrogen gas was evolved in the presence of proton and platinum catalyst in the liquid phase. Methyl viologen absorbed on cellulose was reduced photochemically in the solid phase sensitized by Ru(II)-$(bPy)_3$, using Na_2EDTA as a reducing agent (148). Kurimura et al. reported that the $Ru(bPy)_3^{2+}$-photosensitized reduction of MV^{2+} proceeded rapidly in water-swollen iminodiacetic acid-type chelate resin beads, which absorbed both $Ru(bPy)_3^{2+}$ and MV^{2+} (149).

3.4.4.a Photoreduction of methylviologen of cellulose sensitized by tris(2,2'-bipyridyl)ruthenium complex (148)

Absorption of reaction components on cellulose

Methylviologen (MV^{2+}) is recrystallized from a methanol-acetone mixed solvent. Ru(II)$(bPy)_3$ is prepared by refluxing ruthenium trichloride and 2,2'-bipyridyl (1:3 molar ratio) in methanol for 72 h, and recrystallized twice from methanol. A cellulose paper (1 × 3 cm) is dipped into the aqueous solution of MV^{2+}, Ru(II)$(bPy)_3$, and Na_2EDTA for 10 min, removed, and dried in the air for 15 min. The dipping and drying process is repeated and the sample is dried in vacuum at room temperature for 2 days in the dark and in the presence of silica gel. The quantities of the components absorbed are determined from the concentration decrease of the solution after dipping. The concentrations of MV^{2+} and Ru(II)$(bPy)_3$ are measured by the optical densities at 256 and 454 nm. The quantities of the MV^{2+}, Ru(II)$(bPy)_3$, and Na_2EDTA absorbed on the cellulose are 64.4, 1.88, and 92.4 mg/g cellulose, respectively.

Photoreaction and measurement

Light source is a 12-V 100-W halogen lamp. The visible range is obtained by cutoff filters. For monochromatic irradiation, interference is used in combination with appropriate cutoff filters. Light intensities at the illuminated sample surface are 4800 (white light without filter, 807-nm monochromatic light) W/m^2. A cellulose sample is placed in a quartz cell, irradiated, and the transmission spectrum is measured. A blank cellulose paper is placed in a reference cell during the measurement. Photoreaction is carried out in a Thunberg-type quartz cell at 10^{-3} mm Hg. The rate of MV^{2+} accumulation (mol cm^{-2} min^{-1}) is determined from the rate of OD increase at λ_{max} (620 nm) using the relationship, 1 absorbance = 1.07×10^{-8} (spins cm^{-2} cellulose). The change in the fusible transmittance spectrum of a cellulose sample which absorbed MV^{2+}, $Ru(bPy)_3$, and Na EDTA induced by visible-light irradiation is shown in Figure 22.

A promising system for the storage of solar energy is based on the photosensitized conversion of norbornadiene [*147*] to quadricyclane [*148*] as the energy-storage step followed by the catalyzed exothermic reversion of quadricyclane [*148*] to norbornadiene as the energy-release step. Ideal catalysts for this reversion reaction must meet

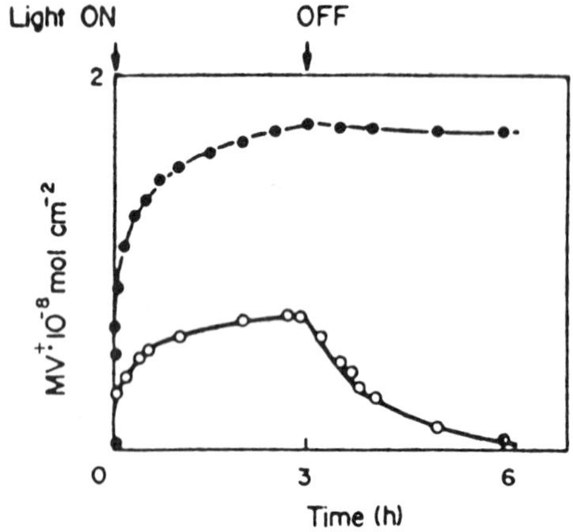

Figure 22. Time dependences of photochemical accumulation and dark oxidation of MV: ●, in vacuo; ○, under air. Irradiated with visible light at 30°C.

stringent criteria, which include long-term stability as well as the ability to effect rapid quantitative conversion with no side reactions. In addition, for an actual energy-storage device based on this principle, the catalyst must be immobilized to prevent dispersion of the catalyst throughout the system. King and Hanes reported an active catalyst for the conversion of quadricyclane to norbornadiene by reacting diphenylphosphinated macroreticular polystyrene with $(CH_3CN)_2PdCl_2$. The activity of this catalyst is less than that of the soluble catalyst $[(C_6H_5)_3P]_2PdCl_2$ by a factor of 2 to 85 (150).

Carboxamide- and sulfonamide-linked polystyrene-anchored cobalt(II) tetraarylporphyrines containing 0.28 to 0.44% cobalt are highly active catalysts for the conversion of quadricyclane to norbornadiene (151). These catalysts gradually lose some of their activity upon repeated recycling. This loss of catalytic activity can be partially restored by treatment with a strong reducing agent such as titanium(III). The apparent activities of these catalysts are increased if the catalyst is finely powdered, thereby suggesting a diffusion limitation in the catalytic activity. The application of these catalysts were proposed for a solar-energy-storage system based on the reversible valence isomerization of norbornadiene to quadricyclane.

3.4.4.b Photosensitized conversion of norbornadiene [*147*] to quadracyclane [*148*] coupled with the polystyrene-anchored Co(II)-tetraarylporpyrin complex [*149*]-catalyzed reversion of quadricyclane to norbornadiene (152)

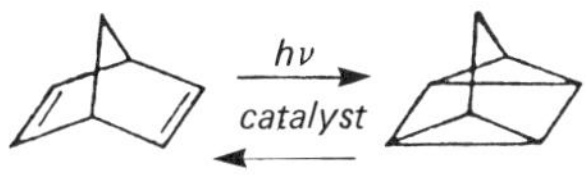

(123)

[*147*] [*148*]

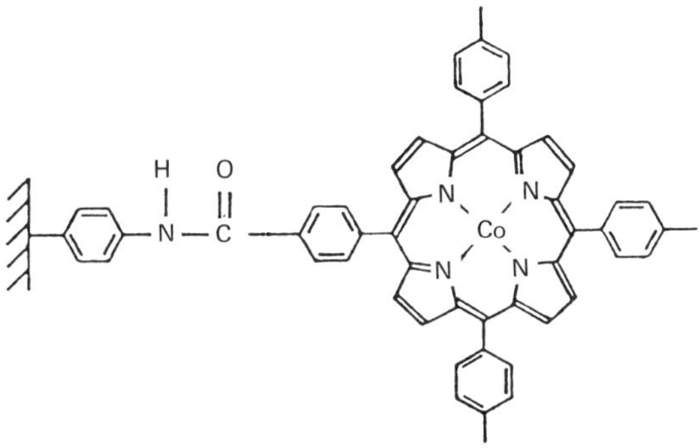

[*149*]

Approximately 0.1 g of polymer-bound cobalt(II)-tetraarylporphyrin is accurately weighed into a 25-mL round-bottomed flask fitted with a magnetic stirring bar, gas inlet adapter, serum cap, and serum cap adapter. The system is subjected to three cycles of evacuation and refilling with nitrogen; thereafter, a nitrogen atmosphere is maintained. During the reaction, the flask is partially immersed in a water bath maintained at 30°C.

Approximately 10 mL of the quadracyclane solution in benzene or xylene (purified by treatment with either potassium metal for at least 6 h or $LiAlH_4$ for 24 h) is introduced with a syringe. The recorded quantity of solution is obtained from the weight difference of the syringe before and after injection. Samples of the reaction solution (20 L) are withdrawn and diluted into 0.2 mL of pentane or carbon disulfide for the gas chromatographic analysis. The initial sample is taken after 60 s with subsequent samples at 200- to 500-s intervals. Normally, eight samples are taken over three to seven half-lives, but occasionally as few as six samples are used.

The recycling studies are conducted in a similar manner. At the end of the reaction period, the solvent is withdrawn through a flat-tipped needle into a syringe and the catalyst washed three times for 15 min per wash with fresh, oxygen-free benzene. The catalyst is dried under a stream of nitrogen at 30°C for 1 h before being used for the next catalytic run.

REFERENCES

1. E. Tsuchida and H. Nishide, Adv. Polym. Sci., 24, 2 (1977).
2. M. Kaneko and E. Tsuchida, Macromol. Rev., 16, 397 (1981).
3. N. Hojo and H. Shirai, Kagaku Sosetsu, 17, 183 (1977).
4. E. Tsuchida, Kagaku Sosetsu, 20, 165 (1977).
5. H. Shirai and N. Hojo, J. Soc. Fiber Sci. Tech., Jpn, 30, 220 (1974).
6. N. Hojo and H. Shirai, in Ion Exchange Resin and Chelate Resin (N. Hojo, Ed.), Kodansha Ltd., Tokyo (1976).
7. T. Saegusa, H. Hirai, and E. Tsuchida, Polymer-Metal Complexes, Kagaku Do-jin, Kyoto (1978).
8. H. Hirai and N. Toshima, Eds., Polymer Complex Catalysts, Gakkai Schuppan Center, Tokyo (1982).
9. N. Hagihara and A. Nakamura, Organometal Polymers, Gakkai Schuppan Center, Tokyo (1982).
10. A. Nakamura, Kagaku Sosetsu, 35, (1982).
11. N. Hojo, H. Shirai, K. Sakata, and M. Fukuda, Kogyo Kagaku Zasshi, 73, 1862 (1972).
12. N. Hojo, H. Shirai, and O. Nobuyoshi, Kogyo Kagaku Zasshi, 74, 1209 (1971).
13. N. Hojo, H. Shirai, and S. Hayashi, J. Polym. Sci. Polym. Symp. 47, 299 (1974).
14. N. Hojo and H. Shirai, Nippon Kagaku Kaishi, 1316 (1972).
15. H. Shirai and N. Hojo, Nippon Kagaku Kaishi, 2223 (1972).
16. H. Shirai, Y. Nio, S. Hayashi, and N. Hojo, Nippon Kagaku Kaishi, 117 (1978).
17. M. Sato, K. Kondo, and K. Takemoto, Makromol. Chem., 179, 601 (1978).
18. M. Sato, K. Kondo, and K. Takemoto, Makromol. Chem., 180, 699 (1979).
19. Y. Nose, M. Hatano, and S. Kambara, Kogyo Kagaku Zasshi, 69, 566 (1966).
20. M. Hatano, Y. Nose, T. Nozawa, and S. Kambara, Kogyo Kagaku Zasshi, 69, 572 (1966).
21. H. Shirai, Nippon Kagaku Kaishi, 390 (1973).
22. N. Hojo, H. Shirai, Y. Kameyama, K. Nagasaki, and S. Ichikawa, Kogyo Kagaku Zasshi, 74, 269 (1971).
23. H. Shirai, M. Yoshida, H. Sakurai, and N. Hojo, Nippon Kagaku Kaishi, 1201 (1973).

24. Y. Kurimura, E. Tsuchida, and M. Kaneko, J. Polym. Sci., A1(9), 3511 (1971).
25. E. Tsuchida, H. Nishide, and M. Takeshita, Makromol. Chem., 175, 2293 (1974).
26. E. Tsuchida, H. Nishide, and Y. Kurimura, Makromol. Chem. 175, 171 (1974).
27. E. Tsuchida, J. Macromol. Sci. Chem., A13, 545 (1979).
28. G. Blauer, Acta Chem. Scand., 17, 8 (1979).
29. S. Yamamoto, T. Nozawa, and M. Hatano, Polymer, 15, 330 (1974).
30. M. Tohjo and K. Shibata, Arch. Biochem. Biophys., 103, 401 (1963).
31. M. Hatano, Kagaku To Kogyo, 18, 926 (1965).
32. E. Tsuchida, K. Honda, and H. Sata, Biopolymers, 13, 2147 (1974).
33. J. R. Miller and G. D. Dorough, J. Am. Chem. Soc., 74, 3997 (1952).
34. W. Scheler, P. Mohr, K. Pommerening, and J. Behlke, Eur. J. Biochem., 13, 77 (1970).
35. J. H. Furrhop, S. Besecke, W. Vogt, J. Ernst, and J. Subramanian, Makromol. Chem., 178, 1261 (1977).
36. E. Collison, F. S. Dainton, F. R. S. B. Mile, S. Tazuke, and P. R. Smith, Nature, 193, 26 (1963).
37. S. Tazuke and S. Okamura, J. Polym. Sci., A1(4), 141, 2461 (1966).
38. Y. Osada, Makromol. Chem., 177, 1259 (1976).
39. Y. Osada, Makromol. Chem., 176, 1893 (1975).
40. Y. Osada, Makromol. Chem., 177, 1273 (1976).
41. C. U. Pittman, Jr., J. C. Lai, D. P. Vanderpool, M. Good, and R. Praclos, Macromolecules, 3, 746 (1970).
42. C. U. Pittman, Jr., and Y. Sasaki, Chem. Lett., 383 (1975).
43. Y. Sasaki, L. L. Walker, E. L. Hurst, and C. U. Pittman, Jr., J. Polym. Sci. Polym. Chem. Ed., 11, 1313 (1973).
44. F. S. Arimoto and A. C. Haven, Jr., J. Am. Chem. Soc., 77, 6295 (1955).
45. C. U. Pittman, Jr., R. L. Voges, and J. Elder, J. Polym. Sci., B9, 191 (1971).
46. C. U. Pittman, Jr., and T. D. Rounsefell, Macromolecules, 9, 936 (1976).
47. C. U. Pittman, Jr., and A. Hirao, J. Polym. Sci. Polym. Chem. Ed., 15, 1677 (1977).
48. V. V. Korshak, S. L. Sosin, and T. M. Frunze, J. Polym. Sci., C22, 849 (1969).
49. I. O. Hartwell and J. C. Bailar, Jr., J. Am. Chem. Soc., 92, 1284 (1970).
50. H. Shirai, K. Kobayashi, Y. Takemae, A. Suzuki, O. Hirabaru, and N. Hojo, Makromol. Chem., 180, 2073 (1979).

51. H. Shirai, K. Kobayashi, Y. Takemae, and N. Hojo, J. Polym. Sci. Polym. Lett. Ed., 17, 661 (1979).
52. E. W. Neuse and L. Bednarik, Macromolecules, 12, 187 (1979).
53. R. W. Kluiber and J. W. Lewis, J. Am. Chem. Soc., 82, 5777 (1960).
54. W. Lautsh, W. Broser, W. Biedermann, and H. Guichtel, Z. Naturforsch., 136, 48 (1957).
55. H. Kamogawa, J. Polym. Sci., B10, 710 (1972).
56. H. Kamogawa, J. Polym. Sci. Polym. Chem. Ed., 12, 2317 (1974).
57. L. D. Rollman, J. Am. Chem. Soc., 97, 2132 (1975).
58. H. Shirai, A. Maruyama, K. Kobayashi, K. Urushido, and N. Hojo, J. Polym. Sci., Polym. Lett. Ed., 17, 661 (1979).
59. H. Shirai, A. Maruyama, K. Kobayashi, N. Hojo, and K. Urushido, Makromol. Chem., 181, 575 (1980).
60. A. Suzuki, H. Shirai, and N. Hojo, Kogyo Zairyo, 17, 73 (1969).
61. D. B. Losee, H. W. Richardson, and W. E. Hatfield, J. Chem. Phys., 59, 3600 (1973).
62. C. Creutz and H. Taube, J. Am. Chem. Soc., 95, 1006 (1973).
63. H. W. Richardson and W. E. Hatfield, J. Am. Chem. Soc., 98, 835 (1976).
64. J. F. Villa and W. E. Hatfield, J. Am. Chem. Soc., 93, 4081 (1971).
65. A. Santoro, A. D. Mighell, and C. W. Reinmann, Acta Crystallogr., B26, 979 (1970).
66. J. Bonner and M. E. Fisher, Phys. Rev., A135, 640 (1964).
67. P. Nanneli, H. D. Gillman, and B. P. Block, J. Polym. Sci., A1(13), 2849 (1975).
68. P. Nanneli, H. D. Gillman, H. G. Monsimer, and S. B. Advani, J. Polym. Sci., A1(12), 2525 (1976).
69. J. T. Wrobleski and D. B. Brown, Inorg. Chem., 18, 498 (1979).
70. N. Hojo, H. Shirai, and A. Suzuki, Kogyo Kagaku Zasshi, 69, 253 (1969).
71. N. Hojo, H. Shirai, and A. Suzuki, Kogyo Kagaku Zasshi, 72, 2040 (1969).
72. N. Hojo, H. Shirai, and A. Suzuki, Kogyo Kagaku Zasshi, 73, 1438 (1970).
73. N. Hojo, H. Shirai, and A. Suzuki, Kogyo Kagaku Zasshi, 73, 2435 (1970).
74. N. M. Rivera and E. M. Engler, J. Chem. Soc. Chem. Commun., 184 (1979).
75. A. Suzuki, K. Ohkawa, S. Kanda, M. Imoto, and S. Watari, Bull. Chem. Soc. Jpn., 48, 2634 (1975).
76. S. Kanda, Nippon Kagaku Zasshi, 83, 560 (1962).
77. S. Kanda, K. Itoh, and T. Nagaito, J. Polym. Sci., C17, 151 (1967).
78. K. Sonogashira, S. Takahashi, and N. Hagihara, Macromolecules, 10, 879 (1977).

79. K. Sonogashira, S. Takahashi, and N. Hagihara, J. Organomet. Chem., 145, 10 (1978).
80. S. Takahashi, M. Kariya, T. Yataka, K. Sonogashira, and N. Hagihara, Macromolecules, 11, 1063 (1978).
81. R. Bannehr, G. Mayer, and D. Wohrle, Polym. Bull., 2, 841 (1980).
82. J. A. Ibers and R. H. Holm, Science, 209, 223 (1980).
83. S. Affrossman and J. P. Murray, J. Chem. Soc., B, 1015 (1966).
84. M. Takeishi, S. Niino, and S. Hayama, Makromol. Chem., 177, 1225 (1976).
85. M. Takeishi, T. Watanabe, S. Fujii, S. Niino, and S. Hayama, Makromol. Chem., 180, 1099 (1979).
86. R. Breslow, D. E. McCure, R. S. Brown, and J. Eisennach, J. Am. Chem. Soc., 97, 194 (1975).
87. I. Tabushi, N. Shimizu, T. Sugimoto, M. Shiozuka, and K. Yamaura, J. Am. Chem. Soc., 99, 7100 (1977).
88. T. Nozawa, Y. Akimoto, and M. Hatano, Makromol. Chem., 158, 21 (1972).
89. I. A. Yamskov, B. B. Berezin, L. A. Belehich, and V. A. Davankov, Makromol. Chem., 180, 799 (1979).
90. T. Takagishi and I. M. Klotz, Biopolymers, 18, 2497 (1979).
91. R. A. Turner, C. F. Huebner, and C. R. Scholz, J. Am. Chem. Soc., 71, 2801 (1949).
92. M. N. Hughes, The Inorganic Chemistry of Biological Processes, John Wiley & Sons, Inc., New York (1972).
93. I. Pecht, A. Levitzki, and M. Anbar, J. Am. Chem. Soc., 89, 1587 (1967).
94. Y. I. Skurlatov, V. Y. Kovner, S. O. Travin, Y. E. Kirsh, A. P. Purmal, and V. A. Kavanov, Eur. Polym. J., 15, 811 (1979).
95. M. Sato, K. Kondo, and K. Takemoto, J. Polym. Sci. Polym. Chem. Ed., 15, 2059 (1977).
96. T. Nozawa and M. Hatano, Makromol. Chem., 141, 21 (1971).
97. T. Nozawa and M. Hatano, Makromol. Chem., 141, 31 (1971).
98. E. Tsuchida and H. Nishide, Polymeric Amine and Ammonium Salts (E. J. Goethals, Ed.), Pergamon Press, Inc., Elmsford, N.Y. (1980), p. 287.
99. H. Nishide, T. Minakata, and E. Tsuchida, J. Mol. Catal., 15, 327 (1982).
100. N. Oguni, S. Shimazu, and A. Nakamura, Polym. J., 12, 891 (1980).
101. B. A. Averill, T. Herskovitz, R. H. Holm, and J. A. Ibers, J. Am. Chem. Soc., 95, 3523 (1973).
102. Y. Nose, M. Hatano, and S. Kambara, Makromol. Chem., 98, 136 (1966).

103. T. Nozawa, M. Hatano, and S. Kambara, Kogyo Kagaku Zasshi, 72, 369 (1969).
104. T. Nozawa, M. Hatano, and S. Kambara, Makromol. Chem., 112, 73 (1968).
105. A. T. Kapanchyan, V. S. Pshezhetski, and V. A. Kavanov, Vysokomol. Soedin., B11, 5 (1969).
106. H. Sigel and G. Blauer, Helv. Chim. Acta 51, 1246 (1968).
107. H. Yamamoto, K. Inouye, and T. Hayakawa, Int. J. Biol. Macromol., 2, 263 (1980).
108. T. Uragami, N. Niwa, and M. Sugihara, Kobunshi Ronbunshu, 39, 669 (1982).
109. H. Sigel, Experimentia, 23, 170 (1967).
110. M. Barteri, M. Farinella, and B. Pispisa, Biopolymers, 16, 2569 (1977).
111. N. Hojo, H. Shirai, Y. Chujo, and S. Hayashi, J. Polym. Sci. Polym. Chem. Ed., 16, 447 (1978).
112. H. Shirai, Y. Nio, S. Hayashi, and N. Hojo, Nippon Kagaku Kaishi, 117 (1978).
113. M. Branca and B. Pispisa, J. Chem. Soc. Faraday Trans. I, 73, 213 (1977).
114. M. Branca, M. E. Marini, and B. Pispisa, Biopolymers, 15, 2219 (1976).
115. V. S. Pshezhetskyi, S. G. Ikryannikov, A. Kuznetsova, and V. A. Kavanov, J. Polym. Sci. Polym. Chem. Ed., 14, 2595 (1976).
116. P. Waldmeier and H. Sigel, Inorg. Chim. Acta, 5, 656 (1971).
117. P. Waldmeier and H. Sigel, Z. Naturforsch., B27, 95 (1972).
118. H. Shirai, A. Maruyama, J. Takano, K. Kobayashi, and N. Hojo, Makromol. Chem., 181, 565 (1981).
119. H. Shirai, A. Maruyama, K. Kobayashi, and N. Hojo, Makromol. Chem., 181, 575 (1980).
120. H. Shirai, A. Maruyama, K. Kobayashi, K. Urushido, and N. Hojo, J. Polym. Sci. Polym. Chem. Ed., 17, 661 (1979).
121. H. Shirai, S. Higaki, K. Hanabusa, and N. Hojo, Makromol. Chem., in press.
122. H. Shirai, S. Higaki, K. Hanabusa, and N. Hojo, J. Polym. Sci. Polym. Lett. Ed., in press.
123. M. Barteri, M. Farinella, and B. Pispisa, Biopolymers, 16, 2569 (1977).
124. H. Shirai, A. Maruyama, M. Konishi, and N. Hojo, Makromol. Chem., 181, 1003 (1980).
125. M. Chanda, K. F. O'Driscoll, and G. L. Rempel, J. Catal., 61, 533 (1980).
126. M. Chanda, K. F. O'Driscoll, and G. L. Rempel, J. Mol. Catal., 8, 339 (1980).

127. B. R. James and G. L. Rempel, Can. J. Chem., 44, 233 (1966).
128. A. Seidell, Solubilities of Inorganic and Metal Organic Compounds, Vol. 1, 4th Ed., Van Nostrand Reinhold Company, Inc., New York (1958).
129. J. H. Schutten and J. Zwart, J. Mol. Catal., 5, 109 (1979).
130. J. Zwart, H. C. Van Der Weide, N. Broker, C. Rummens, and G. C. A. Shutten, J. Mol. Catal., 3, 151 (1977/1978).
131. A. A. Efendiev, T. N. Shankhtakhtinsky, L. F. Mustafaeva, and H. L. Shick, Ind. Eng. Chem. Prod. Res. Dev., 19, 75 (1980).
132. A. Davison and S. S. Wreford, J. Am. Chem. Soc., 96, 3017 (1974).
133. W. D. Bonds, Jr., C. H. Brubaker, Jr., E. S. Chendrasekaran, C. Gibbons, R. H. Grubbs, and L. C. Kroll, J. Am. Chem. Soc., 97, 2128 (1975).
134. K. W. Pepper, H. M. Paisley, and M. A. Young, J. Chem. Soc., 4097 (1953).
135. H. Hirai, Y. Nakao, N. Toshima, and A. Adachi, Chem. Lett., 905 (1976).
136. H. Hirai, Y. Nakao, and N. Toshima, J. Makromol. Sci. Chem., A12, 1117 (1978).
137. E. Tsuchida, M. Kaneko, and H. Nishide, Makromol. Chem., 174, 203 (1973).
138. E. Tsuchida, H. Nishide, and T. Nishihara, Makromol. Chem., 175, 3047 (1974).
139. E. Tsuchida, H. Nishide, and H. Nishikawa, J. Polym. Sci. Polym. Symp. 47, 47 (1974).
140. K. Takemoto, T. Takata, and Y. Inaki, J. Polym. Sci., A1(10), 1061 (1972).
141. K. Ueno, M. Kinoshita, and M. Imoto, Kogyo Kagaku Zasshi, 74, 1921 (1971).
142. Y. Inaki, M. Ishiyama, and K. Takemoto, Makromol. Chem., 160, 127 (1972).
143. Y. Inaki, M. Ishiyama, and K. Takemoto, Angew. Makromol. Chem., 27, 175 (1972).
144. K. Kimura, Y. Inaki, and K. Takemoto, Makromol. Chem., 175, 83 (1974).
145. Y. Inaki, K. Kimura, and K. Takemoto, Makromol. Chem., 171, 19 (1973).
146. M. Kaneko, S. Nemoto, A. Yamada, and Y. Kurimura, Inorg. Chim. Acta, 44, 1289 (1980).
147. M. Kaneko, A. Yamada, S. Nemoto, and M. Yokoyama, Kobunshi Ronbunshu, 37, 685 (1980).
148. M. Kaneko and A. Yamada, Photochem. Photobiol. 33, 793 (1981).
149. Y. Kurimura, M. Nagashima, K. Takato, E. Tsuchida, M. Kaneko, and A. Yamada, J. Phys. Chem., 86, 2432 (1982).

150. R. B. King and R. M. Hanes, J. Org. Chem., 44, 1092 (1979).
151. R. B. King and E. M. Sweet, J. Org. Chem., 44, 385 (1979).

4

The Preparation of Nucleic Acid Analogs

Kiichi Takemoto and Yoshiaki Inaki

Osaka University, Osaka, Japan

4.1 INTRODUCTION

Nucleic acids are known to play an important role in both the replicative and transcriptive functions of genetic codes. DNA, one of the nucleic acids, for example, has a double-stranded structure consisting of two polynucleotide chains twisted about each other in a double helix, which is stabilized by special base-base interactions of nucleic acid bases (Figure 1). The interactions among such purine and pyrimidine families are apparently a common subject of substantial interest among chemists in the fields of organic, macromolecular, and biochemical syntheses (Figure 2).

The functional monomers and polymers containing heterocyclic moieties have received much attention, and numerous studies have been devoted to the preparation and polymerization of these new monomeric species, which may find a number of possible applications as polymeric drugs and other biomaterials.

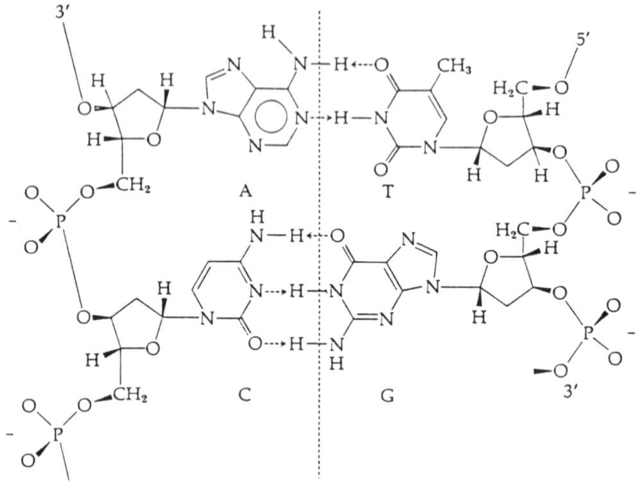

Figure 1

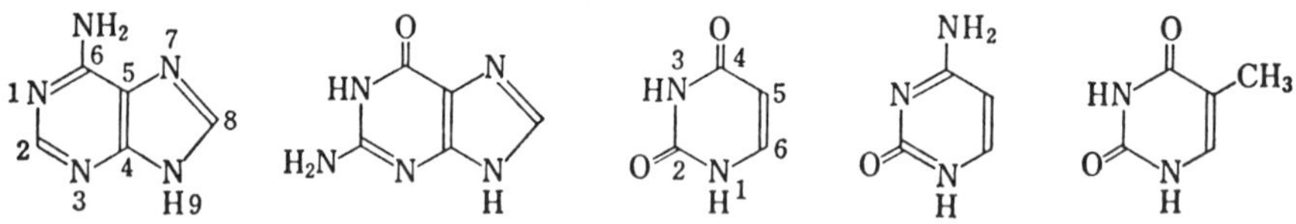

Adenine Guanine Uracil Cytosine Thymine

Figure 2

In this chapter, emphasis is focused particularly on the preparation and polymerization of several series of compounds having nucleic acid bases. Recently, we made a fairly exhaustive survey of the nucleic acid analogs, particularly the specific base-base interactions between the nucleic acid analogs and their biological properties and applications. Reviews have also been written, and these have been referred to in this chapter [1-8, 53, 98, 111, 128].

4.2 POLY-N-VINYL DERIVATIVES OF NUCLEIC ACID BASES

4.2.1 N-Vinyl Compounds of Purines

Poly-9-vinyladenine (*14*) is prepared by free-radical polymerization of the corresponding monomer. The general preparative route to the N-vinyl monomer is illustrated in Scheme 1. Hydroxyethylation of

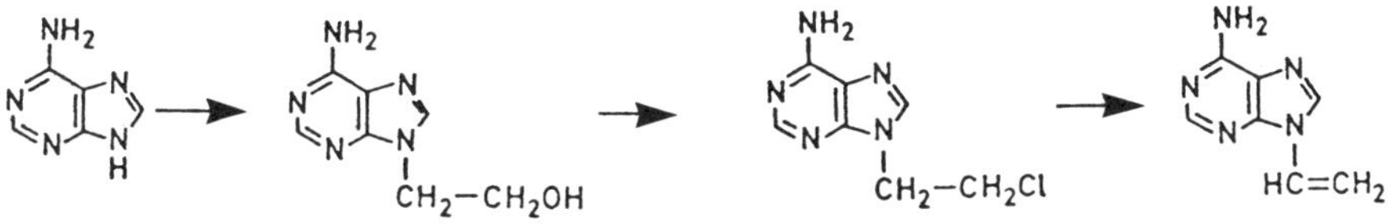

1 2 3

Scheme 1

adenine base can be carried out by treating the base with ethylene carbonate in *N,N*-dimethylformamide in the presence of a trace amount of sodium hydroxide. The hydroxyethyladenine (*1*) thus obtained can readily be chlorinated by thionyl chloride. Finally, the *N*-chloroethyl compound (*2*) is converted to the corresponding N-vinyl monomer (*3*) by treatment with sodium methoxide in dioxane at room temperature. The N-vinyl monomers 9-vinyladenine (*3*) [9], 7-vinyltheophylline (*4*) [12], and 7-vinylguanine (*5*) [12] are synthesized using this procedure.

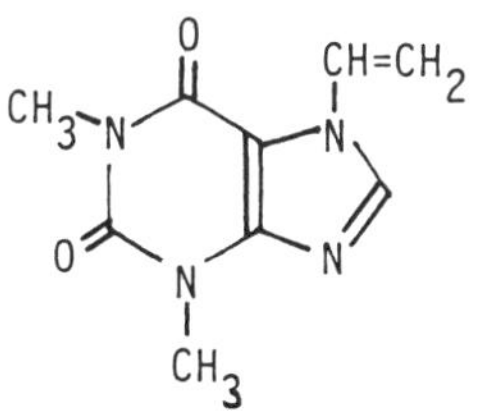

4

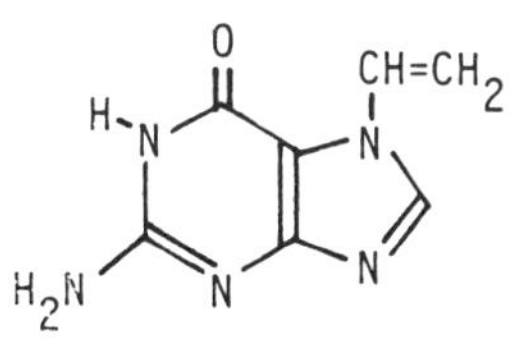

5

9-(2-Hydroxyethyl)adenine (*1*) [9]. To a solution of adenine (13.5 g, 0.01 mol) in dimethylformamide (400 mL) under reflux, ethylene carbonate (9.5 g, 0.01 mol) in dimethylformamide (30 mL) is added dropwise. After the evolution of carbon dioxide ceases, the solvent is removed under reduced pressure. The solid residue is recrystallized from ethanol to yield 13.9 g (79%), mp 238 to 239°C.

9-(2-Chloroethyl)adenine (*2*) [9]. A mixture of 9-(2-hydroxyethyl)adenine (*1*) (5.4 g, 0.036 mol) and thionyl chloride (50 mL) is heated on a steam bath for 40 min. After removal of excess thionyl chloride, the residue is dissolved in water and neutralized with a saturated solution of sodium carbonate to give the product as a precipitate. Recrystallization from ethanol gives colorless needles in 84% yield (5.0 g), mp 204 to 205°C.

9-Vinyladenine (*3*) [9]. To a solution of 9-(chloroethyl)adenine (*2*) (0.60 g, 0.003 mol) in dioxane, a methanolic solution of sodium methoxide, prepared from 4 mL of methanol and sodium (0.4 g), is added. The mixture is stirred for 24 h at room temperature and the solvent is evaporated. The residue is recrystallized from ethanol-benzene to give colorless plates; yield 0.29 g (60%), mp 196 to 197°C.

A vinyl compound containing the hypoxanthine moiety is prepared by a somewhat modified procedure; the chloroethylhypoxanthine (*6*) is obtained from the corresponding adenine derivative (*2*) and is converted into 9-vinylhypoxanthine (*7*) by treatment with sodium methoxide [12] (Scheme 2).

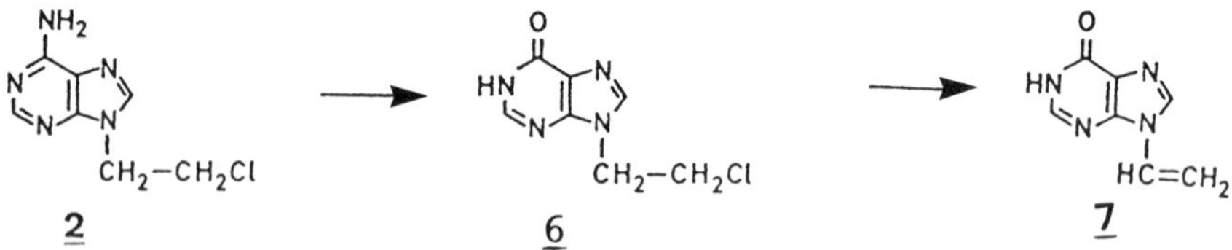

Scheme 2

9-(2-Chloroethyl)hypoxanthine (*6*) [12]. A solution of 9-(2-chloroethyl)adenine (*2*) (1.5 g) in concentrated hydrochloric acid (10 mL) is cooled with stirring to 0°C. Sodium nitrite (0.6 g) in water (20 mL) is added dropwise over 30 min. After the reaction mixture is stirred for a further 2 h at room temperature, it is evaporated to dryness. The residue is dissolved in 28% aqueous ammonium hydroxide (10 mL), and the solution is again evaporated to dryness. Finally, the residue is extracted with ethanol to yield 0.83 g (53%), mp 262°C, after recrystallization from ethanol.

9-Vinylhypoxanthine (*7*) [12]. 9-(2-Chloroethyl)hypoxanthine (*6*) is treated with sodium methoxide using the same procedure for 9-vinyladenine to give 9-vinylhypoxanthine (*7*), in 60% yield, mp 260°C.

An alternative method of obtaining the N-vinyl monomers directly from the corresponding purine bases involves a vinyl exchange reaction of vinyl acetate catalyzed by mercury (II) acetate and sulfuric acid. This procedure has been used to prepare 6-chloro-9-vinylpurine (*8*) and 2,6-dichloro-9-vinylpurine (*9*) [113]. In the synthesis of 9-vinyladenine (*3*), the 6-aminogroup is first protected by a benzoyl group (Scheme 3). The benzoyl derivative (*10*) is reacted with vinyl acetate to give N^6-benzoyl-9-vinyladenine (*11*) in 65% yield [138,143, 144]. Deblocking of the benzoyl group by methanolic ammonia gives 9-vinyladenine (*3*).

6-Chloro-9-vinylpurine (*8*) [113]. A solution of concentrated sulfuric acid (0.1 mL) in vinyl acetate (2 mL) is added to a suspension of mercuric acetate (0.5 g) in vinyl acetate (100 mL) to give a clear solution. Powdered 6-chloropurine (2 g) is added followed by vinyl acetate (5 mL). After flushing with nitrogen for 10 min, the flask is cooled and left in a bath at 45 to 50°C with occasional agitation for

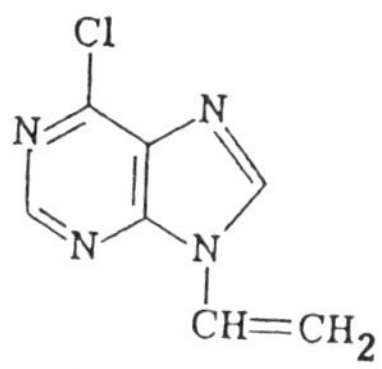

8

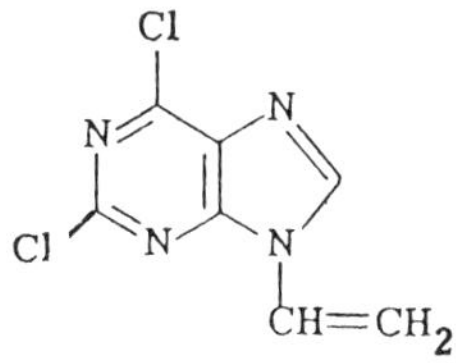

9

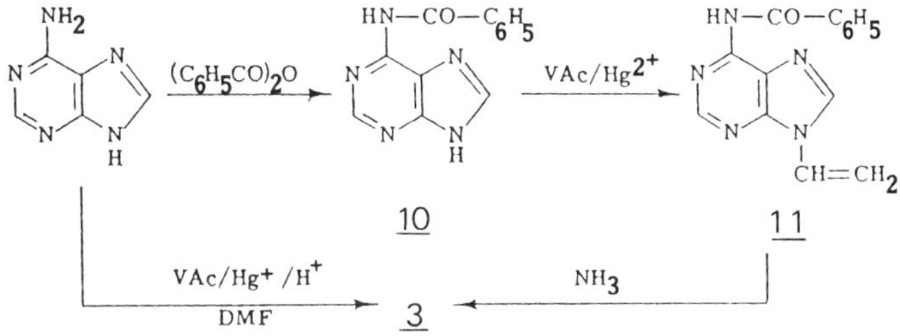

Scheme 3

5 days. Dry Sodium acetate is then added, and the mixture is stirred for 10 min and filtered. The filtrate is evaporated and the residue dissolved in chloroform and the solution is extracted with cold 1 N sodium hydroxide. The chloroform solution is then dried and evaporated. The residue is recrystallized and is purified by sublimation in vacuo to yield 1.6 g (70%), mp 166 to 167°C.

Starting with 6-chloro-9-vinylpurine (*8*), several 6-substituted 9-vinylpurine were prepared in aqueous solution: 9-vinyl derivatives of adenine (*3*) and hypoxanthine (*7*) [117], N^6-mono- and dialkyladenines (*12*), and 6-mercaptopurine (*13*) [13,16] (Scheme 4).

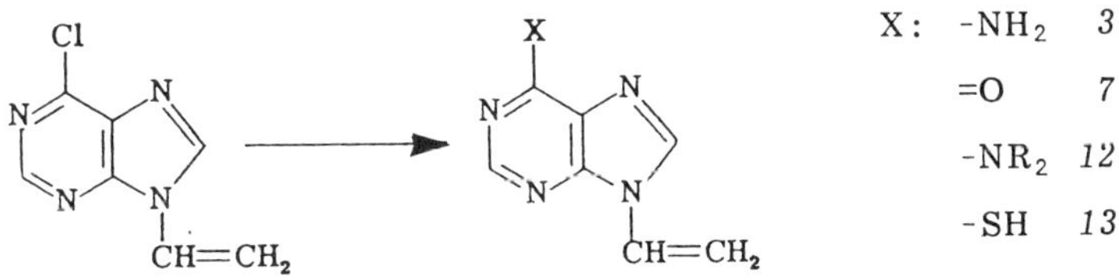

8

Scheme 4

9-Vinyladenine (*3*) [117]. In a stainless steel autoclave, a suspension of 6-chloro-9-vinylpurine (*8*) (1 g) in methanol (25 mL) is saturated with ammonia at 0°C, sealed and heated at 60°C for 48 h. The reaction mixture is then evaporated to dryness. The residue is extracted with boiling toluene, filtered hot, and the filtrate evaporated to half-volume. After cooling, white crystals are collected; yield 660 mg, mp 199 to 200°C.

9-Vinylhypoxanthine (*7*) [117]. 6-Chloro-9-vinylpurine (*8*) (1 g) is suspended in a solution of triethylamine (25 g) in water (500 mL). After stirring for 24 h at room temperature, the homogeneous solution is evaporated in vacuo. The residue is repeatedly recrystallized from water, yielding white leaflets; yield 654 mg, mp 250°C.

6-Mercapto-9-vinylpurine (*13*) [13]. 6-Chloro-9-vinylpurine (*8*) (0.5 g) and an equimolar amount of thiourea in ethanol (20 mL) are stirred for 2 h. Precipitated product is filtered, dissolved in 1 N NaOH, and precipitated with acetic acid to yield 0.37 g (75%), mp 298°C.

The polymerization and copolymerization of the 9-vinyl derivatives of purine were carried out in solvents such as dimethyl sulfoxide, dimethylformamide, and water, using a free-radical initiator [11]. Acrylamide, maleic anhydride, *N*-vinylpyrrolidone, and acrylic acid were used as comonomers [11,147]. Poly-9-vinyladenine is soluble in water and formic acid.

Alternatively, poly-9-vinyladenine (*14*) and poly-9-vinylhypoxanethin (*17*) were prepared by the reaction of poly-6-chloro-9-vinylpurine (*15*) with ammonia or triethylamine aqueous solution (Scheme 5) [115, 117].

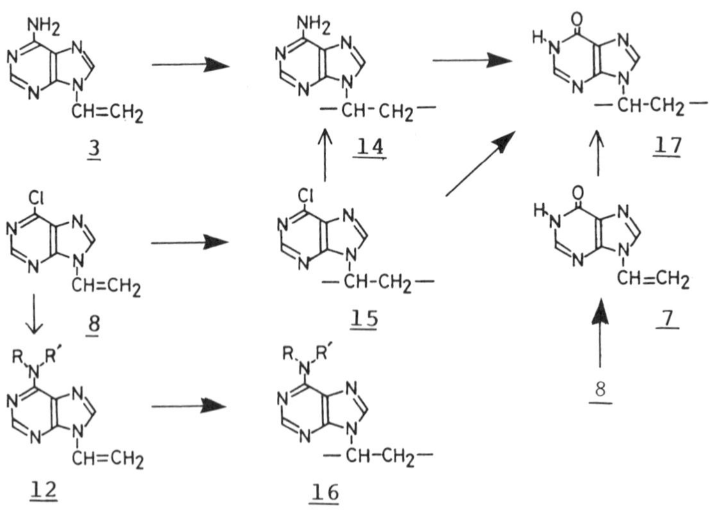

Scheme 5

Poly-6-chloro-9-vinylpurine (*15*) [115]. 6-Chloro-9-vinylpurine (*8*) (1.4 g) is dissolved in dimethylacetoamide (3.5 mL) at 100°C, AIBN (28 mg) is added, and the solution is kept for 1 h at the same temperature. The solvent is evaporated by freeze drying, and the residue is pulverized and mixed with tetrahydrofuran overnight. The insoluble polymer is collected by centrifugation, and dried in vacuo to yield 1.3 g. This polymer is unstable at room temperature and on hydrolysis the product evoked hydrogen chloride. Stored above dry silica gel at 0°C, the polymer does not change even after several months.

Poly-9-vinyladenine from polymer 15 [115]. Polymer *15* (1 g) is suspended in anhydrous methanol (125 mL) and saturated with ammonia at 0°C in a pressure flask. The flask is then closed and the mixture stirred for 2 days at room temperature and 14 days at 36°C. After evaporation in vacuo, the residue is dissolved in 0.05 N hydrochloric acid and neutralized by aqueous ammonia. The suspension is dialyzed against water and separated by centrifugation to give polymer *14* (670 mg). According to the nitrogen analysis, the polymer contains 81% of poly-9-vinyladenine units.

Poly-9-vinylhypoxanthine from polymer 15 [117]. Polymer *15* (140 mg) is dissolved in concentrated hydrochloric acid (25 mL) at room temperature and is stirred for 2 h. The mixture is then evaporated repeatedly from water. The residue is dissolved in 0.1 N NaOH, filtered through sintered glass, neutralized to pH 7, and the precipitated polymer collected by centrifugation and washed with water to yield 107 mg of product.

4.2.2 N-Vinyl Compounds of Pyrimidines

1-Vinyluracil and 1-vinylthymine can be prepared using a procedure similar to that used in the preparation of the N-vinyl compounds of purine [9,113] (Scheme 6). 1-(2-Hydroxyethyl)uracil (*18a*) was prepared by treating uracil or the 2,4-diethoxy derivative with ethylene carbonate and separating the product from the accompanying disubstituted derivative by chromatography [9,113] or recrystalliza-

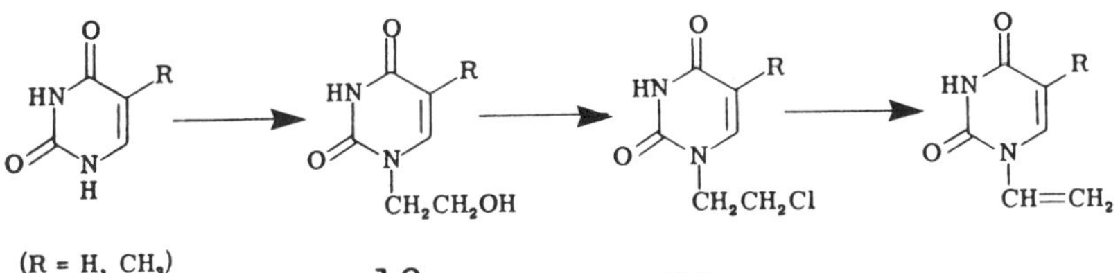

18 19 20

Scheme 6

tion [27]. To avoid disubstitution, bis(*O*-trimethylsilyl)uracil (*21a*) was reacted with 2-bromoethyl acetate to give 1-(2-hydroxyethyl)uracil (*18a*) in high yield [56] (Scheme 7).

The chloroethyl derivative was prepared by chlorination of the hydroxyethyl derivative with thionyl chloride in dioxane [9,113]. Another method that was used to prepare 1-(2-chloroethyl)uracil (*19a*) is the reaction of bis(*O*-trimethylsilyl)uracil (*21a*) with 1,2-dichloroethane, but in low yield (9%) [138] (Scheme 7). Recently,

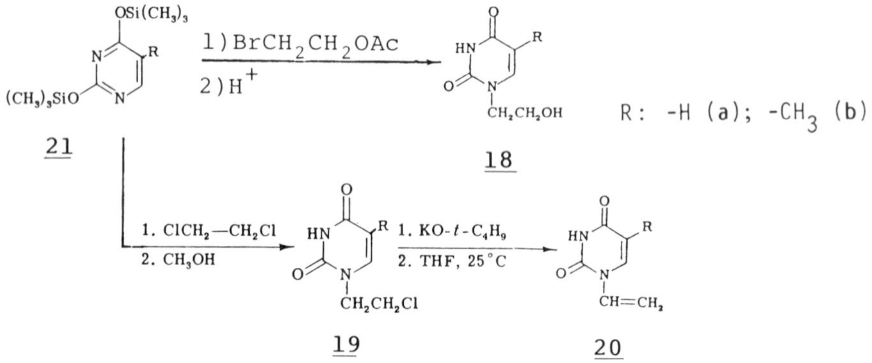

Scheme 7

a convenient method was found to prepare 2-bromoethyluracil (*26*) starting with acrylonitrile with ethanolamine (Scheme 8) [79]. Cyanoethylation of ethanolamine followed by reaction with potassium

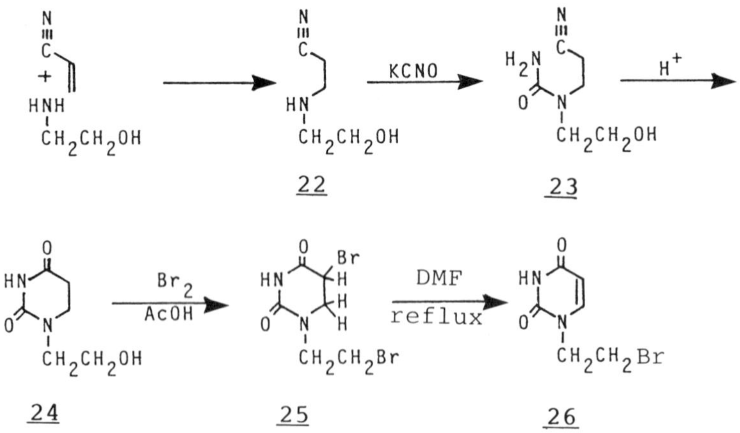

Scheme 8

cyanate gave *N*-cyanoethyl-*N*-hydroxyethylurea (*23*). Cyclization of this compound in aqueous methanolic hydrochloric acid gave 1-(2-hydroxyethyl-5,6-dihydrouracil (*24*). Then bromination in acetic acid, and dehydrobromination, gave 1-(2-bromoethyl)uracil (*26*).

1-Vinyluracil (*20a*) was prepared by dehydrochlorination of 1-(2-chloroethyl)uracil (*19a*) with excess of potassium *t*-butoxide or sodium methoxide in dioxane [9,13]. In these reactions, however, a bicyclic compound is easily formed, especially in the presence of a strong base. For the preparation of 1-vinyluracil, a cyclization reaction of the *N*-vinylurea derivative (*27*) [138] and a vinylation of the silylated uracil with vinyl acetate [140] were reported. These methods, however, need chromatographic purification in the final stage (Scheme 9).

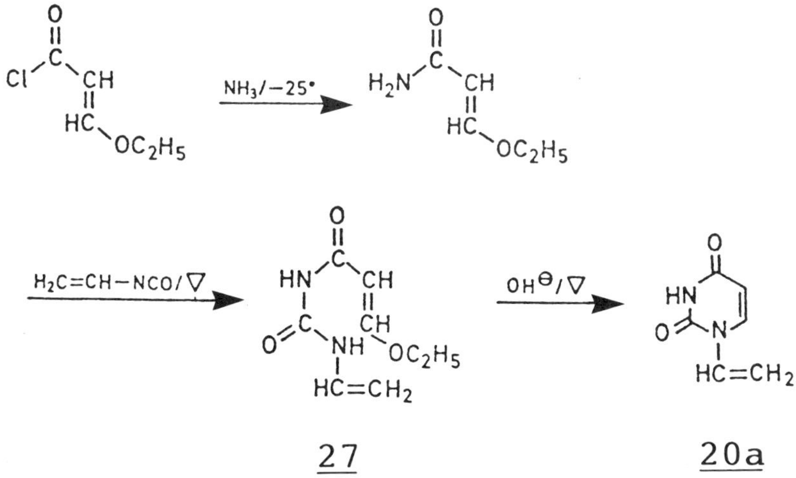

Scheme 9

1-(2-Hydroxyethyl)uracil (*18a*) [56]. Treatment of hexamethyl disilazane (200 mL) with uracil (45 g) in a small amount of dimethylformamide (5 mL) gives 2,4-bis(trimethylsilyloxy)pyrimidine (*22a*) quantitatively; 102 g (99%), bp 116°C/12 mm [201]. This compound is stirred with 2-bromoethyl acetate (95 g) for 10 days at 85°C. Excess methanol (500 mL) is added to the reaction mixture to decompose the silylated compound. The unreacted uracil is filtered off and the filtrate is evaporated to dryness. The residue is recrystallized from ethanol to give the product as colorless needles; yield 48 g (60%), mp 137 to 138°C.

1-(2-Hydroxyethyl)thymine (18b) [48]. A similar method used for the uracil derivative is applied to thymine to give the compound *18b* in 80% yield, mp 179 to 181°C.

1-(2-Chloroethyl)uracil (*19a*) [9]. To a solution of 1-(2-hydroxyethyl)uracil (*18a*) (7.8 g) and a amount of pyridine (1 mL) in dioxane (240 mL), thionyl chloride (16 mL) is added. The mixture is refluxed for 35 min and the solvent is then evaporated to dryness under reduced pressure. Recrystallization of the residue from ethanol gives colorless plates; yield 7.9 g (90%), mp 162 to 163°C.

1-(2-Chloroethyl)thymine (*19b*) [9]. The same conditions for obtaining the chloroethyluracil is used; yield 94%, mp 203 to 205°C.

1-(2-Bromoethyl)uracil (*25*) [79]. Acrylonitrile (55 mL) is added to ethanolamine (42 mL) at 0°C, and the mixture is heated to reflux for 30 min. The viscous oily product obtained is dissolved in 3 N HCl (270 mL); this solution is added dropwise to an aqueous solution (200 mL) of potassium cyanate (70 g). After stirring for 12 h at room temperature, the solvent is evaporated to give an oily residue (*23*). The residue is dissolved in methanol (600 mL), and the precipitated KCl is removed by filtration. To the filtrate is added aqueous methanolic hydrochloric acid (100 mL of concentrated HCl, 300 mL of water, 300 mL of methanol), and refluxed for 3 h. The solvent is then evaporated to dryness. The product is extracted from the residue with hot ethanol, and the filtrate is cooled to give 1-(2-hydroxyethyl)-5,6-dihydrouracil (*24*) as colorless crystals; yield 83 g (85%), mp 143 to 145°C.

The compound *24* (32 g) in acetic acid (50 mL) is refluxed while bromine (10 mL) is added dropwise. After the evolution of HBr gas has ceased, the solvent is removed under reduced pressure. Addition of ethyl acetate to the residue gives crystals of 1-(2-bromoethyl)-5-bromo-5,6-dihydrouracil (*25*) in 60% yield (28 g); mp 200 to 202°C. The product is dissolved in dimethylformamide and is refluxed for 1 h. After evaporation of the solvent, the residue is recrystallized from ethanol to give compound *26*; yield 14 g (70%), mp 152 to 153°C.

1-Vinyluracil (*20a*) [37]. Methanol (10 mL) is added dropwise to a solution of sodium hydride (5 g) and 1-(2-chloroethyl)uracil (5 g) in dioxane (350 mL). The reaction mixture is stirred for 7 h at room temperature, and then the solvent is evaporated under reduced pressure. The residue is recrystallized from ethanol to give colorless needles; yield 2.8 g (71%), mp 180 to 181.5°C.

1-Vinylthymine (*20b*) [9]. The reaction of 1-(2-chloroethyl)thymine with sodium methoxide under the same conditions as those used for the uracil derivative gives compound *20b*; yield 47%, mp 205 to 207°C.

Various N-vinyl monomers of pyrimidine derivatives are prepared by the vinyl interchange reaction with vinyl acetate, catalyzed by mercuric acetate and sulfuric acid: 1-vinyl-3-methyluracil (*28*) [113, 141], 1-vinyl-6-methyluracil (*29*) [142], 1-vinyl-2-ethoxy-4-pyrimidinone (*30*) [112], and 1-vinyl-4-ethoxy-2-pyrimidinone (*31*) [113].

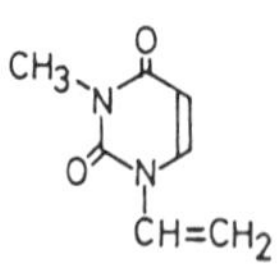

28

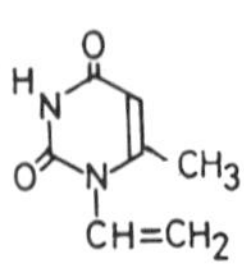

29

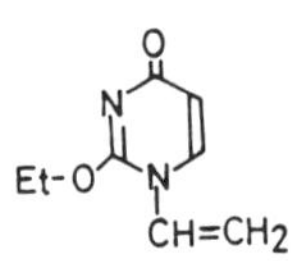

30

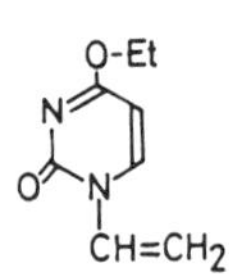

31

1-Vinyl-3-methyluracil (*28*) [113]. This compound is prepared from 3-methyluracil [202] by the same procedure as that used for 9-vinyl-6-chloropurine (*8*); mp 95 to 97°C.

1-Vinyl-6-methyluracil (*29*) [142]. 6-Methyluracil (8 g) is heated at 145 to 150°C with hexamethyl disilazane (20 mL) for 24 h to give 2,4-bis(trimethylsilyl)-6-methyluracil; 4.5 g (27%), bp 97°C/27 mm, mp 72.5 to 74°C. This compound is reacted with vinyl acetate, catalyzed by mercuric acetate and sulfuric acid to yield 2.2 g (73%), mp 198.5 to 200.5°C.

1-Vinyl-2-ethoxy-4-pyrimidinine (*30*) [112]. This compound is prepared from 2-ethoxy-4-pyrimidinone [203] by the vinyl interchange reaction used for 9-vinyl-6-chloropurine (*8*); yield 10%, mp 97 to 99°C.

1-Vinyl-4-ethoxy-2-pyrimidinone (*31*) [113]. This compound is also prepared by a vinyl interchange reaction from 4-ethoxy-2-pyrimidinone [203]; yield 55%, mp 77 to 78°C.

Polymerization and copolymerization of these N-vinyl monomers of pyrimidine derivatives were carried out by free-radical initiators [11, 147]. In the case of the free-radical polymerization of 1-vinyluracil, the formation of substituted dihydrouracil rings occurred via a cyclopolymerization mechanism [141]. However, the solid-state polymerization induced by γ-radiation of the monomer in high concentration and at low temperatures excluded the cyclopolymerization completely [142].

Poly-1-vinyluracil (*32a*) was also prepared by acid hydrolysis of poly-1-vinyl-2-ethoxy-4-pyrimidinone (*33*) [115] or poly-1-vinyl-4-ethoxy-2-pyrimidinone (*34*) [114]. Poly-1-vinylcytosine (*35*) was prepared from poly-1-vinyl-4-ethoxy-2-pyrimidinone (*34*) by the reaction with ammonia [136] (Scheme 10). Alternatively, poly-1-

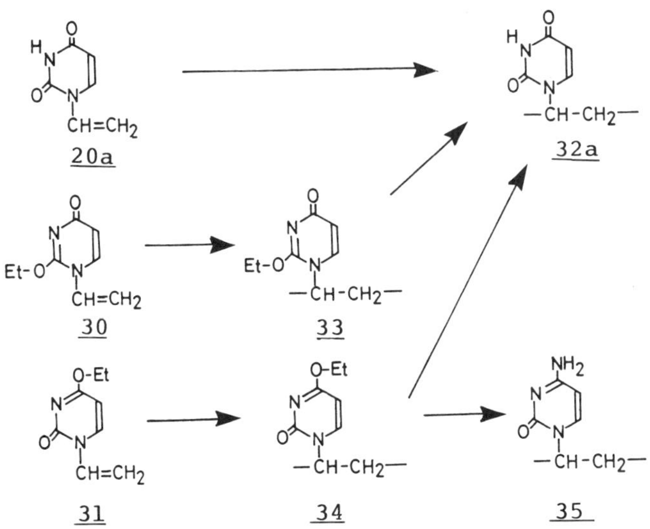

Scheme 10

vinylcytosine (*35*) could be prepared by the polymerization of N^4-acetyl-1-vinylcytosine (*36*) followed by the hydrolysis of the polymer *37*. Preparation of the monomer *36* has been reported, but the polymerization of the monomer is not known [140. Poly-1-vinylthymine (*32b*) can be obtained by the polymerization of 1-vinylthymine monomer (*20b*) (Scheme 11).

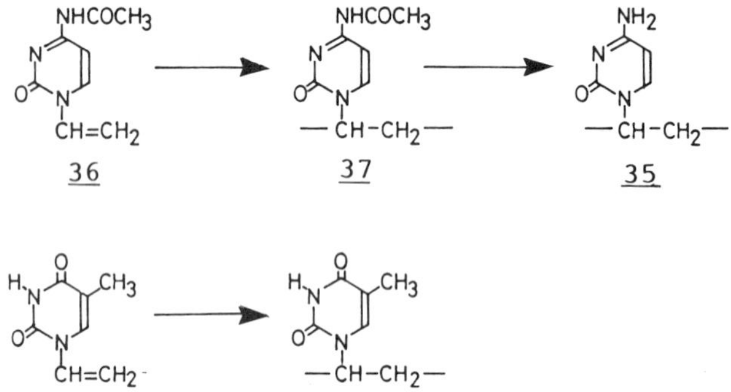

36 37 35

20b 32b

Scheme 11

Poly-1-vinyluracil (32a) from polymer 34 [115]. Monomer *31* is polymerized in dry tetrahydrofuran, initiated by azobisisobutyronitrile at 60°C. After standing overnight at 60°, the resulting cake is ground to a powder and added to dry tetrahydrofuran. The insoluble material is separated by centrifugation and dried to obtain polymer *34*.

Polymer *34* (600 mg) is dissolved in 120 mL of formic acid at room temperature. To this solution is added 40 mL HCl in 80 mL of formic acid dropwise with stirring to prevent precipitation: After 24 h at room temperature the solution is evaporated in vacuo and the residue is dissolved in 30 mL of 1 N NaOH. After centrifugation, poly-1-vinyluracil (*32a*) is precipitated from the supernatant by addition of 1 N HCl to pH 2. The precipitated polymer is collected by centrifugation, washed with water, and dried.

Poly-1-vinylcytosine (35) from polymer 34 [136]. Poly-1-vinyl-4-ethoxy-2-pyrimidinone (*34*) (200 mg) is suspended in 25 mL of methanol and the solution saturated at 0°C with ammonia in a stainless-steel autoclave. The contents is mixed by vibration for 1 h and then heated to 110°C for 70 h. The mixture is evaporated to dryness at 60°C, 0.01 mm overnight, yielding 160 mg of a slightly yellow powder. The degree of conversion is calculated from elementary analyses (87 to 94%).

The intramolecular interaction between nucleic acid bases in the N-vinyl polymers was studied by emission and ultraviolet (UV) spectroscopic studies [29,139,145,146]. The formation of the specific base-paired complexes between the N-vinyl polymers and with polynucleotides [11,16,19,28,118-122,124-134,137], and the biological properties [116,178-193,148,149] were studied.

4.3 POLYACRYLATE AND POLYMETHACRYLATE CONTAINING NUCLEIC ACID BASES

4.3.1 Monomer Synthesis

Methacrylate derivatives of nucleic acid bases were at first prepared by Scheme 12. The hydroxyethyl derivatives of nucleic acid bases

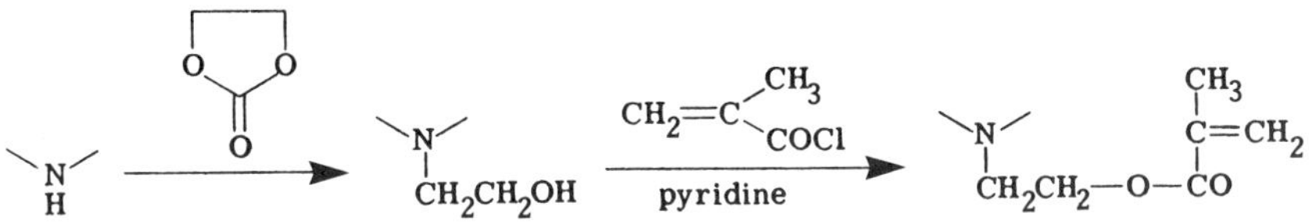

Scheme 12

were treated with methacryloyl chloride in the presence of pyridine to give the methacrylate monomer [10]. In the case of hypoxanthine derivative, the reaction of 9-hydroxyethylhypoxanthine with methacryloyl chloride gives monomer *38* in 30% yield.

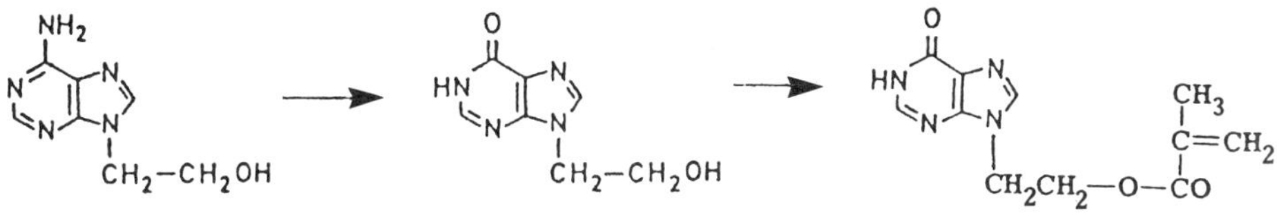

1 38

9-(2-Methacryloyloxyethyl)hypoxanthine (*38*) [12]. To a solution of 9-(2-hydroxyethyl)hypoxanthine (0.18 g) which is prepared from 9-(2-hydroxyethyl)adenine (*1*), in dimethylformamide (10 mL) with a small amount of pyridine (0.1 g), methacryloyl chloride (0.11 g) is added gradually at -50°C. After stirring for 2 h at room temperature, the solvent is removed under reduced pressure. Recrystallization of the residue from acetone gives colorless flakes in 30% yield, mp 192 to 195°C.

The procedure above is somewhat troublesome for both purine and pyrimidine and the total yield of the methacrylate monomers is generally low. In the case of the adenine derivative, the 6-amino group reacts with methacryloyl chloride. A simpler and more convenient method for preparing these methacrylate monomers with substantially higher yields involved reacting the purine and pyrimidine bases with 2-bromoethyl methacrylate (*39*) [35].

For the preparation of the adenine monomer a sodium salt of adenine (*40*) was allowed to react with 2-bromoethylmethacrylate in dimethylformamide (Scheme 13), affording 9-(2-methacryloyloxyethyl)adenine (*41a*) in 70% yield. Similarly, 7-(2-methacryloyloxyethyl)theophylline (*42*) was also prepared.

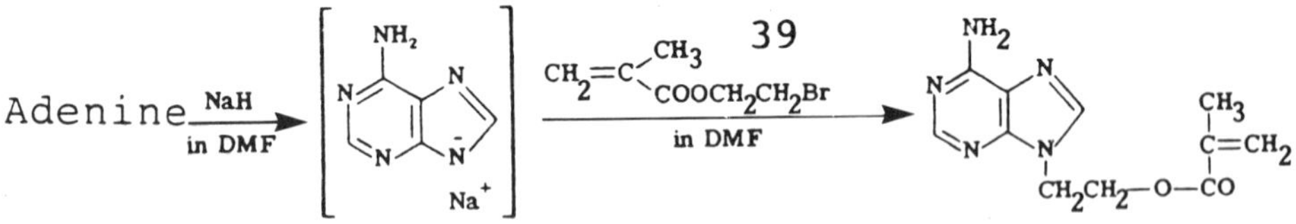

40

41a

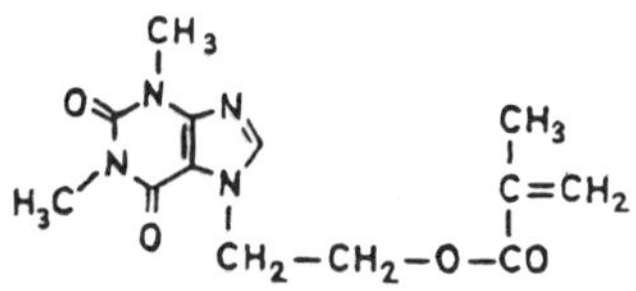

42

Scheme 13

For the pyrimidine derivatives, trimethylsilylated pyrimidine bases (*22*) are reacted with 2-bromoethyl methacrylate (*39*) (Scheme 14). The trimethylsilyl derivative of thymine is reacted with *39* at 60°C to afford 1-(2-methacryloyloxyethyl)thymine (*43c*) in 63% yield. The same reaction for uracil gives 1-(2-methacryloyloxyethyl)uracil (*43a*), but the yield is unexpectedly low (15%).

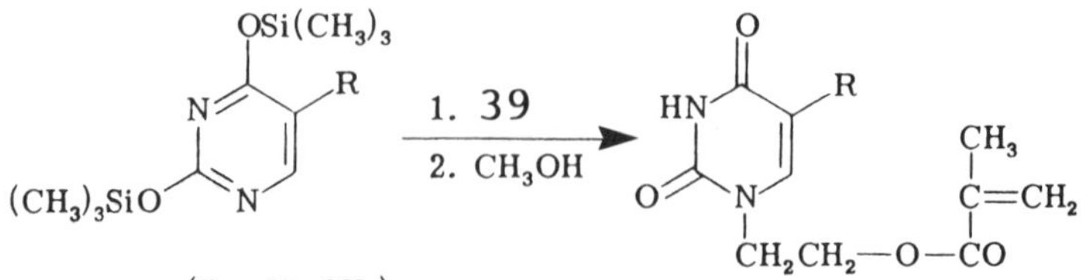

(R = H, CH_3)

21

43

Scheme 14

9-(2-Methacryloyloxyethyl)adenine (*41a*) [35]. A suspension of adenine (1.35 g, 0.001 mol) and sodium hydride (0.48 g, 0.001 mol) in dimethylformamide (100 mL) is stirred at 60°C for 2 h to give sodium salt of adenine (*40*) as a precipitate. To this suspension, 2-bromoethyl methacrylate (*39*) (2.9 g, 0.015 mol) is added, and the mixture is stirred for 24 h at 40°C. The solvent is evaporated under reduced pressure and recrystallization of the residue from ethanol and then water gives colorless needles in 70% yield (1.73 g), mp 201 to 202°C.

1-(2-Methacryloyloxyethyl)thymine (*43c*) [35]. 5-Methyl-bis(trimethylsilyloxy)pyrimidine (2.7 g, 0.01 mol), which is prepared by the reaction of thymine with hexamethyl disilazane, and 2-bromoethyl methacrylate (3.86 g, 0.02 mol) are stirred for 10 days at 60°C. Methanol is then added to the reaction mixture to hydrolyze the silylated compound, and the precipitate is filtered. The filtrate is evaporated and the residue is recrystallized from water to give monomer *43c* as colorless needles; yield 1.5 g (63%), mp 170 to 172°C.

For the preparation of the acrylate derivatives of uracil (*43b*) and thymine (*43d*), the reaction with 2-bromoethyl acrylate or with acryloyl chloride gave only the starting hydroxyethyl derivative because the product is easily hydrolyzed. These acrylate derivatives of uracil and thymine, however, are successfully prepared by the reaction with acrylic anhydride without solvent at 70°C [48]. Also, for the methacrylate derivative of uracil (*43a*), the reaction of 1-(2-hydroxyethyl)uracil (*18a*) and methacrylic anhydride gave excellent results.

1-(2-Methacryloyloxyethyl)uracil (*43a*) [78]. Methacrylic anhydride (2.6 g, 0.02 mol) is added to a solution of 1-(2-hydroxyethyl)uracil (*18a*) (1.7 g, 0.01 mol) in pyridine (10 mL) at 0°C. The mixture is stirred for 24 h at room temperature, then the solvent is evaporated under reduced pressure. The residue is washed with diethyl ether to remove methacrylic anhydride. Recrystallization of the residue from benzene gives the methacrylate monomer *43a* as colorless needles; yield 80%, mp 111 to 112°C.

1-(2-Acryloyloxyethyl)thymine (*43d*) [48]. Acrylic anhydride (2.6 g, 0.02 mol) is added to 1-(2-hydroxyethyl)thymine (*18b*) (1.8 g, 0.01 mol) and stirred for 3 h at 70°C to give a clear solution. The reaction mixture is cooled to 0°C to precipitate the product. The product is filtered and washed thoroughly with diethyl ether and recrystallized from ethanol to give colorless needles; yield 1.9 g (87%), mp 145 to 146°C.

Various 5-substituted uracil derivatives of methacrylate have been prepared from the 5-bromouracil derivative [79] (Scheme 15). Starting with acrylonitrile and ethanolamine, 1-(2-hydroxyethyl)-5,6-dihydrouracil (*24*) was prepared and treated with excess bromine in acetic acid to afford 1-(2-bromoethyl)-5-bromouracil (*44*), followed by the reaction with sodium acetate to give 1-(2-acetoxyethyl)-5-

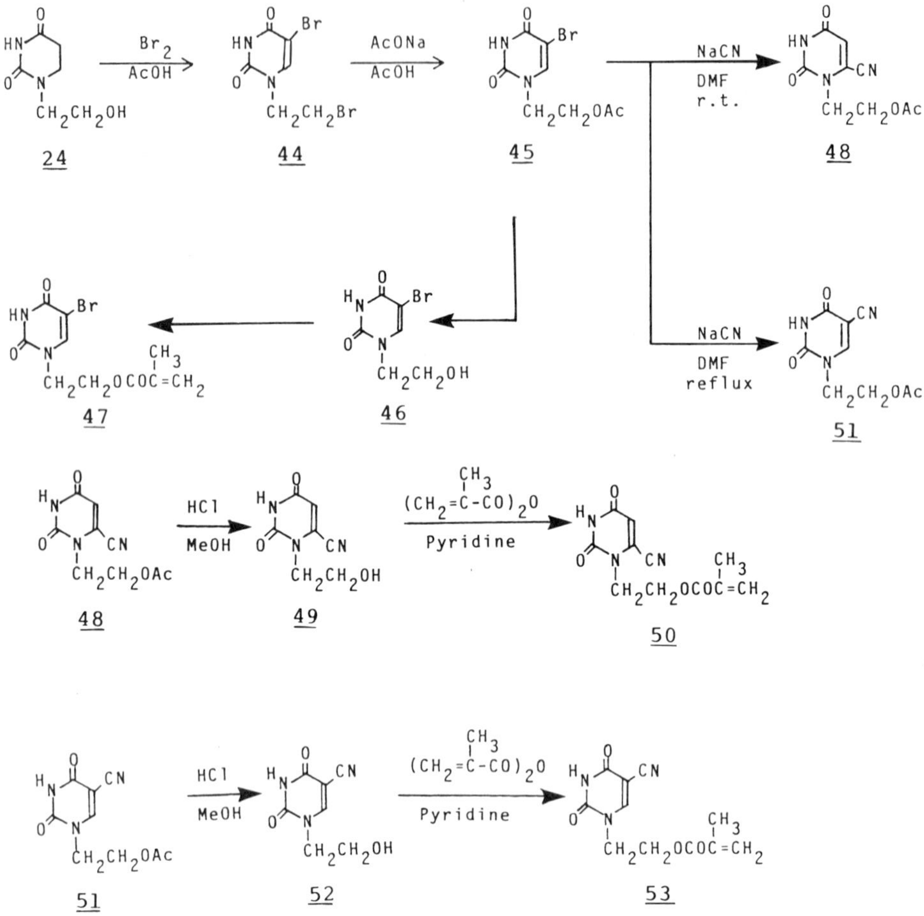

Scheme 15

bromouracil (*45*). Hydrolysis of compound *45* gave 1-(2-hydroxyethyl)-5-bromouracil (*46*) which was reacted with methacrylic anhydride to give 1-(2-methacryloyl)-5-bromouracil (*47*). From compound *45*, 5-cyano- (*50*) and 6-cyano- (*53*) derivatives are prepared by reaction with sodium cyanide, as shown in Scheme 15.

1-(2-Hydroxyethyl)-5-bromouracil (*46*) [79]. Bromine (20 mL, 0.4 mol) is added dropwise to a solution of 1-(2-hydroxyethyl)-5,6-dihydrouracil (*24*) (32 g, 0.2 mol) in acetic acid (50 mL) with reflux, until the color of bromine does not disappear. The solvent is removed by distillation. Recrystallization of the residue from ethanol gives 1-(2-bromoethyl)-5-bromouracil (*44*) as colorless needles almost quantitatively. Compound *44* (12 g, 0.04 mol) is dissolved in a solution of acetic acid (50 mL) and sodium acetate (6.6 g, 0.08 mol) and the mixture is refluxed for 5 h. The solvent is evaporated and the residue extracted with ethyl acetate. Recrystallization from ethyl acetate gives 1-(2-acetoxyethyl)-5-bromouracil (*45*) in 83% yield (9.2 g), mp 174 to 178°C.

Hydrolysis of the acetate derivative (*45*) by hydrochloric acid or aqueous sodium hydroxide gives 1-(2-hydroxyethyl)-5-bromouracil (*46*) almost quantitatively.

1-(2-Hydroxyethyl)-6-cyanouracil (*49*) [79]. 1-(2-Acetoxyethyl)-5-bromouracil (*45*) (17.4 g, 0.063 mol) and sodium cyanide (3.3 g, 0.067 mol) in dimethylformamide (100 mL) are stirred for 24 h at room temperature. The solvent is removed under reduced pressure, and the residue is recrystallized from ethanol to give 1-(2-acetoxyethyl)-5-cyanouracil (*48*) as colorless needles; yield 8.2 g (59%), mp 206 to 209°C.

This compound (*48*) is hydrolyzed to give 1-(2-hydroxyethyl)-6-cyanouracil (*49*), which is reacted with methacrylic anhydride to give 1-(2-methacryloyloxyethyl)-6-cyanouracil (*50*); yield 83%, mp 192 to 194°C.

1-(2-Hydroxyethyl)-5-cyanouracil (*52*) [79]. Compound *45* is is reacted with 2 molar equivalents of NaCN in dimethylformamide for 24 h at 100°C. After the solvent is removed, the residue is recrystallized from ethanol to give 1-(2-acetoxyethyl)-5-cyanouracil (*51*). Hydrolysis of *51*, followed by reaction with methacrylic anhydride, gives the methacrylate containing 5-cyanouracil (*53*).

4.3.2 Polymerization and Polymer Reactions

Poly-9-(2-methacryloyloxyethyl)adenine (*54a*), poly-1-(2-methacryloyloxyethyl)uracil (*55a*) and -thymine (*55c*) as well as the corresponding polyacrylates (*54b*, *55b*, and *55d*) were prepared by free-radical polymerization of the corresponding monomers [11] (Scheme 16). Polymers containing adenine are soluble in dimethyl sulfoxide, ethylene glycol, and acidic aqueous solution (below pH 3), while the polymers having uracil and thymine moieties are soluble in dimethyl sulfoxide, dimethylformamide, and alkaline aqueous solution (above pH 10). The solubility of these polymers in organic solvents is increased by alkylation on nucleic acid bases [78]. By methylation of uracil or thymine, the polymers containing these pyrimidine bases are soluble in chloroform and chlorobenzene [78,82].

Alternatively, the polymethacrylates containing nucleic acid bases can be prepared by the reaction of poly-2-bromoethylmethacrylate (*56*) with sodium salts of nucleic acid bases. By this procedure, stereoregular polymers containing nucleic acid bases were obtained

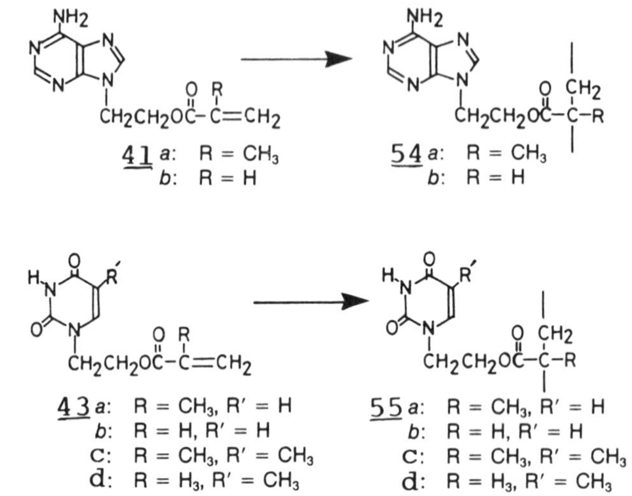

Scheme 16

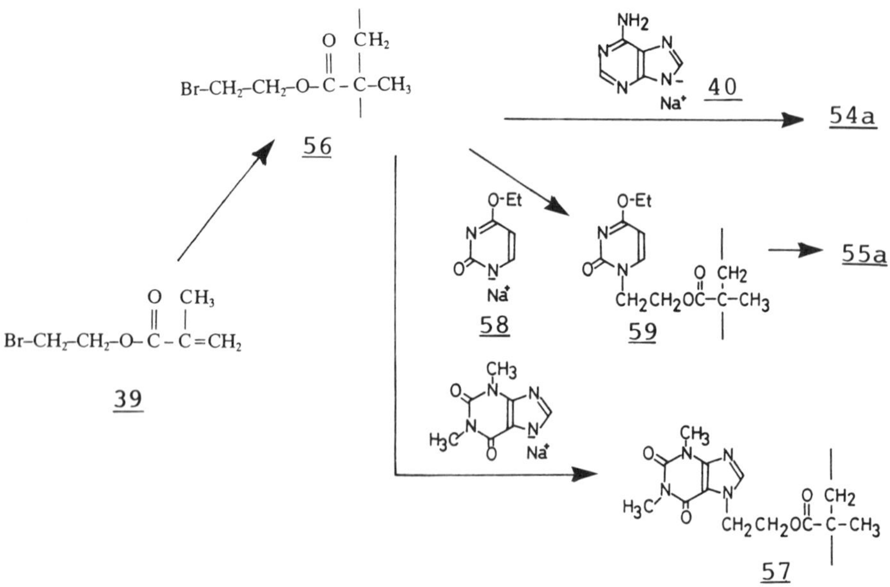

Scheme 17

(Scheme 17) [34]. The synthesis of stereoregular polymers having pendant nucleic acid base starts with 2-bromoethyl methacrylate (*39*), which was polymerized with anionic initiators at -78°C to yield poly-2-bromoethyl methacrylate (*56*) of high isotacticity and syndiotacticity (Table 1). These stereoregular polymers were then reacted with nucleic acid bases to give high yields of stereoregular polymers containing the corresponding base residues.

The reaction of the polymer *56* with sodium salt of adenine (*40*) or theophylline affords the polymers *54a* or *57*. In the case of uracil, the sodium salt of 4-ethoxy-2-pyrimidone (*58*) was reacted with polymer *56*, followed by acid hydrolysis of the resulting polymer *59* to give the polymer *55a*. The results of the polymer reactions are listed in Table 2.

Stereoregular poly-9-(2-methacryloyloxyethyl)adenine (*54a*) [34]. The sodium salt of adenine (*40*) (3 mmol) and the stereoregular poly-2-bromoethyl methacrylate (*56*) (1 mmol) in Table 1 in dimethyl sulfoxide (30 mL) are stirred for 1 week at room temperature. Then the reaction mixture is poured into 0.1 N sodium hydroxide aqueous solution (300 mL) to give the polymer *54a* as a precipitate. The polymer contains 90% adenine units as determined by elemental analysis.

Stereoregular poly-1-(2-methacryloyloxyethyl)uracil (*55a*) [34]. The sodium salt of 4-ethoxy-2-pyrimidinone (*58*) [113] (3 mmol) and the stereoregular polymer *56* (1 mmol) in dimethylformamide (30 mL) are stirred for 1 week at room temperature. Then the reaction mixture is poured into excess methanol to give poly-1-(2-methacryloyloxyethyl)-4-ethoxy-2-pyrimidinone (*59*) as a precipitate. The polymer (200 mg) is dissolved in 9 N hydrochloric acid (80 mL) and stirred for 2 days

Table 1. Anionic Polymerization of β-Bromoethyl Methacrylate

Initiator	Solvent	Time (h)	Conversion (%)	Tacticity[a] (%) I	H	S	$M_n \times 10^{-4}$ [b]
n-C_4H_9Li	Toluene	3	80	75	15	10	2.9
C_6H_5MgBr	Toluene	48	70	81	11	8	3.1
$LiAlH_4$	Ether	0.25	60	88	7	5	1.6[c]
$(C_2H_5)_2AlNPh_2$	Toluene	72	90	4	12	84	1.9

[a]Measured by NMR (100 MHz, at 60°C in $CDCl_3$).
[b]Measured with membrane osmometer (at 37°C in toluene).
[c]Measured with vapor pressure osmometer (at 57°C in DMF).

Table 2. Results of the Polymer Reactions of *56* with the Sodium Salts of Nucleic Acid Bases

Tacticity of *31*	Nucleic acid base in the polymer	Conversion[a] (%)	λ_{max} of polymers[b] (nm)
Isotactic	Adenine	91	267
Syndiotactic	Adenine	90	267
Isotactic	Uracil (OC_2H_5)	88	265[c]
Syndiotactic	Uracil (OC_2H_5)	86	265[c]
Isotactic	Theophylline	90	276[d]
Syndiotactic	Theophylline	95	276[d]

[a]Measured by elementary analysis (from N value).
[b]Measured in DMSO solution.
[c]Values obtained for poly-MAOU.
[d]Measured in chloroform solution.

at room temperature. Then the solution is diluted with water (300 mL) and neutralized with aqueous sodium hydroxide solution under cooling to give polymer *55a*. This polymer contains 88% uracil units from elemental analysis.

Polymethacrylate and polyacrylate derivatives of uracil and thymine (*55a* to *55d*) can also be prepared by the reaction of the cyclic derivatives of uracil and thymine (*60*) with polymethacrylic or polyacrylic acids [54] (Scheme 18). Cyclic derivatives of uracil and thymine (*60*) were obtained in high yield from the corresponding 1-(2-chloroethyl)pyrimidine (*19*) by dehydrochlorination with 1,8-diazabicyclo(5,4,0)undec-7-ene (DBU) in dimethylformamide solution [52,56]. The structure of the cyclic compound was confirmed by spectroscopy. Recently, it was found that colorless needles of the cyclic compound precipitated on the addition of an equimolar amount of KOH in hot methanol into a hot solution of 1-(2-bromoethyl)uracil (*26*) in methanol [81].

Reactions of the cyclic compound (*60*) with various kinds of organic reagents at room temperature gave a series of pyrimidine derivatives without any catalyst [52]. One of the typical reactions involves esterification with acetic acid in bulk to afford the corresponding acetate. When the reaction is applied to polymeric acids, such as poly(acrylic acid), poly(methacrylic acid), and poly(L-glutamic acid), the polymers containing uracil or thymine are obtained in high yield [54,81]. Copolymers such as poly(acrylic acid-co-1-(2-acryloyloxyethyl)uracil) are also obtained successfully [54].

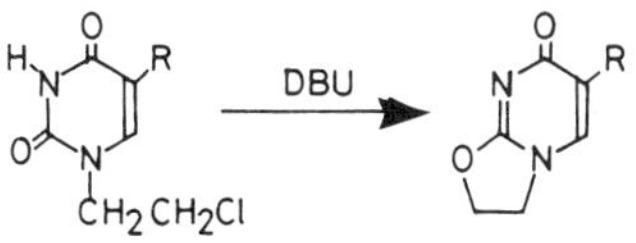

19 a: R = H
b: R = CH_3

60 a: R = H
b: R = CH_3
c: R = Br

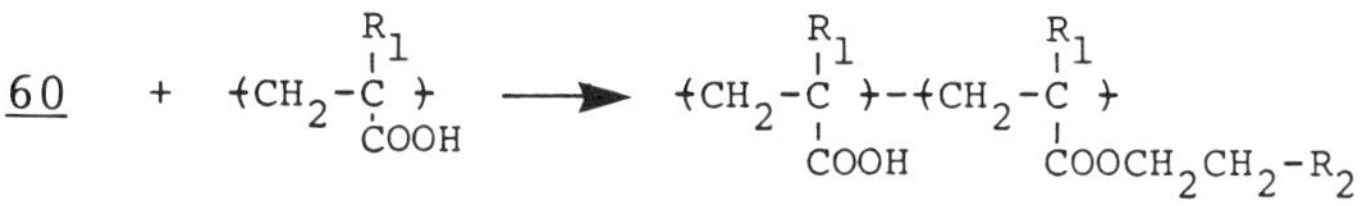

R_1=H, CH_3; R_2= Uracil, Thymine
61

Scheme 18

1, 2-O-Ethanouracil; oxazolidino(2, 3, a)-γ-pyrimidone (60a) [52]. To a solution of 1-(2-chloroethyl)uracil (*19a*) (3.5 g, 0.02 mol) in dimethylformamide (10 mL), 1,8-diazabicyclo[5,4,0]undec-7-ene (3 ml, 0.02 mol) is added at 0°C. The product precipitates within a few minutes, and the reaction mixture is stirred overnight at room temperature to complete the reaction. After filtration, the product is recrystallized from ethanol to give colorless needles in 80% yield (2.2 g), mp > 300°C.

1, 2-Ethanothymine (60b) [52]. This compound is prepared from 1-(2-chloroethyl)thymine (*19b*) by the same procedure as that used for ethanouracil; yield 88%, mp > 300°C.

1, 2-O-Ethano-5-bromouracil (60c) [81]. To a solution of 1-(2-bromoethyl)-5-bromouracil (*44*) in hot methanol, equimolar KOH in hot methanol is added. By cooling the solution, 1, 2-O-ethano-5-bromouracil precipitates as colorless needles; yield 90%, mp 300°C.

Poly-1-(2-methacryloyloxyethyl)thymine (55c) from polyacrylic acid [52]. To a suspension of polyacrylic acid (1.44 g, 20 mmol) in dimethylformamide (30 mL), 1, 2-O-ethanouracil (*60a*) (3.4 g, 20 mmol) is added, and the suspension is stirred for 10 days at 80°C. The suspension becomes a clear solution after few days. The solution is concentrated under reduced pressure, and excess acetone is added to the solution to give the polymer *55c* as a precipitate. Reprecipitation from dimethylformamide-acetone gives the polymer in quantitative yield. The polymer contains almost 100% thymine units.

4.3.3 Properties of the Polymers

Detailed spectroscopic and NMR studies on the behavior of the monomers in various solvents as well as the tacticity of the polymers were made [33]. The rate of polymerization was found to be affected by the type of bases and solvents used, whereas the stereoregularity of the polymers appears not to be affected by these parameters. Activation parameters, $\Delta\Delta H^{\ddagger}$ and $\Delta\Delta S^{\ddagger}$ values for the polymerization of different methacryloyloxy-type monomers in various solvents were estimated. The results reveal that for the polymerization of the monomers having adenine as the side chain in dimethyl sulfoxide, syndiotactic placement appears to be favored by the additional enthalpy of activation required for isotactic placement [38].

To find out how the specific base-base interactions occurring between complementary nucleic acid bases can be utilized for free-radical polymerization systems, radical copolymerization of methacrylate-type monomers bearing complementary nucleic acid bases in different solvents was carried out. The copolymerization was found to proceed predominantly under the influence of base-base pairing between adenine and uracil or thymine rings [27]. The rate acceleration is caused by association of growing polymer end units with the monomer and the reactivity of the complexed monomer is low [42].

Complex formation of stereoregular methacrylate-type polymers containing pendant nucleic acid bases was studied in a dimethyl sulfoxide-ethylene glycol (3:2, v/v) mixed solvent system [44]. The formation of polymer complexes between nucleic acid analogs was found to be influenced by the solvent and temperature. The stereoregularity of the polymer is also important to the complex formation. The ability to form the complex of atactic polymers is higher than with stereoregular polymers and is dependent on the stereoregularity of the polymer containing uracil (Table 3).

Template polymerization is known in replicative biopolymer synthesis. It seems to be one of the most attractive studies if template polymerization can be followed by suitable monomer-polymer pair formation with complementary nucleic acid bases. The free-radical template polymerization of the methacrylate type monomers in the presence of the template polymer with complementary nucleic acid

Table 3. Hypochromicity of Poly-MAOA-Poly-MAOU[a] (1:1) (%)

	Poly-MAOA		
Poly-MAOU	Atactic	isotactic	Syndiotactic
Atactic	21	10	6
Isotactic	18	5	11
Syndiotactic	0	2	2

[a]Poly-MAOA; *54a*; poly-MAOU; *55a*.

bases was studied [31,43]. The polymerization of the methacrylate containing adenine was found to be accelerated by the addition of increasing amounts of template polymer. This fact suggests that complex formation between a growing chain and a template polymer causes acceleration of the polymerization in the presence of the template polymer by repressing the termination reaction or stabilizing the transition state.

The formation of thymine photodimers by UV irradiation onto nucleic acid is well known and the photodimerization of nucleic acid, its model compounds, and pyrimidine bases has been reported by many groups, except the work on the synthetic polymers containing thymine bases, which has hitherto been unknown. Detailed studies have been made on the photodimerization of thymine or uracil bases for the polymethacrylate-containing pyrimidine bases [46,48,61,62,63,65, 79,82]. The photopolymers are useful as image- and information-processing materials as well as photoresists that can be used for integrated-circuit and large-scale integration (Figure 3).

4.3.4 Polymers Containing Nucleosides

In connection with the synthesis of vinyl and related monomers containing nucleic acid bases, those containing nucleoside structures are also of interest. Acrylates [160-162] and methacrylate [195-197] derivatives of nucleosides such as adenosine (*62*), uridine (*63*), thymidine (*64*), guanosine (*65*), and cytosine (*66*) were prepared. The reaction of acrylic or methacrylic anhydrides with protected nucleoside in the presence of pyridine was found to be an effective procedure. The methacrylate derivative of glucopyranosylthymine was prepared by the reaction with methacryloyl chloride [194].

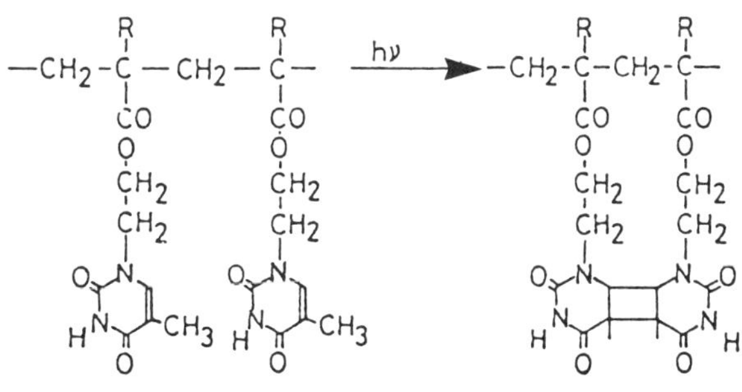

Figure 3. Photodipolymerization of poly-MAOT and poly-AOT.

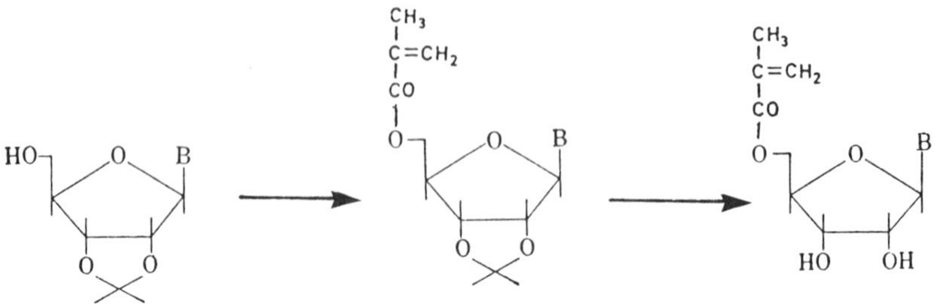

67 68

Base:

Adenine	(*62*)
Uracil	(*63*)
Thymine	(*64*)
Guanosine	(*65*)
Cytosine	(*66*)

Acrylate derivatives of nucleosides are homopolymerized, and copolymerized with acrylamide in aqueous solution using persulfate initiator [160]. The homopolymers are insoluble in water but the copolymers are soluble in water. The copolymers contain about 10% nucleoside units. Another way of obtaining polyacrylate having nucleoside is to react nucleosides with polyacrylic acid or its derivative. A convenient way of doing this is to react ribonucleosides with periodate and then to react the nucleoside dialdehydes so produced with polyacrylic acid hydrazide (Scheme 19). Stereoregular polymethacrylic acid is also used for the reaction. The polymers so formed contained about 10% base residue [166,176].

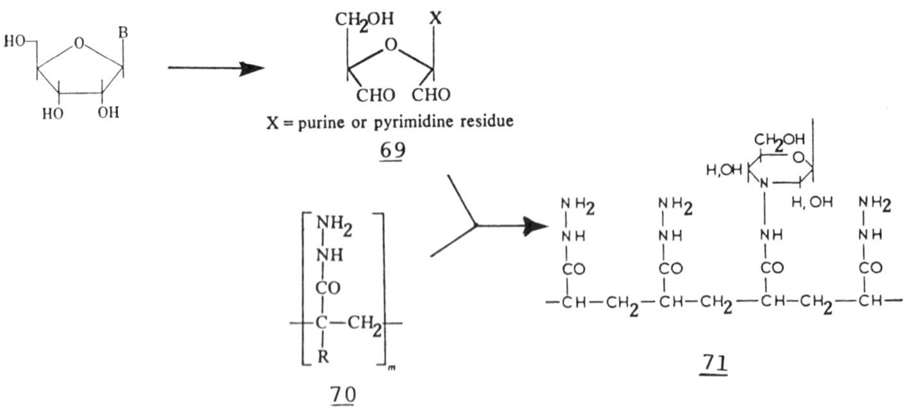

X = purine or pyrimidine residue

69

70

71

Scheme 19

5'-O-Acryloyladenosine (*62*) [167]. Acrylic anhydride (0.45 mL) is added slowly to a solution of 2',3'-*O*-isopropylideneadenosine (1.1 g) in pyridine (100 mL) and the solution is kept at room temperature for 18 h. The solvent is then removed in vacuo. The resultant syrup is dissolved in ethanol-chloroform (1:19) and fractionated on a column of silica gel in the same solvent, to give 5'-*O*-acryloyl-2',3'-*O*-isopropylideneadenosine as a white solid (350 mg, mp 124 to 125°C). This solid (170 mg) is dissolved in 98% formic acid (15 mL). Water (15 mL) is added and the solution is kept at room temperature for 40 h. The formic acid is removed by coevaporation with water until the pH of the solution increases to 4.5. The solution is then concentrated to dryness in vacuo to give 5'-*O*-acryloyladenosine (*62*) as a white solid (114 mg); mp 72 to 76°C.

5'-O-Acryloyluridine (*63*) [161]. 2',3'-*O*-Isopropylideneuridine (1 g, 3.5 mmol) is dissolved in dry dimethylformamide (20 mL) and cooled to -35°C. Acrylic anhydride (0.4 mL) and pyridine (0.8 mL) are added, and after 1 h at -35°C the mixture is set aside at room temperature for 24 h. The mixture is evaporated to a syrup under reduced pressure and hydrolyzed at 100°C for 2 h with 10% acetic acid (1200 mL). The acetic acid is removed by repeated coevaporation with water and the solution is then concentrated to a syrup. The syrup is fractionated by countercurrent distribution with an ethyl acetate-water mixture. The solution is evaporated to dryness, and the residue is recrystallized from ethanol-hexane to give 5'-*O*-acryloyluridine (290 mg, mp 130 to 132°C).

5'-O-Acryloylguanosine (*65*) [167]. To a solution of 2',3'-*O*-isopropylideneguanosine (2 g) in pyridine-dimethylformamide (80 mL-80 mL), acrylic anhydride (1.65 mL) is slowly added. After 18 h at room temperature the solvent is removed in vacuo at 40°C and the residue is dissolved in chloroform. Separation by chromatography on a column of silica gel in ethanol-chloroform (1:9) gives 5'-*O*-acryloyl-2',3'-*O*-isopropylideneguanosine (640 mg, mp 242 to 247°C). The protecting group is removed by stirring with 98% formic acid for 3 days to give 5'-*O*-acryloylguanosine (*65*) (363 mg, mp 166 to 168° dec.).

5'-O-Acryloylcytidine (*66*) [167]. 4-*N*-Acetyl-2',3'-*O*-isopropylidenecytidine (0.5 g) is dissolved in chloroform (50 mL), and finely ground anhydrous potassium carbonate is added. The suspension is stirred vigorously and after acrylic anhydride (0.5 mL) is added the stirring is continued at room temperature for 3 h. The suspension is then filtered and the filtrate is evaporated to dryness. The residue is dissolved in pyridine and *n*-hexane is added slowly. The first flocculent-like precipitate is filtered off and then more *n*-hexane is added until a slight turbidity is produced. Crystallization occurs slowly at room temperature. The product is filtered off, washed with *n*-hexane to give 4-N-acetyl-5'-*O*-acryloyl-2',3'-*O*-isopropylidenecytidine (0.35 g, mp 108 to 110°C). The blocking group of this compound is removed by stirring in 98% formic acid (24 mL) for 18 h, and then 0.1 N hydrochloric acid (20 mL) for 2 days, to give 5'-*O*-acryloylcytidine, mp 75 to 100°C.

Copolymerization of 5'-O-acryloyluridine with acrylamide [161]. 5'-*O*-Acryloyluridine (*63*) (20 mg) and acrylamide (47.7 mg) are dissolved in 1 M acetate buffer (10 mL, pH 5.2), a freshly prepared solution of ammonium persulfate (1%, 0.4 mL) is added, and the solution is maintained at 60° for 6 h. The solution is exhaustively dialyzed against distilled water and freeze dried to yield a copolymer (60 mg). The product is analyzed for uridine directly by measuring the UV absorption of the copolymer; 7 to 8 uridine residues per 100 total residues is obtained.

Ribonucleoside dialdehyde derivatives of polyacrylic acid (*71*) [166]. The ribonucleoside (0.26 mmol) is dissolved in water (70 mL), sodium metaperiodate (0.30 mmol) added, and the solution left at room temperature for 1 h. Polyacrylic acid hydrazide (156 mg), which is prepared from polyethyl acrylate by reaction with hydrazine hydrate for 10 h at 140°C, is dissolved in water. The two solutions are mixed and the reaction allowed to proceed at room temperature for 24 h. Then the reaction mixtures are exhaustively dialyzed and the solution inside the dialysis bags centrifuged to remove insoluble material. The polymers are stored in solution.

4.4 POLYACRYLAMIDE AND POLYMETHACRYLAMIDE CONTAINING NUCLEIC ACID BASES

4.4.1 Monomer Synthesis

Acrylamide and methacrylamide derivatives of nucleic acid bases are prepared by the reaction of the aminoethyl derivatives with acrylic or methacrylic anhydride (Scheme 20) [22,45,55,72]. The procedures

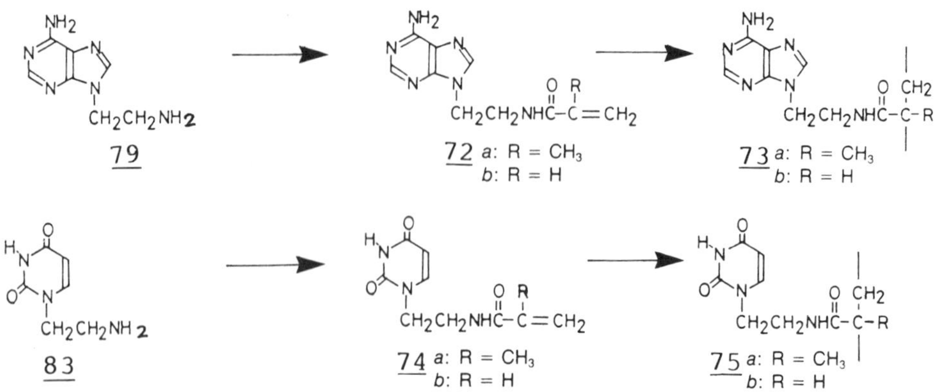

Scheme 20

for the preparation of the 2-aminoethyl derivatives of nucleic acid bases reported gave the product in low yield, or use carcinogenic ethyleneimine [22]. Recently, 2-aminoethyl derivatives of adenine, uracil, and thymine were successfully prepared by the Schmidt-reaction of the carboxyethyl derivatives in high yield [72]. The carboxyethyl derivatives were prepared by the Michelis addition of the bases on ethyl acrylate (Scheme 21), or the reaction with β-propiolactone [25]. The carboxyethyl derivative of hypoxanthine (*84*) was obtained from the carboxyethyl derivative of adenine (*77*), and the methacrylamide derivative (*86*) was prepared using the same procedure as for adenine (*72*) or uracil (*74*) derivatives [72].

9-Carboxyethyladenine (*77*) [204]. To a refluxing suspension of adenine (27 g, 0.2 mol) in ethanol (800 mL) with catalytic amounts of sodium, ethyl acrylate (64 mL, 0.6 mol) is added dropwise. After the suspension becomes a clear solution, the solution is cooled to give the product as a precipitate, and filtered. Recrystallization from ethanol gives 9-carboethoxyethyladenine (*76*) as colorless needles; yield 43.4 g (93%), mp 167 to 168°C. Compound *76* is hydrolyzed with 6 N hydrochloric acid (600 mL) for 3 h at reflux temperatures. The solvent is coevaporated with water, and the concentrated solution is neutralized with a saturated sodium carbonate solution to pH 4 to 5 to give 9-carboxyethyladenine as a white precipitate; yield 37.6 g (98%), mp 284 to 288°C.

9-(2-Aminoethyl)adenine (*79*) [72]. To a solution of 9-carboxyethyladenine (*77*) (21 g, 0.1 mol) in concentrated sulfuric acid (50 mL), sodium azide (13 g, 0.2 mol) is added at 45°C over 5 days with mechanical stirring. Then the mixture is adjusted to pH 5 by concentrated aqueous sodium hydroxide to give a crystalline precipitate. After filtration, the product is recrystallized from water to give 9-(2-aminoethyl)adenine sodium sulfate (*78*) as colorless needles; 16 g, 70%, mp 290 to 292°C. The sulfate salt (*78*) is dissolved in water, and concentrated sodium hydroxide solution is added. Then excess methanol is added to the solution, and the precipitated sodium sulfate is filtered. The filtrate is evaporated, and the residue is recrystallized from ethanol to give 9-(2-aminoethyl)adenine (*79*) as colorless needles; yield 11.7 g (94%), mp 216 to 217°C.

9-(Methacryloylaminoethyl)adenine (*72a*) [72]. To a suspension of 9-(2-aminoethyl)adenine (*79*) (1 g, 5.6 mmol) in pyridine (10 mL), methacrylic anhydride (1.2 mL, 8.4 mmol) is added at 0°C. After stirring for 3 h at room temperature, acetone is added to the reaction mixture to give the product as a precipitate. Filtration and recrystallization from ethanol gives the methacrylate derivative (*72a*) as colorless needles; yield 1.2 g (90%), mp 199 to 200°C.

1-(2-Carboxyethyl)uracil (*81*) [150]. To a suspension of uracil (22.4 g, 0.2 mol) in ethanol (800 mL) with catalytic amounts of sodium,

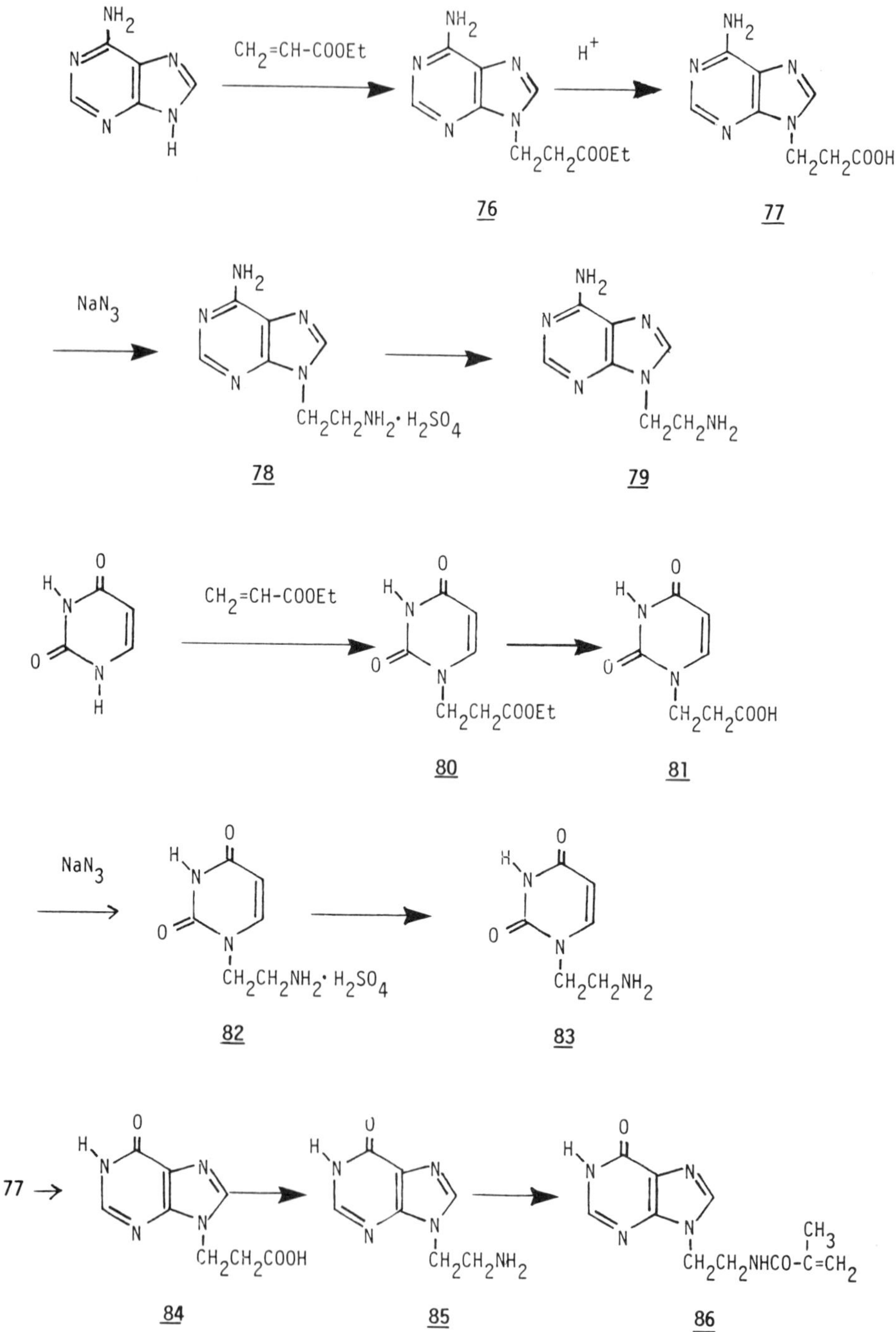
NH2
CH2=CH-COOEt
H+
CH2CH2COOEt
CH2CH2COOH
76
77
NaN3
CH2CH2NH2· H2SO4
CH2CH2NH2
78
79
CH2=CH-COOEt
CH2CH2COOEt
CH2CH2COOH
80
81
NaN3
CH2CH2NH2· H2SO4
CH2CH2NH2
82
83
77 →
CH2CH2COOH
CH2CH2NH2
CH3
CH2CH2NHCO-C=CH2
84
85
86

Scheme 21

ethyl acrylate (43 mL, 0.4 mol) is added dropwise under reflux during 3 h. Then the solvent is evaporated under reduced pressure. Recrystallization of the residue from ethanol gives 1-(2-carboethoxyethyl)uracil (*80*) as colorless needles in 99% yield; 42.8 g, mp 101 to 103°C. Compound *80* (42.8 g) is hydrolyzed with 6 N hydrochloric acid (900 mL) for 3 h under reflux. The solvent is coevaporated with water, and the residue is crystallized from water to give compound *81*; yield 29.2 g (80%), mp 183 to 185°C.

1-(2-Aminoethyl)uracil (*83*) [72]. To a solution of 1-(2-carboxyethyl)uracil (*81*) (18.6 g, 0.1 mol) in concentrated sulfuric acid (50 mL), sodium azide (13 g, 0.2 mol) is added at 45°C over 5 days. The reaction mixture is then poured into excess 2-propanol to give a white solid (*82*). After filtration, the solid is dissolved in concentrated ammonium hydroxide and an excess of methanol is added to the solution. The precipitated salt is removed by filtration, and the filtrate is evaporated to dryness; this step is repeated three times. The residual solid is recrystallized from ethanol to give the aminoethylated uracil (*83*) as colorless crystals; yield 17 g (90%), mp 173 to 175°C.

1-(2-Methacryloylaminoethyl)uracil (*74a*) [72]. To a solution of 1-(2-aminoethyl)uracil (*83*) (1 g, 6.4 mmol) in pyridine (10 mL), methacrylic anhydride (1.2 mL, 8.4 mmol) is added at 0°C, and stirred for 3 h at room temperature. The solvent is evaporated under reduced pressure, and the residue is washed with diethyl ether. Recrystallization from water gives colorless needles of *74*; yield 1.38 g (96%), mp 229 to 230°C.

4.4.2 Polymerizations and Properties of the Polymers

Polyacrylamide and polymethacrylamide derivatives of adenine (*73a* and *73b*) and uracil (*75a* and *75b*) are prepared by the usual free-radical polymerization of the corresponding monomers [72] (Scheme 20). Hydrogen-transfer polymerization of the acrylamide derivatives was tried in dimethyl sulfoxide, dimethylformamide, and pyridine solution with potassium t-butoxide as catalyst. The β-alanine-type structure of the polymer was confirmed by NMR spectra [55].

The formation of the polymer complexes between polymethacrylamide derivatives of nucleic acid bases were studied under conditions similar to the polymethacrylate derivatives. The polymethacrylamides containing complementary nucleic acid bases form a stable 1:1 polymer complex in a dimethyl sulfoxide-ethylene glycol mixed solvent system. Polymethacryloylamino-type derivatives have a structure similar to that of the polymethacryloyloxy-type polymers, with a minor difference in the side-chain structure. However, the ability to form polymer complexes is substantially higher for polymethacryloylamino derivatives than for the polymethacryloyloxy type (Table 4 [72], Table 5 [44]).

Table 4. Hypochromicity of the Systems of Methacryloylamino-type Derivatives[a]

	Poly-MAEA	Di-MAEA	MAEA
Poly-MAEU	66	5	0
Di-MAEU	9	4	0
MAEU	7	0	0

[a]MAEA, *72a*; MAEU, *74a*.

Table 5. Hypochromicity of the Systems of Methacryloyloxy-type Derivatives[a,b]

	Poly-MAOA	Di-MAOA	MAOA
Poly-MAOU	21	0[a]	0
Di-MAOU	0	0	0
MAOU	0	0	0

[a] Interaction was observed only at high concentration.
[b]MAOA, *41a*; MAOU, *43a*.

4.5 POLYETHYLENEIMINE DERIVATIVES OF NUCLEIC ACID BASES

In recent years considerable interest has been generated in the use of natural and synthetic macromolecules for interferon inducers [205]. It is also known that interferon inducers can prevent or regress certain tumors. For example, the synthetic double-stranded complex of poly I and poly C is effective as an interferon inducer; however, it has a high level of toxicity [206]. Similarly, polyethyleneimine has been shown to be highly effective against ascite tumors. The high toxicity of this polymer was eliminated, without significant reduction of its activity, by grafting monomeric serine and/or histidine residue to the polymer. Recently, new model polymers with a polyethylene-imine backbone and nucleic acid base derivatives as pending side chains have been prepared [150]. These polymers are interesting as interferon inducers and antitumor agents.

4.5.1 Grafting of Nucleic Acid Base Derivatives

Polyethyleneimine can be alkylated by 9-(2-chloroethyl)adenine (*2*) and 1-(2-chloroethyl)thymine (*19b*), respectively, in dimethylformamide [104]. The content of nucleic acid base is determined by elementary analysis to be about 20 to 50% for the adenine derivative (*87*), and 60% for the thymine derivative (*88*).

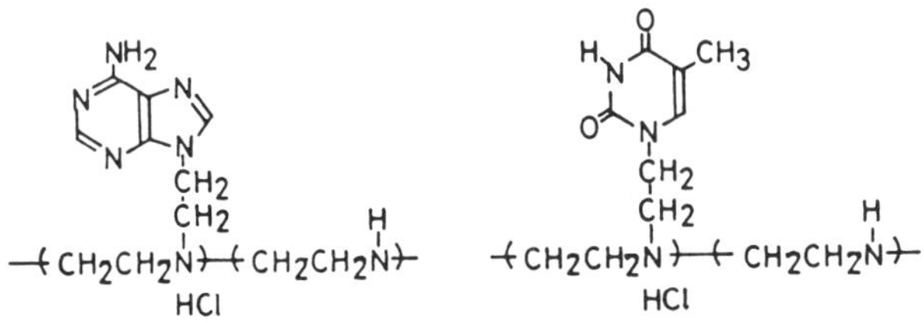

87 88

Synthetic routes to the graft polymers of polyethyleneimine containing adenine units by an amide bond are shown in Scheme 22 [150]. One route is the reaction of polyethyleneimine (*89*) with

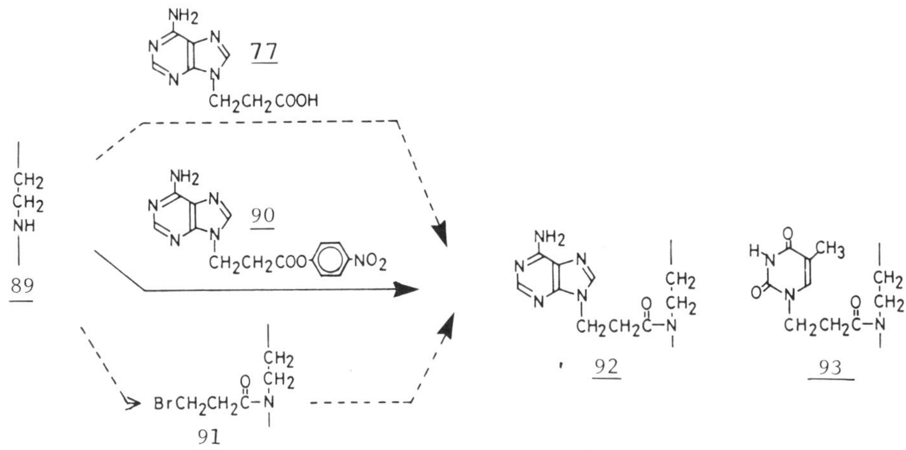

Scheme 22

9-(2-carboxyethyl)adenine (*77*) by a condensation reagent such as carbodiimide or by the activated ester method. The other route involves reaction of the functional graft polymer (*91*) with the sodium salt of adenine (*40*). The activated ester method is most successful.

The *p*-nitrophenyl esters were prepared from the carboxyethyl derivatives by the reaction of *p*-nitrophenyl trifluoroacetate (Scheme 23). As an activated ester, pentachlorophenyl esters (*134* and *140*) were preferred.

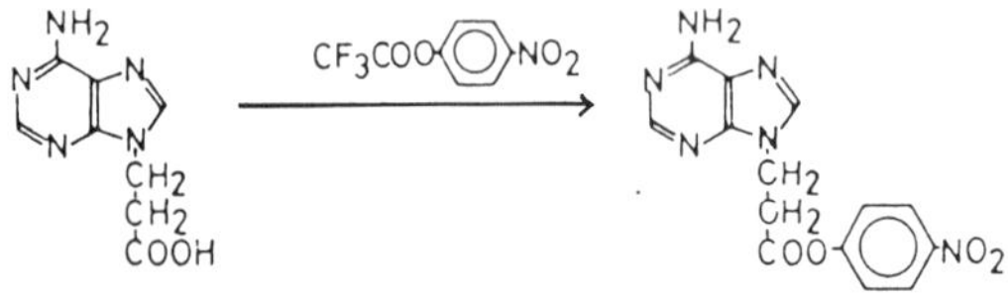

77 90

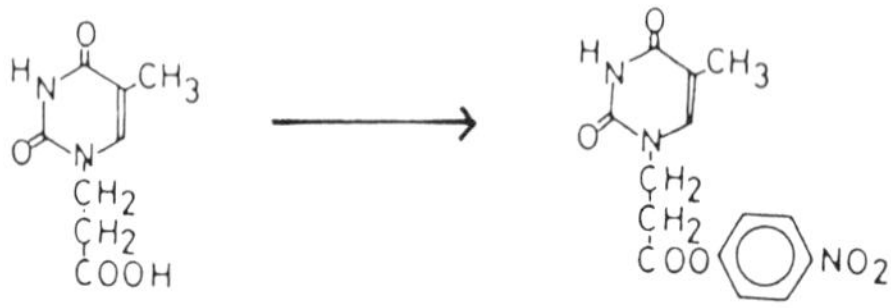

81 94

Scheme 23

The reaction of polyethyleneimine with the *p*-nitrophenyl ester (*90*) was carried out in dimethyl sulfoxide at 25°C in the presence of imidazole. The polymer obtained, adenine-PEI (*92*) is soluble in water below pH 2, in acidic solvents such as formic acid, dimethyl sulfoxide, and ethylene glycol, and insoluble in concentrated hydrochloric acid, dimethylformamide, and other common organic solvents. The amount of nucleic acid base units in the polymers was determined by spectroscopy, elementary analysis, and hydrolysis of the polymer. The adenine units in the polymer is about 60%. The low content of adenine units in the polymer is caused by instability of the *p*-nitrophenyl ester of the adenine derivative (*90*). Stability of the activated ester of the adenine derivative can be improved by protection of the 6-amino group of adenine base. The adenine units can then be grafted on the polymer almost quantitatively [76].

The synthetic way to the graft polymer containing thymine, thymine-PEI, is the same as that used for the adenine derivative. The thymine derivative was grafted almost quantitatively onto polyethyleneimine. The polymer (*93*) is soluble in water above pH 9, dimethyl sulfoxide, and dimethylformamide.

p-Nitrophenyl ester of 9-(2-carboxyethyl)adenine (*90*) [150]. *p*-Nitrophenyl trifluoroacetate (24 g, 0.10 mol) is added to a suspension of 9-(2-carboxyethyl)adenine (77) (16 g, 0.08 mol) in dry pyridine

(500 mL). The mixture is stirred for 2 days at 25°C to give a clear solution. After vacuum distillation of pyridine at 60°C to complete dryness, the residual solid is washed with benzene. The fine white product is filtered and washed with benzene to give the *p*-nitrophenyl ester (*90*) in neutral form; 19 g (80% yield), mp 179 to 181°C. This product can be used for the reaction with polymers without further purification.

Adenine-PEI (*92*) [150]. To a suspension of linear polyethyleneimine (*89*) [207] (0.22 g, 5 mmol) in dimethyl sulfoxide (10 mL), *p*-nitrophenyl ester of 9-(2-carboxyethyl)adenine (*90*) (2 g, 6 mmol) is added in the presence of imidazole (1 g). The solution becomes clear after 30 min. After additional stirring for 2 days at 25°C, the reaction mixture is poured into an excess of dimethylformamide to precipitate the polymer. The polymer is purified by reprecipitation from dimethylsulfoxide-dimethylformamide, washed with dimethylformamide and acetone, and vacuum dried to give adenine-PEI (*92*) (0.72 g).

p-Nitrophenyl ester of 1-(2-carboxyethyl)thymine (*94*) [150]. To a suspension of 1-(2-carboxyethyl)thymine (*81b*) (19 g, 0.1 mol) in dry pyridine (400 mL), *p*-nitrophenyl trifluoroacetate (23 g, 0.1 mol) is added. After stirring for 2 days at 25°C, the pyridine is removed by vacuum distillation at 60°C to complete dryness. Recrystallization of the residue from ethanol gives the *p*-nitrophenyl ester *94* in 95% yield (30 g), mp 179 to 180°C.

Thymine-PEI (*93*) [150]. The *p*-nitrophenyl ester *94* (2 g, 6 mmol) is added to a suspension of linear polyethyleneimine (*89*) (0.22 g, 5 mmol) in dimethyl sulfoxide (6 mL) with imidazole (1 g). The suspension becomes a clear solution within 5 min. After additional stirring for 2 days at 25°C, the polymer is precipitated by adding ethyl ether. The polymer is washed with water, reprecipitated from dimethylformamide-ethyl ether, and washed with acetone to produce thymine-PEI in 97% yield (0.68 g).

In addition to the linear polyethyleneimine, branched polyethyleneimines having three kinds of molecular weight; Dow-1000, Dow-400, and Dow-18 are used as the main chain. The results of the grafting reactions are shown in Table 6.

The interactions of the graft polymers of adenine and thymine on linear and branched polyethylenimine with polynucleotides, poly A, and poly U were studied [151]. From Job plots of the adenine-PEI/thymine-PEI system at pH 7.4, the formation of the complex was determined to have a 1:1 stoichiometry with a hypochromicity of about 16%. Adenine-PEI and poly U (polyuridylic acid) also form a 1:1 complex at neutral pH and even at pH 1.7. Poly A (polyadenylic acid) and thymine-PEI, however, form 1:9 (A:T) complex.

Nucleic acid analogs with a polyethylenimine backbone and nucleic acid base-containing pendants have been characterized by means of

Table 6. Base Content in the Graft Polymers

Polyethylenimine	Base	Base content[a] (unit mol %)
Linear	Adenine	63
	Thymine	100
Dow-1000	Adenine	35
	Thymine	52
Dow-400	Adenine	37[b]
	Thymine	73
Dow-18	Adenine	51
	Thymine	62

[a]From UV spectra: adenine, 275 nm; 285 nm.
[b]From elemental analysis

360-MHz ^1H NMR as well as 90- and 20-MHz ^{13}C NMR and UV spectra [156,157]. Interactions and photochemical reactions were studied for oligoethyleneimine derivatives [64,67-69].

4.5.2 Grafting of Asymmetric Nucleic Acid Base Derivatives

Model polymers of polynucleotides with linear polyethylenimine backbone and optically active nucleic acid base derivatives as pending side chain were synthesized [153]. Synthesis of the 2-propionic acid derivatives 2-(adenin-9-yl)propionic acid (*98*) and 2-(thymin-1-yl)propionic acid (*96*) was achieved by treatment of sodium salt of the corresponding nucleic acid base with ethyl-2-bromopropionate in dimethylformamide, followed by hydrolysis with hydrochloric acid (Scheme 24).

The enantiomeric pairs of thymine derivative are optically resolved with quinine. The optical resolution of corresponding adenine derivatives is extremely difficult mainly because of the poor stability of the salt with resolving reagent.

These optically active thymine derivatives are grafted onto polyethylenimine through amide bond by direct coupling with diethylphosphoryl cyanide to give optically active polymers (*103*). In the case of the adenine derivative, the *p*-nitrophenyl active ester method is used (Scheme 25).

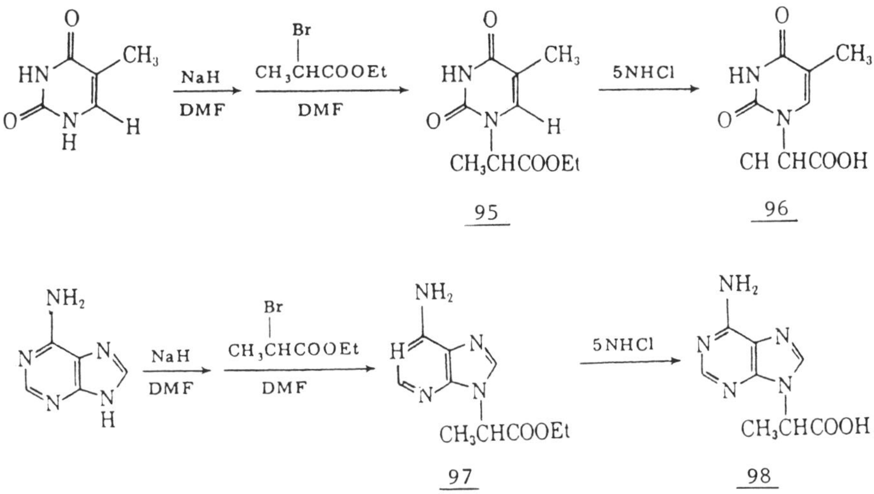

Scheme 24

2-(Thymine-1-yl)propionic acid (96) [153]. A suspension of thymine (37.8 g, 0.3 mol) and sodium hydride (15 g, 0.6 mol; 50% dispersion in mineral oil) in dimethylformamide (750 mL) is stirred at 50°C for 2 h. To the resulting suspension of sodium thymine is added ethyl-2-bromopropionate dropwise over 3 h under vigorous stirring at 50°C. After additional stirring for 2 h, the solvent is evaporated under reduced pressure. The residue is stirred for 3 h with chloroform and the insoluble salt is removed by filtration. The filtrate is evaporated to dryness and recrystallized from carbon tetrachloride to give colorless flakes of ethyl-2-(thymin-1-yl)propionate (*95*); yield 53 g (78%), mp 109 to 111°C.

The ester (50 g, 0.22 mol) is hydrolyzed with 5 N hydrochloric acid (700 mL) for 5 h under reflux. After evaporation of the solvent, the residue is recrystallized from methanol-water (2:1) to give the propionic acid derivative (*96*); yield 36 g (82.6%), mp 264 to 265°C.

2-(Adenin-9-yl)propionic acid (98) [153]. The adenine derivative (*98*) is prepared by the same procedure as that used for the thymine derivative (*96*). The ethyl ester *97* is obtained by recrystallization from carbon tetrachloride-chloroform (1:1) in 67.4% yield, mp 166 to 167°C. The ester is hydrolyzed and recrystallized from methanol-water (1:1) to give compound *98*; yield 80%, mp 308 to 310°C.

Optical resolution of 2-(thymine-1-yl)propionic acid [153]. The 1:1 salt of the propionic acid derivative of thymine and quinine is formed in methanol and the less soluble (-) salt is isolated by 10 recrystallizations from acetone-water (40:1). The partly resolved (+) salt is recovered from the combined filtrates of the fractional recrystallization. Small amounts of (-) salt are removed from the

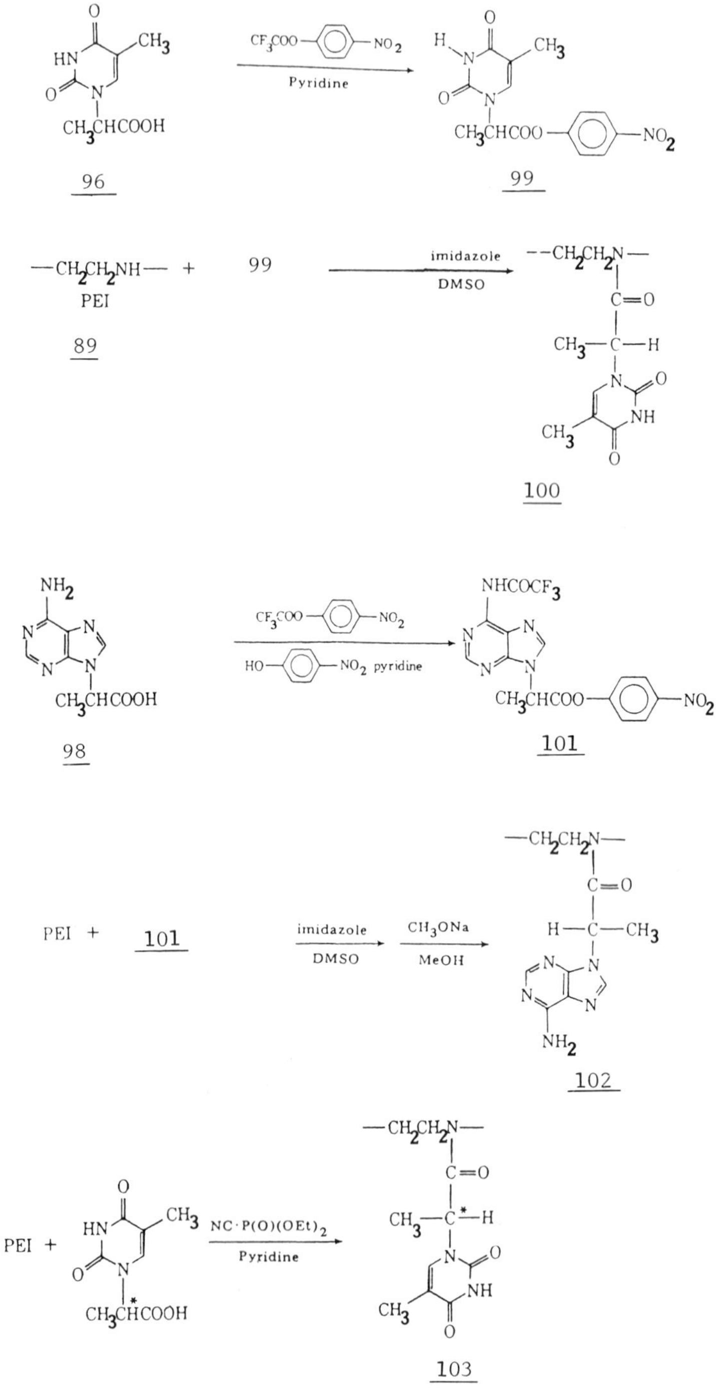
96
99
Pyridine
PEI
89
imidazole
DMSO
100
98
pyridine
101
MeOH
102
103

Scheme 25

crude (+) salt by recrystallization from acetone seeded with pure (-) salt crystal until pure (+) salt is recovered from the filtrate. Treatment of the salts with aqueous sodium hydroxide followed by acidification with hydrochloric acid yielded optically active 2-(thymin-1-yl)propionic acids.

Grafting of optically active thymine derivatives [153]. Optically active 2-(thymin-1-yl)propionic acid (1.23 g, 6.2 mmol) and linear polyethylenimine (0.215 g, 5 mmol) are dissolved in pyridine (23 mL) at 0°C. To this solution is added diethylphosphoryl cyanide [208] (2.65 g, 16.3 mmol) dropwise over 1.5 h under stirring. The reaction mixture is stirred overnight at room temperature and more diethylphosphoryl cyanide (0.41 g, 2.5 mmol) is added at one time. After 2 h the solvent is evaporated at 30°C to dryness. The residue is dissolved into chloroform-methanol (1:1) and passed through a basic alumina column. Evaporation of the solvents leaves the polymer as a brittle material in the flask in 42% yield.

Optical properties of linear polyethylenimine containing optically active 2-(thymin-1-yl)propionyl group as grafted pendant were investigated by UV and circular dichroism (CD) and NMR and compared to those of the related monomer and dimer model compounds [152,154]. A CD spectrum of the polymer in neutral aqueous solution suggests that the polymer exists in some ordered conformation (at least locally). The effects of complementary base pairing on the CD spectra have also been studied by using a linear polyethylenimine containing 2-(adenin-9-yl)propionyl grafts.

4.5.3 Grafting of Amino Acid and Nucleic Acid Base

The carboxyethyl derivatives of nucleic acid bases are further grafted onto the monomeric graft polymers of L-α-amino acid on linear polyethylenimine [150,155]. The synthetic route to the graft polymers containing L-α-alanine and nucleic acid base is shown in Scheme 26. *N*-Carbobenzyloxy-L-α-alanine and linear polyethylenimine (*89*) were condensed by dicyclohexylcarbodiimide (DCC) to give polymer *104* (Cbz-Ala-PEI). The protecting group of polymer *104* was removed by HBr gas in acetic acid to give the graft polymer *105* (HBr-Ala-PEI). Polymer *105* was grafted almost quantitatively as determined by elemental analysis and the NMR spectrum. The L-α-alanine grafted polymer, HBr-Ala-PEI, was further reacted with *p*-nitrophenyl ester (*90*) or *94*, according to the procedure above. From the NMR and UV spectra the polymers obtained were found to contain about 98% adenine units for adenine-Ala-PEI (*106*), and about 100% thymine units for thymine-Ala-PEI (*107*).

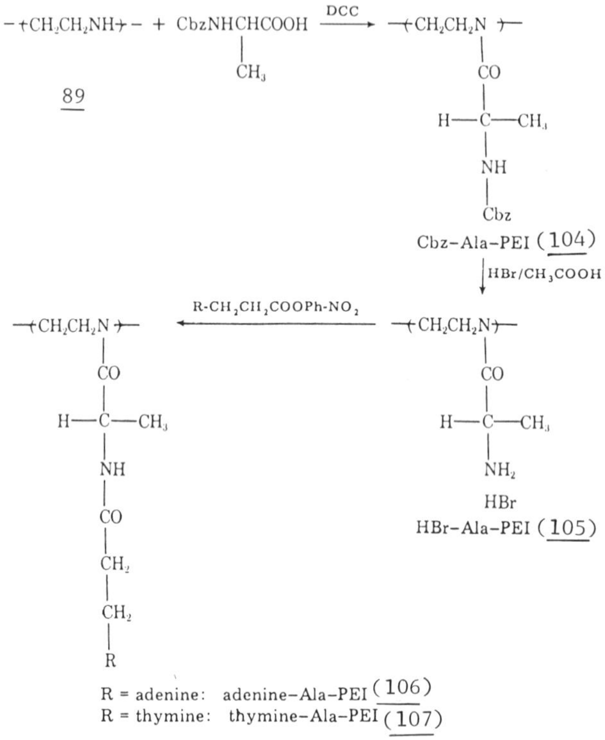

Scheme 26

HBr-Ala-PEI (105) [150]. *N*-Carbobenzyloxy-L-α-alanine (2.5 g, 0.11 mol), linear polyethylenimine (0.43 g, 0.01 mol), and DCC (3.0 g, 0.015 mol) in chloroform (50 mL) are stirred for 5 h at 0°C and for 3 days at 25°C. After dicyclohexylurea is removed by filtration the chloroform solution is precipitated with ethyl ether to give Cbz-Ala-PEI (*104*). This polymer is dissolved in acetic acid and treated with HBr gas for 3 h to precipitate polymer *105*. The mixture is poured into excess acetone, filtered immediately, and washed with acetone. Drying in vacuo gives 1.6 g of HBr-Ala-PEI (*105*) (80% yield).

Adenine-Ala-PEI (106) [150]. To HBr-Ala-PEI (*105*) (0.6 g, 3 mmol) in dimethyl sulfoxide (15 mL) with triethylamine (0.5 ml, 3.1 mmol), *p*-nitrophenyl ester (*90*) (1.0 g, 3 mmol) is added and stirred for 2 days at 25°C. After excess triethylamine is evaporated the reaction mixture is poured into excess methanol. The precipitate is further washed with methanol. The polymer is dissolved in 3N hydrochloric acid and poured into ethanol after evaporation of excess hydrochloric acid with water. The precipitated polymer is washed with ethanol and vacuum dried to give the hydrochloride salt of polymer *106* (HCl-adenine-Ala-PEI); yield 0.64 g. This polymer in aqueous solution is neutralized with 5% sodium carbonate solution to precipitate the neutral polymer adenine-Ala-PEI (*106*).

Thymine-Ala-PEI (*107*) [150]. A solution of HBr-Ala-PEI (*105*) (0.6 g, 3 mmol), *p*-nitrophenyl ester of carboxyethylthymine (*94*) (1.0 g, 3 mmol), and triethylamine (0.5 mL) in dimethyl sulfoxide (15 mL) is stirred for 2 days at 25°C. The solution is poured into methanol and the precipitate is washed and vacuum dried to yield 0.85 g of thymine-Ala-PEI (*107*).

The monomeric graft polymer of L-aspartic acid on linear polyethylenimine; HBr-Asp-PEI (*110*), was obtained by the condensation reaction of linear polyethylenimine with *N*-carbobenzyloxy-*O*-benzyl-L-aspartate (*108*), followed by the removal of the protecting groups. The β-carboxyl group of the aspartic acid unit in HBr-Asp-PEI is again protected by esterification to give polymer *111* (OMe-HBr-Asp-PEI). The NMR spectrum and elemental analysis indicates that the grafting of L-aspartic acid onto polyethylenimine is almost complete. The reaction of OMe-HBr-Asp-PEI (*111*) with *p*-nitrophenyl ester of carboxyethyl-adenine (*90*) gave polymer *112* (adenine-OMe-Asp-PEI), which was treated with hydrochloric acid to yield polymer *113* (HCl-adenine-Asp-PEI). This polymer *113* contains 95% adenine units. The reaction of OMe-HBr-ASp-PEI with *p*-nitrophenyl ester of 1-carboxyethylthymine (*94*) gives polymer *114* (thymine-OMe-Asp-PEI), which contains about 70% thymine units. The synthetic route to the graft polymer that contains L-aspartic acid and the nucleic acid base is shown in Scheme 27.

OMe-Asp-PEI (*111*) [150]. Linear polyethylenimine (0.40 g, 9 mmol) and β-benzyl-*N*-carbobenzyloxy-L-aspartate (4.3 g, 12 mmol) are condensed by DCC (3 g, 15 mmol) in chloroform (50 mL) on stirring for 3 days at 25°C. After filtration the chloroform solution is poured into ethyl ether to give OBzl-Cbz-Asp-PEI (*109*). The polymer is dissolved in dry acetic acid and treated with HBr gas for 5 h to give polymer *110* (HBr-Asp-PEI). The precipitated polymer is filtered and washed with acetone. Polymer *110* is again dissolved in methanol saturated with HBr gas and stirred overnight. The methanol solution is concentrated and poured into ethyl ether and the precipitate is washed with acetone and vacuum dried to give 1.25 g of polymer *111* [OMe-Asp-PEI (HBr salt)].

HCl-adenine-Asp-PEI (*113*) [150]. A solution of OMe-HBr-Asp-PEI (*111*) (0.5 g, 2 mmol) and the *p*-nitrophenyl ester (*90*) (0.66 g, 2 mmol) in dimethyl sulfoxide (10 mL) with triethylamine (0.5 mL) is stirred for 2 days at 25°C. The reaction mixture is precipitated with

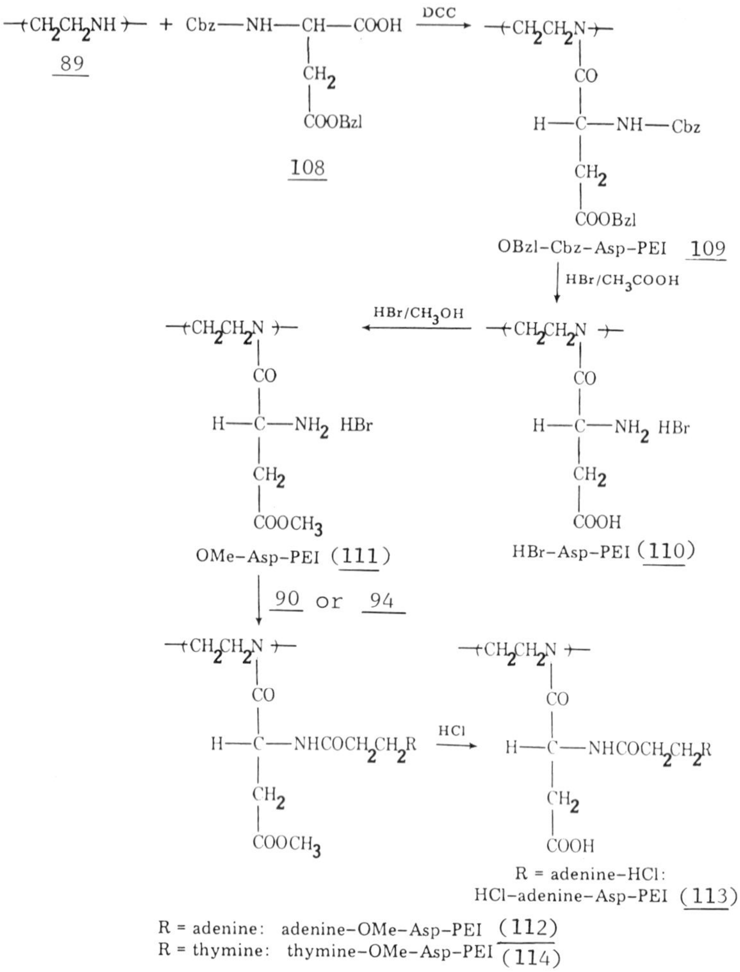

Scheme 27

ethyl ether, washed with acetone and methanol, and dried in vacuo. The polymer obtained, *112* (adenine-OMe-Asp-PEI), is dissolved in 2 N HCl and refluxed for 30 min. After evaporation of excess hydrochloric acid with water the aqueous solution is poured into acetone. The precipitate is washed with acetone and methanol and vacuum dried to give 0.79 g of polymer *113* (HCl-adenine-Asp-PEI).

Thymine-OMe-Asp-PEI (*114*) [150]. *p*-Nitrophenyl ester *94* (0.64 g, 2 mmol) is added to a solution of OMe-HBr-Asp-PEI (*111*) (0.5 g, 2 mmol) in 10 mL of dimethyl sulfoxide with triethylamine (0.5 mL) and stirred for 2 days at 25°C. The solution is precipitated in ethyl ether, washed with acetone and ethanol, and dried in vacuo to give 0.22 g of polymer *114*. The synthetic route to the graft polymer that contained L-serine and nucleic acid base is shown in Scheme 28.

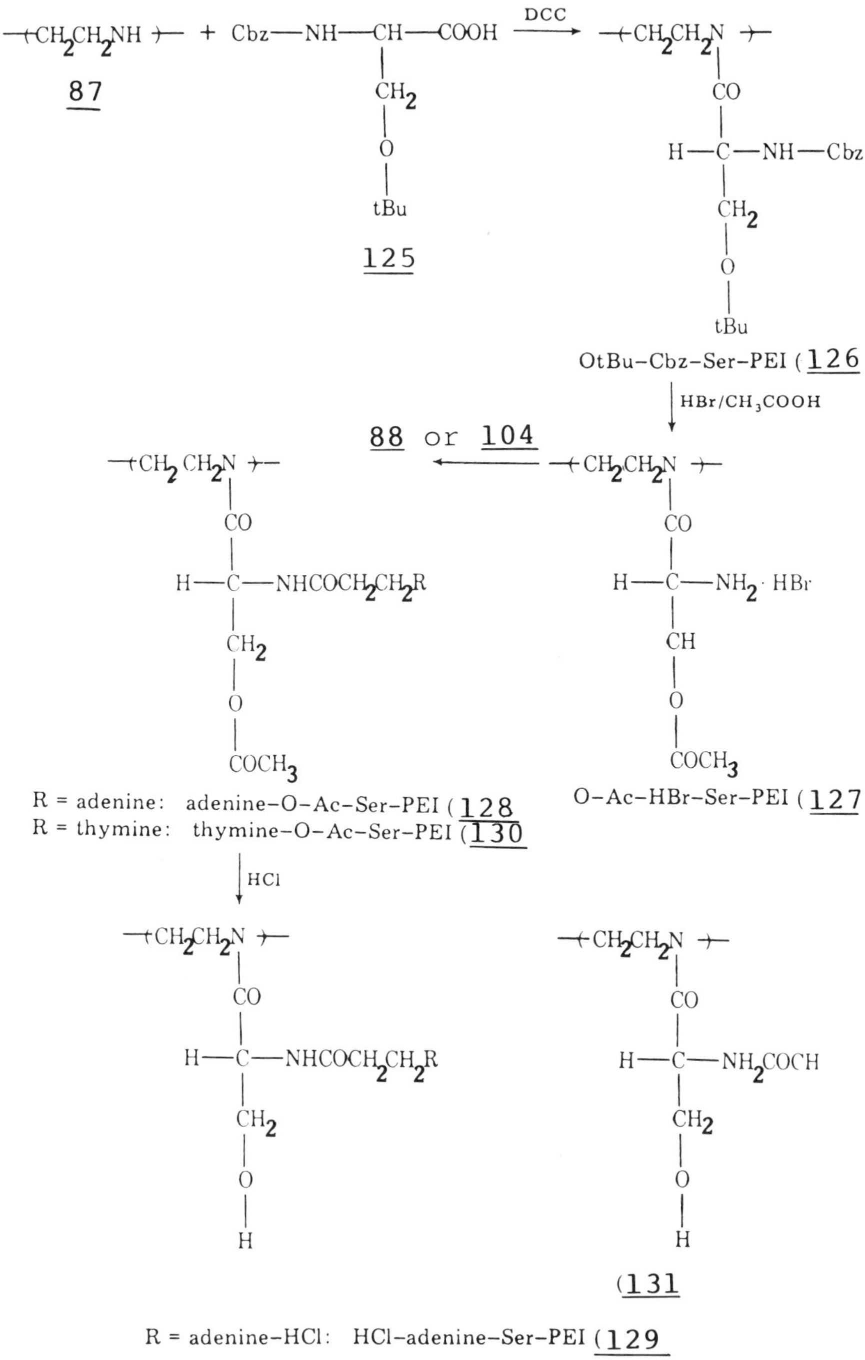
DCC
87
125
OtBu-Cbz-Ser-PEI (126
HBr/CH3COOH
88 or 104
O-Ac-HBr-Ser-PEI (127
R = adenine: adenine-O-Ac-Ser-PEI (128
R = thymine: thymine-O-Ac-Ser-PEI (130
HCl
(131
R = adenine-HCl: HCl-adenine-Ser-PEI (129

Scheme 28

The reaction of polyethylenimine with the protected L-serine *115* (*O*-*t*-butyl-*N*-carbobenzyloxy-L-serine) by DCC gave polymer *116* (OtBu-Cbz-Ser-PEI) in high yield. The protecting groups of OtBu-Cbz-Ser-PEI (*116*) are removed to give OAc-HBr-Ser-PEI (*117*); during the reaction the β-hydroxy group is acetylated. Polymer *117* contains almost 100% L-serine unit; this was determined by elemental analysis and NMR spectroscopy.

Polymer *117* is reacted with the *p*-nitrophenyl ester of adenine derivative *90* and treated with hydrochloric acid to give the polymer containing L-serine and adenine *119* (HCl-adenine-Ser-PEI). The polymer *119*, however, contained 170% adenine units, according to the UV spectrum. This may have been caused by an ester exchange reaction between the *O*-acyl group and the *p*-nitrophenyl ester. In the reaction of OAc-HBr-Ser-PEI (*117*) with the *p*-nitrophenyl ester of 1-(2-carboxyethyl)thymine (*94*), the acyl group moved from *O* to the amino group to give polymer *121*, which contains about 10% thymine units (*120*).

The graft polymer that contains L-histidine and nucleic acid base was prepared by the synthetic route shown in Scheme 29. Linear polyethylenimine is grafted by *N*-carbobenzyloxy-L-histidine, which

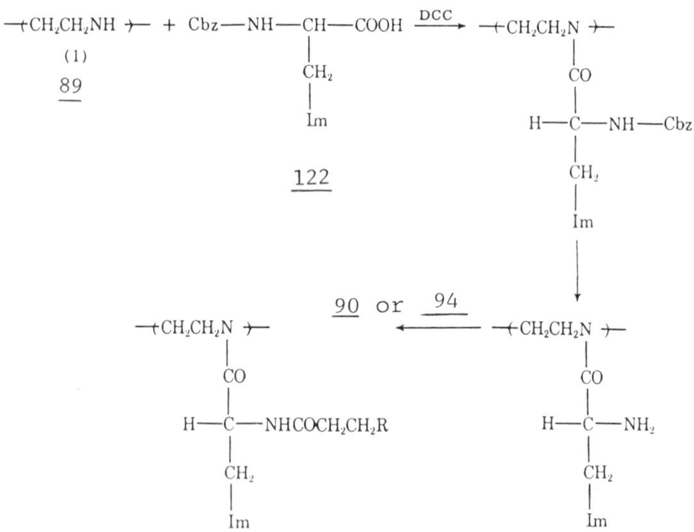

R = adenine–2HCl: HCl–adenine–HCl–His–PEI (126)
R = thymine: thymine–His–PEI (127)

Scheme 29

is followed by removal of the protecting group and by neutralization with triethylamine to give the L-histidine grafted polymer *125* (His-PEI). This polymer contains only 50% histidine units, probably because of the low solubility of compound *122* in chloroform-dimethylformamide mixed solvents.

The reaction of polymer *125*, which contains 50% histidine units, with *p*-nitrophenyl ester *90* gave a copolymer with L-histidine and adenine units. This is a copolymer containing units of *126* and *92*. From the reaction of *125* with *p*-nitrophenyl ester *94* a copolymer with units *127* and *93* was obtained.

Monomeric graft of L-proline onto linear polyethylenimine and further grafting of 2-(thymin-1-yl)propionyl group is shown in Scheme 30 [155]. Direct condensation of *N*-carbobenzyloxy-L-proline (*128*) and polyethylenimine by DCC yielded a graft polymer (*129*); Cbz-Pro-PEI. The N-protecting group of the polymer *129* was removed by HBr gas in acetic acid to give the graft polymer *130* (HBr-Pro-PEI). This polymer was neutralized by Amberlite IR-400 to give polymer *131* (Pro-PEI: 100% graft from NMR). The Pro-PEI was further treated with optically active 2-(thymin-1-yl)propionic acid using diethylphosphoryl cyanide to give polymer *132* [Thy(-)-Pro-PEI].

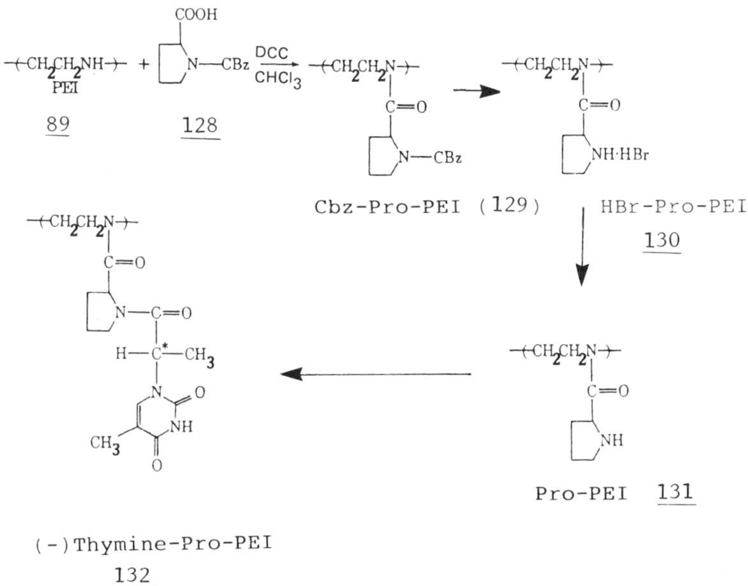

Scheme 30

Pro-PEI (*131*) [155]. To a stirred solution of polyethylenimine (0.86 g, 0.02 mol) and *N*-carbobenzyloxy-L-proline (6.48 g, 0.02 mol) in chloroform (50 mL), DCC (5.36 g, 0.026 mol) in chloroform (20 mL) is added at 0°C. The mixture is stirred at 0°C for 5 h and at room temperature for 2 days. After the resulting dicyclohexylurea is removed by filtration, the chloroform is evaporated to dryness. The residue is dissolved in 20 mL of ethyl acetate and kept at -10°C overnight to precipitate the remaining dicyclohexylurea. After filtration, the filtrate is poured into an excess of diethyl ether to precipitate graft polymer *129* (Cbz-Pro-PEI).

This polymer *129* is dissolved in acetic acid and treated with HBr gas at 5°C for 2 h and at room temperature for 1 h. The mixture is poured into an excess of ether and the resulting solid is washed with acetone. Drying in vacuo gave 3.0 g of HBr-Pro-PEI (*130*). An aqueous solution of the polymer *130* is passed through an Amberlite IR-400 column and the collected elutent is evaporated to dryness to give Pro-PEI (*131*).

(-)Thymine-Pro-PEI (*132*) [155]. To a stirred mixture of optically active (-)-2-(thymin-1-yl)propionic acid (96) (0.991 g, 5 mmol) in pyridine (10 mL) is added diethylphosphoryl cyanide [208] (1.06 g, 6.5 mmol) at 0°C, followed by the addition of Pro-PEI (*131*) (0.588 g, 4.2 mmol) in pyridine (10 mL) over a period of 1 h. After the mixture is stirred at 0°C for an additional 4 h, another diethylphosphoryl cyanide (0.521 g, 3.5 mmol) is added, and the mixture is stirred at room temperature overnight. Pyridine is evaporated to dryness, and the residue is dissolved in chloroform-methanol (1:1) and passed through a basic alumina column. Evaporation of the solvent from the elutent and reprecipitation from chloroform-methanol into excess ether give polymer *132*; yield 0.85 g.

Polyethylenimine derivatives containing amino acid and nucleic acid bases were prepared by another procedure; as shown in Schemes 31 and 32, amino acid derivatives of nucleic acid bases are grafted onto polyethylenimine [69,71,77]. At first, the activated ester of carboxyethyl derivatives of nucleic acid bases is reacted with amino acid ester hydrochloride (*135*) by adding triethylamine. The ethyl ester of the product (*136*) is then hydrolyzed in an alkaline solution to give the amino acid derivative of nucleic acid bases (*137*). From the amino acid derivative the activated ester is then prepared and allowed to react with polyethylenimine to give the graft polymer of amino acid and nucleic acid base onto polyethylenimine (*139*).

Interactions and photochemical reactions of oligo- and polyethyleneimides containing α-alanine or β-alanine, and nucleic acid bases were studied [69,71,80].

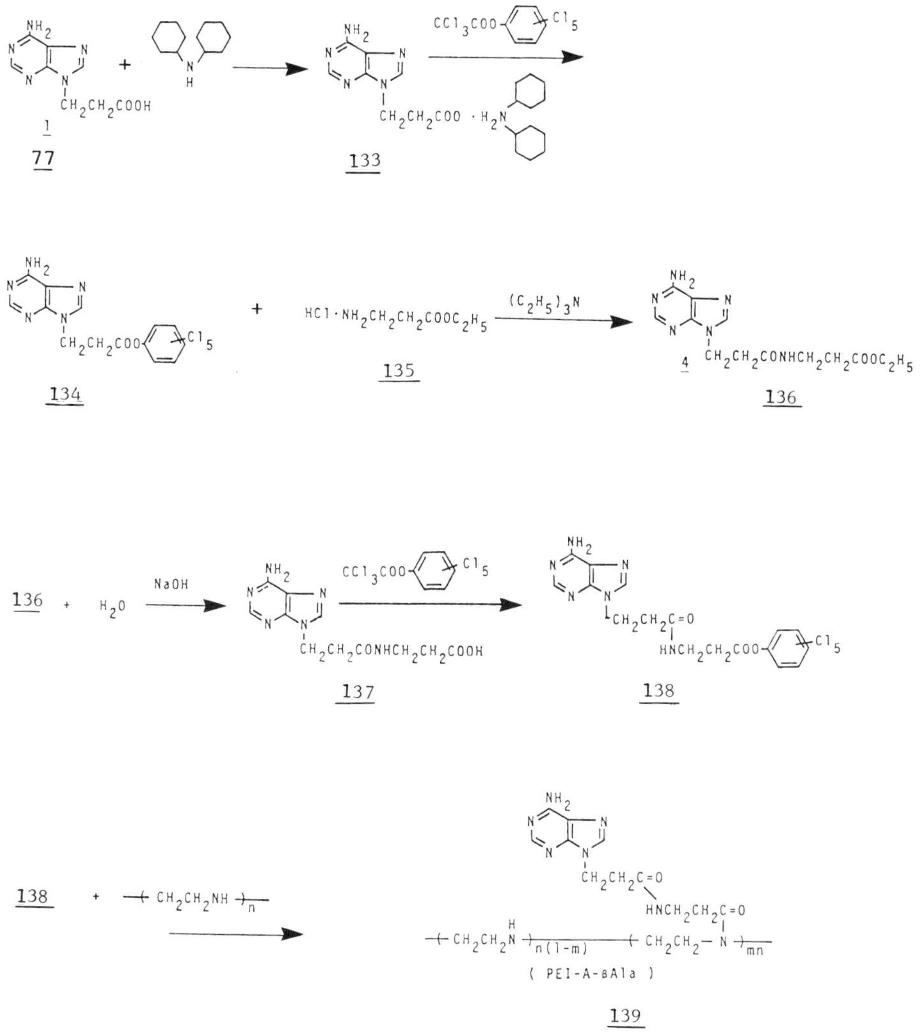

Scheme 31

Adenine-β-Ala-OEt (136) [69]. Dicyclohexylamine (7.28 g, 0.040 mol) is added to a suspension of 9-carboxyethyladenine (77) (8.0 g, 0.039 mol) in 100 mL of methanol-water (1:1), and the mixture is stirred for a few minutes at room temperature to give a clear solution. The solvent is evaporated under reduced pressure, and acetone is added to the residue. The precipitate is filtered and dried to give

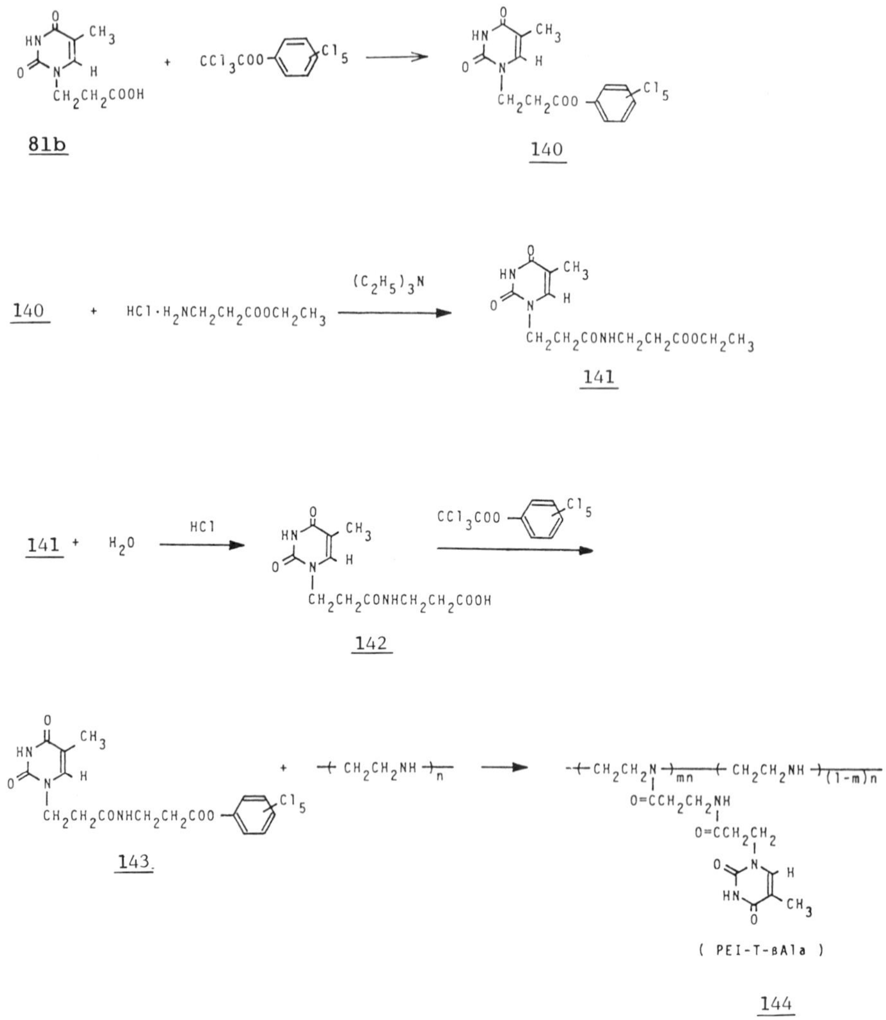

Scheme 32

a dicyclohexylamine salt of 9-carboxyethyladenine *133*; yield 15.1 g (100%).

To a suspension of 9-carboxyethyladenine dicyclohexylamine salt (17.8 g, 0.046 mol) in dimethylformamide (300 mL), pentachlorophenyl trichloroacetate [209] (20.3 g, 0.049 mol) is added and the mixture is stirred for 20 min at 0°C to form a yellow precipitate of pentachlorophenyl ester (*134*).

To this mixture, β-alanine ethyl ester hydrochloride (*135*) (7.0 g, 0.046 mol) and triethylamine (6.63 g, 0.047 mol) are added and stirred for 1 h at 0°C and for 1 day at room temperature. The precipitated dicyclohexylammonium hydrochloride is removed by filtration and the solvent of the filtrate is evaporated. The residue is washed with ethyl acetate to remove triethylammonium trichloroacetate. Recrystallization from ethanol gives colorless needles of adenine-β-Ala-OEt (*136*); yield 11.9 g (84%), mp 210 to 211°C.

Adenine-β-Ala-H (*137*) [69]. The ester (*136*) (13 g, 0.042 mol) is dissolved in 1 N NaOH aqueous solution (45 mL) and stirred for 1 h at room temperature. The solvent is then evaporated to dryness under vacuum and the residue is further dissolved in 2 N HCl solution (50 mL). The pH of the solution is adjusted to 4.5 with a saturated sodium carbonate solution to give a colorless powder. Recrystallization from water gives colorless needles of *137*; yield 8.77 g (75%), mp 282 to 283°C.

Adenine-β-Ala-PEI (*139*) [69]. To a solution of adenine-β-Ala-H (*137*) (2.78 g, 0.01 mol) in dimethyl sulfoxide (40 mL), pentachlorophenyl trichloroacetate (4.40 g, 0.0107 mol) and triethylamine (15 mL) are added and stirred for 20 min at room temperature to give the pentachlorophenyl ester (*138*). To this mixture, linear polyethylenimine (0.43 g, 0.01 mol) is added and stirred for 1 day at room temperature. The solvent is removed under reduced pressure, and the residue is washed with ethanol. Reprecipitation from dimethyl sulfoxide in acetone gives adenine-β-Ala-PEI (*139*) (44% yield). The degree of substitution of the adenine derivative on the polymer is found to be 88% from NMR spectroscopy.

Thymine-β-Ala-OEt (*141*) [69]. 1-(2-Carboxyethyl)thymine (*81b*) (1.9 g, 0.01 mol) is dissolved in dry dimethylformamide (10 mL), and triethylamine (1.4 mL, 0.01 mol) and pentachlorophenyl trichloroacetate (4.5 g, 0.01 mol) are added in that order. After stirring for 20 h at room temperature, the solvent is distilled under vacuum and the residue is washed thoroughly with cold water. Crystallization from acetone-benzene gives colorless needles of *140*; yield 3.3 g (74%), mp 236.0 to 236.5°C.

To a solution of the pentachlorophenyl ester (*140*) (44.7 g, 0.1 mol) in dimethylformamide (200 mL), β-alanine ethyl ester hydrochloride (*135*) (14.6 g, 0.095 mol) is added with triethylamine (13.2 mL, 0.96 mol) and the mixture is stirred for 30 min at room temperature. After the reaction, the precipitate (triethylamine hydrochloride) is removed by filtration. The filtrate is evaporated to dryness under vacuum and the residue is washed thoroughly with ethyl acetate. Recrystallization of the residue from ethanol gives colorless needles (thymine-β-Ala-OEt); yield 18.4 g (65%), mp 171.5 to 172.0°C.

Thymine-β-Ala-H (*142*) [69]. The ester obtained above (*141*) (10 g, 0.034 mol) is dissolved in 1 N NaOH aqueous solution (70 mL) and stirred for 1 h at room temperature. After evaporation of the solvent, the oily residue is acidified by 1 N hydrochloric acid to pH 2.5 to precipitate the product. Recrystallization from water at pH 2.5 gives a colorless product of thymine-β-Ala-H; yield 7.81 g (86%), mp 171.5 to 174.0°C.

Thymine-β-Ala-PEI (*144*) [69]. To a solution of thymine-β-Ala-H (*142*) (2.69 g, 0.01 mol) in dimethyl sulfoxide (30 mL), pentachlorophenyl trichloroacetate (4.4 g, 0.0107 mol) and triethylamine (1.5 mL, 0.011 mol) are added. The mixture is stirred for 20 min at room temperature to form thymine-β-Ala-OPcp (*143*). To this solution, linear polyethylenimine (0.43 g, 0.01 mol) in dimethyl sulfoxide (15 mL) is added and stirred for 1 day at 40°C. After removing the solvent, the residue is washed from ethanol. Repeated reprecipitation from dimethyl sulfoxide into acetone gives thymine-β-Ala-PEI, in 78% yield. The content of the thymine units on the polymer is determined by UV spectroscopy of the hydrolyzed sample of the polymer as 1-(2-carboxyethyl)thymine (95 mol %).

4.6 POLY(α-AMINO ACID) CONTAINING NUCLEIC ACID BASES

4.6.1 Poly(L-lysine) Derivatives

Poly(L-lysine) derivatives containing pendent nucleic acid bases were prepared by two different methods [53,41]: (a) synthesis of the nucleic acid base-substituted L-lysine derivatives followed by their polymerization using the *N*-carboxyamino acid anhydride (NCA) method [40] (Scheme 33), and (b) polymer reaction of the carboxyethyl derivatives of the nucleic acid bases with poly(L-lysine) [39] (Scheme 34).

4.6.1.a NCA Method

The L-lysine derivatives containing nucleic acid bases were prepared as shown in the reaction Scheme 33. The reaction of α,*N*-Cbz-L-lysine (*145*) with the *p*-nitrophenyl ester of the carboxyethyl derivatives is carried out in dimethyl sulfoxide or dimethylformamide solution at 25°C for 2 days. The yield is about 70 to 90% because the amino group of adenine base would not react with the activated ester under the condition used (Table 7).

Poly(L-lysine) derivatives were prepared according to the NCA method. As the solubility of NCAs (*148*) in common organic solvents is low, preparation of the NCAs is unsuccessful by the Fuchs [210] method using phosgene. Consequently, the NCAs was prepared by the Leuchs method [211] using thionyl chloride. As the NCAs are soluble in dimethyl sulfoxide and dimethylformamide but insoluble in common organic solvents, purification by recrystallization is not preferable. The NCAs were washed with dry ethyl ether and allowed to

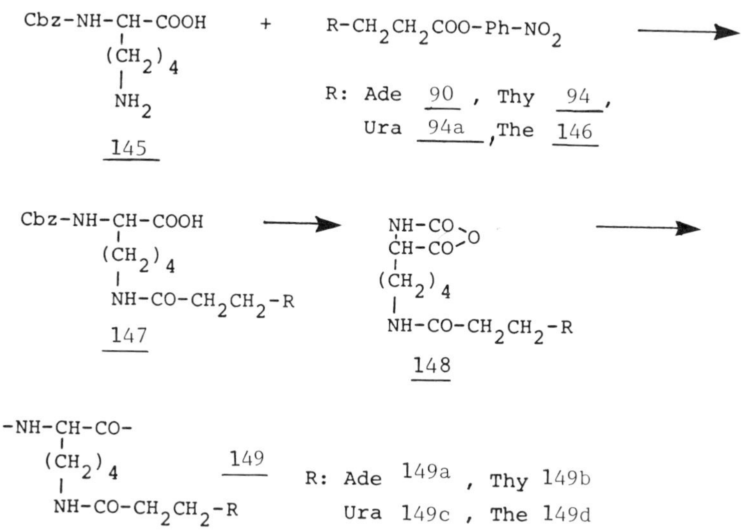

Scheme 33

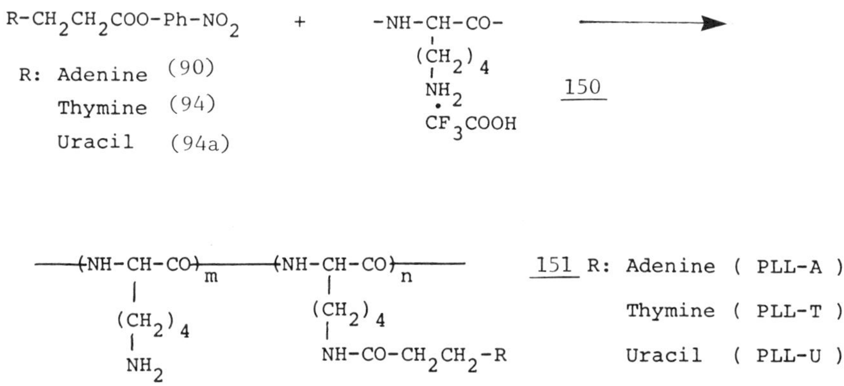

Scheme 34

polymerize in dimethyl sulfoxide solution with triethylamine as the initiator. In the case of the adenine derivative, more than an equimolar amount of triethylamine was consumed by the adenine derivative to initiate the polymerization. The NCAs of the adenine derivative is present as a salt form with hydrogen chloride. The degree of polymerization of polymer *149* was calculated from the terminal amino groups, which are determined by the trinitrobenzene sulfonate method.

Table 7. Synthesis of α,*N*-Cbz-ε,*N*-*R*-L-Lysines (*147*)

No.	R	Solvent[a]	Yield (%)	mp (°C)	UV[b] λ_{max} (nm)	ε_{max}	$[\alpha]_D$[c]
147a	Ade	DMSO	70	125-127	263	13,300	-2.4°
147b	Thy	DMF	87	159-161	271	9,500	-3.2°
147c	Ura	DMF	87	169-171	266	10,000	-2.2°
147d	The	DMSO	69	141-143	276	8,400	-1.0°

[a] Solvents used for the reaction.
[b] In ethanol at 25°C.
[c] In DMSO (c = 2) at 21°C.

a,N-Cbz-ε,N-Ade-L-Lysine (*147a*) [40]. To a suspension of *a,N*-benzyloxycarbonyl-L-lysine (*145*) (2.5 g, 8.9 mmol) in dimethyl sulfoxide (20 mL), *p*-nitrophenyl ester (*90*) (3.3 g, 10 mmol) is added at 25°C with stirring. The solution becomes clear after few hours. After stirring for an additional 2 days, the solvent is removed under reduced pressure. Addition of an excess amount of ethyl acetate to the oily residue gives colorless crystals. Recrystallization from ethanol-ethyl acetate gives the product *147a*; yield 2.9 g (70%), mp 125 to 127°C.

Poly(Ade-L-lysine) (*149a*) [40]. Thionyl chloride (1.0 mL, 14 mmol) is added to Cbz-Ade-Lys (*147a*) (0.6 g, 1.3 mmol) at 0°C and the reaction mixture is gradually heated to 40°C. After the evolution of gas ceases, the mixture is heated to 60°C for 15 min to remove the excess thionyl chloride. The residue is washed repeatedly with dry ethyl ether to afford the NCA derivative of adenine (*148a*). The product is dried in vacuo overnight.

To NCA (*148a*), dissolved in dimethyl sulfoxide (10 mL) is added triethylamine (0.25 mL, 1.8 mmol) as an initiator. The polymerization is carried out at 25°C for 48 h with stirring. After the solvent is removed in vacuo, an excess of ethanol is added to the oily residue to precipitate the polymer. Reprecipitation from dimethyl sulfoxide-ethanol gives polymer *149a*; yield 0.29 g (73%).

Poly(Thy-L-lysine) (*149b*), *poly(Ura-L-lysine)* (*149c*), *and poly(The-L-lysine)* (*149d*) [40]. These polymers are prepared using a procedure similar to that described for preparing poly(Ade-L-lysine) (*149a*). The results are listed in Table 8.

Table 8. Poly(ε,*N*-*R*-L-lysine) (*149*)

No.	R	[I]/[M][a] (mol %)	Yield (%)	UV[b] λ_{max} (nm)	ε_{max}	$[\alpha]_D$[c] (deg)	DP[d]
149a	Ade	1.4	73	266	11,800	-0.8	14.5
149b	Thy	0.4	79	274	8,800	-5.9	10.4
149c	Ura	0.4	74	269	8,300	-3.3	9.5
149d	The	0.4	82	278	8,000	-0.6	7.4

[a]I and M denote initiator and monomer, respectively.
[b]In DMSO solution at 25°C.
[c]In DMSO (c = 1) at 22°C.
[d]Calculated from terminal amino acid units.

4.6.1.b Polymer Reaction Method

The derivatives of poly(L-lysine), which have various amounts of nucleic acid bases, were synthesized by grafting the carboxyethyl derivatives of adenine, thymine, and uracil onto poly(L-lysine) as shown in Scheme 34. Poly(L-lysine) (*150*) used in these reactions is present as a trifluoroacetic acid or hydrobromide salt. The reaction of the *p*-nitrophenyl ester with poly(L-lysine) is carried out in dimethyl sulfoxide in the presence of triethylamine at 25°C. Copolymers with various amounts of nucleic acid bases were prepared by the polymerization of poly(L-lysine) with various molar ratios of *p*-nitrophenyl esters. The nucleic acid base content of these copolymers was determined by UV spectroscopy of hydrolyzed samples.

With thymine and uracil derivatives, graft reactions proceed almost quantitatively, while in the case of the adenine derivative, the activated ester does not react well with poly(L-lysine), and only about 67% adenine units are in the polymer. This may be caused by the instability of the activated ester and could be improved by using the protected adenine derivative (*172*). The results of the grafting reactions are shown in Table 9.

Poly(L-lysine trifluoroacetate) (*150*) [212]. Poly(ε,*N*-trifluoroacetyl-L-lysine) is prepared by the polymerization of ε,*N*-trifluoroacetyl-α,*N*-carboxy-L-lysine anhydride in dried dioxane with triethylamine as an initiator. The molecular weight of the polymer is determined from the intrinsic viscosity in dichloroacetic acid. The trifluoroacetyl protecting group is removed with piperidine in methanol, and the resulted clear solution is evaporated to dryness under vacuum to afford the polymer *150*.

Table 9. Synthesis and Analytical Data of the Copolymers

No.	Base	Sample	Base/PLL[a] (mol/mol)	Mol % of base in the polymer[b]	DP[c]
151a	Ade	PLL-A-53	1.53	53	200
		-67	1.48	67	430
		-31	0.66	31	430
		-15	0.34	15	430
151b	Thy	PLL-T-97	1.50	97	200
		-93	1.44	93	430
		-79	0.80	79	430
		-65	0.65	65	430
		-34	0.35	34	430
		- 8	0.10	8	430
151c	Ura	PLL-U-97	1.51	97	200
		-93	1.45	93	430
		-76	0.87	76	430

[a]Molar ratio of the activated ester (base) to poly(L-lysine) (PLL) in the grafting reaction.
[b]From UV spectra of the hydrolyzed samples.
[c]From the viscosity of the original polymer.

Poly(ε,N-adenine-L-lysine) (*151a*) [39]. To a solution of the polymer *150* (400 mg, 1.7 mmol) in dimethyl sulfoxide (10 mL), *p*-nitrophenyl ester of the carboxyethyl adenine (*90*) (850 mg, 2.6 mmol) and triethylamine (0.3 mL, 2.1 mmol) are added, and the solution is stirred for 3 days at 25°C. The resulting mixture is poured into an excess of ethyl ether to give an oily polymer. The polymer is precipitated with acetone and purified by repeated precipitation from dimethylformamide-ethyl ether or dimethyl sulfoxide-acetone. The polymer is washed with ethyl ether and dried to give a white powder, poly(ε,*N*-adenine-L-lysine) (*151a*); yield 425 mg (79%).

Polypeptides of lysine and ornithine with pending pyrimidine bases were prepared by another method (Scheme 35). Starting with β-methoxymethacryloylisocyanate (*152*) and α,*N*-Cbz-L-lysine or α,*N*-Cbz-L-ornithine, α,*N*-Cbz-amino acids with pyrimidine bases in the side chain (*154*) were synthesized. These protected nucleoamino acids are converted to the corresponding NCAs via the silylester method [200]. The poly(nucleoamino acid) (*158*) obtained by polymerization of the NCAs are characterized by spectroscopy and have DP values in the range 20 to 30°C.

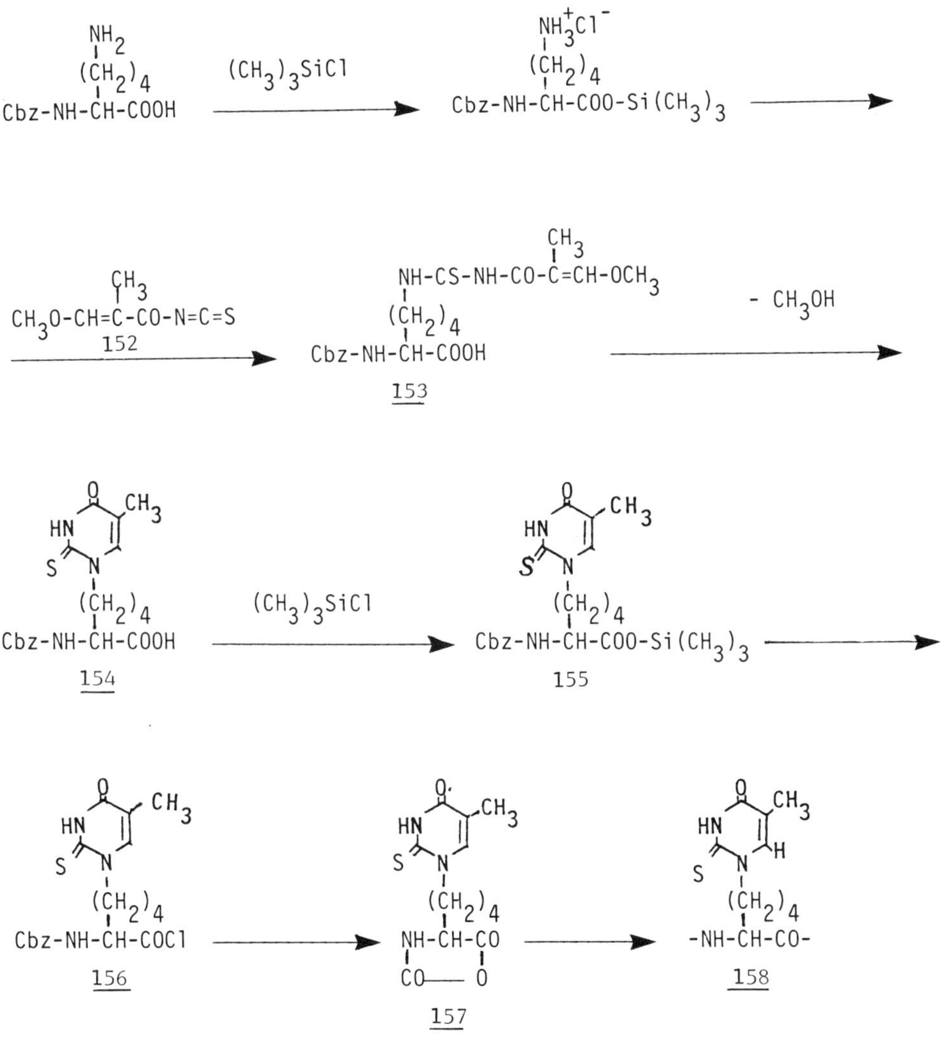

Scheme 35

α,N-Cbz-ε,N-(3-Methoxymethacryloyl thiocarbamoyl)-L-lysine (*153*) [200]. α,*N*-Cbz-L-lysine (28 g, 0.11 mol) and trimethyl chlorosilane (11 g, 0.1 mol) are refluxed in a mixture of dry chloroform (150 mL) and dry acetonitrile (30 mL) until a clear solution of the α,*N*-Cbz-L-lysine trimethylsilyl ester hydrochloride is obtained. At 0°C, 3-methoxymethacryloyl isothiocyanate (*152*) [213] (15.7 g, 0.1 mol) is added to this solution at once and triethylamine (10.1 g, 0.1 mol) dropwise. The reaction mixture is stirred at room temperature for 2 h. After the product is washed with 1 N HCl (100 mL) and water (100 mL), it is dried over sodium sulfate. The dry solution is concentrated and the product crystallized by slow addition of carbon tetrachloride and ligroin under cooling with ice to yield 30.6 g (70%) of *153*, mp 132 to 134°C.

L-α-Benzyloxycarbonylamino-ε-(2-thiothymine-1-yl)caproic acid (*154*) [200]. Compound *153* (43.8 g, 0.1 mol) is stirred in 2 N NaOH (125 mL) at 25 °C for 10 min and at 60°C for 10 min. The clear solution is cooled to 5°C, acidified with ice-cold 5 N HCl to pH 2, and extracted with 300 mL and then 100 mL of ethyl acetate. The combined extracts are washed with water (100 mL), dried over sodium sulfate, and concentrated in vacuo. The product is crystallized by slow addition of carbon tetrachloride and ligroin with cooling to yield 24.3 g (60%), mp 164 to 166°C.

L-α-Amino-ε-thiopyrimidin-1-yl-acid NCA (*157*) [200]. An α,*N*-Cbz-ε,*N*-thiopyrimidin-1-yl-amino acid (*154*) (20 mmol) and triethylamine (2.5 g, 25 mmol) are heated in dry ethylacetate (100 mL) and trimethylchlorosilane (2.4 g, 22 mmol) is added dropwise. After cooling to 0°C, the reaction mixture is filtered under dry conditions and the filtrate concentrated to 50 mL. Freshly distilled thionyl chloride (2.4 g, 20 mmol) is added to the concentrated solution, which is stirred at 20°C for 3 days. The reaction mixture is then cooled to 0°C and the upper layer decanted from the lower viscous portion. The decantate is concentrated and the oily residue is treated with a small amount of dry ethyl acetate and carbon tetrachloride to yield NCA as crystals. The bottom viscous material is treated with 30 mL of dry THF with cooling to crystallize NCA.

Polymerization of NCA [200]. (a) A nucleoamino acid NCA (*157*) (10 mmol) is dissolved in dimethylformamide (20 mL) and *t*-butylamine (or α-aminodiphenylmethane) (0.2 mmol) is added as a 1 M solution in dioxane. The reaction mixture is stirred at 80°C for 10 days, and the polymer precipitated from water.

(b) Nucleoamino acid NCA (10 mmol) is dissolved in dry pyridine and heated to 80°C and after 5 min the evolution of carbon dioxide is observable. After 4 h the polymer is precipitated from water.

Polypeptides with pending nucleosides were prepared by the reaction of polypeptides with reactive nucleoside derivatives. In the first step in this direction, the 5'-OH group of nucleosides is converted to a carboxy function, which is then reacted with polylysine by means of dicyclohexylcarbodiimide; but only a poor yield is obtained. The binding of nucleosides to basic polypeptides via isocyanates was studied as shown in Scheme 36 [199]. The 5'-OH group of 2',3'-*O*-isopropylideneuridine and of N^6-benzoyl-2',3'-*O*-isopropylideneadenosine

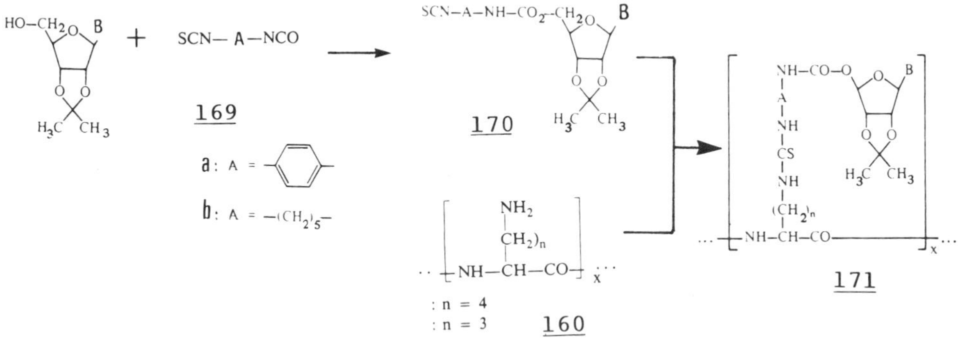

Scheme 36

were attached selectively with the isocyanate group of aromatic and aliphatic α-isocyanato-ω-isothiocyanates (*159*). The isothiocyanate group of the resulting nucleoside derivatives (*160*) was then reacted with the free amino groups of polylysine or polyornithine.

Isothiocyanatonucleosides (*160*) [199]. An aliphatic or aromatic isocyanatoisothiocyanate (0.1 mol), 2',3'-*O*-isopropylidenenucleoside (0.12 mol), and 1,4-diaza[2,2,2]bicyclooctane (0.2 g) are refluxed in dry THF until the isocyanate IR2280-cm^{-1} peak disappears. The solution is concentrated and the residue dissolved in methylene chloride (200 mL). This solution is washed with 1% citric acid solution (100 mL) to remove the catalyst and unreacted nucleosides. The organic layer is dried with sodium sulfate and concentrated. The residue is powdered and dried at 80°C for 48 h; crystallization is never successful (80 to 95% yield).

Nucleopolypeptides (*161*) [199]. Polylysine or polyornithine hydrobromide (10 mmol) is dissolved in dimethyl sulfoxide (30 mL) and triethylamine (1.5 mL). Isothiocyanate-nucleoside (*160*) (11 mmol) is added and the reaction mixture is stirred at room temperature for 2 to 3 days. The polymer is precipitated from cold methanol and extracted five times with methanol to remove triethylamine hydrobromide. Finally, the product is dried at 80°C for 48 h.

The conformations of the poly(L-lysine) derivatives of nucleic acid bases were studied by circular dichroism (CD) and optical rotatory dispersion (ORD) measurements [49,59]. Poly(L-lysine) is known to exist in disordered, helical, and β-conformations depending on the temperature, pH, and the solvent used. The side chain of the polymer has a significant effect on the backbone conformation.

Poly(L-lysine) with 67% adenine units (PLL-A-67) has a helical structure even in acidic aqueous solution, where adenine base in the side chain is positively charged. With an increase in pH the helical conformation of the polymer increased due to release of the electrostatic repulsion between positively charged side chain. Above pH 2.5, the spectra cannot be measured, as the polymer begins to precipitate in aqueous solution.

The thymine polymer derivatives of poly(L-lysine) (PLL-T-93 and PLL-T-79) tend to exist in a helical conformation at neutral pH, while poly(L-lysine) exists in a random coil structure. In contrast the helicity of PLL-Ts decreases with increasing pH because of the electrostatic repulsion between negatively charged thymine bases.

The polymer complex formation between the complementary poly(L-lysine) derivatives was studied by UV spectroscopy in dimethylsulfoxide-ethylene glycol mixed solvent system [51,58]. The complex formation between PLL-A-67 and PLL-T-93 can be clearly observed. The overall stoichiometry of the complex based on the base units was about 1:2 (adenine:thymine). As the adenine-substituted poly(L-lysine) contains 67% adenine units, the stoichiometry based on the polymer chain was about 1:1, and the hypochromicity at this point was 31% (Figure 4).

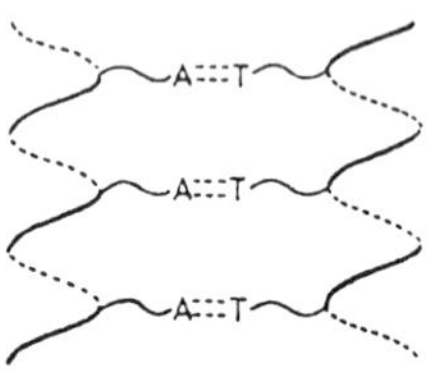

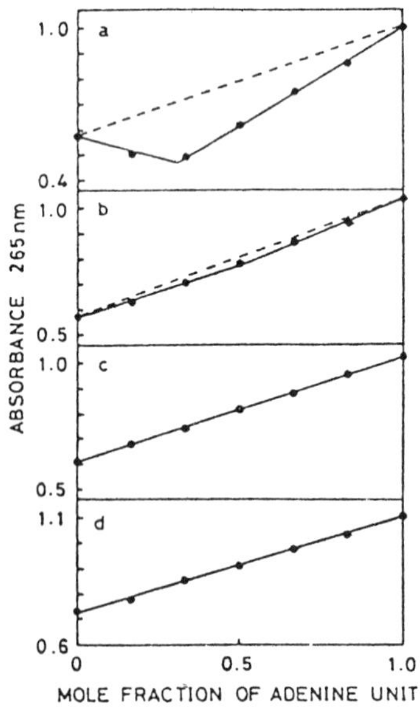

Figure 4. Mixing curves between PLL-A-67 and PLL-Ts in DMSO-EG (3:2 v/v). Absorbance in a 10 mm cell at 25°C. (a) PLL-T-93, (b) PLL-T-79, (c) PLL-T-65, (d) low molecular weight PLL-T.

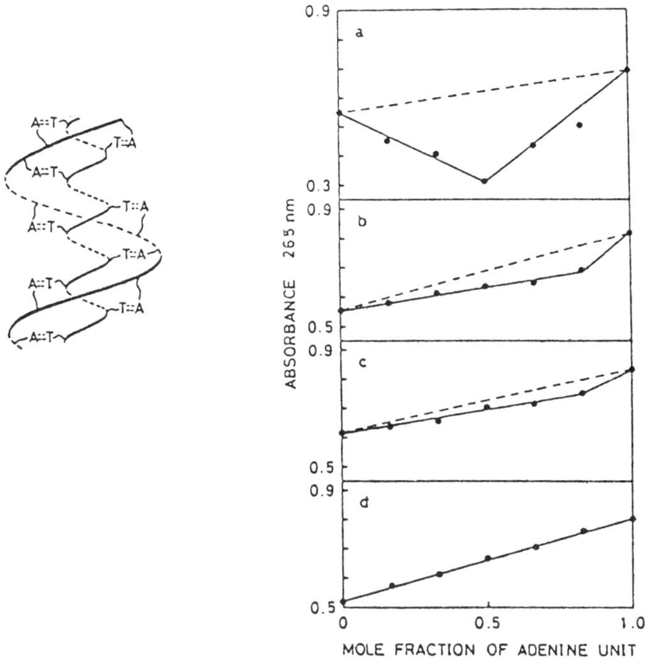

Figure 5. Mixing curves between poly-MAOA and PLL-Ts in DMSO-EG (3/2, v/v). Absorbance at 265 nm was obtained in a 10 mm cell at 25°C. (a) PLL-T-93, (b) PLL-T-79, (c) PLL-T-65, (d) low molecular weight PLL-T.

The stability and stoichiometry of the complex between polymers containing nucleic acid bases are affected by the compatibility of the different base-base distances in the polymers, and also by mutual penetration ability between the main chains. Poly(L-lysine) derivatives and vinyl polymers are apparently incompatible and seem to form unstable complexes. In spite of these considerations, PLL-T-93 with poly(9-methacryloyloxyethyladenine) (*54a*) was found to give a stable polymer complex, when the overall stoichiometry based on nucleic acid bases was 1:1, with a high hypochromicity value (53%). These facts indicate that the polymer complex assumes a double helical structure as illustrated in Figure 5.

The photodimerization reaction of pendant thymine bases in the thymine-containing polylysine derivatives was studied over a wide range of pH in aqueous solution [70]. It was found that the quantum yield of the photodimerization of the pendant thymine bases was affected mainly by the conformation of the polymers in the solution; helix or random coil. It was also found that a conformational change of the polymer chain occurred on photoirradiation. Photodimer formation between thymine bases at the side chain fixed the helical structure of the polymer. The fixed helical structure was stable and not affected by pH changes in the system [83].

4.6.2 Isopoly(L-lysine) Derivatives

Isopoly(L-lysine) (*169*) is a poly(ε-amino acid) in which the ε-amino group of L-lysine is condensed with the α-carboxy group. This produces a chain backbone that is a variant of the one usually seen in proteins with an α-amino group as the side chain. Poly(α-amino acid) is known to form α-helix since it is difficult to form α-helical structure with poly(ε-amino acid). Conformational studies of poly(ε-lysine) suggest that the likelihood of a chain order that is formally similar to the antiparallel pleated-sheet conformation of proteins [214]. It is therefore of interest to study the conformation and the interaction of the poly(ε-lysine) containing nucleic acid bases on the α-amino group.

For the synthesis of isopoly(L-lysine) with pending nucleic acid bases, suitable monomers bearing the nucleic acid base in the side chain was carried out [75]. The L-lysine derivatives containing nucleic acid bases at the α-amino group (*163*) were prepared by condensation of *N*-ε-Cbz-L-lysine (*162*) with the carboxyethyl derivatives of nucleic acid bases using the activated ester method (Scheme 37). After removing of Cbz group, ε-amino acid derivatives of nucleic acid bases (*164*) were obtained. The reaction of the ε-amino acid derivatives with pentachlorophenyl trichloroacetate gave the pentachlorophenyl ester trichloroacetic acid salt (*165*). This activated ester was heated to 130 to 150°C to give the isopoly(L-lysine) derivatives (*166*).

N-ε-Cbz-N-α-9-carboxyethyladenyl-L-lysine (*163a*) [75]. Pentachlorophenyl ester of 9-carboxyethyladenine (*134*) (8.9 g, 20 mmol) in dimethyl sulfoxide (30 mL) are stirred for 3 days at 30°C to give a clear solution. After removal of the solvent under reduced pressure, the residue is washed with acetone by vigorous stirring. Recrystallization from isopropyl alcohol gives product *163a*; yield 3.8 g (42%), mp 177 to 179°C.

N-α-Adenine-L-lysine (*164a*) [75]. To *N*-ε-Cbz-*N*-α-adenine-L-lysine (*163a*) (8.6 g, 18 mmol), HBr saturated acetic acid (70 mL) is

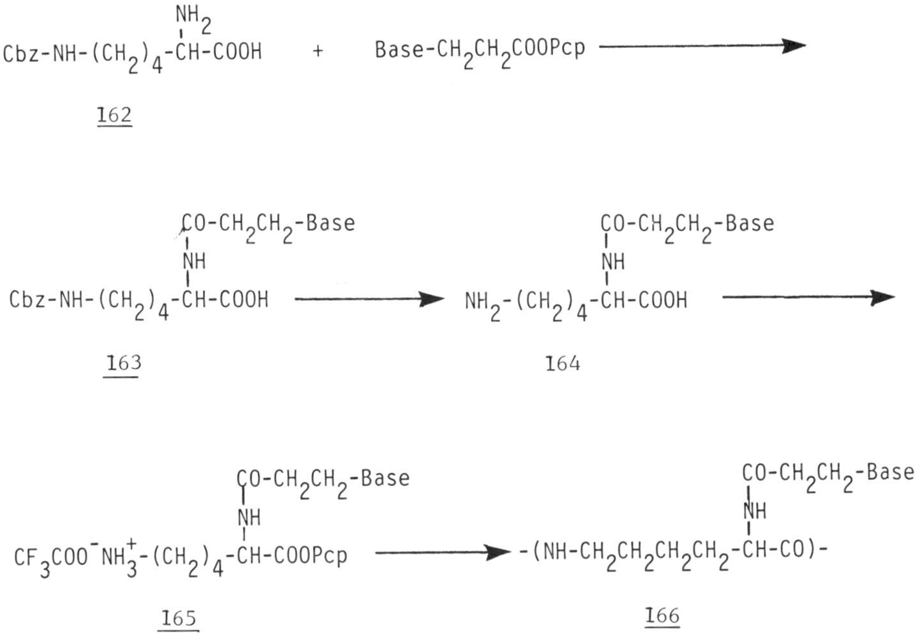

Pcp; Pentachlorophenyl Base: Adenine (**a**)
Uracil (**b**)

Scheme 37

added and stirred for 1.5 h. After evaporation of the solvent, the residue is washed with ethyl ether, dissolved in water, and neutralized with sodium bicarbonate solution to pH 7. The solvent is removed and the residue is recrystallized from methanol to give *164a*; yield 5.1 g (84%), mp 179 to 180°C.

Isopoly(N-α-adenine-L-lysine) (*166a*) [75]. To a suspension of N-α-adenine-L-lysine (*164a*) (400 mg, 1.2 mmol) in dimethylformamide (1 mL), pentachlorophenyl trichloroacetate (490 mg, 1.2 mmol) is added with stirring at 0°C. Continued stirring for 30 min under reduced pressure at room temperature gives a clear solution. To this solution, dry Celite-545 (400 mg) is added and the solvent is removed under reduced pressure. The residue is heated to 130°C under high vacuum for 1 day to sublimate the pentachlorophenol. After that, the residue is washed with water and the polymer extracted with 1 *N* hydrochloric acid (30 mL) from celite. The solution is concentrated and a small amount of methanol is added. The polymer isopoly(N-α-adenine-HCl-L-lysine) (*166a*) is precipitated from acetone in 15% yield (55 mg).

N-ε-Cbz-N-α-Uracil-L-lysine (*163b*) [75]. This compound (*163b*) is prepared from pentachlorophenyl ester of the uracil derivative (*140a*) and N-ε-Cbz-L-lysine (*162*) by the same procedure as that used for the adenine derivative in 88% yield, mp 151 to 153°C after recrystallization from nitromethane.

N-α-Uracil-L-lysine (*164b*) [75]. The Cbz group of compound *163b* is removed with HBr in acetic acid in 82% yield, mp 168 to 170°C (recrystallized from methanol).

Isopoly(N-α-uracil-L-lysine) (*166b*) [75]. From N-α-uracil-L-lysine (*164b*) (400 mg, 1.28 mmol) using a procedure similar to that for preparation of the adenine derivative, the isopolylysine derivative of uracil is obtained. After the reaction, the polymer is separated from celite by 1 N NaOH aqueous solution. Precipitation from methanol with acetone gives isopoly(N-α-uracil-L-lysine) (*166b*) as the sodium salt. The polymer is obtained, by neutralization with hydrochloric acid, in 20% yield (78 mg).

Alternatively, isopoly(L-lysines) containing nucleic acid bases may be prepared by grafting of the nucleic acid base derivatives onto isopoly(L-lysine) (*169*). The N-hydroxysuccinimide ester method was used [198] for the grafting of 2-thiothymine derivative, the isothiocyanate derivatives of nucleosides were used [199] for the grafting of nucleosides onto isopoly(L-lysine). These procedures are similar to those used for the poly(L-lysine) of poly(L-ornithine) derivatives.

Recently, the carboxyethyl derivatives of adenine and thymine were grafted on isopoly(L-lysine) by the activated ester method [76]. Isopoly(L-lysine) was prepared by condensation using the activated ester method (Scheme 38). The p-nitrophenyl ester of N-α-Cbz-L-lysine trifluoroacetic acid salt (*167*) was polymerized by the addition of triethylamine in dimethylformamide to give isopoly(Nα-Cbz-L-lysine)

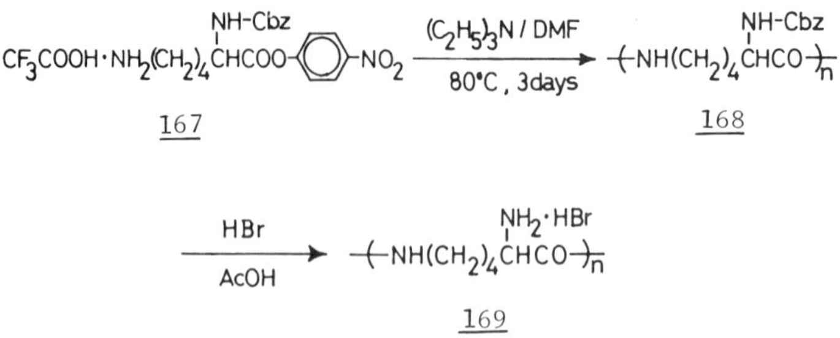

Scheme 38

(*168*). Removing the Cbz group with HBr in acetic acid gave isopoly(L-lysine-HBr) (*169*).

For the grafting of the adenine derivative, the activated ester of 6-amino protected adenine derivative was prepared (Scheme 39). The

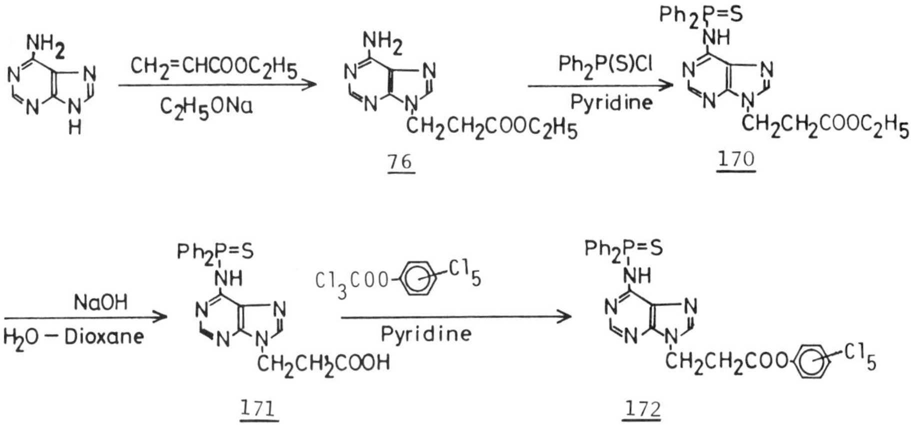

Scheme 39

protected activated ester was found to be more stable than the unprotected ester (*90*). The activated ester *172* is reacted with isopoly(L-lysine) in dimethyl sulfoxide by the addition of triethylamine. The adenine derivative was successfully grafted onto isopoly(L-lysine) to give the polymer containing 92% adenine units (Scheme 40). The thymine derivative is almost quantitatively grafted onto isopoly(L-lysine) by using the activated ester (*140*), which is used for the grafting on polyethylenimine (Scheme 41).

Isopoly(N-α-Cbz-L-lysine) (*168*) [76]. To a solution of N-α-Cbz-L-lysine-OPNP · CF_3COOH (*167*) (32.1 g, 50 mmol) in dimethylformamide (50 mL), triethylamine (10.1 g, 100 mmol) is added and stirred for 1 day at 25°C and for 3 days at 80°C. Then 1 N hydrochloric acid is added to the reaction mixture to precipitate the polymer. The polymer is filtered and washed with 1 N hydrochloric acid and then with acetone to yield 43% (5.6 g) of *168*.

Isopoly(L-lysine-HBr) (*169*) [76]. Isopoly(N-α-Cbz-L-lysine) (*168*) (13.1 g, 50 mmol) is dissolved in HBr/acetic acid (50 mL) and stirred for 5 h at room temperature, 10 h at 60°C, and 2 h at room temperature. To the solution, ethyl ether is added to precipitate the polymer. The solvent is removed by decantation and the polymer washed with acetone. Reprecipitation from water (2 mL) in acetone (500 mL) gives isopoly(L-lysine-HBr) (*169*) in 72% yield (7.6 g).

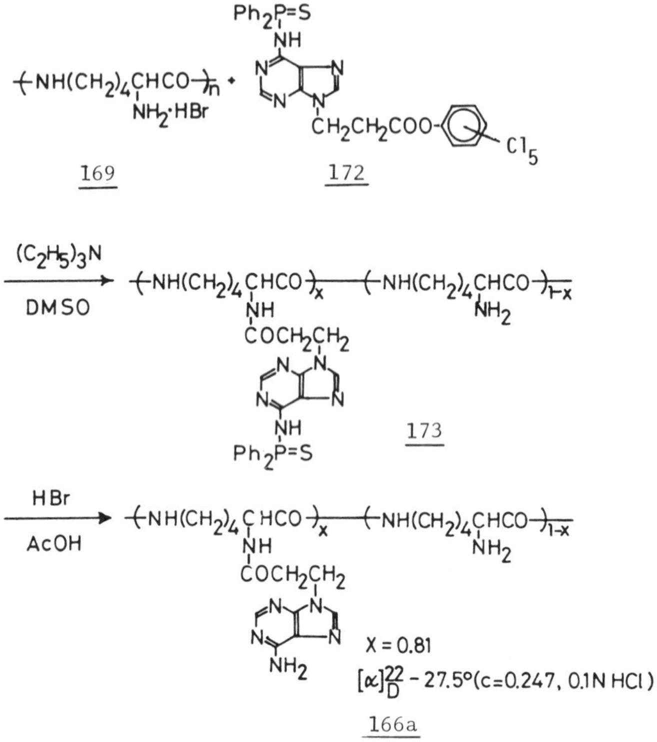

Scheme 40

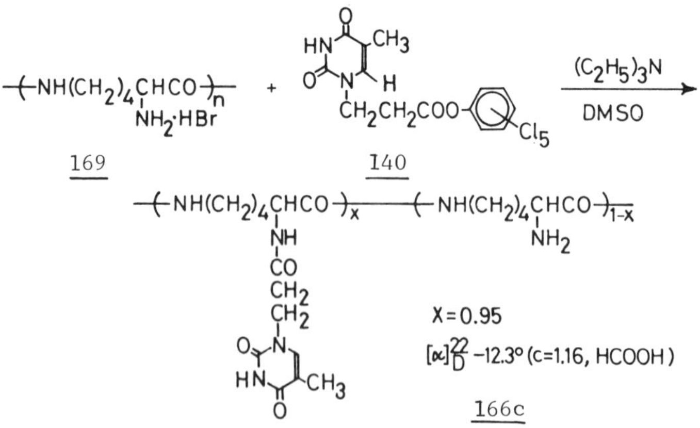

Scheme 41

6-(N-diphenylphosphinothioyl)amino-9-(2-carboethoxyethyl)purine; Ppt-adenine-COOEt (170) [76]. To a suspension of 9-(2-carboethoxyethyl)adenine (*66*) (4.75 g, 20 mmol) in pyridine (20 mL), diphenylphosphinothioyl chloride (8.9 mL, 44 mmol) is added at room temperature and refluxed for 2 h. The solution is cooled to 0°C and the precipitated salt is removed and the filtrate evaporated. To the residue, benzene (200 mL) and water (100 mL) are added and stirred. The benzene solution is separated and washed with water and dried over sodium sulfate. After evaporating the benzene, methanol (200 mL) is added to the residue to precipitate the product as crystals. After filtration, the filtrate is evaporated to give an oily product. Ethyl ether (200 mL) is added to the oily residue and stirred overnight to give the product (*170*) as crystals in 80% yield (7.22 g), mp 117 to 119°C.

Ppt-Adenine-COOH (171) [76]. To a solution of Ppt-adenine-COOEt (*170*) (6.42 g, 1.42 mmol) in dioxane (50 mL), 2 N NaOH aqueous solution (20 mL) is added and stirred for 20 min at room temperature. The solution is concentrated by evaporation to 20 mL and neutralized with 1 N hydrochloric acid to pH 4 to precipitate the product. After filtration, recrystallization from ethanol-water (200:10 mL) gives crystals in 99% yield (5.93 g), mp 213 to 215°C.

Ppt-Adenine-COO-Pcp (172) [76]. Ppt-adenine-COOH (*171*) (0.98 g, 2.3 mmol) is dissolved in pyridine (10 mL) and pentachlorophenyl trichloroacetate [209] (4.6 g, 2.3 mmol) is added. The mixture is stirred for 20 h at room temperature. After evaporation of the solvent, cyclohexane (50 mL) is added to the residue and the insoluble material is removed. The solvent is evaporated and the residue is crystallized from benzene to give product *172* as crystals in 90% yield (1.40 g), mp 122.5 to 123°C.

Isopoly(Ppt-adenine-L-lysine) (173) [76]. To a solution of isopoly(L-lysine-HBr) (*169*) (179 mg, 0.856 mmol) in dimethyl sulfoxide (6 mL), Ppt-adenine-COO-Pcp (*172*) (863 mg, 1.28 mmol) and triethylamine (0.26 mL) are added. The mixture is stirred for 2 days at room temperature. The solvent is evaporated and the residue is washed successively with methanol, ethanol, and acetone. The polymer is purified by reprecipitation from dimethylformamide (4 mL) in acetone. This polymer (*173*) is found by NMR spectroscopy to contain 92% adenine units. The molecular weight of the polymer determined by vapor pressure osmometer method is 11,000.

Isopoly(adenine-L-lysine) (166a) [76]. To a solution of isopoly(Ppt-adenine-L-lysine) (*173*) (182 mg) in acetic acid (2 mL), saturated HBr/acetic acid (2 mL) is added and stirred for 2 days at room temperature. Then ethyl ether is added to the solution to precipitate the polymer. The polymer is washed with benzene and then acetone. Reprecipitated from dimethylformamide with acetone affords isopoly(HBr-adenine-L-lysine) (HBr salt of polymer *166a*).

The polymer is dissolved in water and neutralized by 1 N NaOH aqueous solution to pH 7. The precipitated polymer is filtered and washed with water. Reprecipitation from dimethyl sulfoxide to acetone gives polymer *166a*, isopoly(adenine-L-lysine).

Isopoly(thymine-L-lysine) (*166c*) [76]. To a solution of isopoly(L-lysine-HBr) (*169*) (177 mg, 0.847 mmol) in dimethyl sulfoxide (6 mL), pentachlorophenyl ester of 1-(2-carboxyethyl)thymine (*140*) (378 mg, 0.847 mmol) and triethylamine (0.24 mL, 1.7 mmol) are added and stirred for 2 days at room temperature. The solvent is removed and the residue is washed with ethyl ether and acetone to give the polymer *166c*. The polymer is reprecipitated from dimethyl sulfoxide into acetone. This polymer is found by NMR spectroscopy to contain 95% thymine units.

The conformations of the isopoly(L-lysine) derivatives of nucleic acid bases were studied by circular dichroism (CD) measured [76] in aqueous or in water-ethylene glycol solution. Poly(L-lysine) derivatives containing nucleic acid bases were found to exist in α-helical conformation at neutral pH. However, isopoly(L-lysine) derivatives of adenine and thymine were found to exist in β conformation from the CD spectrum.

The polymer complex formation between the complementary isopoly(L-lysines) was observed by UV and CD spectra in a water-ethylene glycol mixed solvent system. It is suggested from the CD spectra of the polymer complex that the double-helical structure was caused by specific interaction between adenine and thymine bases in the side chain [76].

The photochemical reaction of the isopoly(L-lysine) derivative containing thymine bases was also studied. The rate of the photodimerization reaction of the isopoly(L-lysine) derivative was found to be lower than that of poly(L-lysine) derivative.

4.6.3 Poly(L-glutamine) Derivatives

The preparation of poly(L-glutamate) containing nucleic acid bases was tried [17]. For the theophylline derivative, the L-glutamate derivative was prepared and polymerized by the NCA method to give poly(L-glutamate) containing theophylline [47]. The preparation of L-glutamate derivatives of nucleic acid bases were tried, but the selective esterification of L-glutamic acid at the γ position was difficult [47]. On the other hand, preparation of L-glutamine derivatives of nucleic acid bases was successful [74] (Scheme 42).

Glutamic acid was protected [*N*-α-Cbz-α-Bz-L-Glu (*174*)] and was condensed with 2-aminoethyl derivatives of nucleic acid bases by the activated ester method to give the L-glutamine derivatives (*175*). After the benzyl-group was removed, the NCA of the glutamine

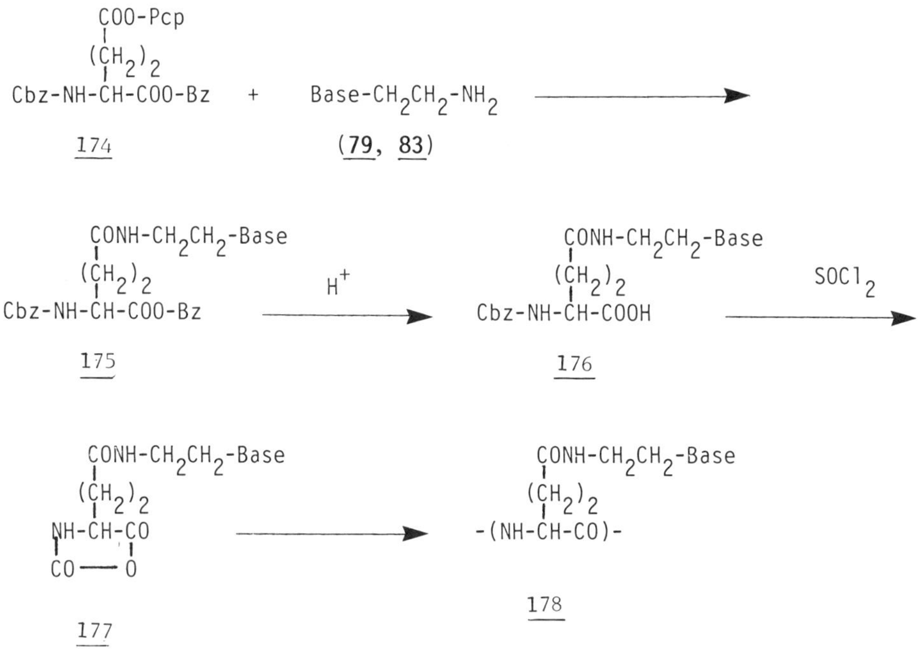

Scheme 42

derivatives were prepared by the reaction with thionyl chloride (*177*). The polymerization of the NCAs (*177*) gave the poly(L-glutamine) derivatives containing nucleic acid bases (*178*).

The poly(L-glutamine) derivatives were also prepared by grafting of the aminoethyl derivatives of nucleic acid bases onto the poly(γ-pentachlorophenyl-L-glutamate), which was prepared from poly(L-glutamic acid) (*180*) with pentachlorophenyl trichloroacetate (Scheme 43).

N-α-Cbz-α-Bz-N-γ-Ade-L-glutamine (*175a*) [74]. Dicyclohexylamine salt of N-α-Cbz-α-Bz-L-glutamic acid (5.5 g, 10 mmol) and pentachlorophenyl trichloroacetate (3.1 g, 10 mmol) in dimethyl sulfoxide (80 mL) are stirred for 1 h at room temperature to give a clear solution of the pentachlorophenyl glutamate (*174*). To this solution, 9-(2-aminoethyl)-adenine (*79*) (2.0 g, 11 mmol) is added and stirred for 24 h at room temperature. The solvent is removed under vacuum and the residue is dissolved in ethyl acetate and washed with water. The solution is concentrated by evaporation to give the product as crystals. Recrystallization from ethyl acetate gives product *175a* in 60% yield (3.0 g), mp 143 to 144°C.

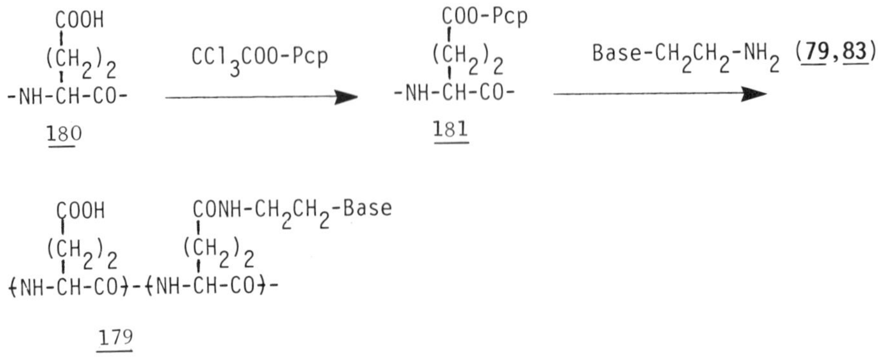

Base: Adenine (a) Cbz: Carbobenzoxy Pcp: Pentachlorophenyl
Uracil (b) Bz: Benzyl

Scheme 43

N-α-Cbz-N-γ-Ade-L-glutamine (*176a*) [74]. Compound *175a*, *N*-α-Cbz-α-Bz-*N*-γ-Ade-L-glutamine (6 g, 13 mmol), is dissolved in acetone and 1 N NaOH (11.3 mL) is added and stirred for 3 h. The solution is concentrated by evaporation and neutralized with 10% citric acid to pH 5 to yield crystals. Recrystallization from ethanol-water gives product *176a* in 60% yield (3.0 g), mp 234.0 to 234.5°C.

Poly(N-γ-Ade-L-glutamine) (*178a*) [74]. Thionyl chloride (2 mL) is added to *N*-α-Cbz-*N*-γ-Ade-L-glutamine (*176a*) (800 mg) and stirred for about 15 min at 40°C to give the acid chloride. The excess thionyl chloride is distilled off. The residue is dissolved in acetic anhydride and heated to 60°C. The acetic anhydride is then removed under vacuum. The residue is recrystallized from acetic anhydride in ethyl ether to give the NCA (*177a*).

The NCA is dissolved in dimethylformamide (10 mL), and triethylamine (300 μl) is added as an initiator. After stirring for 3 days at room temperature, the solution is added to methanol to give polymer *178a* as a precipitate.

N-α-Cbz-α-Bz-N-γ-Ura-L-glutamine (*175b*) [74]. This compound is prepared using 1-(2-aminoethyl)uracil (*83*) by the same procedure as that used for the adenine derivative (*175a*) in 87% yield, mp 126 to 127°C.

N-α-Cbz-N-γ-Ura-L-glutamine (*176b*) [74]. Hydrolysis of *N*-α-Cbz-α-Bz-*N*-γ-Ura-L-glutamine (*175b*) (7.1 g, 14 mmol) is carried out by 1 N NaOH in acetone-water for 3 h at room temperature. The solution is concentrated and acidified with 10% citric acid to pH 3. The oily product is extracted with ethyl acetate and dicyclohexylamine is added to the solution to precipitate the salt. The dicyclohexylamine salt is recrystallized from isopropanol. The salt is then acidified by hydrochloric acid and the product is extracted with acetone. The solvent is evaporated and the residue is recrystallized from methanol-ethyl ether to give *N*-α-Cbz-*N*-γ-Ura-L-glutamate in 77% yield (4.5 g), mp 95 to 96°C.

Poly(N-γ-Ura-L-glutamine) (*178b*) [74]. The uracil derivative (*176b*) is reacted with thionyl chloride by the same procedure as that used for the adenine derivative *177a* to give the NCA *177b*. The polymerization procedure of the uracil derivative *177b* is similar to that for the adenine derivative.

Poly-L-glutamine derivative by grafting [77]. To a suspension of poly(L-glutamic acid) (*180*) (6.1 mmol) in dimethyl sulfoxide (30 mL), pentachlorophenyl trifluoroacetate (6.8 mmol) is added and stirred for 3 days at 60°C to give a clear solution of poly(pentachlorophenyl-L-glutamate) (*181*). To the solution, 2-aminoethyl derivative of adenine or thymine is added and stirred for 3 days at 60°C. The solution is added to methanol to precipitate the polymer [poly(L-glutamine) derivatives of adenine or thymine (*179*)].

4.6.4 Polyalanine Derivatives

L-Willardiine (*184*) is a naturally occurring compound, which is an amino acid having uracil as the side chain [215,216]. Analogs of willardiine—alanine derivatives of nucleic acid bases such as *185* to *188*—were prepared and were polymerized by the NCA or the activated ester method [21,24,135,164,173,174]. These willardiine analogs are prepared by treating the corresponding nucleic acid bases with bromoacetaldehyde diethyl acetal to give *182*. The acetal derivative *182* is hydrolyzed to give the aldehyde *183*, which is converted to D,L-alanine derivative (*184*) by the Strecker synthesis (Scheme 44). The optical resolution of thymine derivative is reported using fractional crystallization of the α-methylphenylethylamine salt of the *N*-α-formyl derivative.

Acetal derivatives of nucleic acid bases (*182*) [164]. The base is suspended in dimethylformamide (25 mL/g base), dry potassium carbonate (1 mol portion), 1-bromo-2,2-diethoxyethane (1 eq mol) added and the mixture heated to 130°C with stirring for 14 h. The solution is concentrated in vacuo to dryness and the uracil derivative is extracted with chloroform from aqueous solution in 60% yield by trituration with ether. Evaporation gave a crystalline solid, mp 89°C. The 4-*N*-acetylcytosine derivative is also extracted with chloroform, deacetylated by treatment with methanol ammonia overnight at room

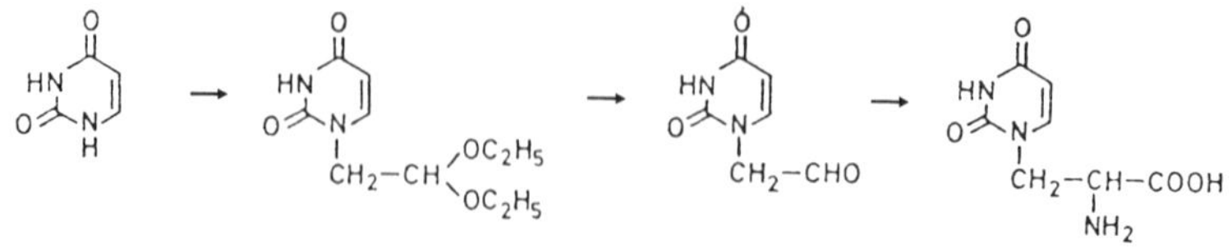

182 183 184

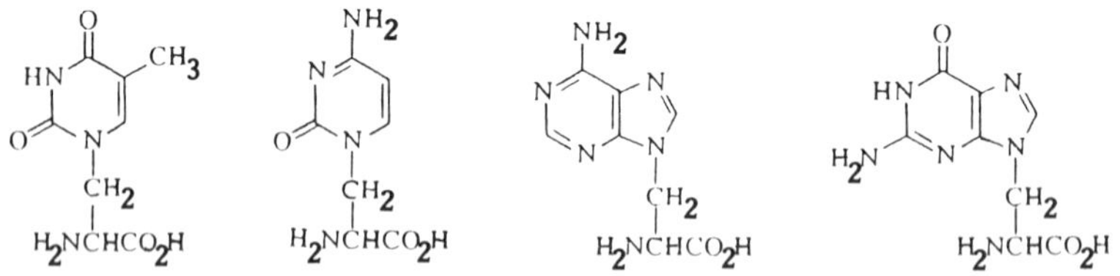

185 186 187 188

Scheme 44

temperature to give 1-(2,2-diethoxyethyl)cytosine in 57% yield, mp 232 to 234°C. The adenine derivative, similarly obtained, is crystallized from ethanol to give 9-(2,2-diethoxyethyl)adenine in 47% yield, mp 212°C.

Aldehyde derivatives of nucleic acid bases (*183*) [164]. Hydrolysis of the uracil derivative is carried out with ammonium chloride solution for 3.5 h on a steam bath, in 83% yield, mp 213 to 214°C. The thymine and adenine derivatives are hydrolyzed with 1 N HCl at 100°C for 1 h. The thymine derivative is obtained in 79% yield, dec. > 300°C. The adenine derivative is obtained as the hydrochloride dihydrate in 69% yield, mp > 300°C. The cytosine derivative is hydrolyzed with 4N HCl at 100°C for 1 h, but the aldehyde is not isolated but converted directly into the amino acid.

Alanine derivatives of nucleic acid bases (*184*) [164]. D,L-Willardiine is obtained by the Strecker synthesis solution of the aldehyde (*183*) (2.03 g) in water (7 mL) containing aqueous ammonia (1.8 mL), potassium cyanide (0.935 g), and ammonium chloride (0.79 g) is heated at 55 to 60°C for 5 h. The solution is mixed with 10 N hydrochloric acid (25 mL) and treated under reflux for 1 h and then set aside overnight. The solution is evaporated to a small volume, mixed with 10 N HCl (15 mL), and heated under reflux for 3 h. The solution is evaporated. The residue is dissolved in water, neutralized to pH 4 with aqueous ammonia, and set aside. A crystalline precipitate of the amino acid; D,L-willardine (*184*), is obtained; yield 1.58 g, mp 205 to 209°C. D,L-alanine derivatives of adenine (*187*) (53% yield), cytosine (*186*) (49% yield), and thymine (*185*) are also obtained by the same procedure.

Polymerization of the willardiine analogs by the NCA method [24]. The willardiine analog is suspended in dioxane and treated with phosgen for 4 h at 40 to 50°C. The excess phosgen is removed by purging with dry nitrogen and the solvent evaporated at 50°C. The residue is recrystallized to give the NCA. The NCA is dissolved in dimethylformamide and triethylamine is added as an initiator and stirred for 2 days at room temperature. Then the solution is added to ethyl ether to precipitate the polymer. The adenine derivative is obtained in 62% yield and the uracil derivative in 13% yield.

Polymerization by the mixed anhydride method [174]. β-(Thymine-1-yl)alanine (*185*) (0.1 g) is suspended in ethylene glycol (25 mL) and the suspension stirred with warming for 12 h. The solution is then cooled to -15°C, *N*-methylmorpholine (47 mg) and 2-methylpropan-1-yl chloroformate (68 mg) are added with stirring, and the resulting white suspension stirred at -15°C for 2 min. It is allowed to warm to room temperature and stirred for 180 h after the solution is dialyzed exhaustively against distilled water, it is freeze dried to give the polymer, 10 mg (mW 1900 to 4000).

4.7 OTHER SYNTHETIC POLYMERS CONTAINING NUCLEIC ACID BASES

4.7.1 Polystyrene Derivatives

Polystyrene derivatives containing nucleic acid bases were prepared by the polymerization of the styrene derivatives of nucleic acid bases or the grafting of the nucleic acid base derivatives onto the chloromethylated polystyrene.

Styrene derivatives—thymine and uracil propionic acid vinylanilides (*190*)—were successfully prepared from the carboxyethyl derivatives of thymine and uracil with 4-aminostyrene by using dicyclohexylcarbodiimide (DCC) as a condensating agent (Scheme 45). The same reaction, however, could not be applied in the case of adenine and cytosine because these bases have an amino group which can participate in the reaction, affording undesirable DCC adducts [30]. Therefore, 3-bromopropionic acid 4-vinylanilide (*191*), which is prepared by condensating 4-aminostyrene (*189*) with 3-bromopropionic acid, was allowed to react with the sodium salt of adenine in dimethylformamide solution [32]. The adenine derivative (*192*) was obtained in 38% yield, and the cytosine derivative (*193*) in 42% yield.

The styrene derivatives were polymerized with a free-radical initiator in dimethyl sulfoxide or dimethylformamide at 60°C. These monomers can also be copolymerized with styrene under similar conditions. The polymers obtained are soluble in dimethyl sulfoxide with molecular weights of about 20,000.

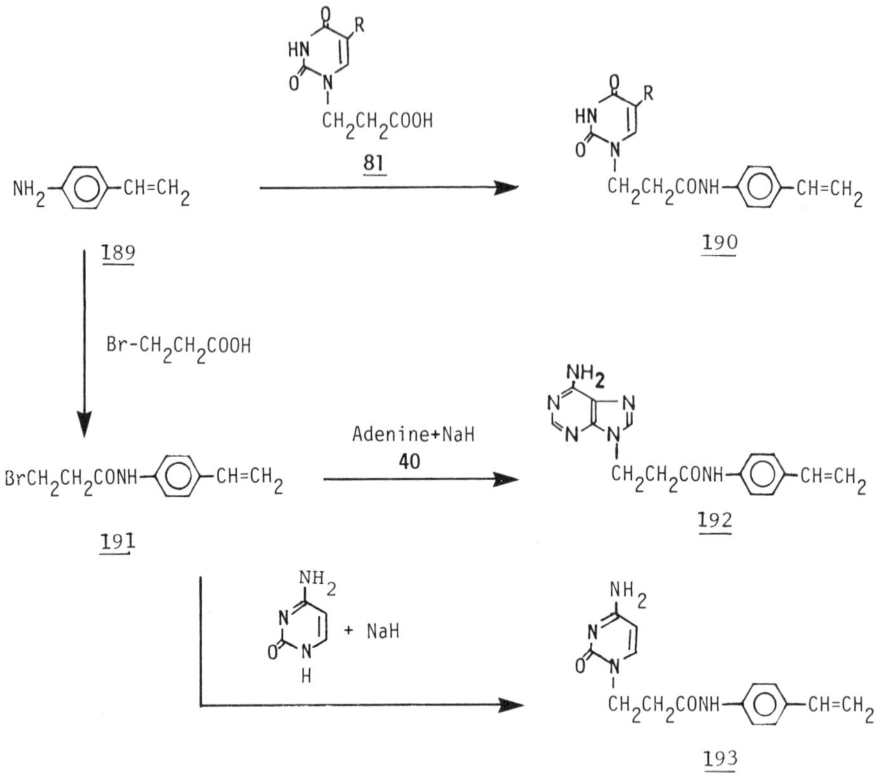

Scheme 45

3-(Uracil-1-yl)propionyl-4-vinylanilide (*190a*) [30]. To a solution of 1-(2-carboxyethyl)uracil (*81*) (5.52 g, 30 mmol) in dimethylformamide (250 mL), DCC (6.18 g, 30 mmol) is added and stirred for 2 h at room temperature. Then 4-aminostyrene (*189*) (3.57 g, 30 mmol) in dimethylformamide (50 mL) is added to the solution and stirred for 4 days. Precipitated dicyclohexylurea is filtered off and the filtrate is evaporated under reduced pressure. The residue is recrystallized from ethanol to give colorless needles in 40% yield, mp 233 to 234°C. The thymine derivative (*190b*) is prepared by a similar procedure in 25% yield, mp 218 to 220°C.

(3-Bromopropionyl)-4-vinylanilide (*191*) [32]. To a solution of 3-bromopropionic acid (1.53 g, 10 mmol) in methylene chloride (10 mL), DCC (2.60 g, 10 mmol) is added and stirred for 2 h at room temperature. Then 4-aminostyrene (*189*) (1.19 g, 10 mmol) is added and stirred for 24 h. The precipitated dicyclohexylurea is filtered off

and the filtrate is condensed under reduced pressure. The residue is recrystallized from methylene chloride to give colorless needles in 70% yield, mp 127 to 128°C.

3'-(Adenin-9-yl)propionyl-4-vinylanilide (*192*) [32]. To a suspension of adenine (1 g, 7 mmol) in dimethylformamide (80 mL), sodium hydride (50%, 0.53 g) is added and stirred for 2 h at 60°C. After cooling to 30°C, 3'-bromopropionyl-4-vinylanilide (1.88 g, 7.4 mmol) dissolved in dimethylformamide (20 mL) is added dropwise and stirred further for 48 h. Insoluble by-product is removed and the filtrate is evaporated. The residue is washed with water (100 mL) and recrystallized from ethanol to give colorless needles in 38% yield, mp 247 to 248°C. The cytosine derivative (*193*) is prepared by the same procedure and recrystallized from methanol in 42% yield, mp 265°C.

Polystyrene resins with pendant nucleic acid bases (*194*) were prepared by the grafting of the corresponding bases onto the chloromethylated polystyrene with potassium carbonate in dimethyl sulfoxide [66]. They are found to be suitable for high-performance liquid chromatography separating effectively nucleic acid base derivatives [60].

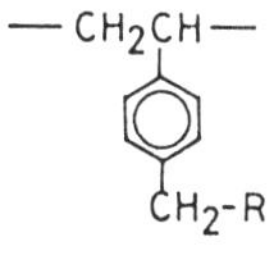

R: Adenine
Uracil

194

Grafting onto polystyrene resin (*194*) [60]. Chloromethylated polystyrene (24% chloromethylated, 2% cross-linked, 100 to 200 mesh, 1 g) is stirred with adenine (270 mg) or uracil (240 mg) in dimethyl sulfoxide in the presence of potassium carbonate (140 mg) at 25 to 30°C for 168 h. The polymer is filtered, washed with aqueous sodium carbonate solution, followed by water and methanol, and dried in a vacuum oven. These polymers contain about 20% base units as determined from elementary analysis.

4.7.2 Polyvinylpyridine Derivatives

Cationic model compounds of polynucleotides such as *195* and *196* are prepared by the Menschutkin reaction of poly(4-vinylpyridine) with 9-(2-chloroethyl)adenine (*2*) and 1-(2-chloroethyl)thymine (*19b*). The interaction between the model compounds and polynucleotide and between the model compounds themselves were studied [105-111].

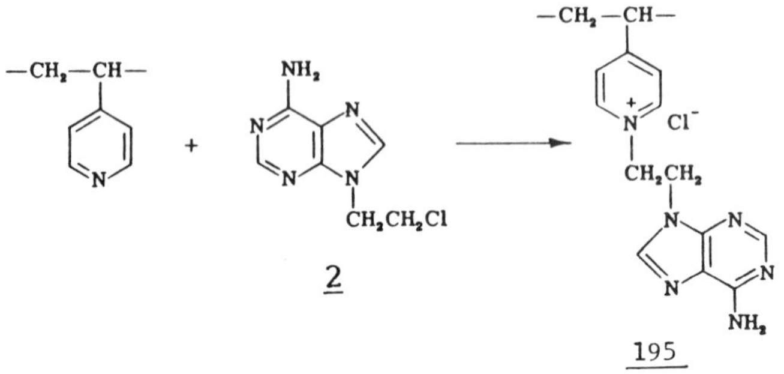

Grafting onto polyvinylpyridine (195 and 196) [111]. Poly(4-vinylpyridine) (6 g) is quaternized by reacting 9-(2-chloroethyl)adenine (2) (12.5 g) in dimethylformamide (40 mL) for 3 days at 85°C. The precipitated product is thoroughly washed with the same solvent and then with dioxane to give a slightly yellow powder. The degree of quaternization and the yield is 85 to 97%.

The thymine derivative (196) is prepared by heating 1-(2-chloroethyl)thymine (19b) (11.5 g) and poly(4-vinylpyridine) (6 g) in dimethylformamide (13 mL) at 85°C for 24 h. The solution is poured into dioxane to precipitate the polymer. Yield 70%; degree of quaternization 82%.

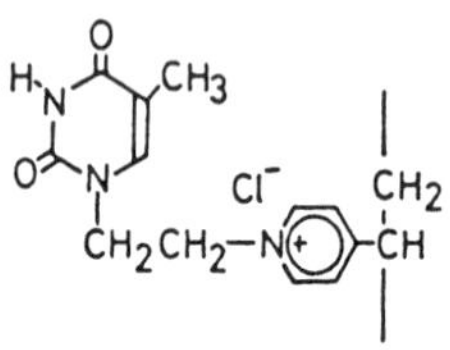

196

4.7.3 Poly(vinyl alcohol) Derivatives

Poly(vinyl alcohol) derivatives containing nucleic acid bases were prepared by grafting of the phosphate derivatives using dicyclohexylcarbodiimide [87,94-96]. The amount of uracil, thymine, and hypoxanthine attached to the polymer was 50 to 60 mol %, and that of adenine was about 10%. Interactions of these polymers with DNA in aqueous solution were studied [97] (Scheme 46). Nucleic acid bases were also grafted onto cellulose [23] and chitin [50].

β-Dihydrogenphosphatoethyl derivatives (197) [87,94]. The hydroxyethyl derivative of nucleic acid bases (5 g) is mixed with trimethyl phosphate (45 mL) and cooled to 0°C. To this solution, phosphorus oxychloride (5 g) is added and stirred for 15 h. After the addition of water (10 mL) the mixture is stirred for 2 h at room

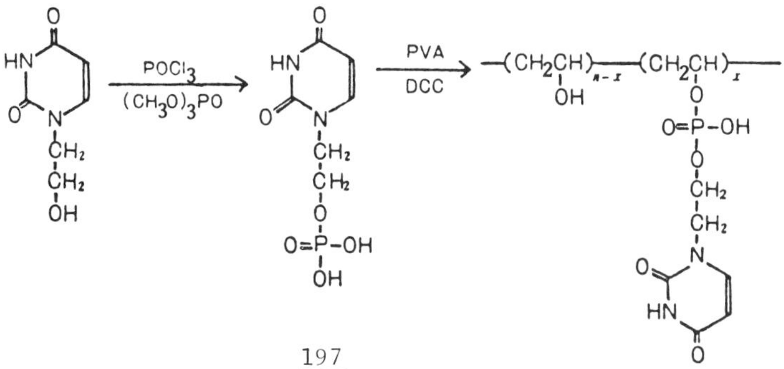

197

198

Scheme 46

temperature. Evaporation of water below 50°C under reduced pressure, followed by the addition of a mixture of ethanol-ether (7:3), yields a white precipitate, which is crystallized from water to give colorless needles. Yield: adenine (70%), hypoxanthine (65%), thymine (40%), uracil (50%), cytosine (55%).

Poly(vinyl alcohol) derivatives (198) [87,94]. To a solution of poly(vinyl alcohol) (0.22 g) and the phosphate derivative (*197*) (2 g) in dimethylformamide (12 mL), DCC (40 g) is added and the mixture is refluxed for 24 h. The reaction mixture is poured into a large volume of water and allowed to stand at room temperature overnight. The mixture is filtered and the filtrate is evaporated under reduced pressure. The residue is dissolved in a minimum amount of water and poured into a mixture of ethanol and ether (7:3) to precipitate the polymer. Yield and base contents: adenine, 42 mg (12%); hypoxanthine, 80 mg (52%); thymine, 87 mg (68%); uracil, 52 mg (68%); cytosine, 60 mg (59%); adenosine, 40 mg (12%).

4.7.4 Polyvinylamine Derivatives

Model polymers of polynucleotides with stereoregular poly(vinylamine) backbones and an optically active nucleic acid base derivative as a pending side chain were prepared [158]. The grafting of (+) and (-)-2-(thymine-1-yl)propionic acid to poly(vinylamine) prepared either by hydrolysis of poly(vinyl acetoamide) or poly(vinyl-*t*-butyl carbamate) was more difficult than in the case of polyethyleneimine. This may be due to a combination of the low solubility and steric factors of polyvinylamine. Polyvinylamine forms a complex with oximes such as ethyl-2-hydroxyiminocyanoacetate (EHICA), which activates the amino group of polyvinylamine; it becomes soluble in polar solvents and gives a higher percent of graft. The optically active carboxylic acid derivative of thymine were grafted onto the polyvinylamine-EHICA complex through amide bonds by direct coupling with sulfonic acid esters of hydroxybenzotriazoles to give an optically active graft polymer *200* (Scheme 47). The coupling agents 1-(methansulfonyl)-4-

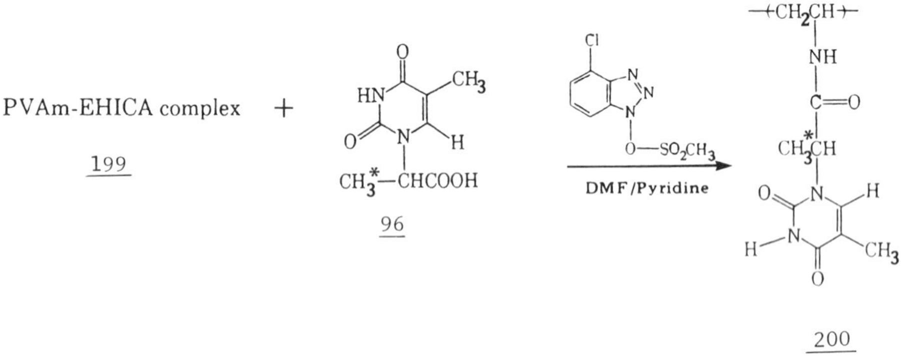

Scheme 47

chlorobenzotriazole (MSCBT) and 1-(4-chlorobenzenesulfonyl)-4-chlorobenzotriazole (CBSCBT) are superior to diethylphosphoryl cyanide (DEPC) [153] for retarding racemization. The optical properties of the polyvinylamine derivatives of thymine and the interaction with poly(adenylic acid):poly A, were investigated [159].

Polyvinylamine-EHICA complex (*199*) [158]. To a solution of polyvinylamine (0.86 g, 20 mmol) [217-219] in methanol (100 mL) is added ethyl-2-hydroxyimino cyanoacetate (EHICA) [220] (2.42 g, 20 mmol) under stirring. The solution turns yellow upon complex formation. This solution is stirred at room temperature for 1 h. After the solution is concentrated to about 10% of the original volume, the residual viscous material is triturated in ether. The resulting pale yellow powder is washed thoroughly with ether to yield 2.29 g (83%) of the complex.

Grafting onto polyvinylamine [158]. To an ice-cooled solution of polyvinylamine-EHICA complex (0.35 g, 2.5 mmol) and (-)-2-(thymine-1-yl)propionic acid (*96*) (0.549 g, 3 mmol) in dry dimethylformamide-pyridine (15:3 mL) is added 1-(methansulfonyl)-4-chlorobenzotriazole (MSCBT) (0.248 g, 2.5 mmol) [221]. The reaction mixture is stirred overnight at 0°C and another portion of (-)-2-(thymin-1-yl)propionic acid (*96*) (0.239 g, 1.5 mmol) and MSCBT (0.371 g, 1.5 mmol) is added at 0°C. After 24 h, the solvents are evaporated at 30°C in vacuo. The residue is dissolved in chloroform-methanol (1:1) and poured into an excess of ethanol-methanol (1:1), filtered, and washed with solvent. The precipitate is redissolved in dimethylformamide (10 mL) and treated with strong basic ion-exchange resin overnight. The solution is filtered and reprecipitated from ether to yield 0.418 g (75%) of polymer.

4.7.5 Polyester and Polyamide Derivatives

The preparation of polyesters containing nucleic acid bases such as *201* was attempted by condensation of the diol and the dicarboxylic acid derivatives of nucleic acid bases [18,99-101]. Similarly, preparations of the polyphosohoric esters were tried [15,20,26,84-86,88-93]. The condensation of some polyamides containing nucleic acid bases were also tried with the dicarboxylic acid derivatives of nucleic acid bases and various diamines (*202*) [36,102,103]. Unfortunately, high-molecular-weight polymers were never obtained by these polycondensation reactions, only oligomers.

Base
|
CH_2
|
$-O-CHCH_2-O-CO-(CH_2)_n-CO-$

201

Base
|
$-CO-CH-CO-NH-(CH_2)_n-NH-$

202

Analogs of oligo- and polynucleotides with structures similar to *203*, in which the nucleosides are joined by acetate ester linkages, were synthesized and shown to interact with natural polynucleotides and in some cases to inhibit the binding of tRNA to ribosomes in the presence of messenger RNA and to interfere with protein biosynthesis in a cell-free system [163,165,168-171,175,177]. Attempts to prepare the polymer containing an oxyacetamide linkage (*204*) were unsuccessful because of the formation of a lactam [172].

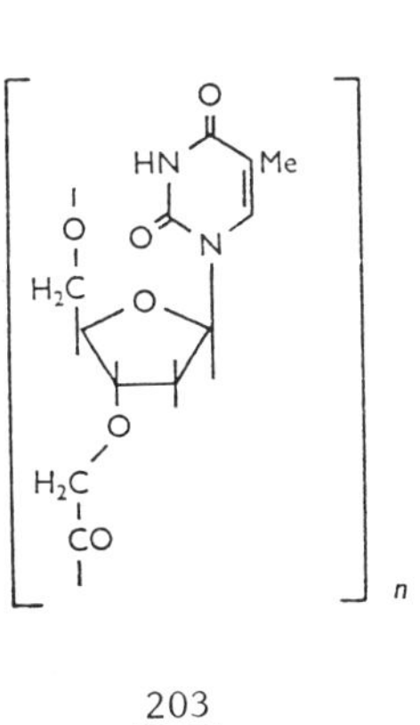

203

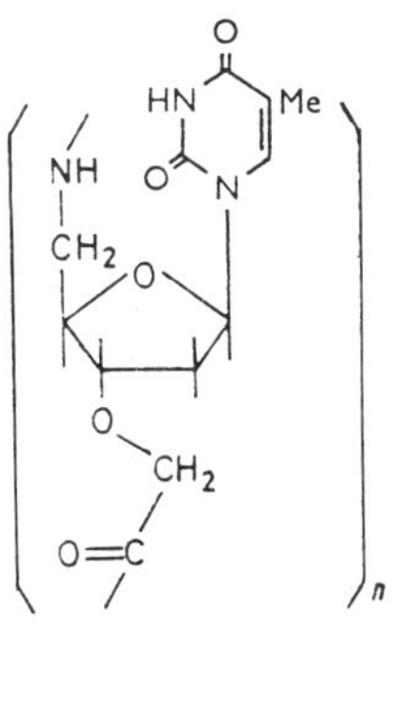

204

The reaction of the disodio derivative of 5'-*O*-tritylthymidine with sodium chloroacetate gave 3'-*O*-carboxymethyl-5'-tritylthymidine (*205*) as the major product. The acid hydrolysis of (*205*) gave 3'-*O*-carboxymethylthymidine (*206*). Polymerization of *206* in the presence of *205* (to act as chain terminator) by the action of dicyclohexylcarbodiimide and removal of the terminal trityl group gave poly(3'-*O*-carboxymethylthymidine) 40% with a molecular weight of 4000 (*203*) (Scheme 48). Polymers of adenosine and uridine were prepared by the same procedure.

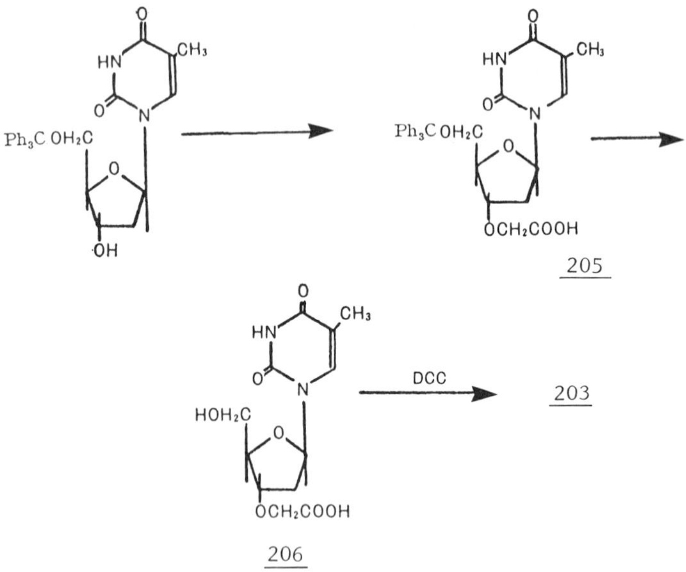

Scheme 48

3'-O-Carboxymethyl-5'-O-tritylthymidine (*205*) [163]. 5'-O-Tritylthymidine (monobenzene adduct; 1.12 g, 2 mmol) is dried in vacuo (P_2O_5) and dissolved in dry dimethyl sulfoxide (10 mL). Sodium hydroxide (105 mg, 4 mmol) is added and the solution is stirred with the exclusion of moisture for 1 h and then set aside for 2 h at room temperature. Dry sodium chloroacetate (233 mg, 2 mmol) is added and the mixture is shaken at room temperature for 5 h and then set aside for 3 days. The mixture is then diluted with water (15 mL) and the pH adjusted to 7. The white precipitate is collected and dissolved in ethanol (5 mL). The solution is evaporated to dryness and the last traces of dimethyl sulfoxide are removed by repeated evaporation to dryness from aqueous ethanol. The residue is dried in vacuo and then extracted with hot chloroform-ethanol (1:3) (3 × 60 mL). The combined extracts are cooled and the white crystals are collected (400 mg) and recrystallized from chloroform-ethanol (1:3) to give 3'-*O*-carboxymethyl-5'-*O*-tritylthymidine as the sodium salt monohydrate.

3'-O-Carboxymethylthymidine (*206*) [163]. 3'-*O*-Carboxymethyl-5'-*O*-tritylthymidine (*205*) (300 mg) is hydrolyzed with acetic acid-water (4:1) at 100°C for 20 min. The triphenylmethanol produced is extracted with hot *n*-heptane and the crude product (150 mg) is isolated from the aqueous acid layer. Crystallization from ethanol gives 3'-*O*-carboxymethylthymidine (*206*) as the sodium salt monohydrate.

Poly(3'-O-carboxymethylthymidine) (*203*) [163]. 3'-*O*-Carboxymethylthymidine (*206*) (sodium salt; 130 mg, 0.4 mmol) and 3'-*O*-carboxymethyl-5'-*O*-tritylthymidine (*205*) (sodium salt; 22 mg, 0.04 mmol) are dissolved in pyridine-water (1:1) and the solution is passed down a column of Zeo Karb 225 (pyridinium form). The column is eluted with pyridine-water (1:1) (100 mL) and the eluent is evaporated to dryness. The resulting oil is evaporated to dryness several times from dry pyridine to give a glassy product which is dried in vacuo. After dissolving in dry pyridine (0.5 mL), dicyclohexylcarbodiimide (300 mg) is added. A clear solution is initially obtained that becomes cloudy and viscous. After 5 min a precipitate forms. After 30 min, dry pyridine (0.3 mL) is added and the mixture is shaken at room temperature for 48 h. Water (6 mL) is added and the suspension is shaken for 24 h. The solvents are removed by evaporation at 30°C and the residue is evaporated several times from water and dissolved in acetic acid-water (4:1) (25 mL). The solution is heated at 100°C for 10 min and then extracted with hot *n*-heptane (3 × 20 mL). The aqueous layer is evaporated to dryness at 30°C and the residue is dissolved in warm dimethylformamide (25 mL) and dialyzed against six changes of dimethylformamide (300 mL) for 6 days. A portion of this solution (40 mL) is then dialyzed exhaustively against distilled water and the resulting suspension is freeze dried. The solid residue is dried at room temperature to give poly(3'-*O*-carboxymethylthymidine) (*203*) (58 mg) as a white solid.

4.7.6 Polymers Containing Nucleic Acid Bases in the Main Chain

Polyamide containing adenine base in the main chain was obtained as a by-product after treating 9-(2-carboxyethyl)adenine (*77*) with phosphorus pentachloride and then adding triethylamine in dimethylformamide [14]. Polymerization of the cyclic derivatives of thymine and uracil gave polymers containing pyrimidine bases in the main chain [37,52,73]. As mentioned above, the cyclic derivatives of uracil [ethanouracil (*60a*)] and thymine [ethanothymine (*60b*)] undergo a variety of reactions to give many valuable derivatives of uracil and thymine. The ethanouracil and ethanothymine monomers were polymerized with cationic and anionic initiators by a unique ring-opening process. The polymerization proceeds by ring opening with isomerization of the pyrimidine ring to give polymers in which pyrimidine rings are connected with ethylene between N-1 and either N-3 or O-4 (*207* and *208*) in the pyrimidine ring (Scheme 49). The structure of these

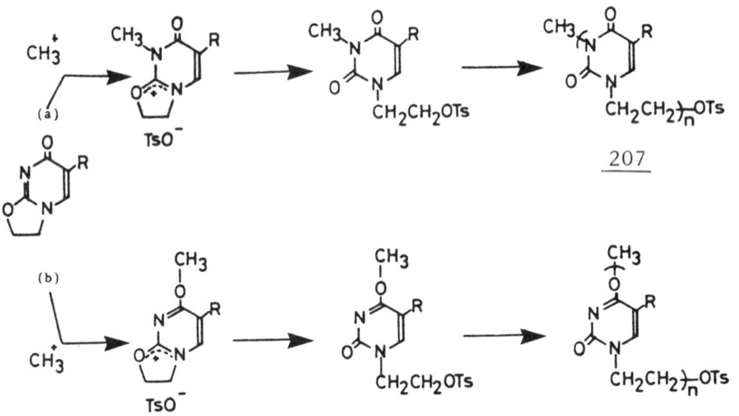

207

208

Scheme 49

polymers was determined by NMR, UV, IR, and hydrolysis of the polymer. Alternatively, polymers in which the pyrimidine bases (*207*) were connected with ethylene between N-1 and N-3 were prepared by successive alkylation of 1-(2-chloroethyl)uracil (*19a*) or -thymine (*19b*) with base such as potassium carbonate in dimethyl sulfoxide at high temperature.

The polymerizations of ethanouracil and ethanothymine were carried out in acetonitrile at 50 to 80°C with methyl *p*-toluenesulfonate as an initiator. The polymer was obtained in 20 to 80% yield, and the molecular weight was 6000 to 9000.

Since the polymer obtained by methyl *p*-toluenesulfonate is a living polymer, it is possible to form a block copolymer containing uracil and thymine units. After the polymerization of ethanouracil, the polymer was washed with the same solvent and a solution of ethanothymine added without initiator to give a block copolymer. Analysis of the NMR spectra of the block copolymer reveals that 15 units of ethanothymine had been added to about 25 units of ethanouracil polymer.

Ethanouracil and ethanothymine were also polymerized by anionic initiators such as sodium methoxide, potassium *t*-butoxide, and *n*-butyl lithium in acetonitrile at 30 to 70°C. These anionic polymerizations gave polymers in which the pyrimidine bases were connected with ethylene between N-1 and N-3 or O-2 in the pyrimidine ring.

Poly(1,2-O-ethanouracil) (*207*) [52]. A glass vessel with a magnetic stirrer is flame dried under dry nitrogen. To this vessel, 1,2-*O*-ethanouracil (*60a*) (690 mg, 5 mmol) and methyl *p*-toluenesulfonate (18 mg, 0.1 mmol) in acetonitrile (20 mL) are added. The monomer dissolves immediately and the polymer precipitates on the wall of the vessel as the polymerization proceeds. After the reaction has proceeded with stirring for 72 h at 70°C, the polymer is precipitated from methanol in 78% yield (538 mg). Cationic polymerization of 1,2-*O*-ethanothymine (*60b*), and anionic polymerizations of the cyclic monomers are carried out according to the procedure described above.

Block copolymer of 1,2-O-ethanouracil and 1,2-O-ethanothymine [52]. 1,2-*O*-Ethanouracil (*60a*) (173 mg, 1.25 mmol) is polymerized with methyl *p*-toluenesulfonate (9.4 mg, 0.05 mmol) in acetonitrile at 70°C using the procedure described above. After polymerization for 96 h, the solvent is removed by decantation, and the polymer is washed with acetonitrile. To this polymer, 1,2-*O*-ethanothymine (*60b*) (190 mg, 1.25 mmol) in acetonitrile (5 mL) is added and the mixture stirred for 5 days at 70°C. The block copolymer is precipitated from methanol; yield 285 mg (79%).

Poly(1-(2-chloroethyl)uracil) (*208*) [52]. To a solution of 1-(2-chloroethyl)uracil (*19a*) (1.74 g, 0.01 mol) in dimethyl sulfoxide (10 mL) is added anhydrous potassium carbonate (1.60 g, 0.01 mol), and the mixture is stirred for 15 h at 90°C. The reaction mixture is poured into excess water to precipitate the polymer; yield 1.28 g (92%). The molecular weight of the dimethylformamide soluble portion is about 1500.

REFERENCES

1. K. Takemoto, J. Macromol. Sci. Rev. Macromol. Chem., C5, 29 (1970).
2. M. Imoto and K. Takemoto, Synthesis, 173 (1970).
3. K. Takemoto, J. Polym. Sci. Polym. Symp., 55, 105 (1976).
4. K. Takemoto, in Polymeric Drugs (L. G. Donaruma and O. Vogl, Eds.), Academic Press (1978), p. 103.
5. K. Takemoto, Y. Inaki, and M. Akashi, J. Macromol. Sci. Chem., A13, 519 (1979).
6. K. Takemoto and Y. Inaki, Adv. Polym. Sci., 41, 1 (1981).

7. K. Takemoto, in Bioactive Polymeric Systems (C. G. Gebelein, Ed.), Plenum Publishing Corporation, New York (1983).
8. K. Takemoto, Makromol. Chem., Suppl., 12, 293 (1985).
9. N. Ueda, K. Kondo, M. Kono, K. Takemoto, and M. Imoto, Makromol. Chem., 120, 13 (1968).
10. K. Kondo, H. Iwasaki, N. Ueda, K. Takemoto, and M. Imoto, Makromol. Chem., 120, 21 (1968).
11. K. Kondo, H. Iwasaki, K. Nakatani, N. Ueda, K. Takemoto, and M. Imoto, Makromol. Chem., 125, 42 (1969).
12. K. Kondo, H. Iwasaki, N. Ueda, K. Takemoto, and M. Imoto, Makromol. Chem., 125, 298 (1969).
13. K. Takemoto, F. Kawakubo, and K. Kondo, Bull. Chem. Soc. Jpn., 44, 1718 (1971).
14. K. Kondo, M. Miyata, and K. Takemoto, Bull. Chem. Soc. Jpn. 44, 2554 (1971).
15. N. Ueda, T. Kawabata, and K. Takemoto, J. Heterocycl. Chem., 8, 827 (1971).
16. K. Takemoto, F. Kawakubo, and K. Kondo, Makromol. Chem., 148, 131 (1971).
17. K. Kondo, Y. Murata, and K. Takemoto, Tech. Rep. Osaka Univ., 22, 785 (1972).
18. K. Kondo, K. Kuwata, and K. Takemoto, Makromol. Chem., 160, 341 (1972).
19. F. Kawakubo, K. Kondo, and K. Takemoto, Makromol. Chem., 169, 37 (1973).
20. T. Kawabata, N. Ueda, and K. Takemoto, Chem. Lett., 55 (1973).
21. M. Miyata, K. Kondo, and K. Takemoto, Tech. Rep. Osaka Univ., 23, 339 (1973).
22. K. Kondo, Y. Hisaoka, and K. Takemoto, Chem. Lett., 125 (1973).
23. N. Ueda, Y. Nakatani, S. Terada, K. Kondo, and K. Takemoto, Tech. Rep. Osaka Univ., 23, 713 (1973).
24. K. Takemoto, H. Tahara, A. Yamada, Y. Inaki, and N. Ueda, Makromol. Chem., 169, 327 (1973).
25. K. Kondo, T. Sato, and K. Takemoto, Chem. Lett., 967 (1973).
26. Y. Inaki, T. Renge, and K. Takemoto, Tech. Rep. Osaka Univ., 24, 365 (1974).
27. K. Takemoto, M. Akashi, and Y. Inaki, J. Polym. Sci. Polym. Chem. Ed., 12, 1861 (1974).
28. M. Akashi, Y. Inaki, and K. Takemoto, Tech. Rep. Osaka Univ., 25, 253 (1975).
29. Y. Inaki, T. Renge, K. Kondo, and K. Takemoto, Makromol. Chem., 176, 2683 (1975).
30. K. Kondo, T. Sato, Y. Inaki, and K. Takemoto, Makromol. Chem., 176, 3505 (1975).

31. M. Akashi, H. Takada, Y. Inaki, K. Kondo, and K. Takemoto, Nucleic Acids Res. Symp. Ser., 2, 101 (1976).
32. K. Kondo, Y. Ohbe, and K. Takemoto, Makromol. Chem., 177, 3461 (1976).
33. M. Akashi, Y. Inaki, and K. Takemoto, Makromol. Chem., 178, 353 (1977).
34. Y. Inaki, H. Takada, K. Kondo, and K. Takemoto, Makromol. Chem., 178, 365 (1977).
35. M. Akashi, Y. Kita, Y. Inaki, and K. Takemoto, Makromol. Chem., 178, 1211 (1977).
36. K. Kondo, S. Miki, and K. Takemoto, Polym. J., 9, 429 (1977).
37. M. Akashi, H. Futagawa, Y. Inaki, K. Kondo, and K. Takemoto, Nucleic Acids Res. Symp. Ser., 3, 7 (1977).
38. M. Akashi, Y. Inaki, and K. Takemoto, J. Macromol. Sci., A12, 619 (1978).
39. T. Ishikawa, Y. Inaki, and K. Takemoto, Polym. Bull., 1, 85 (1978).
40. T. Ishikawa, Y. Inaki, and K. Takemoto, Polym. Bull., 1, 215 (1978).
41. T. Ishikawa, Y. Shigeno, T. Takahara, Y. Inaki, K. Kondo, and K. Takemoto, Nucleic Acids Res. Symp. Ser., 5, 279 (1978).
42. M. Akashi, Y. Kita, Y. Inaki, and K. Takemoto, J. Polym. Sci. Polym. Chem. Ed., 17, 301 (1979).
43. M. Akashi, H. Takada, Y. Inaki, and K. Takemoto, J. Polym. Sci. Polym. Chem. Ed., 17, 747 (1979).
44. M. Akashi, T. Okimoto, Y. Inaki, and K. Takemoto, J. Polym. Sci. Polym. Chem. Ed., 17, 905 (1979).
45. K. Kondo, S. Tanioku, and K. Takemoto, Polym. J., 11, 81 (1979).
46. Y. Kita, T. Uno, Y. Inaki, and K. Takemoto, Nucleic Acids Res. Symp. Ser., 6, 63 (1979).
47. Y. Shigeno, K. Kondo, and K. Takemoto, Polym. Bull., 1, 599 (1979).
48. Y. Kita, Y. Inaki, and K. Takemoto, J. Polym. Sci. Polym. Chem. Ed., 18, 427 (1980).
49. T. Ishikawa, Y. Inaki, and K. Takemoto, J. Polym. Sci. Polym. Chem. Ed., 18, 949 (1980).
50. Y. Sakuma, Y. Inaki, and K. Takemoto, Tech. Rep. Osaka Univ., 30, 319 (1980).
51. T. Ishikawa, Y. Inaki, and K. Takemoto, J. Polym. Sci. Polym. Chem. Ed., 18, 1847 (1980).
52. Y. Inaki, H. Futagawa, and K. Takemoto, J. Polym. Sci. Polym. Chem. Ed., 18, 2959 (1980).
53. Y. Inaki, T. Ishikawa, and K. Takemoto, Modification of Polymers, ACS Symp. Ser., 121, 359 (1980).

54. Y. Kita, H. Futagawa, Y. Inaki, and K. Takemoto, Polym. Bull., 2, 195 (1980).
55. K. Kondo, S. Tanioku, and K. Takemoto, Makromol. Chem. Rapid Commun., 1, 303 (1980).
56. Y. Inaki, H. Futagawa, and K. Takemoto, Org. Prep. Proced. Int., 12, 275 (1980).
57. M. Miyata, S. Senda, and K. Takemoto, Tech. Rep. Osaka Univ., 30, 591 (1980).
58. Y. Inaki, T. Ishikawa, S. Sugita, and K. Takemoto, J. Polym. Sci. Polym. Lett. Ed., 18, 725 (1980).
59. Y. Inaki, T. Ishikawa, and K. Takemoto, Nucleic Acids Res. Symp. Ser., 8, 137 (1980).
60. K. Kondo, T. Horiike, and K. Takemoto, J. Macromol. Sci. Chem., A16, 793 (1981).
61. Y. Kita, T. Uno, Y. Inaki, and K. Takemoto, J. Polym. Sci. Polym. Chem. Ed., 19, 477 (1981).
62. Y. Kita, T. Uno, Y. Inaki, and K. Takemoto, J. Polym. Sci. Polym. Chem. Ed., 19, 1733 (1981).
63. Y. Kita, Y. Uno, Y. Inaki, and K. Takemoto, J. Polym. Sci. Polym. Chem. Ed., 19, 2347 (1981).
64. Y. Inaki, Y. Suda, Y. Kita, and K. Takemoto, J. Polym. Sci. Polym. Chem. Ed., 19, 2519 (1981).
65. Y. Kita, T. Uno, Y. Inaki, and K. Takemoto, J. Polym. Sci. Polym. Chem. Ed., 19, 3315 (1981).
66. K. Kondo, K. Hyodo, T. Horiike, and K. Takemoto, Nucleic Acids Res. Symp. Ser., 10, 125 (1981).
67. Y. Inaki, Y. Sakuma, Y. Suda, and K. Takemoto, J. Polym. Sci. Polym. Chem. Ed., 20, 1917 (1982).
68. Y. Sakuma, Y. Inaki, and K. Takemoto, J. Polym. Sci. Polym. Chem. Ed., 20, 3431 (1982).
69. Y. Sakuma, Y. Inaki, and K. Takemoto, J. Polym. Sci. Polym. Chem. Ed., 22, 2061 (1984).
70. Y. Suda, Y. Inaki, and K. Takemoto, J. Polym. Sci. Polym. Chem. Ed., 21, 2813 (1983).
71. Y. Sakuma, Y. Inaki, and K. Takemoto, Nucleic Acids Res. Symp. Ser., 11, 269 (1982).
72. Y. Inaki, S. Sugita, T. Takahara, and K. Takemoto, J. Polym. Sci. Polym. Chem. Ed., 24, 3201 (1986).
73. E. Mochizuki, Y. Inaki, and K. Takemoto, Polym. Prepr. Jpn., 29, 120 (1980).
74. Y. Inaki, F. Nakao, and K. Takemoto, in preparation.
75. Y. Inaki, M. Omatsu, and K. Takemoto, in preparation.
76. T. Yamamoto, Y. Inaki, and K. Takemoto, Polym. Prepr. Jpn., 31, 1597 (1982).
77. Y. Inaki, M. Kono, and K. Takemoto, in preparation.

78. S. B. Fang, Y. Inaki, and K. Takemoto, Polym. Prepr. Jpn., 29, 1341 (1980); 30, 402 (1981).
79. Y. Inaki, S. Fukunaga, Y. Suda, and K. Takemoto, J. Polym. Sci. Polym. Chem. Ed., 24, 119 (1985).
80. Y. Inaki and K. Takemoto, Makromol. Chem. Suppl., 14, 91 (1985).
81. E. Mochizuki, Y. Inaki, and K. Takemoto, Nucleic Acids Res. Symp. Ser., 16, 121 (1985).
82. S. Hozumi, Y. Suda, Y. Inaki, and K. Takemoto, Polym. Prepr. Jpn., 32, (1983).
83. Y. Suda, Y. Inaki, and K. Takemoto, J. Polym. Sci. Polym. Chem. Ed., 22, 623 (1984).
84. T. Seita, K. Yamauchi, M. Kinoshita, and M. Imoto, Makromol. Chem., 148, 321 (1971).
85. T. Seita, K. Yamauchi, M. Kinoshita, and M. Imoto, Bull. Chem. Soc. Jpn., 45, 926 (1972).
86. T. Seita, K. Yamauchi, M. Kinoshita, and M. Imoto, Makromol. Chem., 154, 255 (1972).
87. T. Seita, K. Yamauchi, M. Kinoshita, and M. Imoto, Makromol. Chem., 154, 263 (1972).
88. T. Seita, K. Yamauchi, M. Kinoshita, and M. Imoto, Makromol. Chem., 164, 7 (1973).
89. T. Seita, M. Konoshita, and M. Imoto, Makromol. Chem., 166, 15 (1973).
90. T. Seita, M. Kinoshita, and M. Imoto, Bull. Chem. Soc. Jpn., 46, 1572 (1973).
91. T. Seita, K. Takahashi, M. Kinoshita, and M. Imoto, Makromol. Chem., 172, 19 (1973).
92. T. Seita, M. Kinoshita, and M. Imoto, J. Macromol. Sci. Chem., A7, 1297 (1973).
93. T. Seita, K. Yamauchi, M. Kinoshita, and M. Imoto, Bull. Chem. Soc. Jpn., 46, 1563, 1572 (1973).
94. T. Seita, K. Yamauchi, M. Kinoshita, and M. Imoto, Makromol. Chem., 164, 15 (1973).
95. T. Seita, M. Kinoshita, and M. Imoto, Bull. Chem. Soc. Jpn., 46, 1762 (1973).
96. T. Seita, M. Kinoshita, and M. Imoto, Makromol. Chem., 164, 345 (1973).
97. T. Seita, M. Kinoshita, and M. Imoto, Bull. Chem. Soc. Jpn., 46, 3310 (1973).
98. M. Kinoshita, K. Yamauchi, and M. Imoto, Prog. Polym. Sci. Jpn., 7, 63 (1974).
99. M. Hattori, H. Takai, and M. Kinoshita, Makromol. Chem., 178, 3211 (1977).
100. M. Hattori, H. Takai, and M. Kinoshita, Makromol. Chem., 179, 905 (1978).

101. M. Hattori and M. Kinoshita, Makromol. Chem., 180, 325 (1979).
102. M. Hattori and M. Kinoshita, Makromol. Chem., 180, 2323 (1979).
103. M. Hattori, N. Nakagawa, and M. Kinoshita, Makromol. Chem., 181, 2325 (1980).
104. T. Shimizu, Y. Konishi, and A. Murakami, Makromol. Chem., 178, 2581 (1977).
105. T. Shimizu, A. Murakami, and K. Yamana, Nucleic Acids Res. Symp. Ser., 5, 379 (1978).
106. T. Shimizu, A. Murakami, and Y. Kinishi, Bull. Chem. Soc. Jpn., 51, 821 (1978).
107. T. Shimizu, A. Murakami, and Y. Konishi, J. Chem. Soc. Perkin Trans. 1, 20 (1979).
108. T. Shimizu, and A. Imai, Bull. Chem. Soc. Jpn., 49, 349 (1976).
109. T. Okubo and N. Ise, Polym. J., 4, 675 (1973).
110. T. Okubo, K. Ban, and N. Ise, Makromol. Chem., 175, 49 (1974).
111. T. Okubo and N. Ise, Adv. Polym. Sci., 25, 135 (1977).
112. J. Pitha and P. O. P. Ts'o, J. Org. Chem., 33, 1341 (1968).
113. J. Pitha, J. Org. Chem., 35, 903 (1970).
114. J. Pitha, P. M. Pitha, and P. O. P. Ts'o, Biochim. Biophys. Acta, 204, 39 (1970).
115. P. M. Pitha and J. Pitha, Biopolymers, 9, 965 (1970).
116. J. Pitha and P. M. Pitha, Science, 172, 1146 (1971).
117. J. Pitha, P. M. Pitha, and E. Stuart, Biochemistry, 10, 4595 (1971).
118. F. Reynolds, D. Grunberger, J. Pitha, and P. M. Pitha, Biochemistry, 11, 3261 (1972).
119. P. M. Pitha and J. Pitha, Nature New Biol., 240, 78 (1972).
120. P. M. Pitha, N. M. Teich, D. R. Lowy, and J. Pitha, Proc. Nat. Acad. Sci. USA, 70, 1204 (1973).
121. P. M. Pitha and J. Pitha, J. Gen. Virol., 24, 385 (1974).
122. P. M. Pitha, J. Pitha, and W. P. Rowe, Virology, 63, 568 (1975).
123. J. Pitha, Anal. Biochem., 65, 422 (1975).
124. J. Pitha, J. Org. Chem., 40, 3296 (1975).
125. J. Pitha, Macromolecules, 9, 771 (1976).
126. J. Pitha and S. H. Wilson, Nucleic Acid Res., 3, 825 (1976).
127. J. Pitha, Cancer Res., 36, 1273 (1976).
128. J. Pitha, Polymer, 18, 425 (1977).
129. L. Noronha-Blob, V. E. Vengris, P. M. Pitha, and J. Pitha, J. Med. Chem., 20, 356 (1977).
130. P. M. Pitha and J. Pitha, J. Pharmacol. Exp. Ther., A2, 247 (1978).
131. L. Noronha-Blob and J. Pitha, Biochim. Biophys. Acta, 519, 285 (1978).
132. J. Pitha, S. H. Wilson, and P. M. Pitha, Biochem. Biophys. Res. Commun., 81, 217 (1978).

133. V. E. Vengris, P. M. Pitha, L. L. Sensenbrenner, and J. Pitha, Mol. Pharmacol., 14, 271 (1978).
134. H. J. Chou, J. P. Foehlich, and J. Pitha, Nucleic Acids Res., 5, 691 (1978).
135. M. Draminski and J. Pitha, Makromol. Chem., 179, 2195 (1978).
136. P. M. Pitha and A. M. Michelson, Biochim. Biophys. Acta, 204, 381 (1970).
137. J. Pitha, M. Akashi, and M. Draminski, in Biomedical Polymers (E. P. Goldberg and A. Nakajima, Eds.). Academic Press (1980), p. 271.
138. H. Kaye, J. Polym. Sci., B7, 1 (1969).
139. H. Kaye, J. Am. Chem. Soc., 29, 5777 (1970).
140. H. Kaye and S. H. Chang, Tetrahedron, 26, 1369 (1970).
141. H. Kaye, Macromolecules, 4, 147 (1971).
142. H. Kaye and S. H. Chang, Macromolecules, 5, 397 (1972).
143. H. Kaye and H. J. Chou, Makromol. Chem., 158, 161 (1972).
144. H. Kaye, Macromolecular Synthesis, Coll. Vol. 1 (J. A. Moore, Ed.), John Wiley & Sons, Inc., New York (1977), p. 521.
145. H. Kaye and S. H. Chang, J. Macromol. Sci., Chem. A7, 1127 (1973).
146. H. Kaye and H. J. Chou, J. Polym. Sci. Polym. Phys. Ed., 13, 477 (1975).
147. P. E. Olson, P. Fraher, S. Boguslawski, and M. P. Mertes, Biochemistry, 14, 4892 (1975).
148. S. Boguslawski, P. E. Olson, and M. P. Mertes, Biochemistry, 15, 3536 (1976).
149. L. Maggiora, S. Boguslawski, and M. P. Mertes, J. Med. Chem., 20, 1283 (1977).
150. C. G. Overberger and Y. Inaki, J. Polym. Sci. Polym. Chem. Ed., 17, 1739 (1979).
151. C. G. Overberger, Y. Inaki, and Y. Nanbu, J. Polym. Sci. Polym. Chem. Ed., 17, 1759 (1979).
152. Y. Morishima and C. G. Overberger, J. Macromol. Sci., A13, 573 (1979).
153. C. G. Overberger and Y. Morishima, J. Polym. Sci. Polym. Chem. Ed., 18, 1247 (1980).
154. C. G. Overberger and Y. Morishima, J. Polym. Sci. Polym. Chem. Ed., 18, 1267 (1980).
155. C. G. Overberger and Y. Morishima, J. Polym. Sci. Polym. Chem. Ed., 18, 1433 (1980).
156. C. G. Overberger and A. G. Ludwick, J. Polym. Sci. Polym. Chem. Ed., 20, 2123 (1982).
157. C. G. Overberger and A. G. Ludwick, J. Polym. Sci. Polym. Chem. Ed., 20, 2139 (1982).
158. C. G. Overberger and S. Kikyotani, J. Polym. Sci. Polym. Chem. Ed., 21, 525 (1983).

159. C. G. Overberger and S. Kikyotani, J. Polym. Sci. Polym. Chem. Ed., 21, 541 (1983).
160. F. Cassidy and A. S. Jones, Eur. Polym. J., 2, 319 (1966).
161. M. G. Boulton, A. S. Jones, and R. T. Walker, J. Chem. Soc., C, 1216 (1968).
162. A. S. Jones, M. K. A. Khan, and R. T. Walker, J. Chem. Soc., C, 1454 (1968).
163. M. H. Halford and A. S. Jones, J. Chem. Soc., C, 2667 (1968).
164. M. T. Doel, A. S. Jones, and N. Taylor, Tetrahedron Lett., 27, 2285 (1969).
165. M. D. Edge, and A. S. Jones, J. Chem. Soc., C, 1933 (1971).
166. M. G. Bouldon, A. S. Jones, and R. T. Walker, Biochim. Biophys. Acta, 246, 197 (1971).
167. M. J. Cooper, R. S. Goody, A. S. Jones, J. R. Tithensor, and R. T. Walker, J. Chem. Soc., C, 3183 (1971).
168. G. J. Cowling, A. S. Jones, and R. T. Walker, Biochim. Biophys. Acta, 254, 452 (1971).
169. M. D. Edge, A. Hodgson, A. S. Jones, and R. T. Walker, J. Chem. Soc. Perkin Trans. 1, 1991 (1972).
170. M. D. Edge, A. Hodgson, A. S. Jones, M. MacCoss, and R. T. Walker, J. Chem. Soc. Perkin Trans. 1, 290 (1973).
171. A. S. Jones, M. MacCoss, and R. T. Walker, Biochim. Biophys. Acta, 294, 365 (1973).
172. M. J. Gait, A. S. Jones, and R. T. Walker, J. Chem. Soc. Perkin Trans. 1, 1684 (1974).
173. M. T. Doel, A. S. Jones, and R. T. Walker, Tetrahedron, 30, 2755 (1974).
174. J. D. Buttrey, A. S. Jones, and R. T. Walker, Tetrahedron, 31, 73 (1975).
175. R. C. Bleaney, A. S. Jones, and R. T. Walker, Nucleic Acids Res., 2, 699 (1975).
176. A. S. Jones, A. F. Markham, and R. T. Walker, Tetrahedron, 32, 2361 (1976).
177. A. S. Jones, Int. J. Biol. Macromol., 1, 194 (1979).
178. S. Hoffmann, H. Schubert, and W. Witkowski, Z. Chem., 11, 345 (1971).
179. S. Hoffmann, H. Schubert, and W. Witkowski, Z. Chem., 11, 465 (1971).
180. S. Hoffmann, H. Schubert, and W. Witkowski, Z. Chem., 14, 98 (1974).
181. S. Hoffmann, W. Witkowski, A. Gyulbudagyan, and H. Schubert, Z. Chem., 14, 475 (1974).
182. S. Hoffmann, W. Witkowski, and H. Schubert, Z. Chem., 14, 14 (1974).

183. S. Hoffmann, W. Witkowski, and H. Schubert, Z. Chem., 14, 154 (1974).
184. S. Hoffmann, W. Witkowski, and A. M. Ladhoff, Z. Chem., 16, 227 (1976).
185. S. Hoffmann, Z. Chem., 19, 241 (1979).
186. S. Hoffmann and W. Witkowski, Nucleic Acids Res., Suppl. 1, 137 (1976).
187. S. Hoffmann and W. Witkowski, Nucleic Acids Res., 4, 221 (1978).
188. S. Hoffmann, W. Witkowski, and K. Washke, Z. Chem., 16, 484 (1976).
189. S. Hoffmann, W. Witkowski, P. Vener, and H. Robler, Z. Chem., 16, 402 (1976).
190. S. Hoffmann, W. Witkowski, K. Grade, and C. Schonheit, Z. Chem., 16, 322 (1976).
191. S. Hoffmann, W. Witkowski, K. Washke, and S. R. Washke, Z. Chem., 17, 61 (1977).
192. S. Hoffmann and W. Witkowski, Z. Chem., 17, 101 (1977).
193. S. Hoffmann and W. Witkowski, Am. Chem. Soc. Symp. Ser., 74, 178 (1978).
194. H. Schott, G. Greber, and L. Bucsis, Makromol. Chem., 136, 303 (1970).
195. H. Schott and G. Greber, Makromol. Chem., 136, 307 (1970).
196. H. Schott and G. Greber, Makromol. Chem., 144, 333 (1970).
197. H. Schott and G. Greber, Makromol. Chem., 145, 11 (1971).
198. H. R. Kricheldorf and M. Fehrle, Polym. Bull., 6, 21 (1981).
199. H. R. Kricheldorf and M. Fehrle, Makromol. Chem., 181, 2571 (1980).
200. H. R. Kricheldorf and M. Fehrle, Biopolymers, 21, 2097 (1982).
201. T. Nishimura and I. Iwai, Chem. Pharm. Bull., 12, 352 (1964).
202. D. J. Brown, E. Hoerger, S. F. Mason, J. Chem. Soc., 211 (1955).
203. Q. E. Hilbert and E. F. Jansen, J. Am. Chem. Soc., 57, 552 (1935).
204. E. P. Lira and C. W. Huffman, J. Org. Chem., 31, 2188 (1966).
205. E. DeClerq and T. C. Merigan, Annu. Rev. Med., 21, (1970).
206. M. Absher and W. R. Stinebring, Nature, 223, 715 (1969).
207. T. Saegusa, H. Ikeda, and H. Fujii, Polym. J., 3, 35 (1972).
208. A. Takamizawa, Y. Sato, S. Tanaka, Yakugaku Zasshi, 85, 298 (1965).
209. M. Fujino and C. Hatanaka, Chem. Pharm. Bull., 16, 929 (1968).
210. F. Fuchs, Ber., 55, 2943 (1922).
211. H. Leuchs, Ber., 39, 857 (1906).

212. M. Sela, R. Arnon, and I. Jacobson, Biopolymers, 1, 517 (1963).
213. G. Schaw and R. N. Warrener, J. Chem. Soc., 153 (1958).
214. D. R. S. Kushwaha, K. B. Mathur, and D. Balasubramanian, Biopolymers, 19, 219 (1980).
215. J. H. Dewar and G. Shaw, J. Chem. Soc., 583 (1962).
216. A. P. Martinez and W. W. Lee, J. Org. Chem., 30, 317 (1965).
217. D. J. Dawson, R. D. Gless, and R. E. Wingard, Jr., J. Am. Chem. Soc., 98, 5996 (1976).
218. R. Hart, Makromol. Chem., 32, 51 (1959).
219. A. R. Hughes and T. St. Pierr, Macromol. Synth., 6, 31 (1977).
220. M. Conrad and A. Schulz, Ber., 42, 735 (1909).
221. M. Ito, H. Nojima, J. Notani, D. Hagiwara, and K. Takai, Bull. Chem. Soc. Jpn., 51, 3320 (1978).

5

Photoresponsive Polymers

Masahiro Irie

Osaka University, Osaka, Japan

5.1 INTRODUCTION

There are many different approaches to the application of basic photochemical reactions to polymer chemistry. If attention is focused only on the molecular weight change of the starting material to the end product, photochemical reactions in polymer systems may be expressed as follows:

$$a \rightarrow P_n \quad (1)$$

$$P_n + P_m \rightarrow P_{n+m} \quad (2)$$

$$P_n \rightarrow P_{n-m} + P_m \quad (3)$$

$$a + P_n \rightarrow P_n' \quad (4)$$

$$P_n \rightarrow P_n' \quad (5)$$

In this scheme, a and P are low-molecular-weight compounds and polymers. Reactions 1 to 3 represent photopolymerization, photocross-linking and photodegradation, respectively. Reaction 4 is photochemical modification of polymers (or grafting).

Photocuring (or photohardening) of polymers is an area of considerable importance to the paint and printing industries, and the process includes photopolymerization of monofunctional monomers, Reaction 1, and photocross-linking of polymers, Reaction 2. Because of the practical nature of the subject, most research has been directed to the modification of polymers using known photocross-linkable groups and screening photocurable polymer compositions so as to meet the requirements of these industries. An understanding of the detailed reaction mechanisms in polymer systems is extremely limited, as the science is very complicated. The cross-linking may be achieved by direct irradiation of the polymer or monomers themselves, but is usually enhanced by the addition of photosensitizers, which makes the situation far more complicated.

Photocross-linkable polymers are classified as follows: (a) those bearing functional groups which cross-link themselves under photoirradiation, (b) those mixed with cross-linkable compounds, and (c) those polymerized, in the presence of polyfunctional monomers. In all cases, sensitizers are effective in enhancing the reactions. Several examples of the key structures of the photocross-linkable groups are listed in Table 1. These groups are introduced as side chains to the polymer and the intermolecular reaction of these groups causes cross-linking of the polymer chains. Cross-linkers and polyfunctional monomers contain multireactive centers in one molecule, such as acryloyl, allyl, mercapto or epoxy groups. Photohardening of the polymers is brought about by the formation of polymer networks with the aid of reactions of these compounds. Progress in this field has been reviewed extensively [1-3].

The effect of light on the polymer is not always beneficial for cross-linking or grafting, since it often leads to degradation of the polymer (Reaction 3), especially in the presence of atmospheric oxygen. Therefore, it is necessary to find methods of preventing such breakdown, particularly for outdoor applications. From this point of view, many types of stabilizers and antioxidants have been developed. An important application of photodegradation is to use the process for making positive photoresists. Both areas have been recently reviewed in great detail [4-6].

Reaction 5 is the subject of this chapter, photostimulated reversible property changes of polymers without any change in the molecular weights. Physical and chemical property changes of the polymers and the polymer solutions are caused by conformational changes in the polymers. Polymers that show property changes in response to light are known as photoresponsive polymers [7,8]. These polymers may be of great interest as molecular analogs of photoreceptors in biological systems [9]. In addition, they could find applications in constructing photoactive devices such as photosensors or information storage devices.

Table 1. Photocross-linkable Functional Groups

Name	Structure
Cinnamic ester	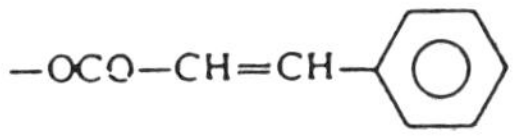
Benzalacetophenone (Chalcone)	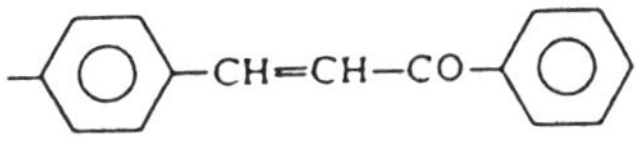
Styrylpyridinium	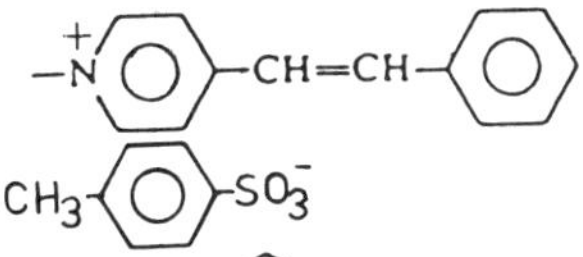
1- or 1,2-Disubstituted maleimide	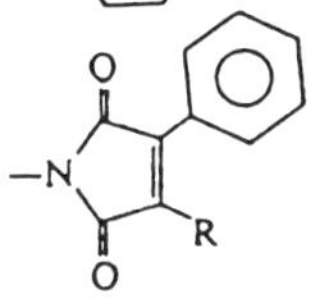 (R = H, alkyl, CN, or halogen)
Cumarin	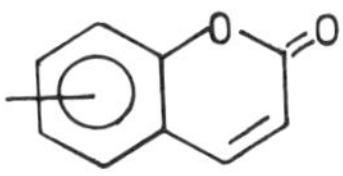
p-Azidobenzoic ester	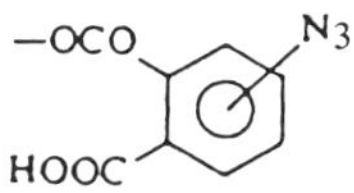
Acid azide	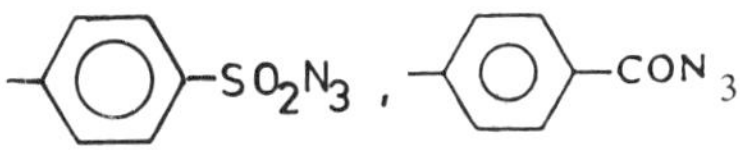
Quinone azide	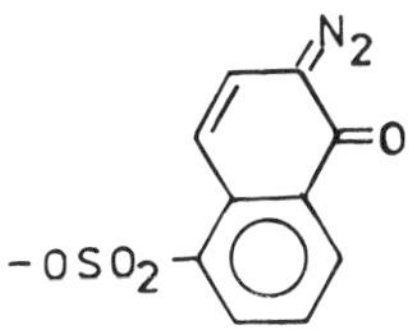

5.2 BASIC PRINCIPLE

A photoresponsive polymer is a special polymer that shows a conformational change in response to light. Therefore, the basic principle of the molecular design is to construct a polymer system having chromophores that can transfer light energy into a change in conformation. The chromophore should show a property change during isomerization large enough to cause a conformational change in the polymer chain.

Many photosensitive molecules are known to be transformed under photoirradiation into other isomers that are reversible to the initial state either thermally or photochemically:

$$A \underset{h\nu',\ \Delta}{\overset{h\nu}{\rightleftharpoons}} B$$

These isomerization processes are named photoisomerizations [10] and are classified into several categories depending on the reaction mechanisms involved, as shown in Table 2.

The isomerizations are always accompanied by some physical and chemical property change in the chromophores. Among such changes, the following properties seem to be useful as a driving force to induce a conformational change of the polymer chain: (a) a change in geometrical structure, (b) a change in dipole moment, and (c) the generation of charges.

Although many chromophores are known to undergo physical property changes under photoirradiation, the number of useful photoactive molecules for photoresponsive polymers are limited. Among the many compounds available, the following chromophores have been used for photoresponsive polymers.

1. Azobenzene

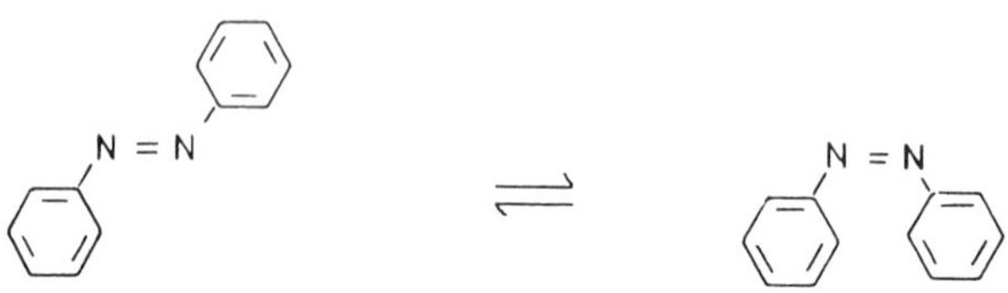

2. Stilbene

3. Spirobenzopyran

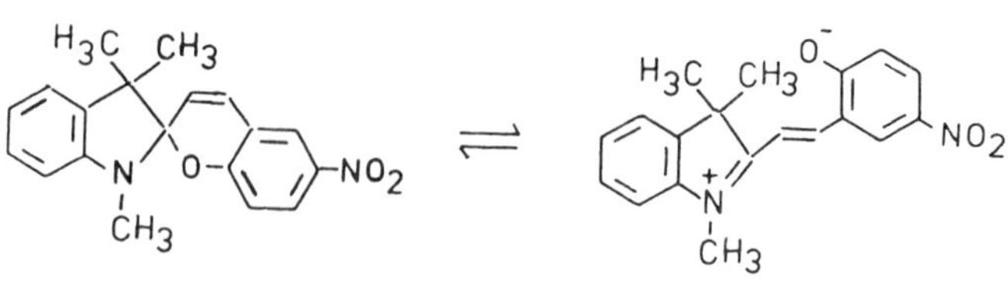

Table 2. Photoisomerizations

Cis-trans isomerization

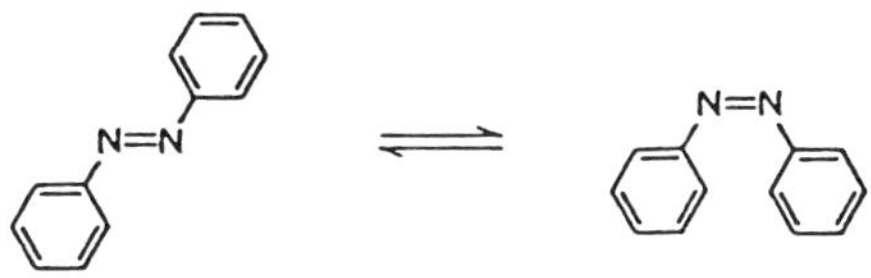

Radical Formation

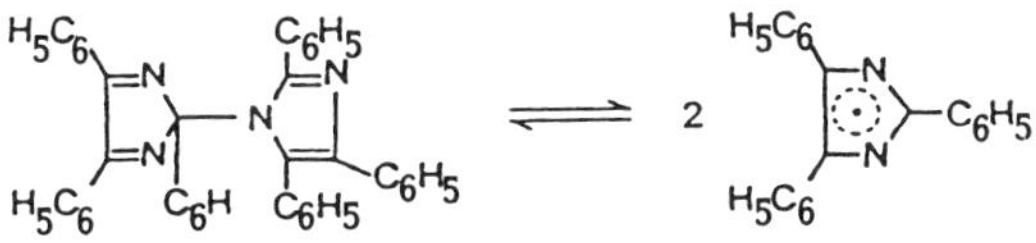

Ring-formation and ring-cleavage

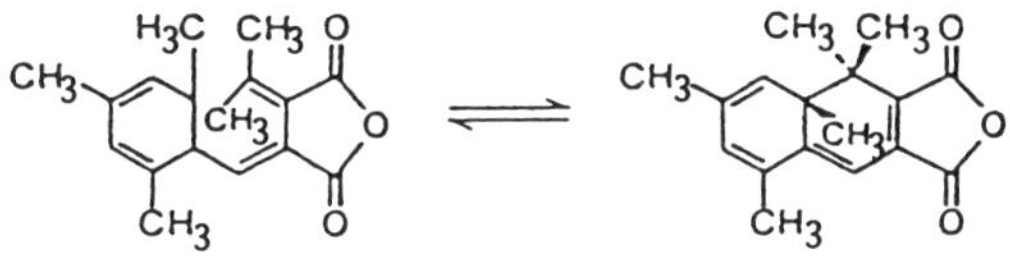

Ionic dissociation

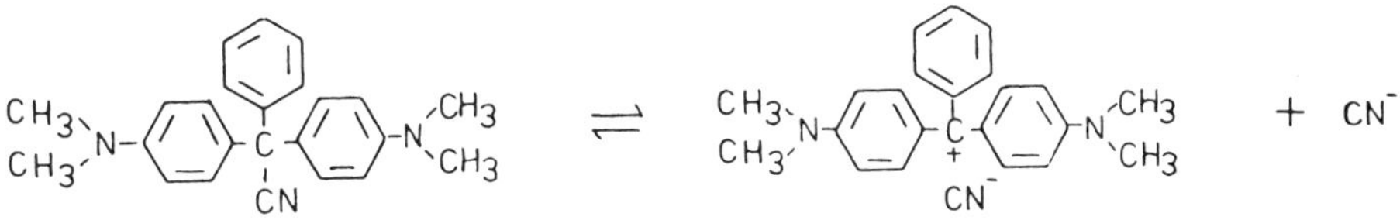

Hydrogen transfer

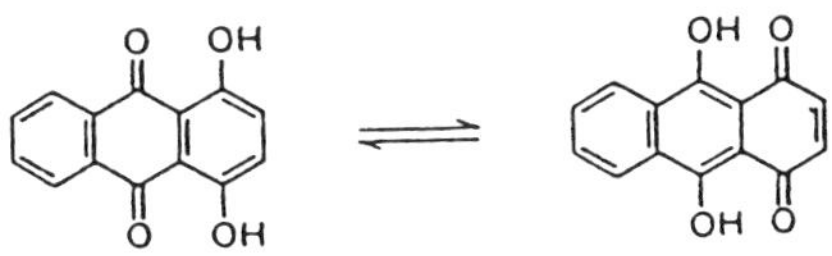

4. Triarylmethane derivatives

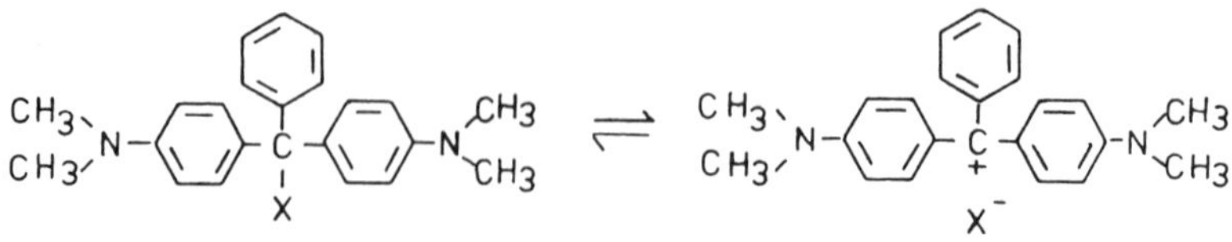

5.3 AZOBENZENE

Azobenzene is the most frequently used chromophore. It undergoes isomerization from the trans to the cis form under ultraviolet irradiation ($300 > \lambda > 400$ nm); the cis form can return thermally or photochemically to the trans form. Summarized in Table 3 are the photoresponsive polymers reported having azobenzene groups.

Physical property changes associated with the isomerization are as follows:

1. *A change in the absorption spectrum.* Trans to cis isomerization causes a decrease in the intense absorption at 320 nm due to π-π^* transition and an increase in absorption at 440 nm due to n-π^* transition (Figure 1) [10].
2. *A change in geometry.* The distance between the 4 and 4' carbons is shortened from 9.0 Å to 5.5 Å by isomerization from the trans form to the cis form [11].
3. *A change in dipole moment.* The dipole moment changes from 0.5 D (trans) to 3.1 D (cis) [12].

5.3.1 Polymer Synthesis [13-18,27]

Azobenzene residues can be introduced into the backbone as well as pendant groups of the polymer by several methods. Representative examples of polymer synthesis having azobenzene in the backbone are as follows.

Example 1

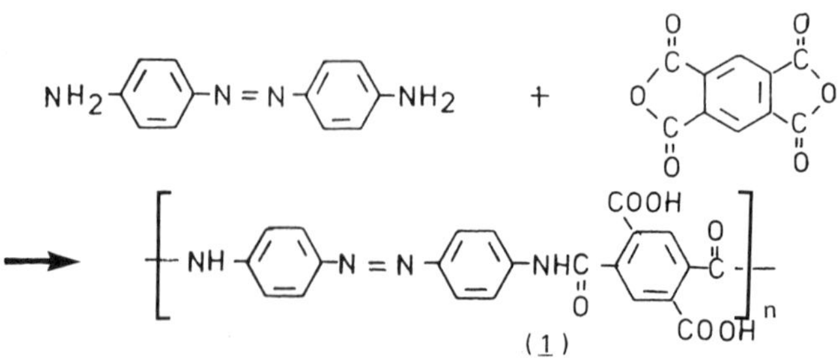

(1)

A solution consisting of 0.636 g of *N,N'*-diaminoazobenzene, 0.654 g of pyromellitic dianhydride, and 14 mL of purified dry *N,N*-dimethylacetamide was stirred for 20 h at 50 to 60°C. The reaction mixture was poured into a large volume of ethanol, yielding a fibrous precipitate. The polymer was purified by repeated reprecipitation from *N,N*-dimethylacetamide solution with ethanol and then dried in a vacuum oven [13].

Example 2

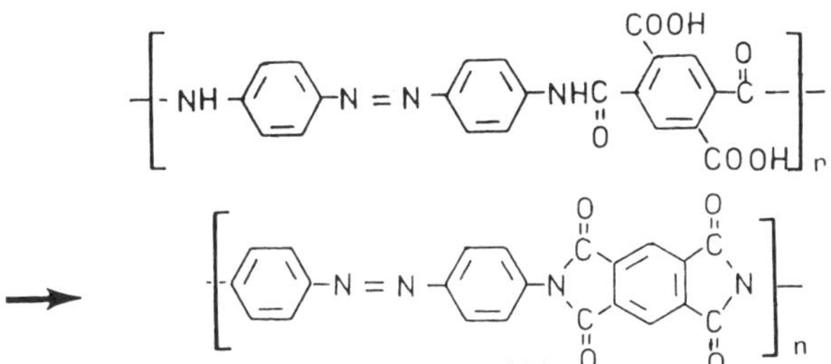

(2)

Heating the polyamide 1 to 200°C for 2 h produced polyimide 2 [14].

Example 3

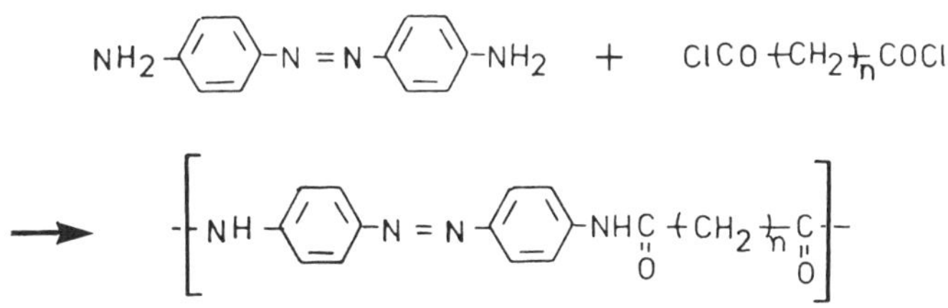

(3)

Equimolar amounts of 4,4'-diaminoazobenzene and $ClCO(CH_2)_nCOCl$ were introduced into a mixed solvent of hexamethylphosphoramide and *N*-methyl-2-pyrrolidone (2:1 by volume) at -10°C and stirred for 5 h. The polymer was washed several times with water, with acetone-water (1:1 by volume) and finally with acetone.

Table 3. Photoresponsive Polymers Having Azobenzene Chromophores

No.	Structure	Reference
(1)	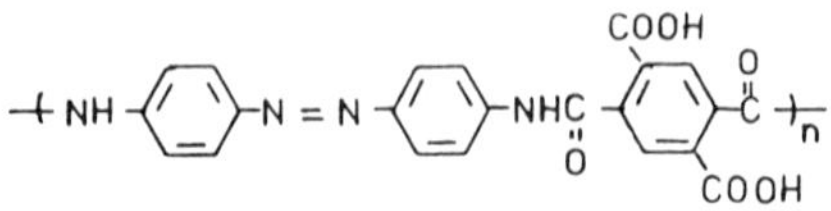	[13]
(2)	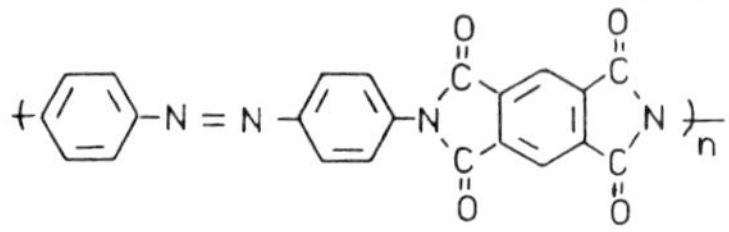	[14]
(3)	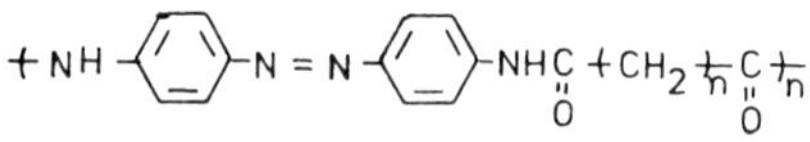	[13]
(4)	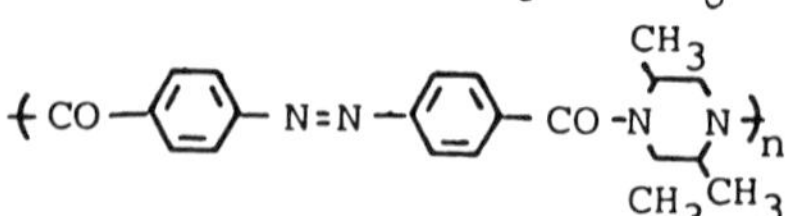	[18]
(5)	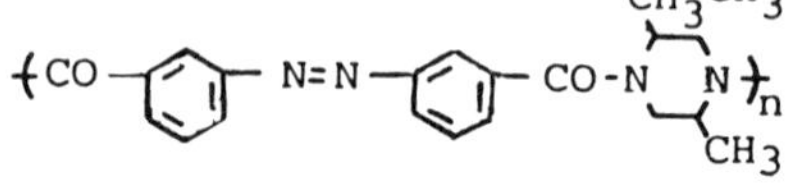	[18]
(6)	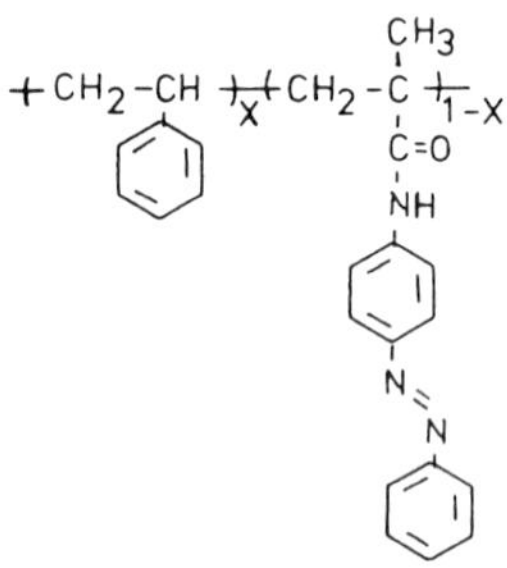	[27]
(7)	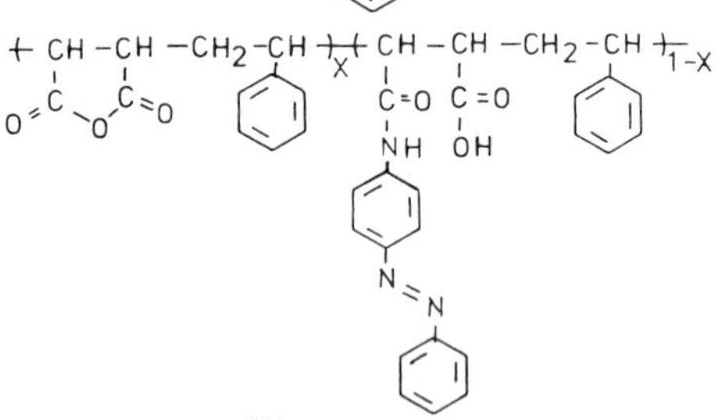	[16]
(8)	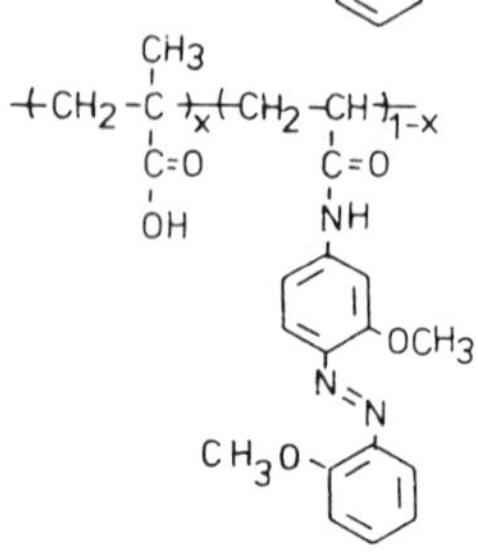	[19]

(9) 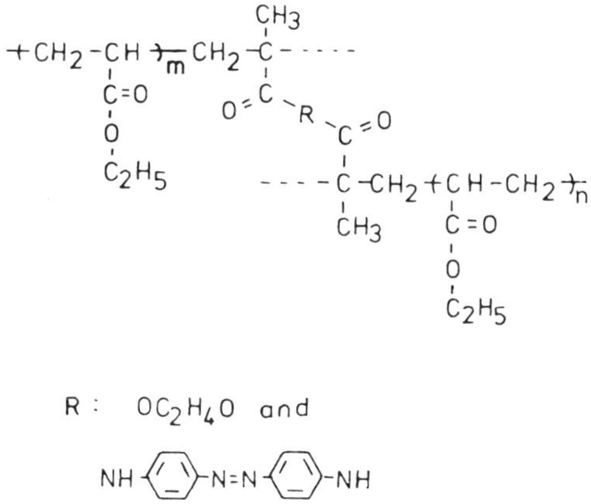 [17]

Example 4

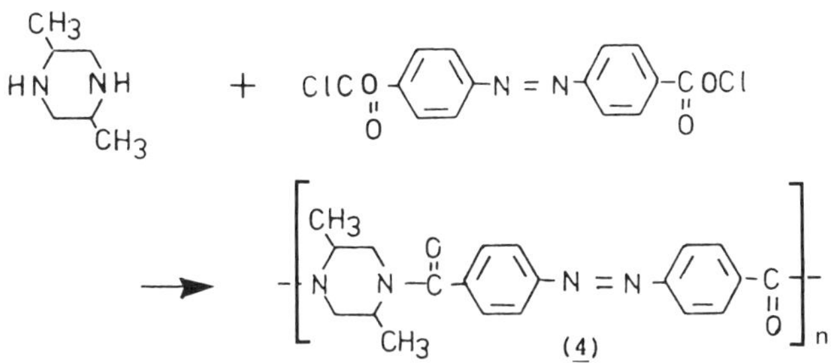

A solution of the amide (0.006 mol) was made in a suitable volume of water (45 mL); to this solution was added sodium carbonate (0.01 mL). The solution was poured into a blender and an emulsion was formed by adding dichloromethane (13 mL) and stirring at high speed. To this emulsion was added, as quickly as possible with continued stirring, a solution of azodiacidic chloride (0.005 mol) in dichloroethane (27 mL). Rapid stirring was continued for 5 to 7 min and then the polymer was isolated.

Polymers having azobenzene residues in the side groups were made as follows.

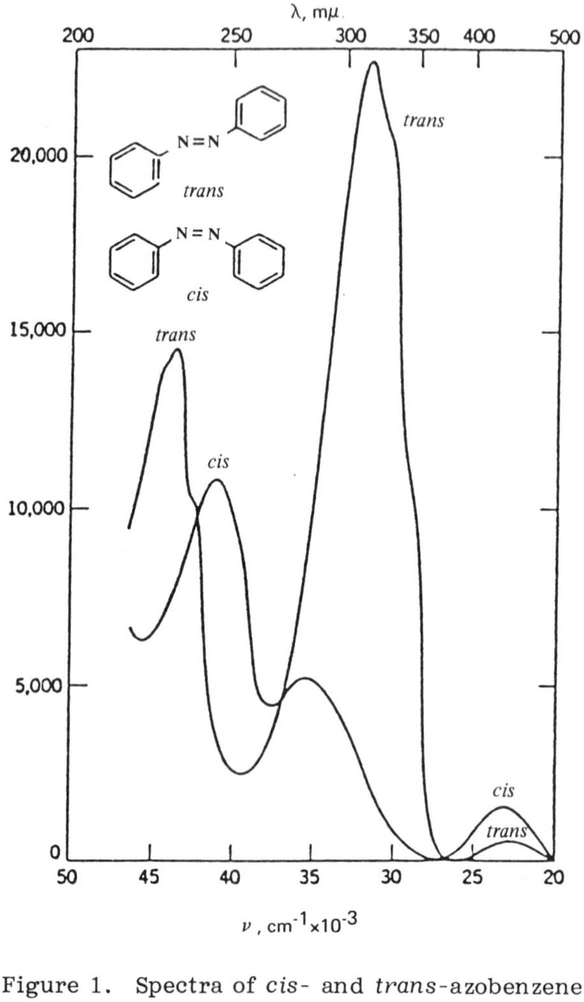

Figure 1. Spectra of *cis*- and *trans*-azobenzene in 95% ethanol.

Example 5 [27]

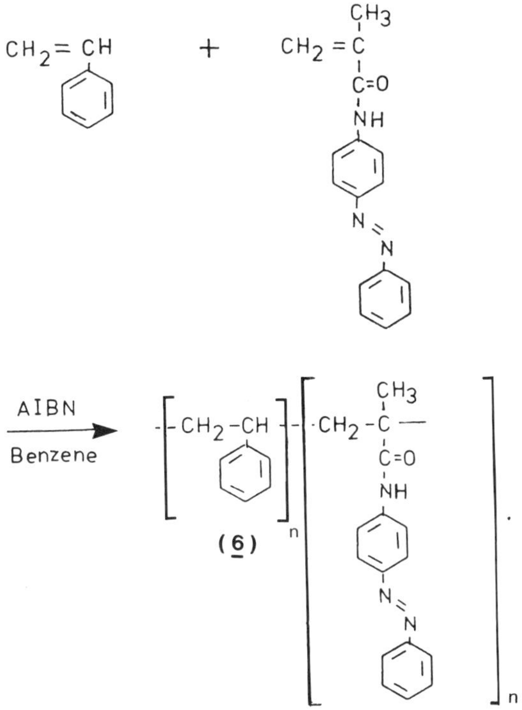

(6)

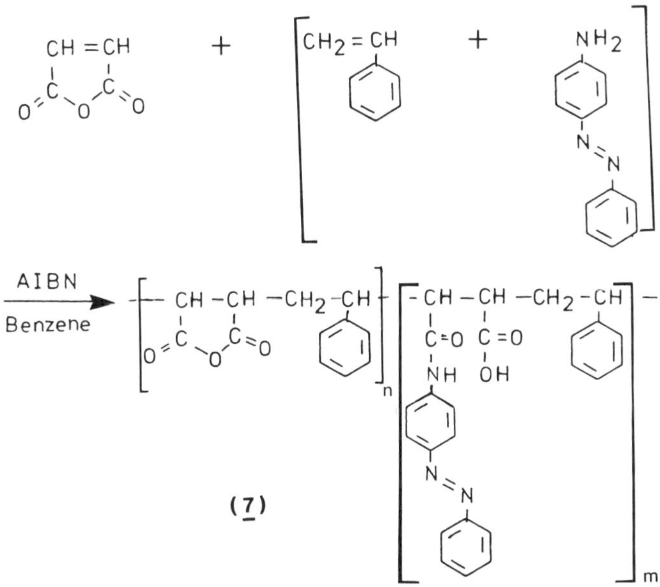

(7)

Polymer networks cross-linked by azobenzene chromophores can be made using 4,4'-di(isobutylaminoazobenzene) as follows [17].

Example 7

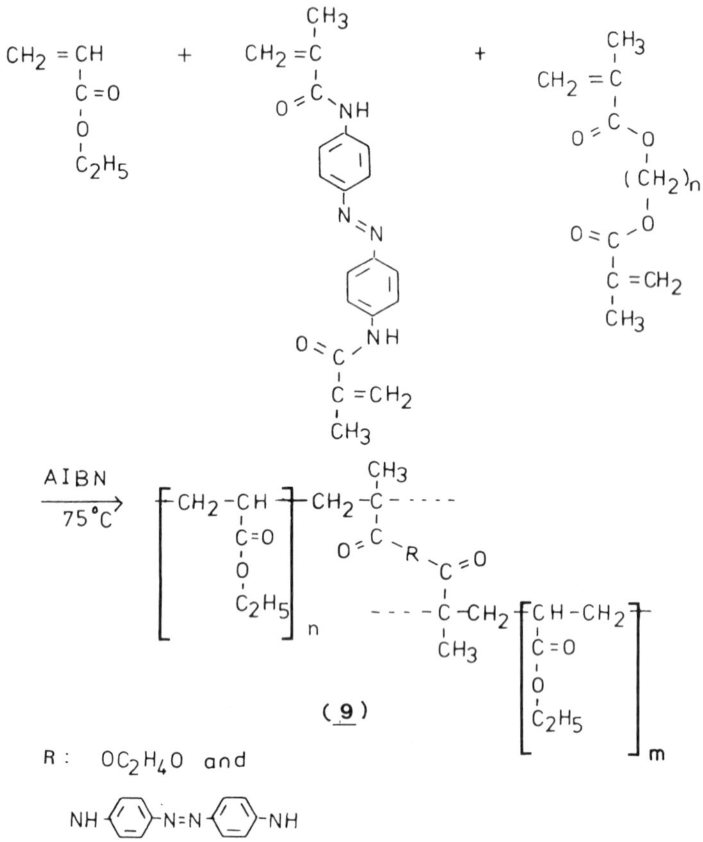

(**9**)

Polymerization was carried out between two glass plates separated by a Teflon spacer ring of 0.15 to 0.5 mm thickness. The system was slowly heated up to 75°C for 4 h and kept at this temperature for 2 days.

5.3.2 Photoresponsive Behavior [13]

5.3.2.a Spectral response [13,17]

Shown in Figure 2 is a representative spectral change of polymer 1 under ultraviolet irradiation. The intense absorption at 380 nm decreased when the *N*,*N*'-dimethylacetamide solution of polymer 1 was irradiated with ultraviolet light ($410 > \lambda > 350$ nm), while the band at 510 nm increased.

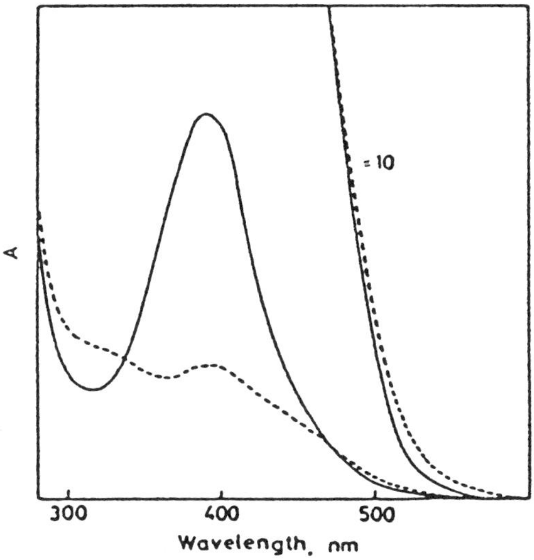

Figure 2. Absorption spectra of polyamide 1 in *N,N*-dimethylacetamide. —, in the dark before irradiation; ---, after irradiation with ultraviolet light (410 > λ > 350 nm).

The intense band at 390 nm was ascribed to the trans form of the amino-substituted azobenzene in the backbone of the polymer chain and the band at 510 nm to the cis form. The interconversion of the 390-nm band to the 510-nm band shows photoisomerization of azobenzene residues in the chain backbone.

After removing the ultraviolet light, the absorption spectrum returned to the initial spectrum in the dark. The return was accelerated by visible light (λ > 470 nm) irradiation. This photoresponsive behavior indicates that the isomerization of azobenzene takes place reversibly in the backbone of polymer chains [13].

Similar spectral changes have been reported when the chromophore is embedded in the poly(ethyl acrylate) network as cross-linking agents [17]. Small deviations from the behavior in solution are due to suppression of the thermal isomerization rate, a slight increase in the apparent activation energy, and a decrease in the ratio of cis isomer in the photostationary state.

5.3.2.b Viscosity [13,18]

A polymer having photoisomerizable unsaturated linkage in the backbone is suspected to change its conformation under photoirradiation. This idea was realized using polyamide having azobenzene residues in the backbone (polymer 1) [13]. In Figure 3 is shown the solution

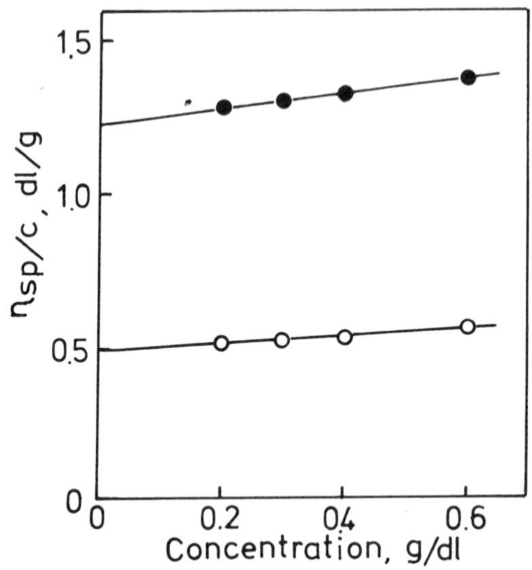

Figure 3. Viscosity of polyamide 1 in *N,N*-dimethylacetamide at 20°C. ●, in the dark before irradiation; ○, under irradiation with ultraviolet light (410 > λ > 350 nm).

viscosity of the polymer in *N,N'*-dimethylacetamide in the dark as well as during irradiation with ultraviolet light (410 > λ > 350 nm). The intrinsic viscosity during irradiation was 60% lower than the viscosity in the dark. The absence of a concentration effect suggests that the photodecrease arises from the conformational change of the polymer chain and not from interchain interaction.

On alternate irradiation of ultraviolet and visible light, the viscosity was changed as much as 60%, as shown in Figure 4. The content of the trans azobenzene form estimated from the absorption spectrum, was related to the viscosity on alternate irradiation with ultraviolet and visible light. The similarity of behavior suggests that the photodecrease and photoincrease in viscosity are directly correlated with the isomerization of azobenzene residues in the polymer backbone. The isomerization from trans to cis form kinks the polymer chain, resulting in compact conformation. The compact conformation returns to the initial extended conformation either thermally or by visible-light irradiation.

Similar experiments have been carried out with polyamides having flexible piperazine residues in the backbone, poly(3,3'-azodibenzoyl-trans-2,5-dimethylpiperazine) and poly(4,4'-azodibenzoyl-trans-2,5-piperazine) (polymers 4 and 5) [15,18]. In each case there was a reduction in viscosity upon irradiation. The change was, however, very small (less than 10% at 0.538 g/dL) compared with that of the system mentioned above. Moreover, no difference was detected between the intrinsic viscosities measured in the dark and under irradiation. The result suggests that the photodecrease of viscosity of polymers 4 and 5 arises from interchain interaction and not from intramolecular conformational change. The absence of noticeable photoviscosity effects by these polymers is possibly due to inclusion of flexible segments in the polymer chain.

The polymer having azobenzene pendant groups would also be expected to show a photoviscosity effect. This has been done, for the first time, by Lovrien [19]. The polymer examined was poly(methacrylic acid) containing azo groups as side-chain groups (8). The negative charges on the polyelectrolyte tend to expand the polymer due to electrostatic force, while the trans forms of the side chains generate hydrophobic attractive force, which tends to contract the polymers. On irradiation, the hydrophobic force is lessened due to the isomeriza-

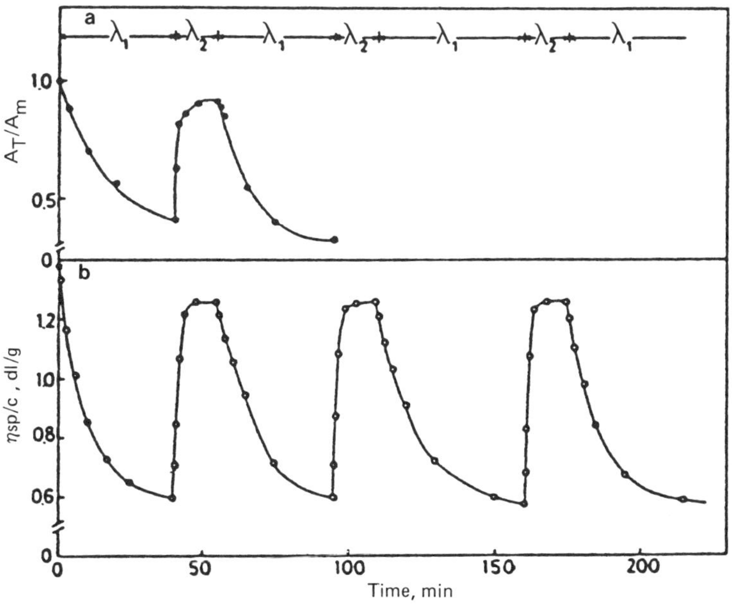

Figure 4. Changes of (a) content of the trans form of azobenzene residues in the polyamide 1 backbone and (b) viscosity of polyamide 1 in N,N-dimethylacetamide on alternate irradiation with ultraviolet ($410 > \lambda_1 > 350$ nm) and visible ($\lambda_2 > 470$ nm) light at 20°C. Concentration of the polymer is 0.9 g/dL.

tion from the trans form to the cis form and the polymer chain expands, bringing about an increase in viscosity. The increase in flow time observed in this system was approximately 1%.

Recently, Matějka and Dušek examined the photoviscosity effect of polymers having pendant azobenzene groups in organic solvents [16]. They used copolymers of maleic anhydride with styrene containing covalently bound aminoazobenzene (polymer 7). For a polymer with a very high content of azobenzene groups (72 mol %), photodecrease in viscosity by as much as 1 to 8% in THF and 24 to 30% in 1,4-dioxane was observed.

The effect may be explained by the following mechanism; radiation induces a strong dipole moment in the azo bond and these dipoles may change intrapolymer and polymer-solvent interactions. The smaller photoviscosity effect in THF solution compared with that in 1,4-dioxane can be explained by the higher dielectric constant of the former, which may suppress the intrapolymer dipole-dipole interaction.

5.3.2.c Photomechanical effect

A photomechanical effect involves a reversible contraction and expansion of an oriented rubbery film upon irradiation based on the change in the dimension of chromophores embedded in the polymer networks. This effect was first described for a partly crystalline polyimide (polymer 2) having azo-aromatic chromophores in the backbone by Agolini and Gay [14]. They reported a contraction of the polymer film at 200°C of 0.5% by ultraviolet irradiation, which in principle could cause isomerization of azobenzene chromophores from trans to cis form. A correlation between the isomerization and the film contraction was not obtained. At this high temperature, however, the content of cis isomer in the photostationary state is believed to be very low because of rapid back reaction from cis to trans form. At the same time, the rapid thermal reaction made it difficult to measure any correlation between the contraction and the isomerization. It seems that the isomerization may not be the sole origin of the contraction of this system.

Eisenbach tried to measure the photomechanical effect for a poly(ethyl acrylate) network having azobenzene chromophores as cross-linking agents (polymer 9) [17]. When the network is stretched, the azobenzene chromophores were suggested to be preferentially oriented parallel to the stretching axis. Upon irradiation of such an oriented sample with ultraviolet light, the configurational change of the azo chromophores is expected to cause a change in the conformation of an adjacent chain segment. A typical photomechanical behavior observed for the poly(ethyl acrylate) rubber at room temperature is shown in Figure 5. Irradiation with ultraviolet light induced trans→cis isomerization and contraction took place. By changing the wavelength of irradia-

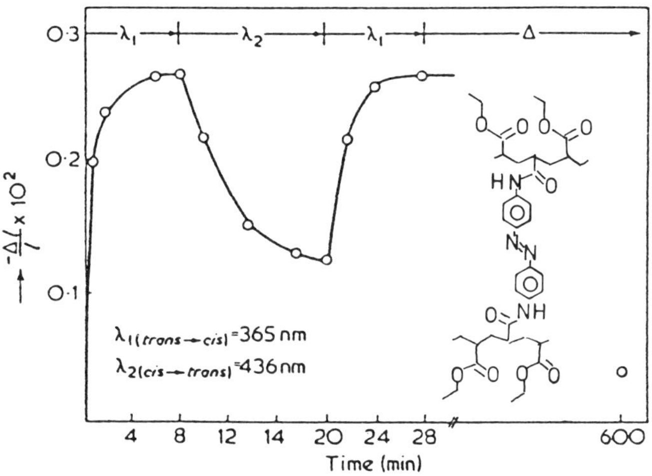

Figure 5. Photomechanical behavior of poly(ethyl acrylate) network with azo-aromatic cross-links upon irradiation; molar ratio monomer/ cross-linking agent: 100; elongation about 150%.

tion to the visible region, the cis form changed to the trans form and partial length recovery was observed. This contraction-recovery cycle could be repeated many times and the average photomechanical effect was 0.15 to 0.25%.

The problem of determining to what extent the photomechanical effect is caused by local heating effects and by real photochemical effects always arises when this process is observed. The reverse reaction, the partial length recovery by visible-light irradiation, indicates that part of the photomechanical effect for the poly(ethyl acrylate) network described above is due to the photochemical isomerization.

Photomechanical effects have also been examined for easily deformed swollen gels. This approach was first tried by Van der Veen and Prins for a system consisting of water-swollen membranes of poly(2-hydroxyethyl methacrylate) cross-linked with 1.1 vol % ethylene glycol dimethacrylate in the presence of Chrysophenine G [20].

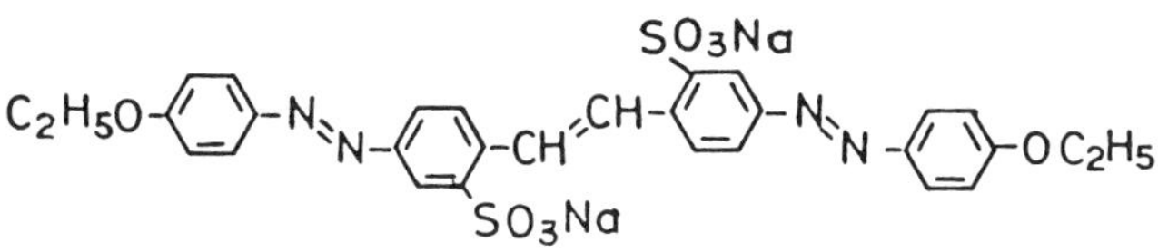

The principle of photoregulation of the swelling due to the mixing of photoisomerizable chromophore is the existence of an equilibrium interaction between the photoisomerizable compounds and the polymer network.

A similar photo effect on the swelling of cross-linked poly(methacrylic acid) (PMA) was observed by the photoisomerization of 4-phenylazophenyltrimethyl ammonium ions [21].

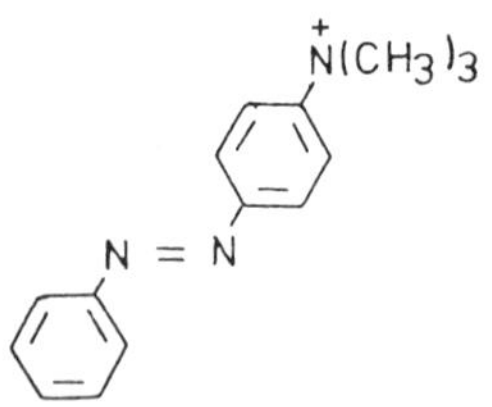

An increased cis percentage of the chromophore at a pH in the range of the conformational unfolding of the PMA led to increased ionization and thus increased degree of swelling (Figure 6). Membrane C, which has the lowest cross-linking density, showed the largest effect. The efficiency of the conversion of light energy into mechanical work performed by the membrane can be calculated to be about 3×10^{-7}.

A covalently bound photoisomerizable chromophore is expected to give a more direct effect on the deformation of the swollen gels. Matějka et al. studied cross-linked systems having azo side groups: copolymers of 2-hydroxyethyl methacrylate and azonaphthol methacrylate, P(HEMA-MAN), swollen in water and maleic anhydride-styrene copolymers with covalently bound aminoazobenzene, P(MAH-STY-AAB), swollen in diethyl phthalate [22]. They carefully estimated the contribution of thermal effect during the photomechanical conversion.

The cross-linked copolymer P(MAH-STY-AAB) swollen in diethyl phthalate was irradiated at the elongation of 1.25%. The irradiation caused a reversible increase in the force by 1% [Figure 7(a)]. Force change was much faster than the isomerization reaction and the response correlated well with the change in the temperature inside the gel. The correlation suggests that the decisive role played in the contraction process is local heating due to light absorption, not the photoisomerization of the photochromes. On intense photoirradiation, the sample is heated by the absorption of radiation even in the case of perfect external thermostatic control. The rise in temperature inside the gel by irradiation was experimentally determined to be 1.2° K. Using an interference filter ($\lambda = 370$ nm) it is possible to minimize the heating effect. The slow process of force change in Figure 6 (b) may be ascribed to trans-cis isomerization even though the effect was less than 1%.

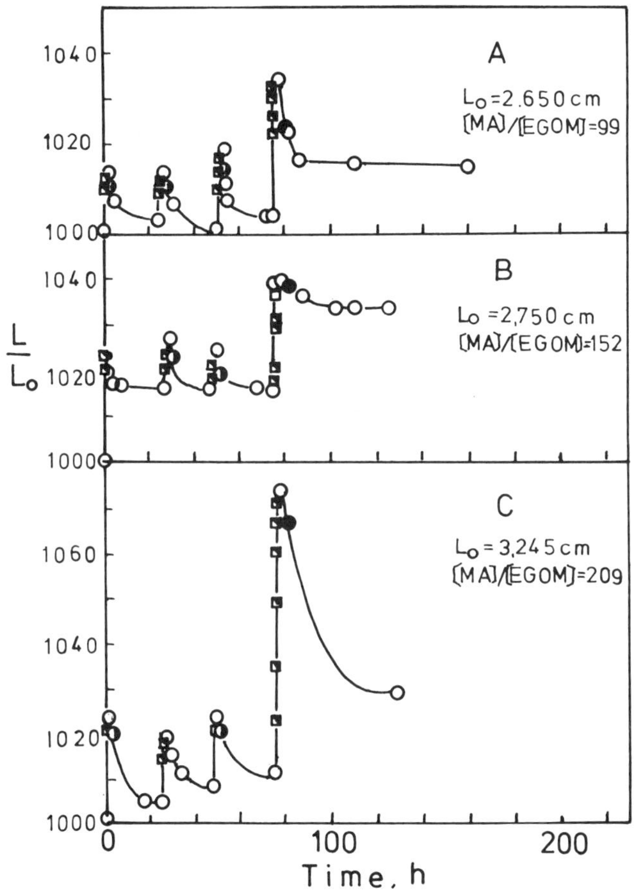

Figure 6. Photomechanochemical cycling of membranes with a monomer-to-dye ratio of 140 at 25 ± 0.05°C in 1/15 M phosphate buffer at pH 5.4. Dilatation is caused by 5 min of irradiation at 320 to 370 nm to a 80% cis photostationary state (□). Contraction occurs after 30 min of irradiation at 400 to 480 nm to a 50% trans state (○). In the last cycle longer irradiations are used (● corresponds to 3 h at 400 to 480 nm).

The large contribution to the heat effect on photomechanical contraction shown above strongly suggests that many examples reported to have photomechanical effects should be reexamined to confirm the extent of the real photomechanical effect. Furthermore, by considering the theory of elasticity of gaussian networks, the following factors should also be considered for a constant deformed sample length.

1. At a small elongation (less than the inversion elongation value, $\lambda < \lambda_{inv}$), temperature increase causes a decrease in force.
2. At a large elongation ($\lambda > \lambda_{inv}$), an increase in the modulus with temperature causes an increase in the contractive force.

The value of elongation in the experiment should always be taken into consideration, together with λ_{inv}, when the photomechanical effect is discussed.

Photoelongation of polyamide containing stilbene chromophores in the polymer backbone may be explained by the effect described above in the case of a small elongation [23]. The solution viscosity of this polymer system decreased upon trans→cis photoisomerization, while a mechanochemical study showed a decrease in force upon photoirradiation, which causes trans→cis isomerization.

5.3.2.d Surface pressure of monolayers [18]

Blair et al. have carried out film balance measurements on monolayers of photochromic polymer 4 and 5. Figure 8 shows an example of the pressure-area curve for polymer 4. A significant reduction in area per monomer unit was observed on changing from dark to light conditions and this was interpreted as being the result of trans-to-cis isomerization of azobenzene chromophores; the more extended conforma-

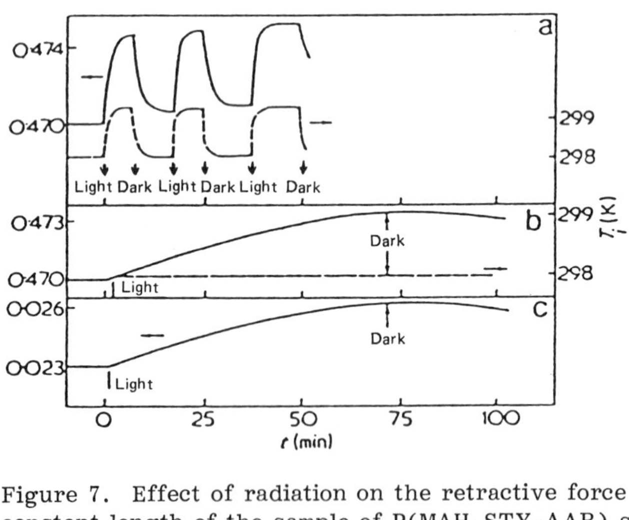

Figure 7. Effect of radiation on the retractive force f (in newtons) at constant length of the sample of P(MAH-STY-AAB) swollen in diethyl phthalate and on the temperature inside the sample, T_i. —, Retractive force; ---, temperature. (a) $\lambda = 1.25$ without interference filter. (b) $\lambda = 1.25$ with interference filter. (c) $\lambda = 1.05$ without interference filter.

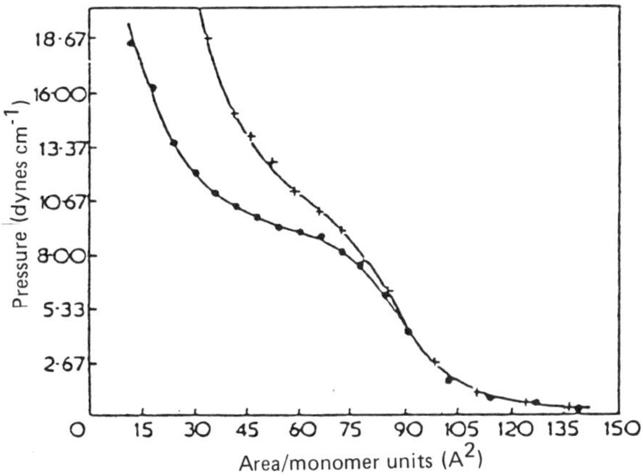

Figure 8. Pressure/area isotherms for polymer 4. +, Dark, •, under UV irradiation.

tion of the trans isomer would produce a larger apparent area. The difference between the pressure-area curves of the meta and para polyamide was interpreted on the assumption that the meta-azo polymer chain may have helical structure while the para-azo polymer has a linear structure.

5.3.2.e Wettability [24]

During the course of isomerization from trans to cis isomer, azobenzene shows an increase in the dipole moment from 0.5D to 3.1 D [12]. The change of dipole moment has been applied for controlling the wettability of polymer surface. Figure 9 shows wettability change on the surface of poly(*p*-phenylazoacrylanilide, PPAAn) on ultraviolet irradiation. The contact angle for water (cos θ) increased from 0 (hydrophobic nature of polystyrene) to 0.2 (slightly hydrophilic nature of poly-acrylate). The increase in hydrophilic nature correlated well with the isomerization of azobenzene groups from trans to cis form. After removing the light, the contact angle again decreased to zero in the dark; the recovery was accelerated by visible-light irradiation.

5.3.2.f Adsorption ability [25,26]

A change in the hydrophilic nature of the polymer surface on irradiation of poly(*p*-phenylazoacrylanilide) or its copolymer with 2-hydroxyethyl methacrylate (HEMA) could control adsorption-desorption behavior of proteins or organic substances onto the polymer. Adsorption of lyso-zyme onto the copolymer of *p*-phenylazoacrylanilide and HEMA was found

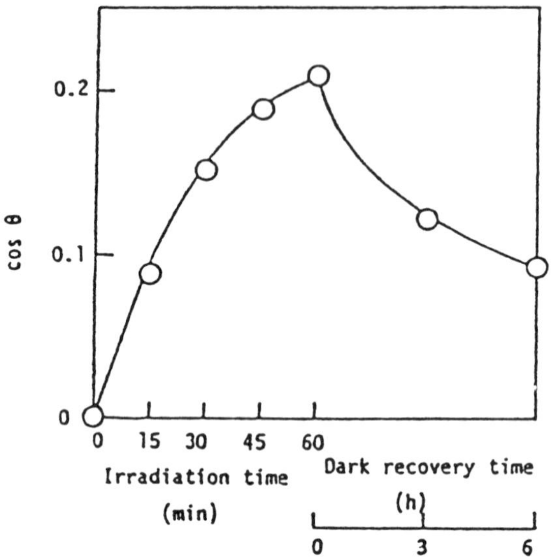

Figure 9. Photo- and thermo-induced wettability change on the surface of PPAAn film: θ, contact angle for water; photo intensity, 0.83 mW/cm^2.

to decrease from 4.6 cg to 1.8 cg per gram of adsorbent on ultraviolet irradiation, which induces the isomerization from trans to cis form [25]. Table 4 summarizes the results of other proteins. Considering the similarity in molecular weight of the three proteins, the difference in adsorption behavior is probably due to the difference in the surface characteristics of these proteins. The decrease in adsorption ability upon ultraviolet irradiation is attributable to the reduction of hydrophobic interaction between the protein and the polymer by the appearance of hydrophilic cis-form azobenzene on the surface.

The photostimulated change in the adsorption nature of a polymer may be applied to adsorption chromatography [26]. Figure 10 shows adsorption of Cephalosporin C on the photoresponsive polymeric column and desorption by irradiation with ultraviolet light. On passing an aqueous solution of Cephalosporin C through the column in the dark, the absorbance of the effluent became equal to that of the solution. When it was irradiated with ultraviolet light, the absorbance of the effluent increased, indicating the desorption of Cephalosporin C from the adsorbent as a result of hydrophobicity change. This result suggests that isolation and purification may be possible using this adsorbent in a system in which water is the only solvent.

Table 4. Photoinduced Change in the Capacity of Proteins to be Adsorbed onto a Photoresponsive Polymeric Adsorbent at 25°C[a]

Proteins	Adsorption amount (cg/g adsorbent)		MW	Average Hydrophobicity[b]
	Dark	Light		
Cytochrome *c*	7.05	5.90	12,400	1050
Lysozyme	4.60	1.83	14,300	890
α-Lactalbumin	3.80	2.00	14,400	1050

[a] $[\text{Proteins}]_{free}$ = 1.20 g/L.
[b] These values have been calculated for proteins from amino acid compositions.

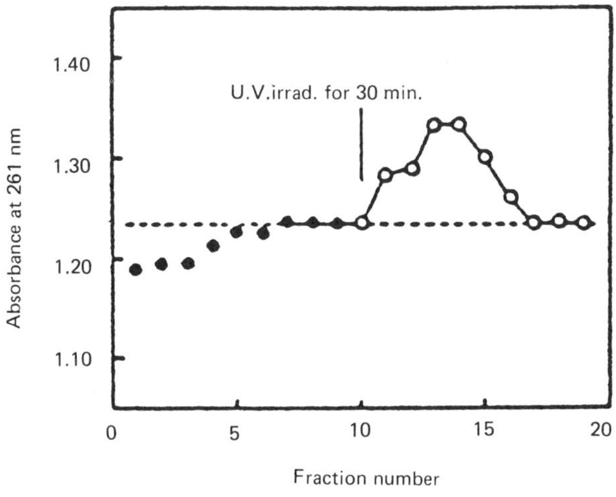

Figure 10. Chromatography of Cephalosporin C on cross-linked polystyrene coated with poly(*p*-phenylazoacrylanilide) at 25°C. Flow rate, 1 mL min^{-1}; fraction volume, 3 mL; $[\text{Cephalosporin C}]_{in\ feed}$ = 0.2 mmol dm^{-3}. ●, in the dark; ○, after irradiation.

5.3.2.g Solubility [27]

The solubility in cyclohexane of polystyrene with a small number of azobenzene pendant groups (ca. 6 mol % of monomer unit) (polymer 6) was found to change reversibly upon irradiation with light of specific wavelength; ultraviolet light (410 > λ > 350 nm) caused precipitation of the polymer, while visible light (λ > 470 nm) resolubilized it (Figure 11). The change of solubility was unambiguously induced as a consequence of photoisomerization of the azobenzene chromophores from trans to cis form.

Important physical property changes of azobenzene associated with the isomerization are a geometrical structural change and an increase in a dipole moment, as previously described. These property changes, however, do not affect the solubility of azobenzene itself in cyclohexane. Precipitation of azobenzene was not observed under ultraviolet irradiation even in a saturated cyclohexane solution of azobenzene. Polystyrene is necessary for the precipitation to take place.

In cyclohexane (theta solvent for polystyrene) intermolecular interaction between a polymer and solvents is in balance with the intra- and interpolymer interaction. The dipole moment increase and the geometrical structural change of pendant groups by ultraviolet irradiation are considered to alter the balance of polymer-solvent and polymer-polymer interactions. The introduction of the nonpolar trans form of azobenzene

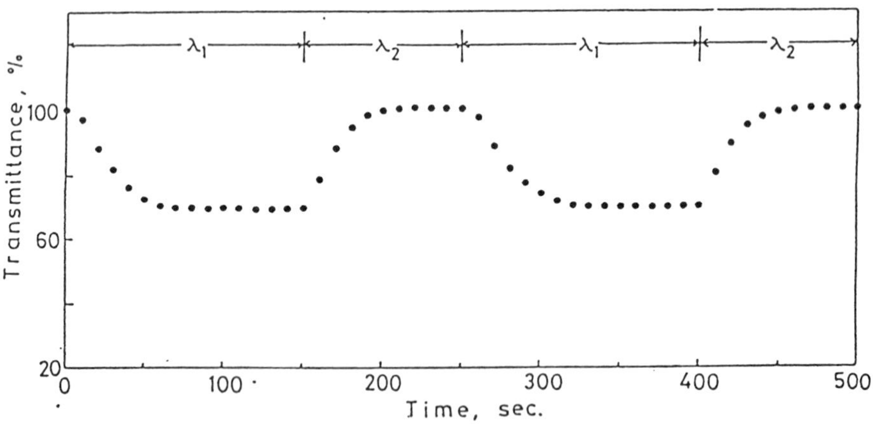

Figure 11. Changes of transmittance at 650 nm of cyclohexane solution containing PS-A-6.1 on alternate irradiation with ultraviolet (410 > λ_1 350 nm) and visible λ_2 > 470 nm) light at 30°C. Concentration of the polymer was 0.046 g/dL.

into polystyrene pendant groups scarcely affects the polymer-solvent interaction in cyclohexane, while the photogenerated polar cis form tends to decrease the polymer-solvent interaction. Therefore, upon ultraviolet irradiation, the polymer-solvent interaction decreases considerably until the polymer precipitates. Molecular-weight-distribution measurements of the precipitate and the soluble polymer revealed that the molecular weight of the precipitate was twice that of the soluble polymer.

The precipitation behavior observed in this system is similar to the fractional precipitation of a polymer from a solution by addition of a miscible nonsolvent. Cyclohexane is a good solvent for the polystyrene with *trans*-azobenzene pendant groups, while it is a poor solvent for photoirradiated polystyrene having *cis*-azobenzene chromophores. The fractionation was mainly controlled by the molecular weight and not by the content of the azobenzene groups.

5.3.2.h pH and conductivity [13]

Photostimulated conformational change of polymer chains gives rise not only to viscosity change but also to other solution property changes. In Figure 12 is shown the response of conductivity of an *N*,*N*'-

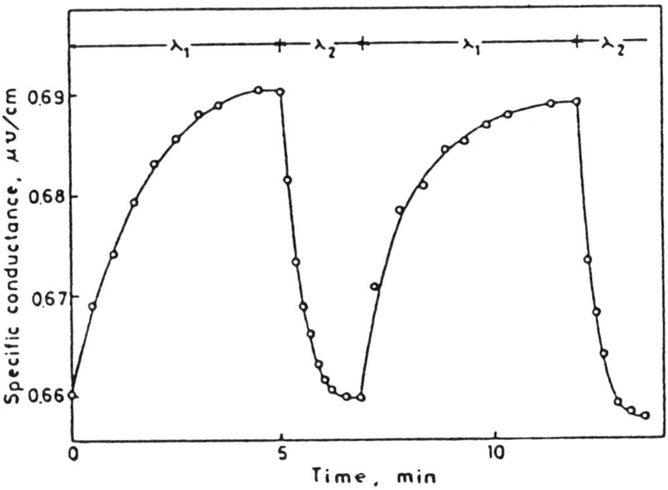

Figure 12. Changes of specific conductance of an *N*,*N*-dimethylacetamide solution of polyamide 1 at 25°C on alternate irradiation with ultraviolet (410 > λ_1 > 350 nm) and visible (λ_2 > 470 nm) light. Concentration of the polymer is 0.04 g/dL.

dimethylacetamide solution containing polymer 1 under alternate irradiation with ultraviolet and visible light. The specific conductivity was increased with ultraviolet irradiation and decreased with visible irradiation. The amount of change was about 4%. The dissociation equilibrium of amide-substituted terephthalic acid residues in the polymer backbone is influenced by a change of microenvironment due to conformational change of the polymer chain. Dissociation of the acid is stimulated in the compact conformation, while the extended conformation depressed the dissociation.

The change in the equilibrium constant for the dissociation of the acid in organic solvent suggests the possibility of pH control in aqueous systems. Measurement of the pH of an aqueous solution of polyamide 1 revealed that pH decreased under ultraviolet irradiation and returned to the initial value under visible irradiation.

5.3.2.i Dynamics of conformational change [28]

It is indispensable to know the responsive rate, when constructing photoresponsive devices using the photoresponsive polymers described above. The most convenient way to find the response time is to measure the conformational change of the polymer chain in solution. Direct measurement of the conformational change was carried out using a flash photolysis method combined with a light-scattering detection system by Irie and Schnabel [28]. Light-scattering intensity change gives information on how fast the conformational change occurs. They measured the time dependence of light-scattering intensity change after a single pulse (530 nm, 20 ns), which induces isomerization from the cis to the trans form of azobenzene chromophores along the polymer backbone, and followed the unfolding process of polyamide 1. The polyamide was found to change its conformation from compact to extended form by visible-light irradiation (see Section 5.3.2.b) [13].

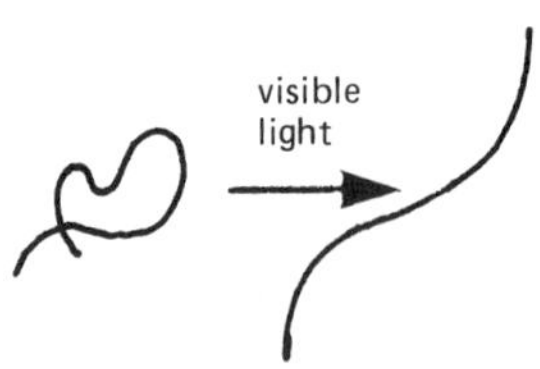

Summarized in Table 5 are the conformational relaxation times of the polyamide in several solvents. In solvents such as *N*,*N*'-dimethylacetamide and *N*,*N*'-dimethylformamide, the unfolding process proceeds rather quickly, while in poor solvents the unfolding is retarded. The response time was found to be in the range 10^{-4} to 10^{-3} s.

Table 5. Conformation Relaxation of Polymer 1 Chains Subsequent to Cis to Trans Isomerization of Azobenzene Groups in the Backbone

Solvent	$\tau_{1/2}$[a] (s)
N,*N*-dimethylacetamide	5.8×10^{-4}
N,*N*-dimethylformamide (DMF)	4.7×10^{-4}
4:1[b] DMF-ethanol	7.2×10^{-4}
3:2[b] DMF-ethanol	11.0×10^{-4}
4:1[b] DMF-water	8.8×10^{-4}

[a]Half-lives of the decrease of the light-scattering intensity at 514 nm and 22°C.
[b]Volume ratio.

5.4 SPIROBENZOPYRAN

Spirobenzopyran undergoes ring opening on ultraviolet irradiation with the production of intensely colored merocyanine; the merocyanine can return thermally or photochemically to colorless spiropyran. Physical property change associated with the isomerization of spirobenzopyran to merocyanine are (a) a change in absorption spectrum, (b) a change in dipole moment, and (c) a geometrical structural change.

5.4.1 Polymer Synthesis

Spirobenzopyran residues can be introduced as pendant groups to polymers by copolymerization of 1,3,3-trimethylindolino-6'nitro-8'-[(methacryloxy)methyl]spirobenzopyran with other monomers. The synthesis is as follows.

Example 8

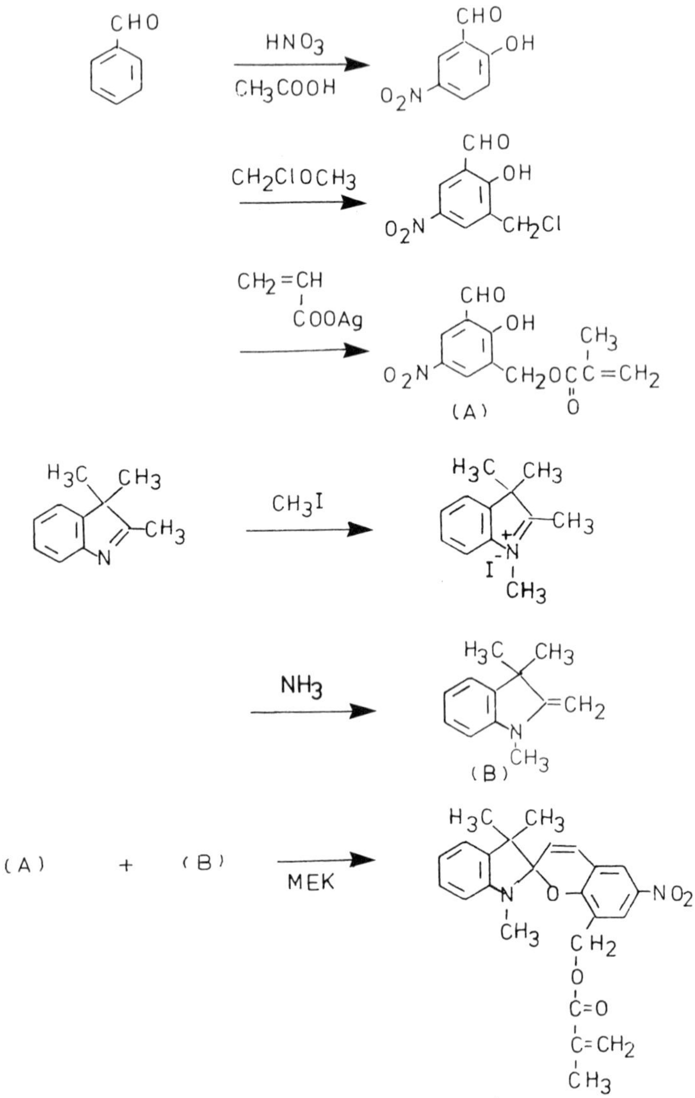

(12)

Benzospiropyran polymer networks having spirochromophores as cross-linking agents were made by copolymerization of ethylacrylate with variable amounts of bis-photochrome dimethacrylate, 1,1'-(α,α'-

p-xylyl)-bis[3',3'-dimethyl-8-methacryloxymethyl-6'-nitro-spiro(2H-1-benzopyran-2,2'-indoline)] as follows [29,34].

Example 9

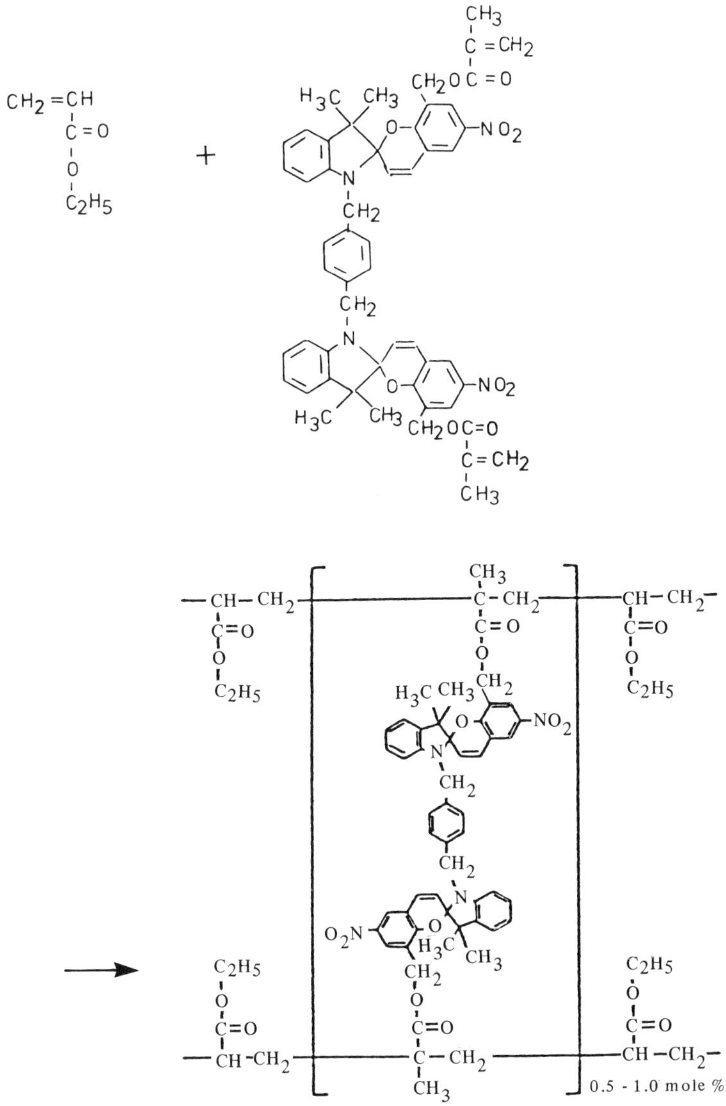

(13)

The copolymerizations were carried out in benzene solution (50 vol %) in the presence of diisopropyl peroxydicarbonate as initiator on the surface of glass or mercury. A representative condition is 100 mmol of ethylacrylate, 0.66 mmol of bis-photochrome dimethacrylate, and 0.096 mmol of diisopropyl peroxydicarbonate. Summarized in Table 6 are the photoresponsive polymers having spirobenzopyran chromophores.

Table 6. Photoresponsive Polymers Having Spirobenzopyran Chromophores

No.	Structure	References
(10)	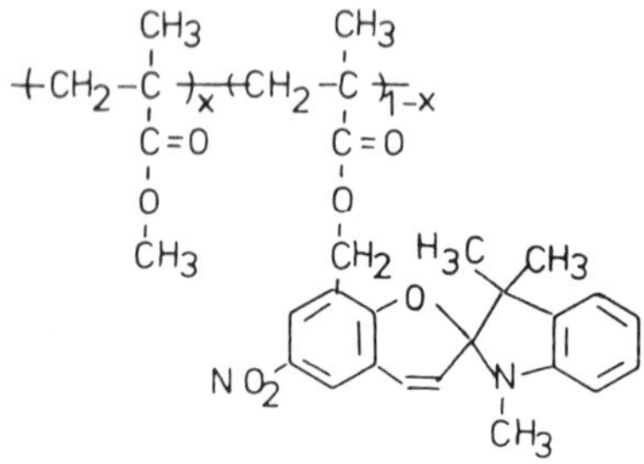	[30]
(11)	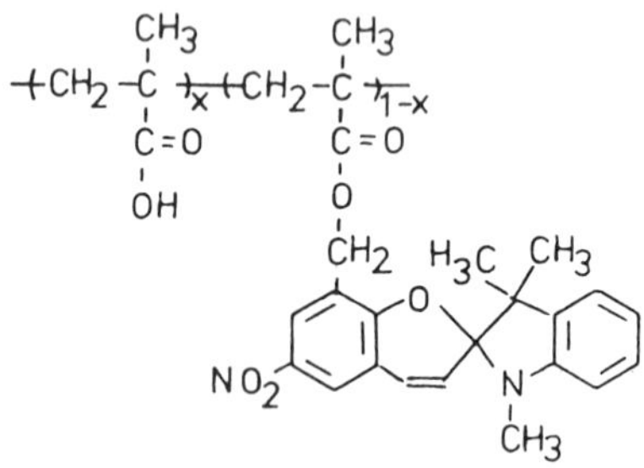	[32, 33]
(12)	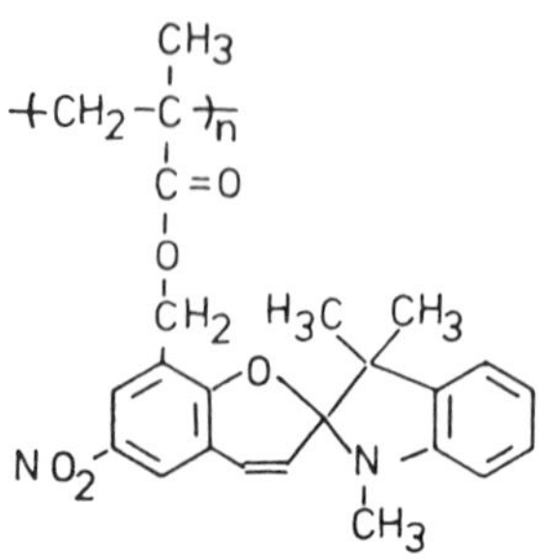	[24]

No.	Structure	References
(13)	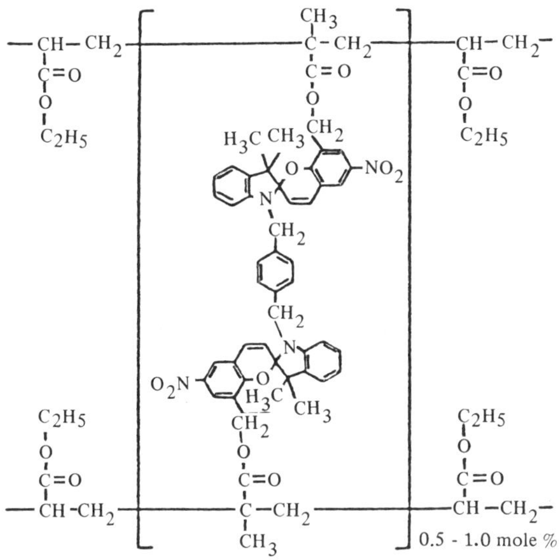	[29,34]

5.4.2 Photoresponsive Behavior

5.4.2.a Spectral response [30]

Spectra of polymer 10 in benzene in the dark and after irradiation by ultraviolet light ($\lambda > 310$ nm) are shown in Figure 13. Before irradiation, the benzene solution is colorless; the solution becomes blue on ultraviolet irradiation. On removing the light, the intense color disappears after several minutes.

In polar solvents, such as alcohols, colored merocyanine forms even in the dark and is in equilibrium with colorless spiropyran (Figure 14). Photoexcitation of the merocyanine with visible light in alcohols brings about ring closure, with the production of colorless spiropyran; the red color disappears on irradiation with visible light. Ultraviolet light enhanced the absorption around 530 nm. The increase in absorption corresponds to the ring-opening reaction of residual spiropyrans. The intense absorption at 530 nm generated by ultraviolet-light irradiation returned to the thermal equilibrium intensity after removal of the light.

In general, as the polarity of the solvent is increased, thermal equilibrium between spiropyrans and merocyanines shifts to the formation of the merocyanines. In addition, the visible absorption maximum

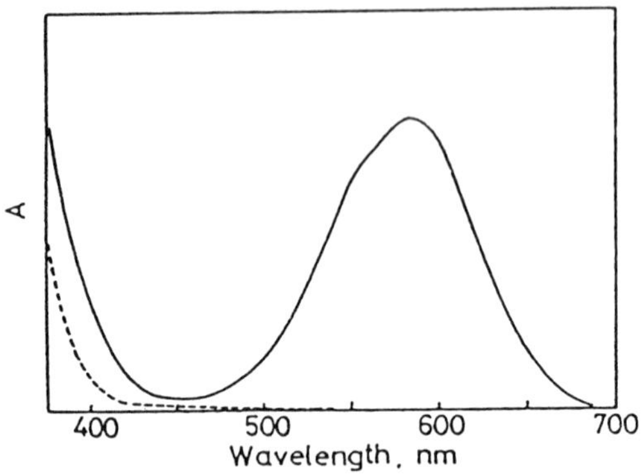

Figure 13. Absorption spectra of merocyanine bonded to poly(methyl methacrylate) in benzene. ---, in the dark; —, after irradiation by ultraviolet light (λ > 310 nm).

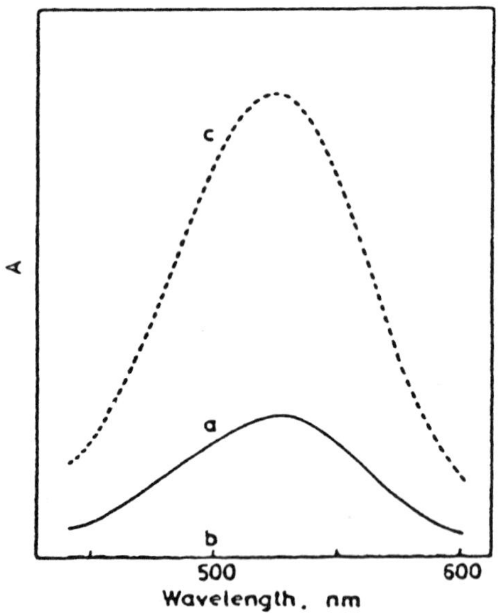

Figure 14. Absorption spectra of merocyanine bound to poly(methacrylic acid) in methanol at 20°C. (a) Under thermal equilibrium conditions. (b) Under photostationary state of visible light (λ > 470 nm) irradiation. (c) Under photostationary state of ultraviolet light (410 nm > λ > 310 nm) irradiation.

of the merocyanine form shifts to a shorter wavelength, the extinction coefficient decreases, and the half-width of the band increases. For example, the blue color (λ_{max} = 590 nm) in benzene changes to red (λ_{max} = 530 nm) in methanol. These spectrum changes give us clear information concerning geometrical structures of the spirobenzopyrans, whether they are in less polar spiropyran form or in the polar open-ring merocyanine form as well as information regarding their micro-environment.

5.4.2.b Viscosity [30, 32, 33]

A change in dipole moment due to isomerization from spiropyran to the merocyanine form would be expected to alter intramolecular interaction of polymer chains when the chromophores are incorporated into the polymer pendant groups or backbone groups. The change of intra-molecular interactions in polymer systems induces a conformational change in the polymer chain and can be detected by viscosity measurement. A viscosity change by photoirradiation of the spirobenzopyran system was observed for the first time in benzene for poly(methyl methacrylate) with spirobenzopyran pendant groups by Irie et al. [30]. The intrinsic viscosity in benzene during irradiation was 17% lower than the viscosity in the dark, while no difference was detected in dichloroethane (Figure 15). The constancy of the viscosity in dichloro-ethane demonstrates that the photoviscosity effect is inherent to the polymer and not caused by a thermal effect.

The viscosity during photoirradiation in benzene returns to the initial value in less than 3 min at 20°C after the light is removed. In Figure 16 is shown the recovery of the viscosity together with the fading behavior of the absorption at 590 nm attributed to the merocyanine form. The good correlation between viscosity and the absorption intensity change due to the merocyanine form unambiguously indicates that the isomerization gives rise to the conformational change in poly(methyl methacrylate) chains.

The increase of viscosity is dependent on the polarity of the solvent; 17% in benzene, 13% in dioxane, 5% in tetrahydrofuran, 3% in chloroform, and 1% in dichloroethane, respectively. The solvent effect suggests that the shrinkage of the polymer chain is caused primarily by specific intramolecular solvation of the ester side groups of the poly(methyl methacrylate) chain of the photogenerated polar merocyanines, in competition with solvation by the solvent.

The concept of preferential intramolecular solvation was experimentally supported by examining the solvent effect on the absorption maximum of the merocyanine form of the pendant groups. The reciprocal of the maximum wavelength of the merocyanine bound to the polymer and the low-molecular-weight analog was plotted against the Dimroth E_{T30} value [31] of the solvent as shown in Figure 17. The absence

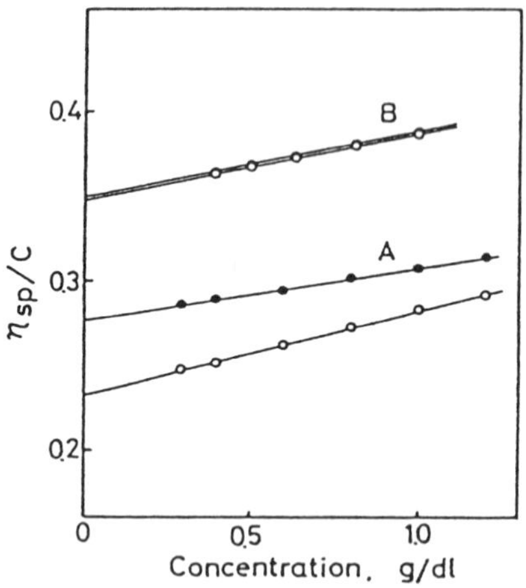

Figure 15. Viscosity of poly(methyl methacrylate) having spirobenzopyran units (polymer 10, x = 0.87) at 30°C. (A) In benzene: •, in the dark; ○, under irradiation (λ > 310 nm). (B) In dichloroethane: •, in the dark; ○, under irradiation (λ > 310 nm).

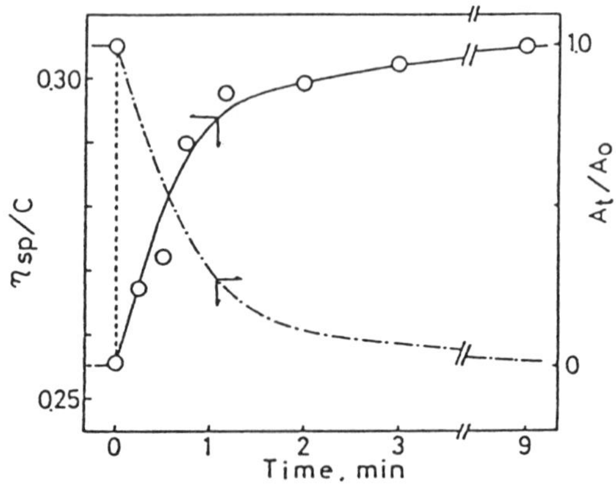

Figure 16. Recovery of the viscosity of poly(methyl methacrylate) having spirobenzopyran units (polymer 10, x = 0.87) and fading of the absorption at 590 nm due to merocyanine formed in benzene at 20°C in the dark. Concentration of the polymer is 1.0 g/dL.

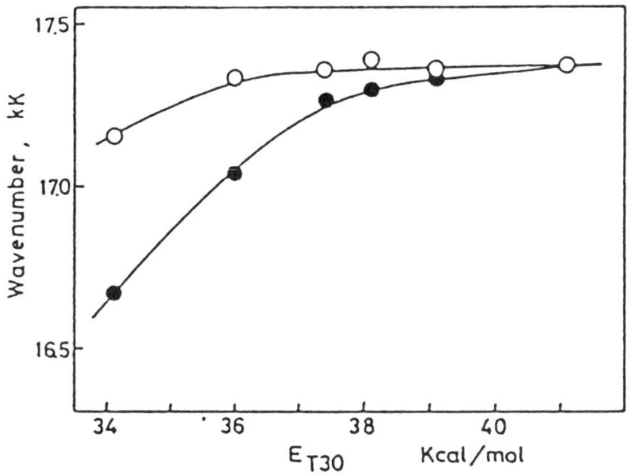

Figure 17. Correlation of absorption maxima (○) of merocyanine bonded to poly(methyl methacrylate) and (●) of 1,3,3-trimethylindolino-6'-nitro-8'-(dimethylcarboxymethyl)spirobenzopyran with the Dimroth polarity parameter E_{T30} (kcal/mol) for a series of solvents.

of a strong solvatochromism in the polymer spectrum implied that the photogenerated merocyanines are preferentially solvated by the ester groups of the same polymer chain and are shielded from the effect of bulk solvent. The stabilization energy arising from self-solvation (around 1.4 kcal/mol in benzene) is considered to be the origin of the shrinkage of the polymer chain.

Poly(methacrylic acid) having spirobenzopyran pendant groups is soluble in polar solvents such as methanol or water when the content of spirobenzopyran is less than 20 mol %. Methanol solution of the polymer has a weak red color under thermal equilibrium conditions, which indicates the presence of merocyanine form in equilibrium with spiropyran form. Visible irradiation completely bleached the absorption (merocyanine → spiropyran); after the light was removed the absorption gradually reappeared in the dark (spiropyran → merocyanine). Ultraviolet irradiation caused enhancement of the absorption around 530 nm (spiropyran → merocyanine); the intense absorption again returned to the thermal equilibrium intensity after removal of the light (merocyanine → spiropyran). Together with these isomerizations, a methanol solution of the polymer showed a viscosity change, as shown in Figure 18 [32]. Visible-light irradiation caused an increase in viscosity, while ultraviolet light irradiation caused a decrease in viscosity. By eliminating the dark period, the viscosity can be changed as much as 40% by alternate irradiation of visible and ultraviolet light, as shown in Figure 19.

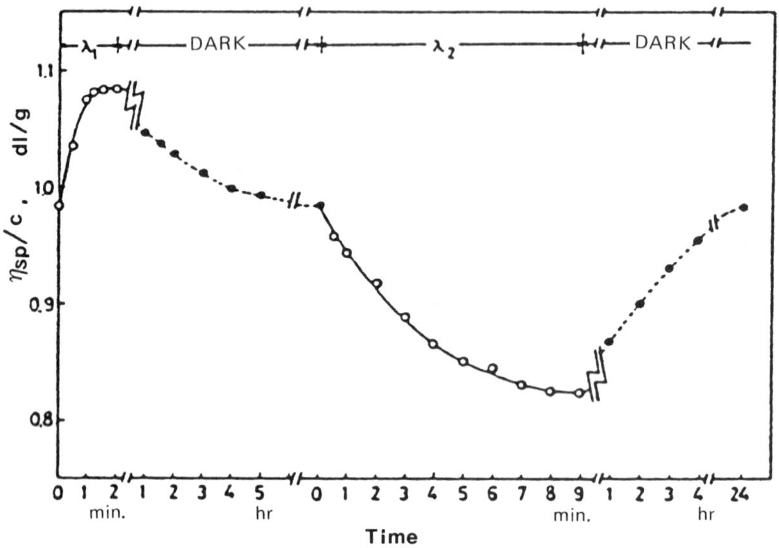

Figure 18. Changes of the viscosity of a methanol solution of poly(methacrylic acid) having spirobenzopyran pendant groups (polymer 11, x = 0.926) on visible light (λ_1 > 470 nm) irradiation and ultraviolet light (410 nm > λ_2 > 310 nm) irradiation at 20°C. Concentration of the polymer is 0.25 g/dL.

The merocyanine has a zwitterion structure. The destruction of the zwitterion structure of the merocyanine pendant groups by visible-light irradiation is considered to alter the balance of the intrachain interactions. The electrostatic attractive forces between the zwitterions and the charges on carboxylic acid residues of the polymer chain tend to contract the polymer chain, while the charges on the carboxylic acid residues, at the same time, generate repulsive forces which tend to expand the polymer. In the dark before irradiation, the attractive forces overcome the repulsive forces and the polymer chain has a contracted conformation. Upon visible-light irradiation, the attractive forces considerably decrease because of the disappearance of the zwitterion structure, unbalancing the system until the electrostatic forces cause the polymer chain to expand. Under a photostationary state of visible-light irradiation, the conformation of the polymer is mainly determined by the repulsive force between the charges on carboxylic acid residues. A similar conformational change by poly(methacrylic acid) with spirobenzopyran pendant groups has been reported in an aqueous system [33].

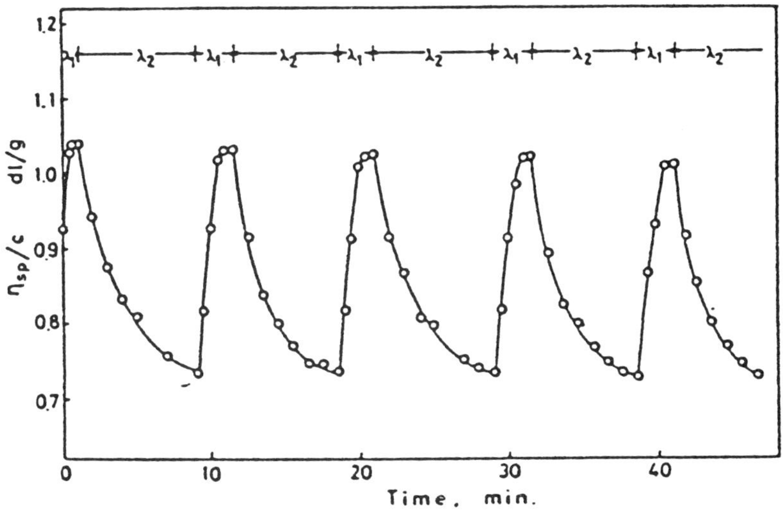

Figure 19. Changes in the viscosity of a methanol solution of poly(methacrylic acid) with spirobenzopyran pendant groups (polymer 11, x = 0.911) on alternate irradiation with visible (λ_1 > 470 nm) and ultraviolet (410 nm > λ_2 > 310 nm) light at 20°C. Concentration of the polymer is 0.25 g/dL.

5.4.2.c Photomechanical effect

The photomechanical effect of stretched rubbery networks cross-linked by isomerizable spirobenzopyran chromophores (polymer 13) has been examined by Smets et al. [29, 34]. The effect was studied under constant stress at constant temperature by following the length contraction as a function of time of irradiation. On irradiation, 2% contraction has been observed; in the dark, length recovery occurs. The photomechanical effect increased with decreasing temperature and showed a maximum at optimum stress depending on M_c value of the network. The light-dark cycles could be repeated several times and were reversible.

Initially it is assumed that the contraction of the rubber is attributed to an entropy increase of the polymer chain due to the higher flexibility of the open-ring merocyanine compared to parent ring-closed spiropyran [34]. This interpretation, however, could hardly explain the following experimental results: (a) Only a minor decrease (ca. 3%) of optical density due to merocyanine form was observed even after the length recovery was completed; (b) the rate of contraction was higher when the film had been previously colored than for the first irradiation cycle; and (c) dark decoloration kinetics indicated an activation energy of 26.8 kcal/mol, while length recovery rate measurements was only 6 to 7 kcal/mol. It was therefore suggested that the contraction is related to photochemical interconversion of photogenerated merocyanine isomers.

On light absorption, the most stable isomer is converted to a less stable isomer, resulting in a change of the conformation of the polymer networks [35,36]. This interpretation is partly supported by the similarity of the action spectrum of contraction and the absorption spectrum of the merocyanine form of the photoresponsive poly(ethyl acrylate) rubber.

Local temperature increase resulting from light absorption would also be expected to contribute to the shrinkage of this stretched rubber, similar to that observed in the contraction behavior of swollen gels having azobenzene chromophores. Such an effect could result from insufficient heat exchange in the solid polymer sample and be dependent on the specific heat capacity and the density of the polymer. The experimental difficulty in measuring simultaneously "in situ" temperature increase of the rubber and contraction under irradiation always leaves an uncertainty in the interpretation of the photocontraction mechanism. There is no easy way to discriminate the heat effect and the real photomechanical effect originating from isomerization when the observed contraction is less than a few percent. At least half of the total contraction is believed to be attributable to the local heating effect.

5.4.2.d pH

Photocontrol of pH value in aqueous solution has been tried using poly(methacrylic acid) having spirobenzopyran pendant groups [33]. In Figure 20 is shown the response of pH of an aqueous solution of the polymer under the influence of visible-light irradiation ($\lambda > 470$ nm). The pH value increased on exposure to visible-light irradiation and returned to the initial value in the dark when the light was removed. The response of the pH value was in good correlation with the isomerization of the spirobenzopyran side groups.

The hydrophobic nature of the merocyanine is considered to be weak compared with that of the spirobenzopyran form because of its zwitterionic structure. The increase in the hydrophobic nature upon isomerization of the merocyanine form to the spiropyran form causes the polymer to contract, resulting in a change of conformation. The dissociation of the carboxylic acid residue in the side chain is influenced by this conformational change of the main chain, giving rise to the pH change. Although the change of pH value is very small in the present system, improvements may be possible by increasing hydrophobic nature change of the photochromes.

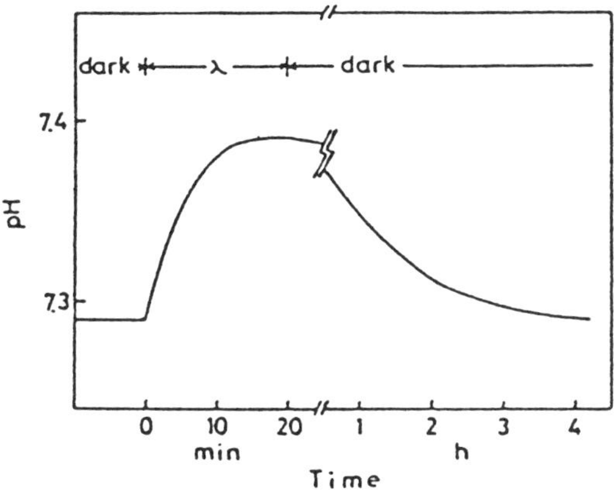

Figure 20. Response of pH value of an aqueous solution of poly(methacrylic acid) having spirobenzopyran units (polymer 11, x = 0.84) under visible light irradiation (λ > 470 nm) and in the dark at 25°C. Concentration of the polymer was 0.08 g/dL.

5.4.2.e Wettability

The wettability change of the surface of polymer 12 has been examined. In the dark, the contact angle for water (cos θ) was close to zero, indicating its hydrophobic nature. On ultraviolet irradiation cos θ increased to 0.2, together with strong coloration. The increase in the hydrophobic nature of the polymer surface is ascribable to the formation of polar merocyanine on the polymer's surface.

5.4.2.f Surface pressure of monolayers

Photoinduced mechanical response of a monolayer consisting of poly(methyl methacrylate) with spirobenzopyran pendant groups was examined at an air-water interface by Gruler et al. [37]. Irradiation with ultraviolet-light pulses isomerized the spiropyran chromophore and led to a relative surface pressure increase up to 10%. The pressure decreased progressively in the dark and a typical decay time was estimated as being approximately 20 min. The decay can be accelerated by use of visible-light pulses, as shown in Figure 21. The whole process can be repeated several times.

5.5 STILBENE

Stilbene is a cis-trans isomerizable chromophore, which has been most extensively studied from the physicochemical point of view [38]. This

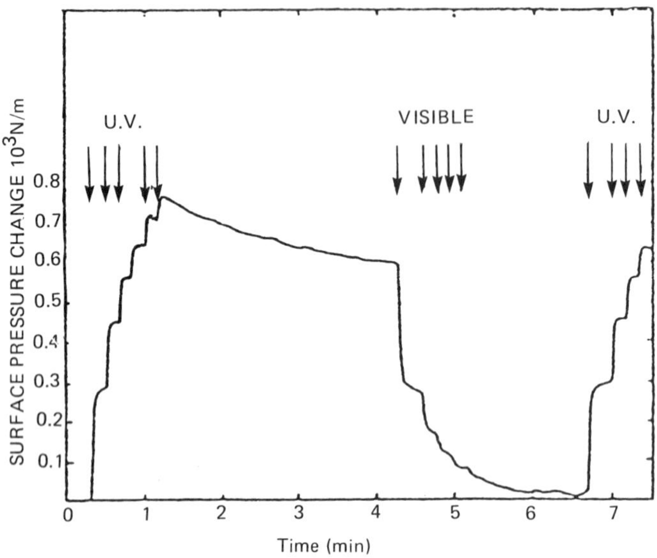

Figure 21. Surface pressure of a spirobenzopyran-methylmethacrylate copolymer monolayer versus time upon irradiation with successive UV and visible-light pulses. Surface concentration = 0.875 mg/m^2, water subphase at pH = 1.0.

molecule shows similar physical property changes as azobenzene during photoisomerization, although the isomerization mechanism is different (azobenzene isomerizes by inversion, whereas stilbene isomerizes by rotation along the unsaturated linkage). By incorporating this molecule into the backbone or as pendant groups of a polymer, similar conformational changes observed for the azobenzene system would be expected; however, very few examples are reported. One of the reasons the molecule was rarely adopted in photoresponsive polymers is that the isomerization is always accompanied by some side reactions, such as cyclization.

5.5.1 Polymer Synthesis

Polymers having stilbene chromophores in the backbone can be made by the following two methods.

Example 10

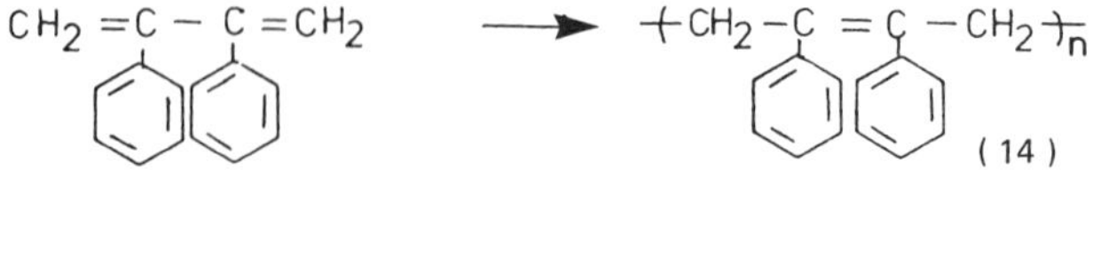

(14)

Poly(2,3-diphenylbutadiene) (cis content about 50%) was synthesized by radical polymerization of 2,3-diphenylbutadiene at 60°C using azobisisobutyronitrile as an initiator. 2,3-Diphenylbutadiene was prepared by the method of Inoue et al. [39]. The monomer was purified by repeated recrystallization from methanol. The polymer was purified by precipitation from benzene solution with methanol.

Example 11

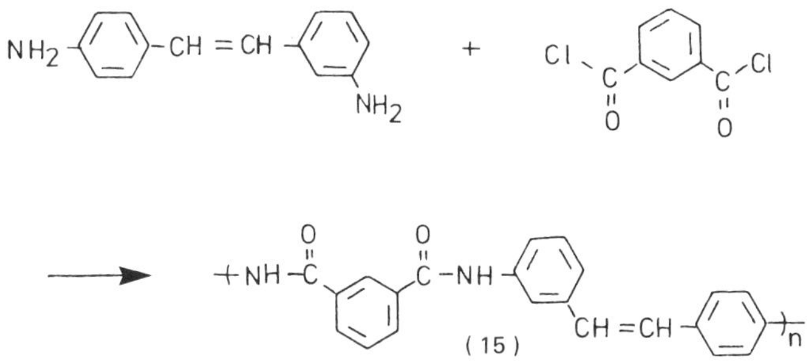

5.5.2 Photoresponsive Behavior

5.5.2.a Spectral response

In dichloromethane poly(2,3-diphenyl butadiene) has an absorption maximum around 250 nm and a tail extending up to 380 nm. On irradiating the aerated solution with 313-nm light, new absorption peaks appeared at 336 and 353 nm due to the formation of phenanthrene rings. Under a deaerated condition, an absorption band around 460 nm was formed in addition to the weak phenanthrene absorption at 353 nm. The absorption is ascribable to dihydrophenanthrene. Dihydrophenanthrene disappeared in the dark in the absence of O_2 according to first-order kinetics ($\tau_{1/2}$ = 4 min at 21°C). The decay is not accompanied by an increase in phenanthrene absorption.

These spectral changes suggest that the following reactions occur with this photoresponsive polymer:

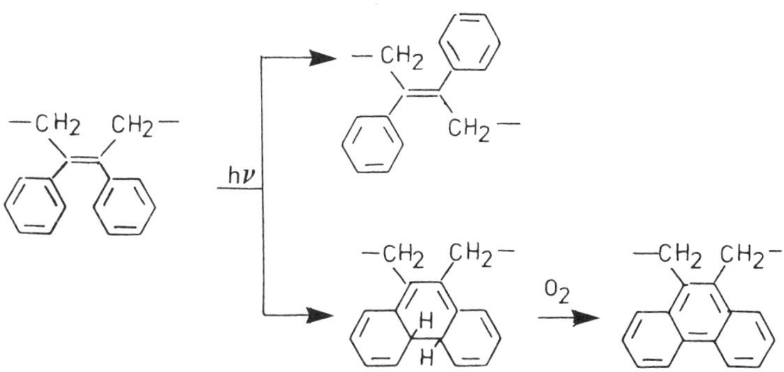

The high efficiency of phenanthrene formation is due to the impeded flexibility of the polymer chain segments. Thus cyclization is favored over cis-trans isomerization. The contribution of photochemical cyclization reaction makes the polymer unsuitable as a reversible photoresponsive system.

5.5.2.b Photomechanical effect

Osada et al. examined the photomechanical energy conversion of polyamide containing stilbene chromophores in the backbone [23]. In solution the viscosity decreased on irradiation, although the recovery was not observed. The film stretched under constant tension, relaxed on irradiation, and returned to the initial stress in the dark. The discrepancy in photoresponsive behavior in solution and in the film state suggests that the photoinduced relaxation in the film is possibly due to a local temperature increase resulting from light absorption as described before. The relatively rapid return to the initial stress after removing the light may well be interpreted by the temperature effect for the present system, where recovery of viscosity was not observed in solution.

5.6 TRIARYLMETHANE DYES

Triarylmethane dyes are well-known photochromic molecules, which dissociate into ion pairs under ultraviolet irradiation, with the production of intensely colored triphenylmethyl cations. The cation thermally recombines with the counterion as follows:

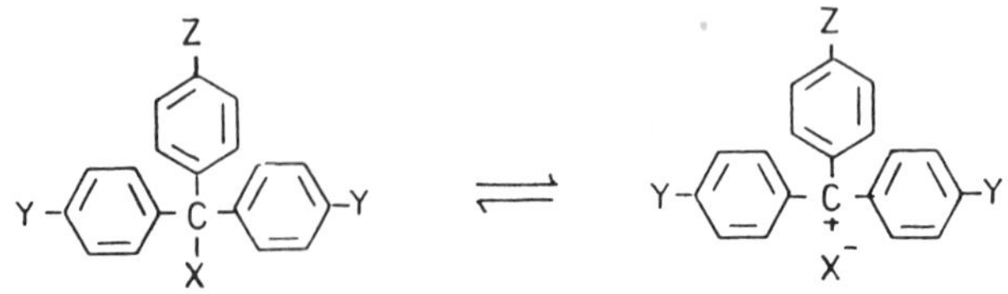

These compounds have long been employed advantageously for their color change on photoirradiation rather than for their chemical property change. According to the reaction scheme shown above, this chromophore functions as a positive charge generator. Photogeneration of positive charges along the polymer chain would expand the polymer chain due to repulsive force among the charges, and the disappearance of the charges in the dark would again shrink the polymer to the initial conformation. This behavior was observed for polystyrene having triphenylmethane leucohydroxide pendant groups [40].

5.6.1 Polymer Synthesis

Triphenylmethane leucohydroxide can be introduced to the pendant groups by a polymer reaction as described below [41].

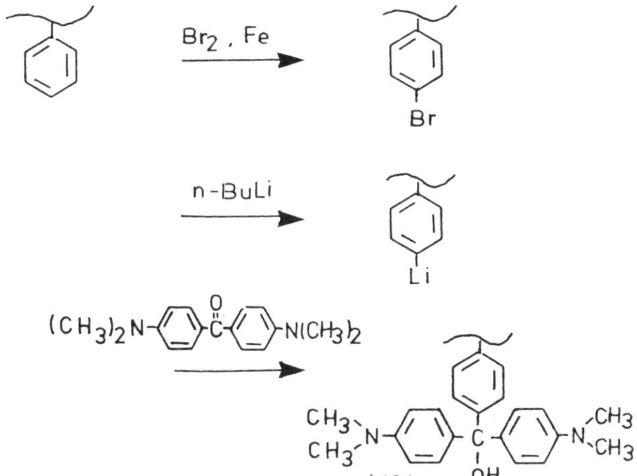

(16)

5.6.2 Photoresponsive Behavior

5.6.2.a Viscosity [40]

Methylene chloride solution containing polystyrene having triphenylmethane leucohydroxide pendant groups became green on ultraviolet irradiation. The appearance of a green color indicates formation of triphenylmethyl cations along the polymer chain.

The increase in absorption at 610 nm was accompanied with an increase in the viscosity of the polymer solution. Coulomb repulsive force between the photogenerated positive charges are considered to cause the increase in solution viscosity. To substantiate the mechanism, a polymer concentration dependence on reduced viscosity was carefully examined before and after irradiation. Before irradiation, the dependence showed normal behavior, while the reduced viscosity showed an anomalous increase at low polymer concentration after irradiation. The increase in the reduced viscosity at low polymer concentration is typical of polyelectrolytes, and this result strongly suggests that the increase in the solution viscosity is caused by the repulsive forces between the photogenerated positive charges.

5.6.2.b Wettability [41]

A wettability change on the surface of polystyrene having triphenylmethane leucohydroxide was measured on ultraviolet irradiation. The contact angle with water (cos θ) was found to increase from 0.2 to 0.8. The change in wettability induced by irradiation is due to the enhanced hydrophilic nature of the film surface brought about by the ionic dissociation of the chromophore.

5.7 MISCELLANEOUS

In aiming at a pronounced photomechanical effect, it may be worthwhile to reconsider the works on "mechanochemistry" studied as early as 1950 [42,43]. A change in dimension of a polyelectrolyte gel was induced by the chemical ionization of the polymer. On addition of strong alkali the thread elongated to about three times the original length, while subsequent addition of acid caused rapid reversion to the original length. Continued research in the field produced deeper understanding of the phenomenon, and many materials were reported to be suitable for conversion of chemical energy into mechanical motion [44].

If ionization is induced by photoirradiation and the lifetime of the ions is long enough to permit the deformation of the polymer film, a similar large effect may be expected. One example of formation of ions by photoirradiation is to use triarylmethane leucohydroxide as described before [45]. Another possibility is to use charge-transfer interaction.

The latter approach has been taken by Aviram [46] in the case of poly(*p*-(*N*,*N*-dimethylamino)-*N*-γ-D-glutamanilide) cross-linked by 2,6-bis(bromomethyl)naphthalene. On irradiation of this polymer in the presence of CCl_4 as an electron acceptor, photoionization occurs.

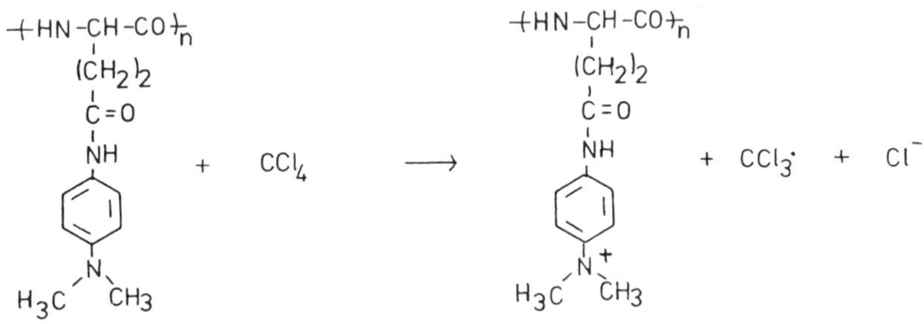

Ionization of the chromophores in the polymer chain caused the deformation of the film up to 35% in each dimension. It should be noted, however, that no attempt was made to reverse the expanded gel back to the original size as the process appears to be irreversible. The photomechanical effect may find applications in the photocopying, printing, actinometry, and other fields.

REFERENCES

1. G. A. Delzenne, Makromol. Chem. Chem. Suppl., 2, 169 (1979).
2. S. P. Pappas, Ed., UV Curing Science and Technology, Technology Marketing Corp., Connecticut (1978).
3. S. P. Panda, J. Sci. Ind. Res., 35, 560 (1976).
4. L. F. Thompson, C. G. Willson, and M. J. Bowden, Ed., Introduction to Microlithography, Am. Chem. Soc. (1983).
5. Z. Osawa, in Development in Polymer Photochemistry, Vol. 3 (N. S. Allen, Ed.), Applied Science Publishers Ltd., Barking, Essex, England (1982), p. 209.
6. W. Schnabel, Polymer Degradation, Hauser, München (1981).
7. M. Irie, in Photophysical and Photochemical Tools in Polymer Science (M. A. Winnik, Ed.), Reidel Pub., Dordrecht (1986), p. 269.
8. M. Irie, in Molecular Models of Photoresponsiveness (G. Montagnoli, and B. F. Erlanger, Ed.), Plenum Pub., New York (1983), p. 291.
9. B. F. Erlanger, Ann. Rev. Biochem., 45, 267 (1976).
10. G. H. Brown, Tech. Chem. (N.Y.), 3 (Photochromism) (1971).
11. a. G. C. Hampson and J. M. Robertson, J. Chem. Soc., 409 (1941).
    b. J. J. de Lange, J. M. Robertson, and I. Woodward, Proc. R. Soc., A 171 398 (1939).
12. D. J. W. Bullock, C. W. N. Cumper, and A. I. Vogel, J. Chem. Soc., 5316 (1965).
13. M. Irie, Y. Hirano, S. Hashimoto, and K. Hayashi, Macromolecules, 14, 262 (1981).
14. F. Agolini and F. P. Gay, Macromolecules, 3, 349 (1970).
15. H. S. Blair and J. E. Riordan, Polymer, 20, 196 (1979).
16. L. Matějka and K. Dušek, Makromol. Chem., 182, 3223 (1981).
17. C. D. Eisenback, Polymer, 21, 1175 (1980).
18. H. S. Blair, H. I. Pague, and J. E. Riordan, Polymer, 21, 1195 (1980).
19. R. Lovrien, Proc. Nat. Acad. Sci. USA, 57, 236 (1967).
20. G. van der Veen and W. Prins, Nature, 230, 70 (1971).
21. J. C. Chung, M. De Sorgo and W. Prins, J. Mechanochem. Cell Motil., 2, 105 (1973).
22. a. L. Matějka, K. Dušek, and M. Iliavský, Polym. Bull., 1, 659 (1979).
    b. L. Matějka, M. Iliavský, K. Dušek, and O. Wichterle, Polymer, 22, 1511 (1981).
23. Y. Osada, E. Katsumura, and K. Inoue, Makromol. Chem. Rapid Commun., 2, 411 (1981).
24. N. Negishi, K. Tsunemitsu, K. Ishinara, I. Shinohara, T. Okano, T. Akaike, and Y. Sakurai, Kobunshi Ronbunshu, 37, 287 (1980).

25. N. Negishi, K. Ishihara, I. Shinohara, T. Okano, K. Kataoka, Y. Sakurai, and T. Akaike, Chem. Lett. 681 (1981).
26. N. Negishi, K. Kitahara, and I. Shinohara, Makromol. Chem. Rapid Commun., 2, 95 (1981).
27. M. Irie and H. Tanaka, Macromolecules, 16, 210 (1983).
28. M. Irie and W. Schnabel, Macromolecules, 14, 1246 (1982).
29. G. Smets, J. Braeken, and M. Irie, Pure Appl. Chem., 50, 845 (1978).
30. M. Irie, A. Menju, and K. Hayashi, Macromolecules, 12, 1176 (1979).
31. K. Dimroth, C. Reichardt, T. Siepmann, and F. Bohlmann, Justus Liebigs Ann. Chem., 661, 1 (1963).
32. A. Menju, K. Hayashi, and M. Irie, Macromolecules, 14, 755 (1981).
33. M. Irie, K. Hayashi, and A. Menju, Polym. Photochem., 1, 233 (1981).
34. G. Smets and G. Evaens, Pure Appl. Chem. Suppl. Macromol. Chem., 8, 357 (1973).
35. O. Chaude, Chem. Phys., 52, 39 (1954).
36. J. Arnaud, C. Wippler, and F. Beaure D'Augeres, J. Chem. Phys., 64, 1165 (1967).
37. H. Gruler, R. Vilanove, and F. Rondelez, Phys. Rev. Lett., 44, 590 (1980).
38. J. Saltiel, J. D'Agostino, E. D. Megarity, L. Metts, K. R. Neuberger, M. Wrighton, and O. C. Zafirion, Org. Photochem., 3, 1 (1973).
39. H. Inoue, M. Helbing, and O. Vogel, Macromolecules, 10, 1331 (1977).
40. M. Irie, and M. Hosoda, Makromol. Chem., Rapid Commun., 6, 533 (1985).
41. N. Negishi, K. Tsunemitsu, and I. Sinohara, Polym. J., 13, 411 (1981).
42. W. Kuhn, B. Hargitay, A. Katchalsky, and E. Eisenberg, Nature (London), 165, 514 (1950).
43. A. Katchalsky and E. Eisenberg, Nature (London), 166, 267 (1950).
44. A. Wasserman, Ed., Size and Shape Change of Contractile Polymers, Pergamon Press Ltd., London (1960).
45. M. Irie, and D. Kungwatchakun, Macromolecules, 19, 2476 (1986).
46. A. Aviram, Macromolecules, 11, 1275 (1978).

6

Electrically Conducting Polymers

Yasuhiko Shirota
Osaka University, Osaka, Japan

6.1 INTRODUCTION

It is well known that solids are usually classified as metals, semiconductors, and insulators from the standpoint of electrical conductivity. Most organic compounds, including polymers, are insulators. Actually, polymers have been used as excellent insulating materials in various ways. Although the phenomenon of photoconduction in organic solids was found for anthracene as early as in 1906 [1], pioneering work on electrical conduction for organic solids was reported from the late 1940s to early 1950s for certain classes of organic compounds. Phthalocyanines have been found to show electrical conductivity [2]. Electrical conduction was found for highly condensed polycyclic aromatic hydrocarbons (e.g., violanthrene) whose molecular structure resembles that of graphite, which is known to be highly conductive [3]. Then it was found that a perylene-bromine charge-transfer complex shows a remarkably higher electrical conductivity [4]. These findings have opened up the research field of organic semiconductors [5]. Since then, studies of organic conductors, including polymers, have made remarkable progress over a period of 30 years from organic semiconductors to organic metals and even to organic superconductors. This field has been the subject of numerous review articles and books [6].

A variety of charge-transfer complexes have been studied in view of electrical conduction. Since a new electron acceptor, tetracyanoquinodimethan (TCNQ), was synthesized in 1960 [7], there have been extensive studies on TCNQ anion radical salts with various cations. Among them, single crystals of a few TCNQ anion radical salts, for example, quinolinium TCNQ complex salt [8], acridinium TCNQ complex salt [9], and *N*-methylphenazinium TCNQ simple salt [10], have been found to show metallic properties in electrical conduction. In 1970 a new electron donor, tetrathiafulvalene (TTF), was synthesized [11], and in 1973 single crystals of a TTF-TCNQ charge-transfer complex were found to exhibit a high room-temperature conductivity with metallic properties. The conductivity increases sharply with a decrease in temperature, with the maximum conductivity approaching that of mercury [12, 13]. These organic metals are characterized by anisotropy in electrical conduction and are termed pseudo-one-dimensional metals. Since then, intensive and extensive studies of pseudo-one-dimensional organic conductors have been made. It was in 1980 that the first known example of organic superconductors was reported: Cation radical salts of tetramethyltetraselenafulvalene (TMTSF), $(TMTSF)_2PF_6^-$, and $(TMTSF)_2ClO_4^-$, prepared by the method of electrocrystallization [14], have been found to transform from metals to superconductors under a pressure of 12 kbar and at atmospheric pressure with transition temperatures of 0.9 K and 1.4 K, respectively [15, 16].

With respect to polymers, a number of polymers (e.g., linear or planar π-conjugated polymers, polymeric charge-transfer complexes, and TCNQ anion radical salts with polymeric cations) have been studied extensively and found to fall into the family of semiconductors. Metallic conductivity in polymers was first found for an inorganic polymer, poly(sulfur nitride), in 1973 [17]. In 1975 this polymer was found to transform into a superconductor below 0.3 K [18]. It was found in 1977 that polyacetylene, which itself is a semiconducting polymer, shows metallic properties when heavily doped with dopants such as halogens [19]. Since then intensive studies on organic polymeric metals have been started. A model for high-temperature polymeric superconductors was proposed in 1964 [20]; however, the synthesis of organic polymeric superconductors has not yet been successful. Potential industrial applications for conducting organic solids are also being actively investigated.

Photoconductivity in organic solids has been studied extensively both theoretically and experimentally on molecular crystals of polycyclic aromatic hydrocarbons such as anthracene. Progress in experimental methods such as drift mobility measurement by means of a time-of-flight method [21-23] and techniques of purification and single-crystal growth of organic compounds has led to a deeper understanding of electrical conduction in organic solids. With the finding that poly(*N*-vinylcarbazole) shows a large photoconductivity and that it can be used as a photoreceptor material in electrophotography, photoconduction of polymers have become the subject of extensive studies. Recent years have witnessed significant advances in both theoretical and experimental studies of photoconduction in disordered organic systems, including polymers.

Photoconducting materials have found practical application as photoreceptors in electrophotography. A number of photoconducting materials have been patented as photoreceptors, and some of them have actually been used. A variety of photoconducting polymers have also been synthesized. Books and review articles on photoconduction are listed in Ref. 24. This chapter deals with organic polymeric semiconductors, polymeric metals, and polymeric photoconductors, attention being focused on the synthesis of these electrically conducting polymers.

6.2 BASIC CONCEPTS IN ELECTRICAL CONDUCTION

6.2.1 Electrical Conductivity

The electrical conduction is classified into ionic conduction and electronic conduction, based on whether the charge carriers are ions or electrons. They are essentially different in that ionic conduction is caused by the transport of matter (i.e., ions), whereas electronic conduction involves the transport of electrons. In general, electronic conduction is enhanced with an increase in pressure, whereas the opposite is the case for ionic conduction. The discussion in this chapter is focused on electronic conduction.

Metals, semiconductors, and insulators are represented in terms of band structure, as shown in Figure 1. For metals a band is about half-filled with electrons and hence no activation energy is necessary for the generation of charge carriers. In contrast, the band is completely filled with electrons for both semiconductors and insulators, and hence activation is required for the transition of electrons from the valence band to the conduction band to generate charge carriers. If the band gap is small enough to generate charge carriers by thermal activation, then solids show a semiconductive property. These are termed intrinsic semiconductors. In most cases, however, electron-donating or electron-accepting impurities participate in the generation of charge carriers. In this case, either holes or electrons play the dominant role in transport. These extrinsic semiconductors are termed p-type or n-type semiconductors, where holes or electrons are the major charge carriers, respectively.

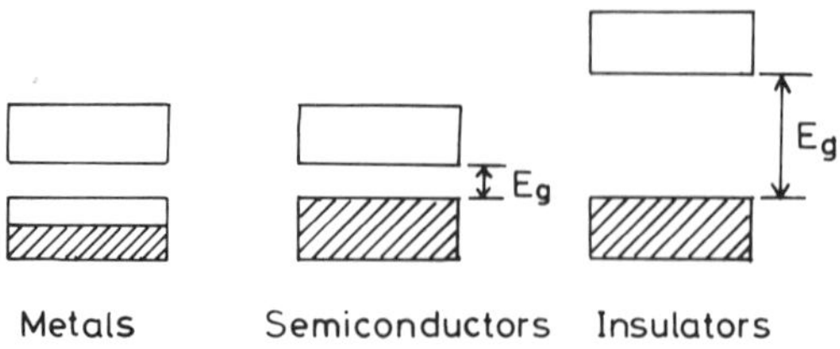

Figure 1. Band structures for metals, semiconductors, and insulators.

The electric current density J [coulomb/(s · cm^2)] is given by

$$J = nev \tag{1}$$

where

n = number of charge carriers per unit volume (cm^{-3})
e = electric charge (coulombs)
v = velocity of charge carriers that move under an electric field (cm/s).

The velocity v is proportional to the strength of the electric field as expressed by

$$v = \mu E \tag{2}$$

where the proportional constant μ [cm^2/(V · s)] is called the drift mobility of charge carriers (i.e., the distance across which charge carriers migrate per second under a unit electric field). The current density is then expressed by

$$J = ne\mu E \tag{3}$$

The magnitude of electrical conductivity is represented in terms of specific conductivity σ (S cm^{-1}) or its reciprocal, specific resistivity, ρ (Ω · cm). The specific conductivity represents the electric current that flows across the unit area (1 cm^2) of electrode under the unit external electric field (1 V/cm) applied to the sample. σ is expressed by

$$\sigma = ne\mu \tag{4}$$

Generally, solids with conductivities of less than 10^{-12} S cm^{-1} are insulators. Solids with metallic properties in electrical conduction generally show conductivities higher than 10^2 S cm^{-1}. Solids with electrical conductivities ranging from 10^{-1} to 10^{2} S cm^{-1} are semiconductors. It should be noted that it is the temperature dependence of electrical conductivity instead of its magnitude that distinguishes metals from semiconductors. While the conductivity increases with a decrease in temperature for metals, the temperature dependence of the electrical conductivity for semiconductors follows the equation

$$\sigma = \sigma_0 \exp\left(\frac{-\Delta E}{kT}\right) \tag{5}$$

The values of σ for compounds described in Section 6.1, together with other typical examples, are given in Figure 2.

6.2.2 Photoconductivity

Photoconduction is observed when the concentration of charge carriers is larger upon photoirradiation than in the thermal equilibrium state. Therefore, photoconduction is usually observed for insulating materials or poor semiconductors.

In the case of photoconduction, the lifetime of photogenerated charge carriers, τ (s), should be taken into consideration. When the light of weak intensity is irradiated, a recombination process of charge carriers is negligible and hence the photostationary concentration of charge carriers n (cm^{-3}) is given by

$$n = n_0 \tau \tag{6}$$

where n_0(cm^{-3} s^{-1}) is the number of charge carriers generated by photoirradiation per second per a unit volume. Therefore, the steady-

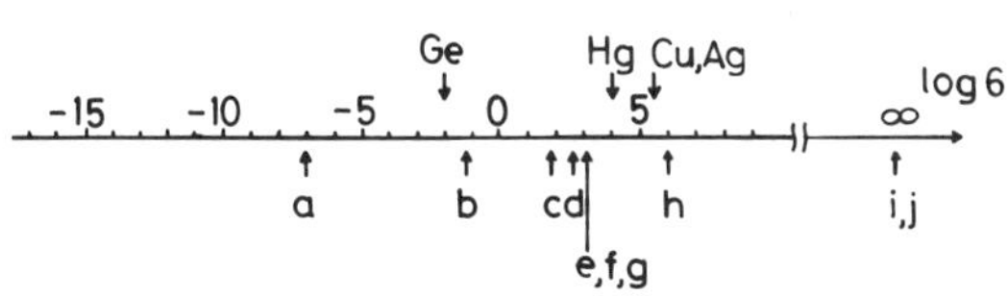

Figure 2. Electrical conductivities of organic and inorganic conductors described in Section 6.1. a, violanthrene; b, perylene · bromine CT-complex; c, quinolinium $TCNQ^{\overline{\cdot}}$ complex salt, N-methylphenazinium $TCNQ^{\overline{\cdot}}$ simple salt; d, TTF · TCNQ CT-complex; e, HMTSF · TCNQ CT-complex; f, AsF_5-doped $(CH)_x$; g, $(SN)_x$; h, graphite · AsF_5 intercalation compound; i, $(TMTSF)_2X$ (below T_c); j, $(SN)_x$ (below T_c); j, $(SN)_x$ (below T_c). TTF, tetrathiafulvalene; HMTSF, hexamethylene tetraselenafulvalene; TMTSF, tetramethyl tetraselenafulvalene.

state photocurrent density J_{ph} [coulombs/(s · cm^2)] is expressed by

$$J_{ph} = n_0 \tau e \mu E \tag{7}$$

when the contact between the electrode and the sample is ohmic.

Now let the intensity of the incident light be I_0 [photons/(cm^2 · s)], the photoabsorption coefficient of the sample be α (cm^{-1}), the quantum yield for the photogeneration of charge carriers be η, the thickness of sample (distance between the two electrodes) be ℓ (cm), and the electrode area be S (cm^2), and the following equation holds:

$$n_0 \tau S \ell = S I_0 (1 - e^{-\alpha \ell}) \eta \tau \tag{8}$$

Then J_{ph} is expressed by

$$J_{ph} = \frac{1}{\ell} I_0 (1 - e^{-\alpha \ell}) \eta \tau e \mu E \tag{9}$$

by introducing Equation 8 into Equation 7. When the light is absorbed on the surface of the sample ($\alpha = \infty$), J_{ph} is represented by

$$J_{ph} = \frac{1}{\ell} I_0 \eta \tau e \mu E \tag{10}$$

The distance across that charge-carrier drift during their lifetimes is termed the range (Schubweg). The range w (cm) is defined as

$$w = \tau \mu E \tag{11}$$

Both the concentration of charge carriers and the drift mobility of charge carriers can be quantitatively evaluated by means of transient measurement using a pulsed light. Usually, the use of blocking electrodes is preferable in handling insulating materials. Either the electric charge induced on the electrode by the drift of charge carriers or the induced electric current that flows the external circuit is measured. In what is called the "time-of-flight method" (Figure 3), a sheet of charge carriers is generated on the surface of samples by pulse excitation, and the transit time t_T (s), which is the time required for charge carriers to drift across the sample thickness to the other electrode under a given electric field, is measured from either an electric charge waveform or a photocurrent waveform (Figure 4). The number of charge carriers that are alive at time t, n(x,t) (cm^{-2}), is given by

$$n(x,t) = n_0 \, e^{-t/\tau} \tag{12}$$

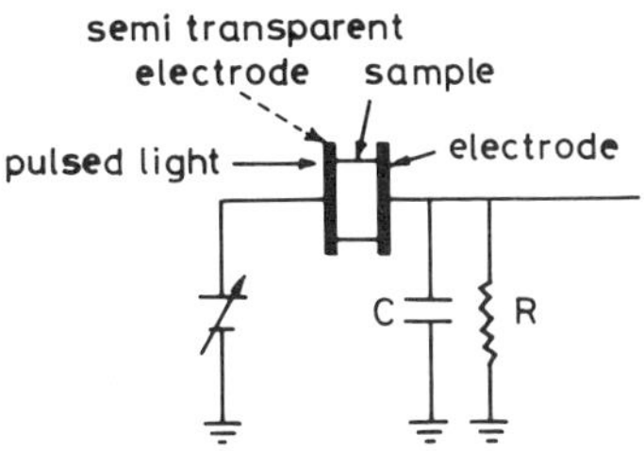

Figure 3. Time-of-flight method.

where n_0 (cm^{-2}) is the number of charge carriers generated on the surface of a sample at time zero (t = 0) and τ (s) is the mean lifetime of charge carriers. The photocurrent density observed at time t, $J_{ph}(t)$ [coulombs/(s · cm^2)] is expressed by

$$J_{ph} = \frac{dq}{dt} = \frac{dx}{\ell} n(x,t)e \frac{1}{dt} = \frac{1}{\ell} e\mu E n_0 e^{-t/\tau} \tag{13}$$

Similarly, the electric charge induced on the electrode at time t is represented by

$$Q(t) = \int_0^t J_{ph}(t)\, dt = \frac{1}{\ell} e\mu E n_0 \tau(1 - e^{-t/\tau}) \tag{14}$$

When $\tau > t_T$, J_{ph} and Q are expressed by Equations 15 and 16, respectively [23].

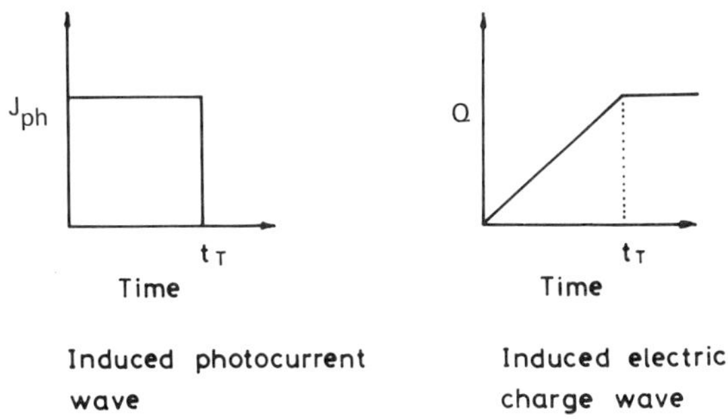

Figure 4. Photocurrent and electric charge waveforms.

$$J_{ph} = \frac{1}{\ell} n_0 e\mu E = \frac{1}{\ell} I_0 \eta e\mu E \tag{15}$$

$$Q = \frac{1}{\ell} n_0 e\mu E t_T = \frac{1}{\ell} I_0 \eta e\mu E t_T \tag{16}$$

That is, the photocurrent expressed by Equation 15 flows until time t_T. If the t_T is measured, drift mobility of charge carriers μ is determined by

$$t_T = \frac{\ell}{\nu} = \frac{\ell}{\mu E} = \frac{\ell^2}{\mu V} \tag{17}$$

6.3 ORGANIC POLYMERIC CONDUCTORS

6.3.1 Survey of Organic Conductors: Semiconductors, Metals, and Superconductors

Organic electrical conductors are largely classified into low-molecular-weight organic compounds and polymers. Low-molecular-weight organic compounds can be classified into the following several classes: (a) highly condensed polycyclic aromatic hydrocarbons that resemble graphite in their structures, (b) charge-transfer complexes, (c) ion radical salts, and (d) metal complexes. Ion radical salts include both anion radical salts of electron acceptors and cation radical salts of electron donors. They are either simple salts where cations and anions are 1:1 in stoichiometric proportions or complex salts which additionally contain neutral acceptor or donor molecules. Usually, complex salts are more conductive than simple salts. In general, a high electrical conductivity is found for charge-transfer complexes, ion radical salts, or partially oxidized metal complexes.

Charge-transfer complexes are formed by the reaction of neutral donor and neutral acceptor molecules. The combination of a strong electron donor with a strong electron acceptor sometimes leads to the formation of ion radical salts. The synthesis of ion radical salts is carried out by metathesis reactions, for example, between quaternary ammonium salts of aliphatic or aromatic amines and metal-TCNQ anion radical salts, or by electrode reactions of neutral donor or acceptor molecules.

Single crystals of charge-transfer complexes are grown by slow cooling of reaction solutions or by slow evaporation of solvents, by diffusion of donor and acceptor molecules in solutions, or by co-sublimation of donor and acceptor molecules. The method of slow diffusion of a donor and an acceptor through a solvent bridge tends to produce large single crystals. Electrode reactions are carried out by either a potentiostatic or a galvanostatic method. In many cases ion radicals generated by electrode reactions instantaneously react with neutral molecules to form highly conductive complex salts with mixed valence interactions, and their single crystals grow on the surface of the electrode. This method is called electrocrystallization.

Many of the charge-transfer complexes that are formed from neutral donors and neutral acceptors take the crystallographic structures of mixed stacks, whereas certain families of charge-transfer complexes and ion radical salts, particularly those containing TCNQ, and ion radical salts prepared by electrocrystallization, take segregated stacks of donors and/or acceptors. Examples of organic semiconductors of each class together with their conductivities are given in Figure 5 and Table 1.

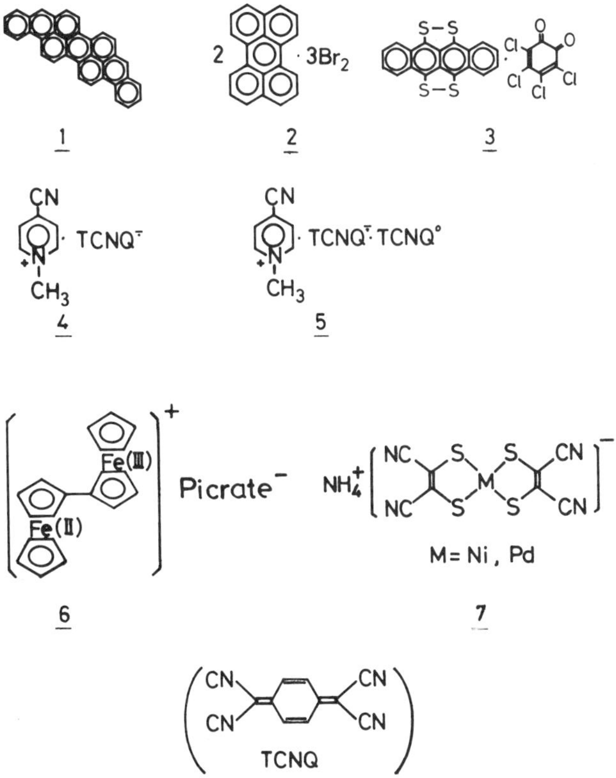

Figure 5. Examples of organic semiconductors.

Table 1. Room-Temperature Conductivities of Organic Semiconductors

Compound	σ_{rt} (S cm^{-1})	Ref.
Violanthrene (1)	10^{-7}	[25]
Perylene-bromine CT complex (2)	10^{-1}	[26]
Tetrathiatetracene-*o*-chloranil CT complex (3)	5×10^{-1}	[27]
4-Cyano-*N*-methylpyridinium-TCNQ anion radical simple salt (4)	10^{-6}	[28, 29]
4-Cyano-*N*-methylpyridinium-TCNQ anion radical complex salt (5)	2×10^{-2}	[28, 29]
Partially oxidized ferrocenylene (6)	10^{-8}	[30]
Ammonium bis(dicyanoethylenedithio) metalate (7)	10^{-4} to 1	[31]

Metallic properties in electrical conduction have been found among compounds that form charge-transfer complexes that make segregated stacks of donor and acceptor molecules, ion radical salts mostly with mixed valence interactions, or partially oxidized metal complexes. They are termed organic metals or synthetic metals. Metallic conduction was first found for single crystals of a few TCNQ anion radical salts as described in Section 6.1. Since the finding of metallic conduction for single crystals of a TTF-TCNQ charge-transfer complex, various modifications of TTF and TCNQ molecules have been carried out, and electrical conductivities of charge-transfer complexes between these modified TTF molecules and TCNQ, TTF and modified TCNQ molecules, and modified TTF and modified TCNQ molecules, have been examined in order to search for new organic metals with better electrical properties than those of the TTF-TCNQ charge-transfer complex and to gain deeper insight into the origin of metallic conduction. The TTF molecule has been modified in several ways: replacement of the sulfur atom by selenium or tellurium, the introduction of substituents, the synthesis of molecules having an isoelectronic structure with TTF, and the synthesis of fused TTF rings. Similarly, the TCNQ molecule has been modified by the introduction of substituents or the extension of π conjugation. Molecules having an isoelectronic structure with TCNQ have also been synthesized. As a result, charge-transfer complexes with electrical properties comparable to or better than those of the parent TTF-TCNQ complex (e.g., the hexamethylene tetraselenafulvalene-TCNQ charge-transfer complex) [32] have been synthesized [33-36]. Synthesis of the derivatives of TTF and TCNQ with isoelectronic structures has enabled the synthesis of organic alloys such as $TSeF_{0.03}TTF_{0.97}TCNQ$ [37] or $TSeF \cdot TCNQ_{0.9}MTCNQ_{0.1}$ [38]. Metallic conduction has also been found for electrochemically prepared ion radical salts and partially oxidized metal complexes. Some examples of organic metals are given in Figure 6 and Table 2.

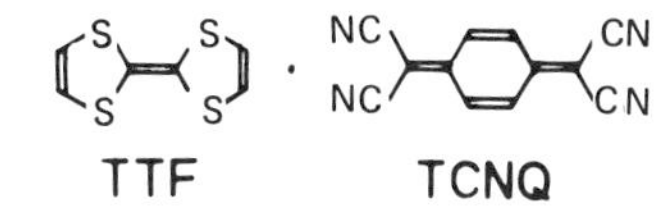

1

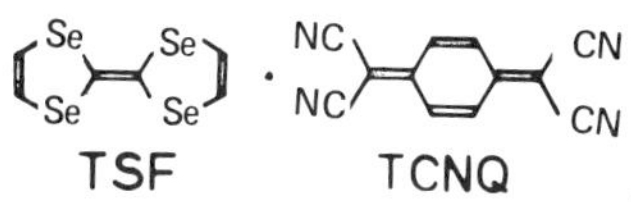

2

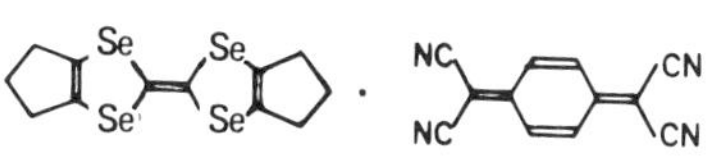

3

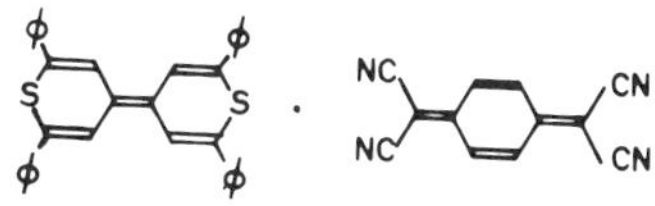

4

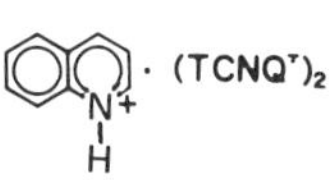

5

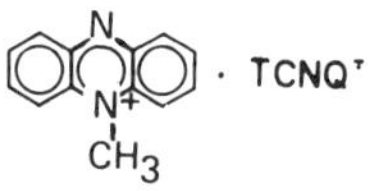

6

$(TTT)_2I_3$

7

$TTT \cdot I_{0.72}$

8

$TTT \cdot (PF_6)_{0.61}$

9

$TTF \cdot Br_{0.70}$

10

$(Peryrene)_2AsF_5$

11

$K_2Pt(CN)_4Br_{0.3} \cdot 3H_2O$

12

$NiPcl_{1.0}$

13

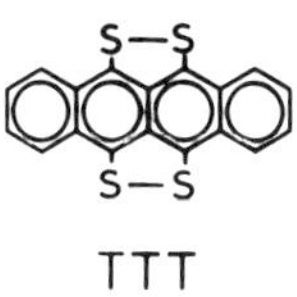

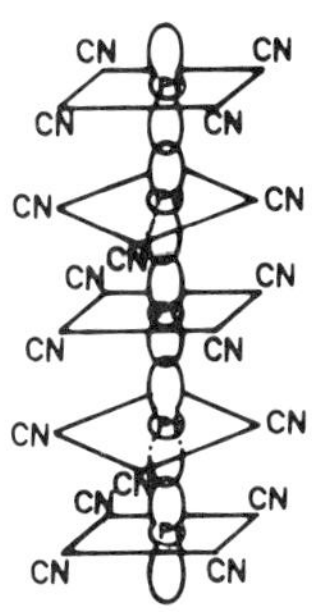

$K_2Pt(CN)_4 \cdot 3H_2O$

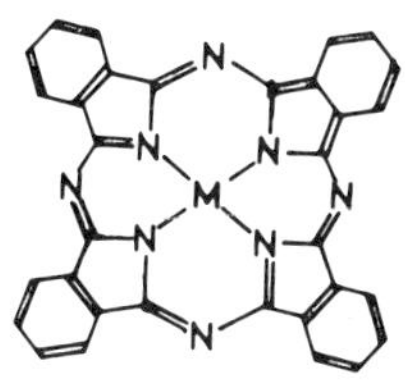

MPc (M=Ni)

Figure 6. Examples of organic metals.

Table 2. Electrical Conductivities of Organic Metals (Single Crystals)

Compound	σ_{rt} (S cm^{-1}); T_c (K)	Ref.
TTF-TCNQ CT complex (1)	$\sigma_{rt} = 500$ $\sigma_{max} = 10^4$ $T_c = 58, 38$	[39]
TSeF-TCNQ CT complex (2)	$\sigma_{rt} = 800$ $T_c = 28$	[37]
HMTSeF-TCNQ CT complex (3)	$\sigma_{rt} = 2 \times 10^3$ $\sigma_{max} = 7 \times 10^3$ $T_c = 16$	[32]
Dithiodipyranylidene-TCNQ CT complex (4)	$\sigma_{rt} = 250 \sim 300$ $T_c = 145$	[40]
Quinolinium-TCNQ anion radical complex salt (5)	$\sigma_{rt} = 50$	[41]
N-Methylphenazinium-TCNQ anion radical simple salt (6)	$\sigma_{rt} = 140$	[42]
TTT cation radical salt (7)	$\sigma_{rt} = 3 \times 10^3$	[43]
TTT cation radical salt (8)	$\sigma_{rt} = 1.1 \times 10^3$	[44]
TTT cation radical salt (9)	$\sigma_{rt} = 760$	[45]
TTF cation radical salt (10)	$\sigma_{rt} = 8.2 \sim 8.6 \times 10^2$	[44]
Perylene cation radical salt (11)	$\sigma_{rt} = 1.2 \times 10^3$	[46]
Partially oxidized potassium tetracyanoplatinate (12)	$\sigma_{rt} = 300$	[47]
Nickel phthalocyanine iodide (13)	$\sigma_{rt} = 350$	[48]

Organic superconductors have been found among compounds of cation radical salts of tetramethyltetraselenafulvalene which were prepared by electrocrystallization. They are transformed from metals into superconductors below transition temperatures under pressure or at an atmospheric pressure [49-51]. Examples of organic superconductors are given in Table 3 [52].

Table 3. Characteristics of $(TMTSF)_2X$ Superconductors

Anion (X^-)	T_{sc} (K)	Pressure (kbar)
PF_6^-	0.9	6.5
AsF_6^-	1.1	12
SbF_6^-	0.4	11
TaF_6^-	1.4	12
ReO_4^-	1.3	9.5
ClO_4^-	1.4	

Source: Ref. 52.

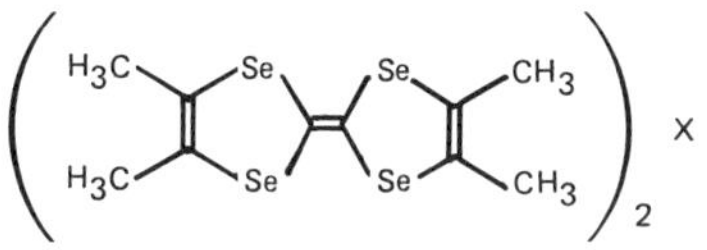

$(TMTSF)_2X$

6.3.2 Survey of Organic Polymeric Conductors: Polymeric Semiconductors and Polymeric Metals

Electrically conducting polymers are classified into several classes: (I) linear π-conjugated polymers, (II) planar π-conjugated polymers, (III) polymeric charge-transfer complexes, (IV) polymeric ion radical salts, (V) metal-containing polymers, (VI) pyropolymers (conjugated polymers), and (VII) others. Examples of polymeric semiconductors are given in Figure 7 [53-71,76-78,110-111,131].

Polymers 1 to 6 are examples of linear π-conjugated polymers [53-58]. Generally, they show low electrical conductivities, being poor semiconductors or almost insulators. However, some linear π-conjugated polymers transform into highly conducting materials when chemically or electrochemically doped, as described later. Polyacetylene, the simplest among the linear π-conjugated polymers, has been shown to be a semiconductor; the conductivity is dependent on the crystallinity [72], the oxygen content [73,74], and the cis-trans content [53]. The electrical conductivity of crystalline films of polyacetylene varies from 10^{-9} S cm^{-1} for the cis isomer to 10^{-5} S cm^{-1} for the trans isomer [53]. On the other hand, poly(phenylacetylene)s are electrically insulators [75]. Polymer 7 is an example of (II) [59].

I Linear π-Conjugated Polymers

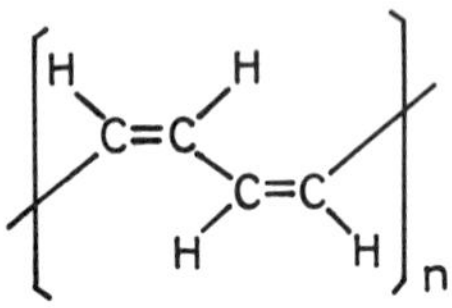

1

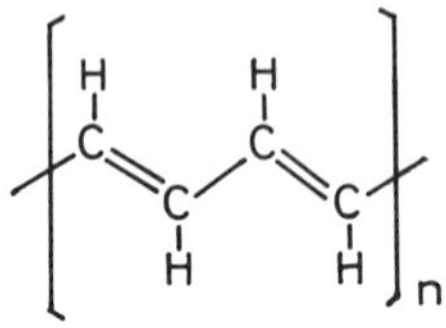

2

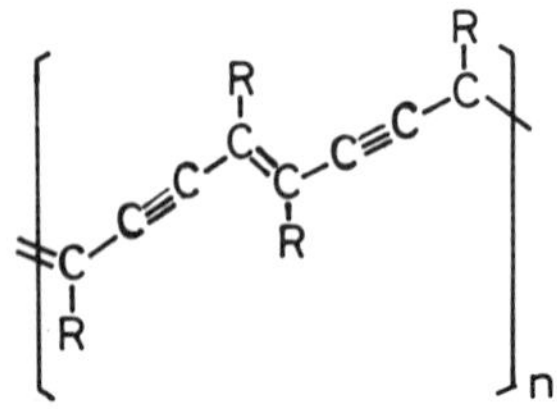

3

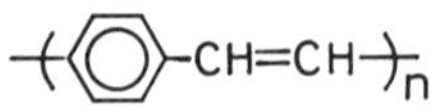

4

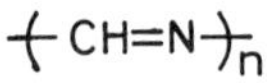

5

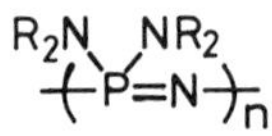

6

II Planar π-Conjugated Polymers

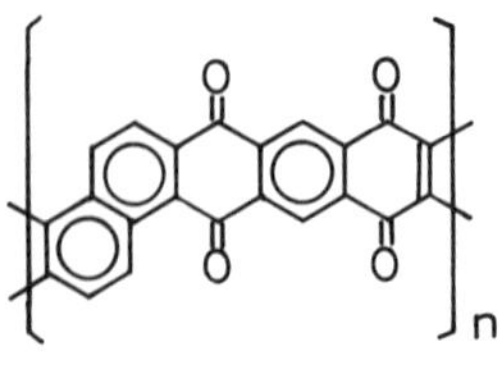

7

Figure 7. Examples of organic polymeric semiconductors.

III Polymeric Charge-Transfer Complexes

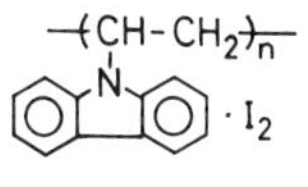

8

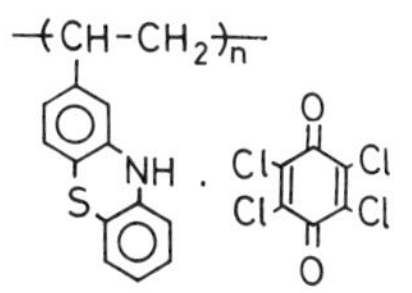

9

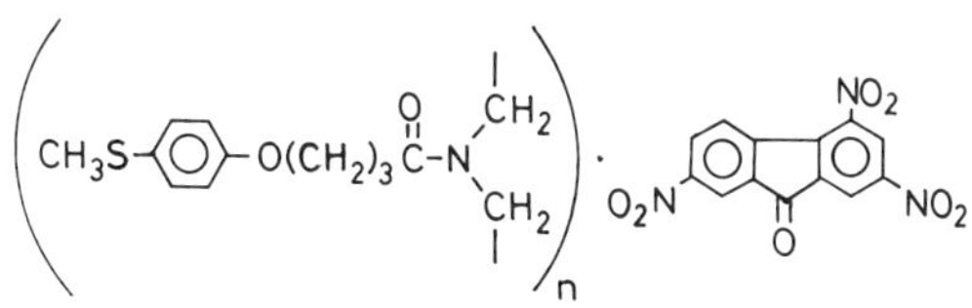

10

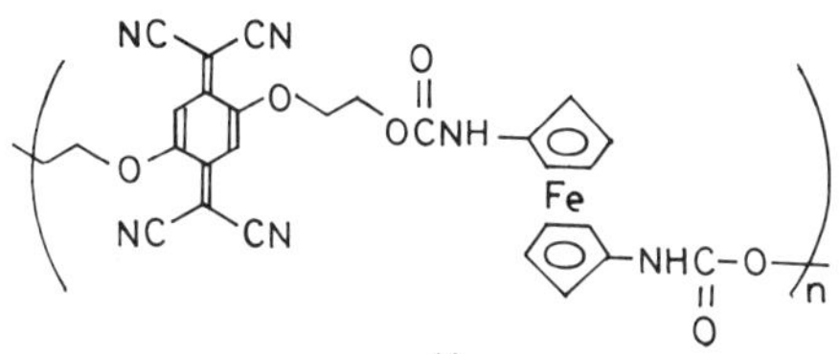

11

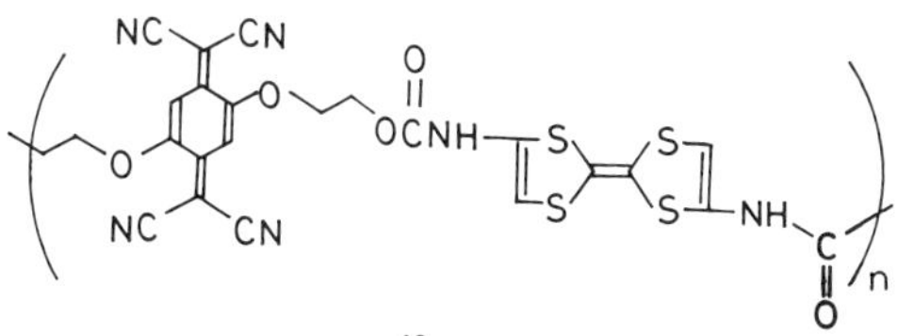

12

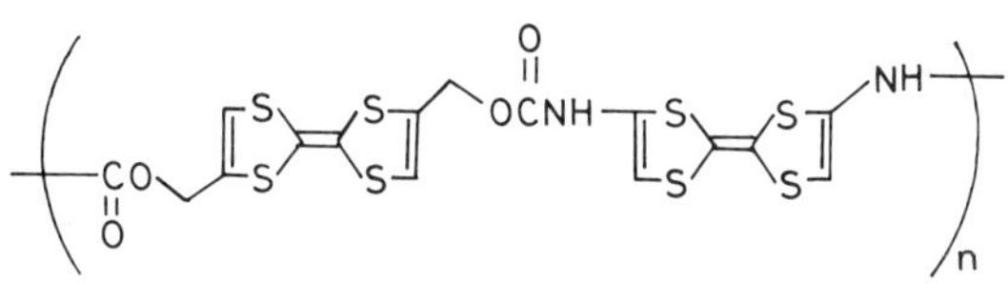

13

Figure 7 (continued)

IV Polycation-TCNQ Anion Radial Salts

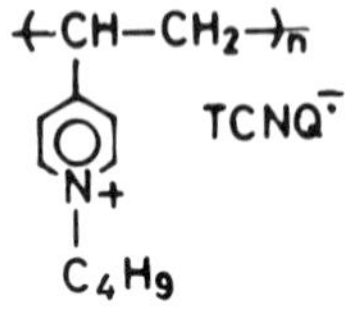

14

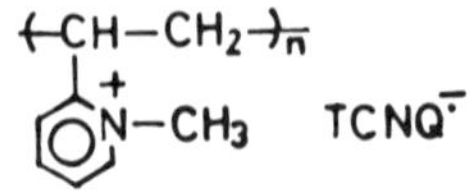

15

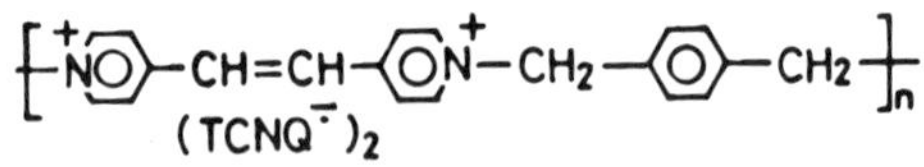

16

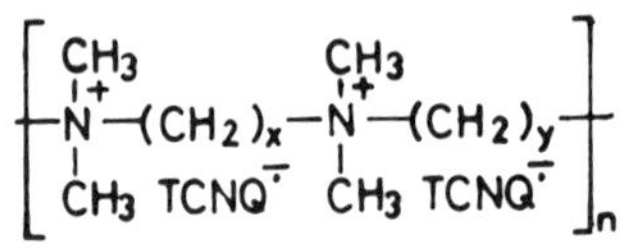

17

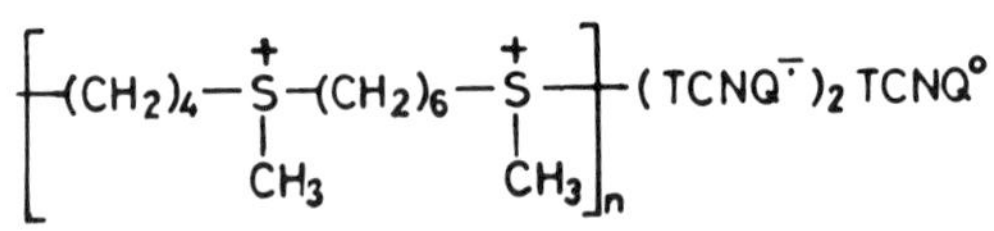

18

Figure 7 (continued)

V Metal-Containing Polymers

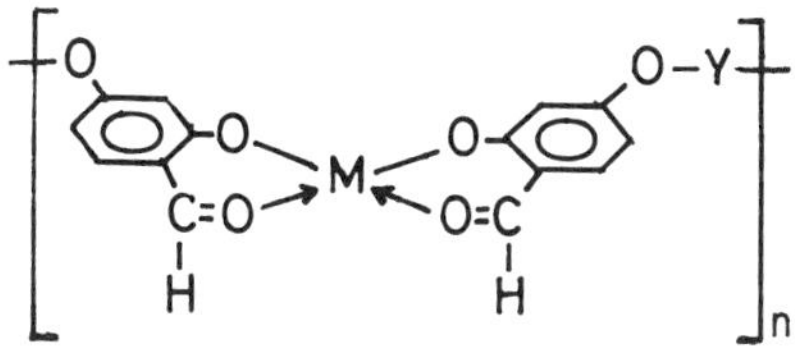

M= Fe^{2+}, Co^{2+}, Ni^{2+}, Zn^{2+}

19

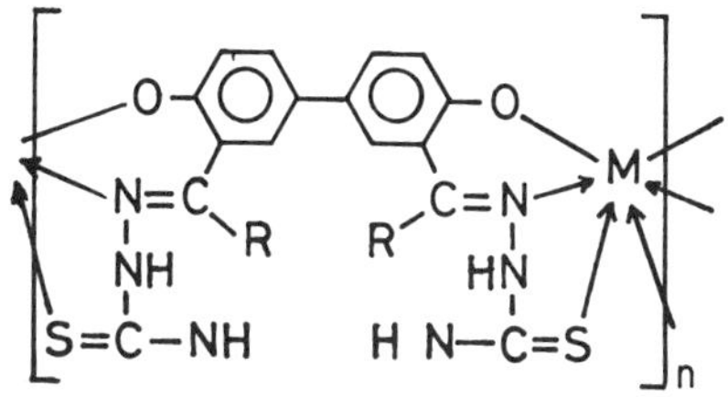

M= Cu^{2+}, Ni^{2+}, Co^{2+}

20

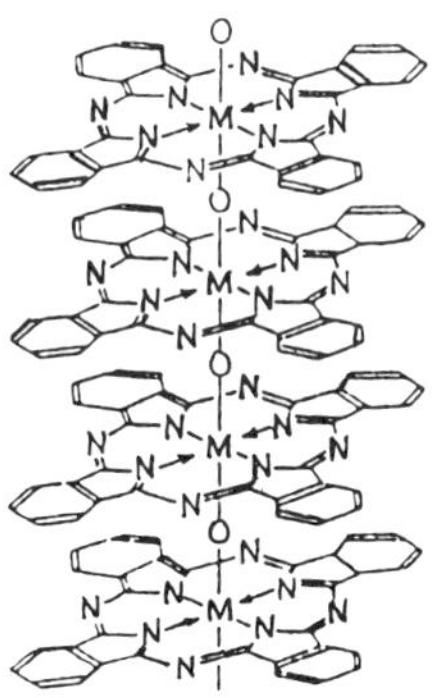

21

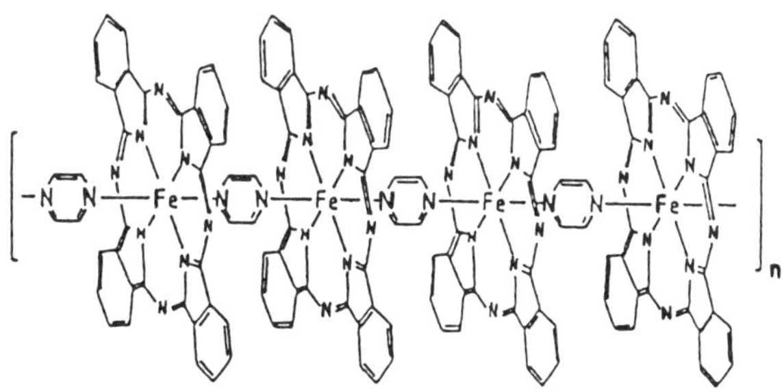

22

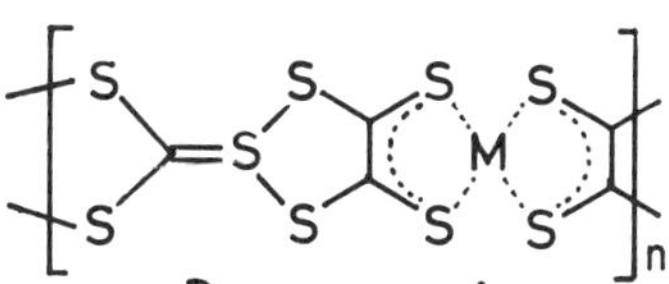

M= Ni, Cu, Pt

23

Figure 7 (continued)

VI Polymer Cation Radical Salt

(Electrochemically Doped Vinyl Polymers)

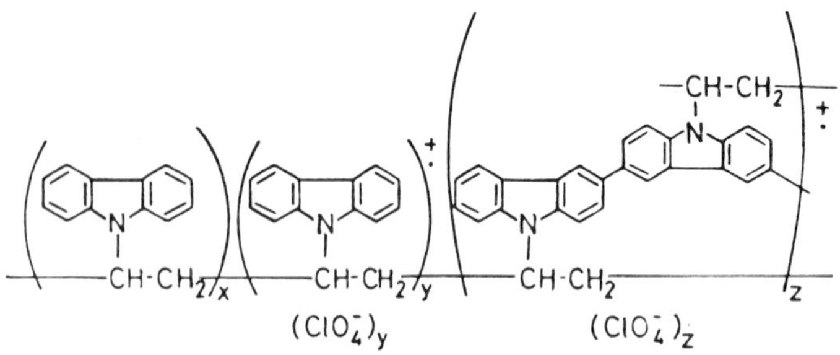

24

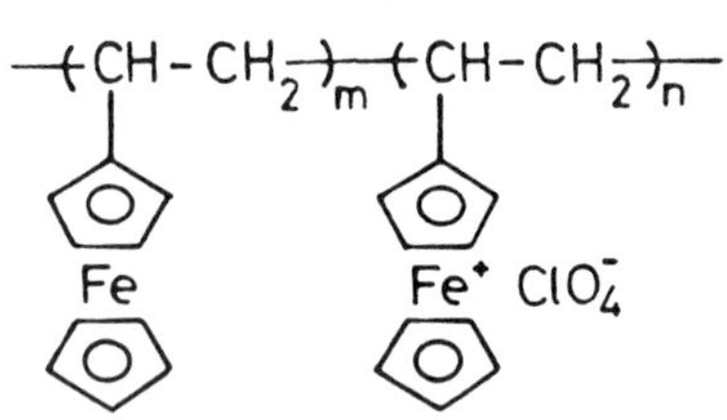

25

Figure 7 (continued)

A few combinations are conceivable for polymeric charge-transfer complexes: (a) polymer (D)-low-molecular-weight compound (A), (b) polymer (D)-polymer (A), and (c) low-molecular-weight compound (D)-polymer (A), where D and A represent an electron donor and an electron acceptor, respectively, and in polymers pendant groups act as D or A. Since available examples of electron acceptor polymers are few in number, many of the polymeric charge-transfer complexes are composed of electron donor polymers and low-molecular-weight electron acceptors. Existing data show that the electrical conductivities are generally low [60,61] for these materials. Room-temperature conductivities of the charge-transfer complexes of poly(*N*-vinylcarbazole) with iodine have been shown to vary from 2×10^{-11}

to 8×10^{-8} S cm^{-1}, depending on the content of iodine [110]. Polymers containing both electron-donating and electron-accepting groups within the polymeric framework also fall into the family of polymeric charge-transfer complexes. Polymers 11 and 12 in Figure 7 are examples of this type [62].

Polymeric ion radical salts include polycation-TCNQ anion radical salts, and cation radical or anion radical salts of polymers that are produced by chemical or electrochemical doping. Polycations are polymeric quaternary ammonium or sulfonium ions which are incorporated either in the main chain or as pendant groups. The electrical conductivity of polycation-TCNQ anion radical salts can be widely varied by controlling the amount of TCNQ. The polycation-TCNQ anion radical salts are characteristic in that they are soluble in organic solvents and can be cast into films. Various kinds of polymeric TCNQ anion radical salts have been prepared and their electrical conductivities have been reported [63-66, 76].

Polymers 19 to 23 in Figure 7 are examples of metal-containing polymers [67-71]. Recently, there has been increasing interest in coordination polymers from the viewpoint that they may be a new class of electrically conducting materials. Some coordination polymers containing Fe^{2+}, Co^{2+}, Ni^{2+}, and Zn^{2+} (e.g., polymers 19 and 20) show semiconducting properties with conductivities on the order of 10^{-6} to 10^{-10} S cm^{-1} [67, 68].

New classes of low-dimensional polymer conductors are being developed. One class of polymers is composed of metallomacrocycles which are covalently linked in a face-to-face orientation (e.g., phthalocyaninato polysiloxane, polygermyloxane, or polystannyloxane). The electrical conductivities observed are about 10^{-8} S cm^{-1} or less but are increased to 10^{-1} S cm^{-1} by chemical doping [69, 77]. Another class of polymers consists of square-planar transition metal complexes which are polymerized via bridging ligands to linear chains, as exemplified by polymer 22 [70, 78].

Kapton is a commercially available polyimide film. It has been reported that pyrolyzed Kapton exhibits high isotropic conductivities which are comparable to those of doped polyacetylene films and that for low-temperature pyrolysis ($T < 650°C$) the concentration of charge carriers predominantly increases, while $T > 650°C$ significant changes occur in the carrier mobility [79].

Chemical doping of polymers results in the formation of charge-transfer complexes or partially oxidized or reduced ion radical salts, which increases their electrical conductivities many orders of magnitude. Like chemical doping, electrochemical doping of polymers has proved to be a useful method of producing highly conducting polymers. Metallic properties have been found for chemically or electrochemically doped linear π-conjugated polymers. Examples of doped polymers are given in Figure 8.

$(CH)_x$ 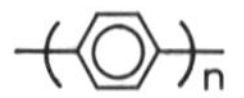 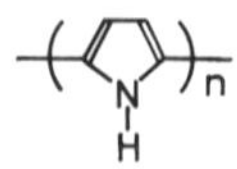

1 2 3

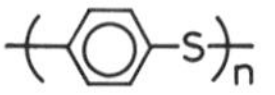 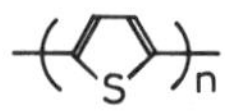

4 5

Figure 8. Examples of doped, highly conductive polymers.

Either *cis*- or *trans*-polyacetylene films can be doped with various electron-accepting species, such as halogens (I_2, Br_2), metal halides (AsF_5, $FeCl_3$), and protonic acids (H_2SO_4, $HClO_4$), and electron-donating species, such as sodium naphthalide, to give p-type and n-type conductors, respectively [80-90]. Electrical conductivities of these doped films increase dramatically with low concentrations of dopants. It has been shown that the concentration dependence of the electrical conductivity and the activation energy suddenly change near 2% dopant levels and that the semiconductor-to-metal transition also occurs near this dopant level in the case of iodine doping [80, 81, 84]. Electrochemical doping of polyacetylene films has also been carried out [91].

Chemical doping of poly(*p*-phenylene) with AsF_5, $SbCl_5$, HSO_4F, $FeCl_3$, or alkali metals increases the conductivity from less than 10^{-14} S cm^{-1}, for the undoped polymer to 10^2 S cm^{-1} for the doped polymer [87, 92, 93].

Highly conducting polypyrrole films have been obtained by anodic oxidation of pyrrole in an organic solution on a platinum electrode [94-98]. The polymer films formed are partially oxidized electrochemically during the polymerization process. The anion part of the supporting electrolyte is incorporated into the polymer as a counter-anion and the ion content is in the range 0.25 to 0.33 per pyrrole ring. The black films, which are thermally stable to temperatures exceeding 300°C, exhibit room-temperature conductivity in the range 10 to 100 S cm^{-1} and metal-like thermopower [95, 96]. Copolymer films similarly prepared from mixtures of pyrrole and *N*-methylpyrrole have conductivities between 5×10^{-3} and 100 S cm^{-1}, depending on the composition of the copolymer [95, 97]. It has been shown that thin films of the metallic polypyrrole on platinum electrode can repeatedly be driven electrochemically between the conducting form (oxidized form) and the insulating form (neutral form) [98-101]. Like polypyrrole, highly conducting polythienylene films have been obtained by the anodic oxidation of thiophene [112].

It has been reported that other π-conjugated polymers, for example, poly(*p*-phenylene sulfide) [102,103], poly(2,5-thienylene) [104], poly(hepta-1,6-diyne) [105], poly(4,4-biphenylene) [106], and poly(*p*-phenylene vinylene) [107], can be doped with iodine or AsF_5 and that electrical conductivities of these doped polymers are significantly high.

Electrical conductivities of metal-containing polymers, for example, polyvinylferrocene, ferrocene-*o*-anisaldehyde condensation polymer [108] or poly(3-vinylbisfulvalenediiron) [109] and phthalocyanato polysiloxane, polygermyloxane, or polystannyloxane [69,77], are also greatly enhanced by chemical doping (i.e., partial oxidation with electron-accepting dopants leading to the formation of mixed valence salts).

Nonconjugated vinyl polymers are transformed into semiconductors by electrochmical doping. It has been shown that electrochemical doping of poly(*N*-vinylcarbazole) and poly(3,6-dibromo-*N*-vinylcarbazole) by means of potentiostatic anodic oxidation produces electrically conducting polymer films with room-temperature conductivities as high as 6×10^{-4} S cm^{-1} [111a]. Similarly, electrochemically doped polyvinylferrocene shows conductivities of about 10^{-5} S cm^{-1} [111b]. The anodic oxidation of *N*-vinylcarbazole in dichloromethane with tetra-*n*-butylammonium perchlorate as the supporting electrolyte deposits an electrically conducting polymer on the surface of the electrode [111c].

6.3.3 Syntheses of Organic Polymeric Conductors: Polymeric Semiconductors and Polymeric Metals

6.3.3.a Synthesis of polyacetylene and substituted polyacetylenes

Polyacetylene, produced by the polymerization of acetylene with Ziegler-Natta catalysts, was first reported in 1958 [113]. Although polyacetylene was obtained in the form of powders, it has since been found that a uniform, free-standing film of polyacetylene can be prepared by using concentrated soluble Ziegler-type catalysts such as a $Ti(OC_4H_9)_4$-$Al(C_2H_5)_3$ system [114,115]. It has been shown that the configuration of polyacetylene prepared at temperatures lower than -78°C is all-cis, whereas all-trans polymers are obtained at temperatures higher than 180°C [114,116]. The films are polycrystalline, consisting of randomly oriented fibrils having diameters of about 200 Å [117-121] or irregularly shaped lamellar-like particles with diameters of a few hundred angstroms [122,123]. The morphology of the polyacetylene films changes with polymerization conditions such as the catalyst system [124,125] or the nature of solvent used [126]. An amorphous polyacetylene film has been prepared by a glow-discharge polymerization technique [127].

Recently, substituted acetylenes have been successfully polymerized to yield linear polymers with high molecular weight. Catalysts such as $ArM(CO)_3$, where Ar is an arene and M is Cr, Mo, W [128,129], WCl_6, or $MoCl_5$ [130,75], have been found to be effective for the polymerization of substituted acetylenes such as phenylacetylene.

6.3.3.b Synthesis of free-standing polyacetylene films [114,115]

A uniform, free-standing polyacetylene film is prepared by introducing acetylene gas into an unstirred concentrated solution of the $Ti(OC_4H_9)_4$-$Al(C_2H_5)_3$ catalyst.

In a typical experiment, 15 mmol of tetrabutoxytitanium and 60 mmol of triethylaluminium, in that order, are added to 60 mL of toluene with hypodermic syringes. After the mixture is stirred for a few minutes at room temperature, a 30-mL aliquot is transferred into a Schlenk-type flask. The flask is cooled in a dry ice-methanol bath and degassed under vacuum. After the system reaches a constant temperature, acetylene is introduced into the flask. Formation of a polyacetylene film is observed immediately after the introduction of acetylene gas. The polymerization is interrupted by evacuating the system. Purification of the film formed is done by removing the catalyst solution remaining under the film with a hypodermic syringe followed by washing repeatedly with toluene or hexane until the solution becomes colorless. The purification is carried out at the temperature of the polymerization in order to prevent cis-trans isomerization of the polymers. The film is dried by blowing room-temperature nitrogen gas on it for a minute and then stored under nitrogen at -30°C.

The polymerization is carried out at temperatures over a wide range (about -100 to 180°C). If polymerization temperatures are lower than 80°C, toluene is used as a solvent; at higher temperatures, *n*-hexadecane is used.

6.3.3.c Chemical doping of polyacetylene films

When polyacetylene films are exposed to the vapor of electron-accepting substances such as halogen, AsF_5, H_2SO_4, and $HClO_4$ [80] or dipped into a solution of dopants such as silver perchlorate or silver tetrafluoroborate in toluene [88], $FeCl_3$ in rigorously dried nitromethane [87], or sodium naphthalenide in THF [83], they become doped with these species, resulting in the formation of the salts of the polymer ion radicals. The level of doping varies with the vapor pressure of dopants, the concentration of dopant in solutions, and the exposure time.

In the case of iodine, a pure polyacetylene film is mounted in a glass vessel to which a bulb containing the halogen is attached. The iodine container is held at a fixed temperature to produce a known vapor pressure of iodine. The halogen content in the final product and at various stages in the reaction is determined from weight uptake and by chemical analysis [84].

It has been reported that AsF_5-doped *cis*-polyacetylene films can be prepared by exposure of acetylene gas to a small amount of AsF_5 in a glass reaction vessel at -75 to -198°C. For example, a polyacetylene film obtained at -75°C with an initial molar ratio of acetylene to AsF_5 of 10 had the composition $[CH_{1.03}(AsF_5)_{0.011}]_X$ and exhibited a room-temperature conductivity of 4.8×10^{-2} S cm^{-1} [132].

6.3.3.d Synthesis of poly(*p*-phenylene vinylene)

Poly(*p*-phenylene vinylene)s and poly(*p*-xylylidence)s have been synthesized by a number of methods, such as the reaction of *p*-xylene-bis(triphenylphosphonium chloride) with terephthalaldehyde in the presence of lithium ethylate (Wittig reaction) [133], the reaction of *p*-xylene tetrabromide with alkyllithium in tetrahydrofuran [55], or by dehydrochlorination of *p*-xylene dichloride with sodium hydride in dimethylformamide [56].

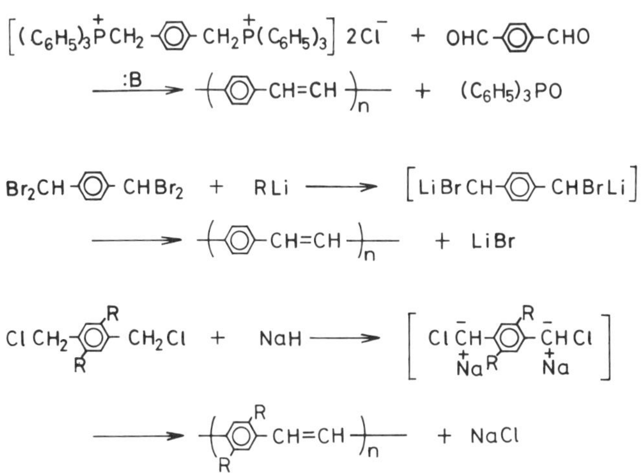

Wittig reaction [133]

p-Xylene-bis(triphenylphosphonium) chloride (20 g, 0.027 mol) which is recrystallized from an ethanol-ether mixture and terephthalaldehyde (3.84 g, 0.329 mol) are stirred in absolute alcohol (200 mol). A solution of Li (0.40 g, 0.058 g atom) dissolved in ethanol (150 mL) is added dropwise under nitrogen over a period of 1 h. An intense orange-red color develops at the point of contact. On mixing, a bright lemon-yellow precipitate forms, which is filtered, washed with alcohol, and dried to give 6.4 g of the poly(*p*-phenylene vinylene), mp 400°C.

Reaction of p-xylene tetrabromide with alkyllithium [55]

p-Xylene tetrabromide (23 g, 0.054 mol) is dissolved in dry tetrahydrofuran (75 mL). To this solution is slowly added methyllithium in ether (0.11 mol, 140 mol) at -15 to -20°C. Brownish-yellow solids precipitate from the reaction solution. After stirring for 4 h, water is added and the solids are filtered, washed with tetrahydrofuran, and dried to give poly(*p*-phenylene vinylene) (5.0 g).

6.3.3.e Synthesis and chemical doping of poly(*p*-phenylene)

Poly(*p*-phenylene) has been synthesized by various procedures, such as oxidative cationic polymerization of benzene with Lewis acid-oxidant systems, electrolytic polymerization, coupling of aromatic halogen compounds, and condensation of aromatic nuclei via organometallic reagents such as aryl Grignard reagents or aryllithium compounds [134].

The catalyst systems effective for the polymerization of benzene include aluminum chloride-cupric chloride, aluminum bromide-cupric bromide, and antimony chloride-cupric chloride [135-138]. Recently, it has been reported that $Cu(AlCl_4)_2$, which is prepared by heating a mixture of aluminum chloride (2.0 g) with copper(II) chloride (1.0 g) at 240°C for 10 h in a glass ampul, reacts with benzene at 20°C to give poly(*p*-phenylene) in a high yield (92% yield for a 2-h reaction) [139].

The polymerization of benzene with aluminum chloride-cupric chloride has been suggested to proceed by oxidative cationic polymerization [135]. However, a cation radical initiation followed by a cationic propagation mechanism has generally been accepted [140]. More recently, a unique mechanism of propagation via coordination of benzene cation radical with benzene nuclei has been presented [141].

Poly(p-phenylene) from benzene-aluminum chloride-cupric chloride-water

The reaction is carried out under nitrogen in a three necked flask equipped with a paddle stirrer. Water (1 mL) is added dropwise to a mixture of benzene (1 mol), aluminum chloride (0.5 mol), and cupric chloride (0.5 mol). The temperature is increased to 35 to 37°C and the reaction allowed to proceed for 30 min at these temperatures during which time HCl gas is evolved. The reaction mixture is then added to water and the polymer is collected by filtration. After treatment with 18 N hydrochloric acid, the product is triturated repeatedly in boiling concentrated hydrochloric acid and then washed with water until the washings are colorless. After two triturations with boiling

2 M sodium hydroxide solution, the polymer is washed with water until the filtrate is colorless. The polymer then is washed repeatedly with distilled water until the filtrate gives a negative test for chloride ion (silver nitrate method). After drying at 120°C for 5 h, the product is a finely divided, light brown solid.

Poly(*p*-phenylene) can be doped with both n- and p-type dopants to provide highly conducting materials [92]. It has also been shown that *p*-phenylene oligomers, such as biphenyl, terphenyl, and *p*-quarterphenyl, form electrically conducting complexes with AsF_5 and that prolonged exposure to AsF_5 causes polymerization of these *p*-phenylene oligomers to give AsF_5-doped highly conducting poly(*p*-phenylene) [142].

6.3.3.f Synthesis and chemical doping of poly(phenylene sulfides) [143,144,102,103]

Commercially available poly(*p*-phenylene sulfide) is prepared by the reaction of *p*-dichlorobenzene with sodium sulfide in *N*-methyl-2-pyrrolidinone at 250°C for 24 h [143]. Poly(*m*-phenylene sulfide) is similarly prepared from 1,3-dichlorobenzene and sodium sulfide [144].

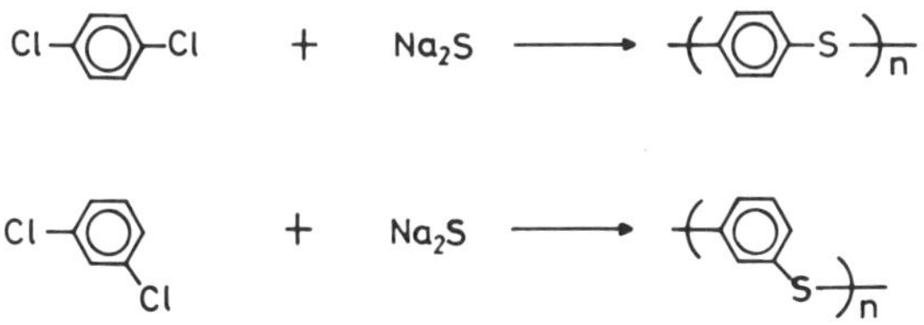

A powder sample of poly(*p*-phenylene sulfide) is doped in various forms as loose powders, pellets prepared from compressed powder, melt-molded plaques, and thin films cast onto quartz windows from diphenyl ether solutions at 230°C [144].

Doping of polymer samples with AsF_5 is carried out in an all-stainless-steel vacuum line employing either stainless steel or glass vessels with Teflon O-ring joints. The sample tubes are first evacuated to 10^{-5} torr, then pressurized with AsF_5 gas to the desired pressure while maintaining the vessel at the desired temperature. After the doping time is complete, the system is evacuated to remove excess AsF_5. Further manipulations with doped samples are performed in a dry box under argon.

6.3.3.g Synthesis of polypyrrole films: electrochemical polymerization and electrochemical doping [94-96]

The polypyrrole films are synthesized galvanostatically on a platinum surface in a single-compartment, two-electrode cell containing 0.1 M

Et_4NBF_4 and 0.06 M pyrrole in 99% aqueous acetonitrile. The films formed are spontaneously doped electrochemically and adhere well to the surface of the electrode. The cell current increases linearly with time over a period of 3 min from zero to a final value. The final current density is usually set to 1 mA cm^{-2}, although values ranging from 0.5 to 1.5 mA cm^{-2} are used. Films of thickness up to 50 μm are prepared in this manner.

Various factors in electrolysis affect the conductivity, adherence, morphology, and flexibility of films formed. Polymerization onto platinum electrodes results in smoother, better-adhering films than when tin oxide electrodes are used. Films formed in acetonitrile are found to have a higher conductivity than those formed in an aqueous solution. The absence of water produces very poorly adhering, nonuniform films while the presence of increasing amounts of water improves adherence. Excessive current densities yield films having very rough surfaces.

6.3.3.h Electrochemical doping of poly(*N*-vinylcarbazole) and poly(3,6-dibromo-*N*-vinylcarbazole) [111a]

Electrochemical doping of poly(*N*-vinylcarbazole) and poly(3,6-dibromo-*N*-vinylcarbazole) is carried out by means of controlled potential electrolysis using a three-electrode system. Polymer films with thickness of 10 to 30 μm are cast from 1 to 3%, w/v, tetrahydrofuran solution onto the surface of a platinum plate electrode and dried in vacuum for 6 h. Anodic oxidation of the polymer films is carried out in acetonitrile with 0.1 M tetra-*n*-butylammonium perchlorate or tetrafluoroborate as a supporting electrolyte at a potential of 1.0 V versus Ag/Ag^+ 0.01 M reference electrode on the basis of cyclic voltammograms of poly(*N*-vinylcarbazole) and poly(3,6-dibromo-*N*-vinylcarbazole). The films turns into a bluish-green color as the electrochemical doping proceeds. When the electrolysis is stopped after an appropriate time, the films that adhere to the electrode are washed with acetonitrile and then with alcohol to remove any supporting electrolyte remaining on the films.

6.3.3.i Synthesis of polymeric charge-transfer complexes

A general method for the preparation of polymeric charge-transfer complexes is to mix equal moles of donor and acceptor in a solvent and evaporate the solvent under nitrogen and then in vacuum [61a].

6.3.3.j Synthesis of polyurethane containing tetrathiafulvalene and tetracyanoquinodimethan units [62]

The polymer is prepared by condensation of 2,5-bis(2-hydroxyethoxy)-7,7,8,8-tetracyanoquinodimethan with 4,4-diisocyanatotetrathiafluvalene.

To a stirred solution of 324.3 mg (1 mmol) of 2,5-bis(2-hydroxyethoxy)-7,7,8,8-tetracyanoquinodimethan (324.3 mg, 1 mmol) and 4,4-diisocyanatotetrathiafulvalene (298.4 mg) in 8 mL of dimethylformamide under an argon atmosphere is added 0.015 mL of dibutyltin dilaurate. The solution gradually becomes viscous and gels over a period of a few hours. After 24 h, a few drops of methanol are added, then after 15 min, an excess of benzene and methanol is added. The black polymer (591 mg) is collected by filtration and dried at 65°C (0.1 mm).

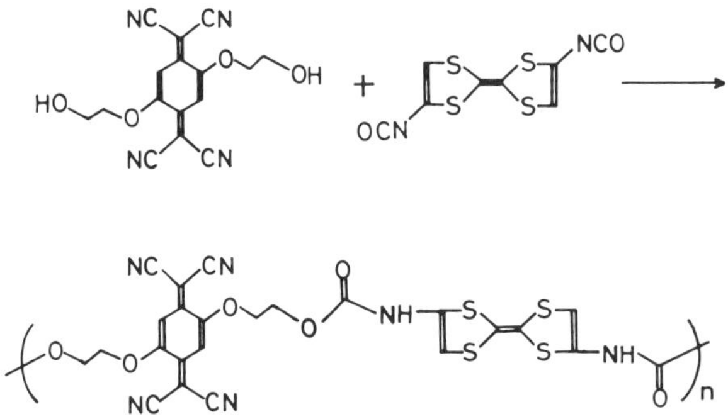

6.3.3.k Synthesis of TCNQ salts of poly(4-vinylpyridine) [63]

A 1-g sample of poly(1-butyl-4-vinylpyridinium iodide), which is prepared by the reaction of poly(4-vinylpyridine) with *n*-butyl iodide in nitromethane at 40 to 45°C for 4 days, is dissolved in a mixture of 100 mL of water and 100 mL of alcohol. Lithium TCNQ (750 mg) is dissolved in 100 mL of water and the two solutions are mixed and stirred for 30 min. The precipitate formed is filtered, washed with alcohol and ether, and dried in vacuum; yield 1.1 g.

Films of complex salt are prepared as follows. Weighed quantities of the simple salt and TCNQ are combined in dimethylformamide, and films are cast from these solutions.

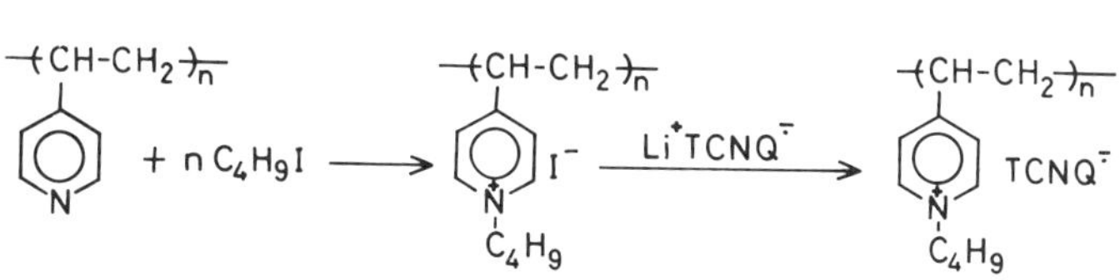

6.3.3.l Synthesis of TCNQ salts of ionene polymers (17) [65]

The ionene bromide is prepared by the reaction of polymethylene bis(*N*,*N*-dimethylamine) with polymethylene dibromide. The ionene bromide is treated with lithium TCNQ in a molar ratio of unit segment to lithium TCNQ of 1:2 to give TCNQ simple salt of the ionene polymer. The complex salt is prepared by treating the ionene bromide with a mixture of lithium TCNQ and neutral TCNQ using a molar ratio of 1:2:2.

6.3.3.m Synthesis and chemical doping of phthalocyaninato polysiloxanes, polygermyloxanes, and polystannyloxanes

Dichlorosilicone, germanium, and tin phthalocyanines [$M(Pc)Cl_2$] are prepared according to the literature and then hydrolyzed in pyridine to produce the corresponding dihydroxides [$M(Pc)(OH)_2$]. Polymerization under high vacuum at 300 to 400°C produces the corresponding phthalocyaninato polysiloxanes, polygermyloxanes, and polystannyloxanes in high yield and purity.

Chemical doping can be carried out by exposure of the powdered phthalocyaninato polysiloxane, polygermyloxane, and polystannyloxane materials to solutions of iodine or to iodine vapor. Depending on the amount of iodine employed, a wide range of dopant stoichiometries are achieved.

6.3.3.n Synthesis of polymeric phthalocyaninato(pyrazine)iron(II) [70]

The reaction of β-iron phthalocyanine (β-PcFe) with an excess of pyrazine (molar ratio 1:5) in boiling chlorobenzene leads immediately to polymeric blue phthalocyaninato(μ-pyrazine)iron(II).

β-PcFe (2 mmol) is slurried in anhydrous chlorobenzene (180 mL) and boiled under reflux with pyrazine (10 mmol) for 24 h. The blue residue is treated by Soxhlet extraction with anhydrous chlorobenzene under N_2 for 3 h to remove phthalocyaninatobis(pyrazine)iron(II). The blue residue in the extraction vessel is washed with light petroleum, filtered, and dried at 80°C; yield 86%.

6.3.3.o Synthesis of chelate polymers of 2,4-dihydroxybenzaldehyde [67]

Bis(2,4-dihydroxybenzaldehyde)Co^{2+} and Ni^{2+} are obtained by the reaction of 2 mol of 2,4-dihydroxybenzaldehyde with 1 mol of cobalt or nickel acetate in water in the presence of sodium acetate for 2 h. The chelate polymers are synthesized from bis(2,4-dihydroxybenzaldehyde) and the diacid chloride, terephthaloyl or isophthaloyl chloride, by interfacial polycondensation. Bis(2,4-dihydroxybenzaldehyde)Co^{2+} and Ni^{2+} (8.33×10^{-3} mol) in an aqueous alkaline solution (1.67×10^{-2} mol in 100 mL of water) reacts with 8.33×10^{-3} mol of terephthaloyl or isophthaloyl chloride in dichloromethane in the presence of sodium lauryl sulfate as a dispersion agent. The mixture is vigorously stirred for 5 to 8 min and the polychelates are isolated by precipitation in acetone. They are purified by repeated extraction with methanol and ethanol. The polychelates are obtained as colored powders, which are insoluble in common organic solvents but partly soluble in dimethylformamide.

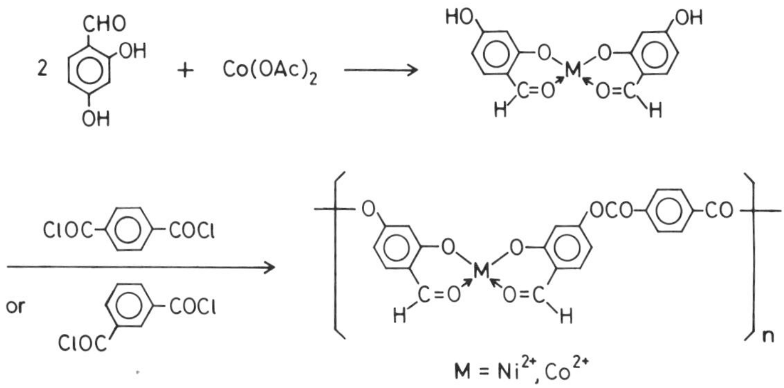

6.3.3.p Synthesis of polymers containing tetrathiafulvalene moiety [145-148]

Interest in tetrathiafulvalene (TTF) and its derivatives, which form charge-transfer complexes with tetracyanoquinodimethan (TCNQ) with metallic conductivity, has expanded to the synthesis of polymers containing the TTF moiety either within the polymer backbone or as the pendant group.

Polyesters, polyurethanes, and polysulfones containing the TTF moiety within the polymer backbone have been prepared by condensation of 2,6(7)-bis(*p*-hydroxyphenyl)tetrathiafulvalene with diacid chlorides, diisocyanates, and disulfonyl chlorides using solution techniques [145].

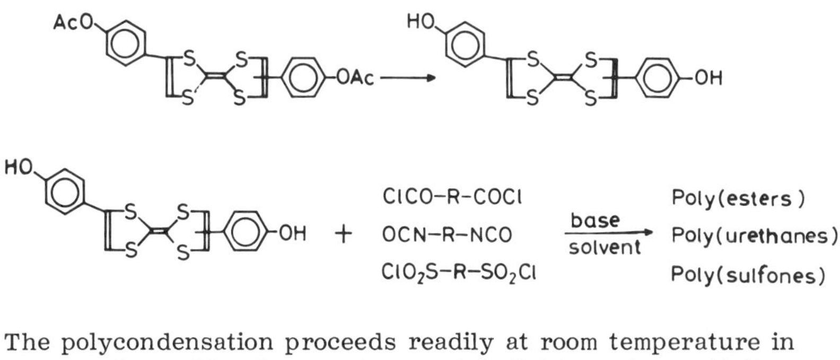

The polycondensation proceeds readily at room temperature in solvents such as chloroform, toluene, 1,2-dichloroethane, *N,N*-dimethylacetamide, tetramethylene sulfone, and hexamethylphosphortriamide as well as in the presence of triethylamine, *N*-ethylmorpholine,

2,6-lutidine, 2-picoline, pyridine, or isoquinoline. The molecular weight of the resulting polymers are low in all cases. They all fail to form charge-transfer complexes with TCNQ and 2,3-dicyano-4,6-dichloro-*p*-benzoquinone (DDQ).

Another type of polymer containing the TTF moiety has been prepared by a coupling process. They form charge-transfer complexes with TCNQ and DDQ [146].

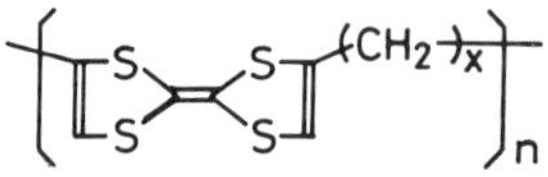

Monomers and the corresponding polymers containing the pendant TTF moiety have also been prepared [147,148].

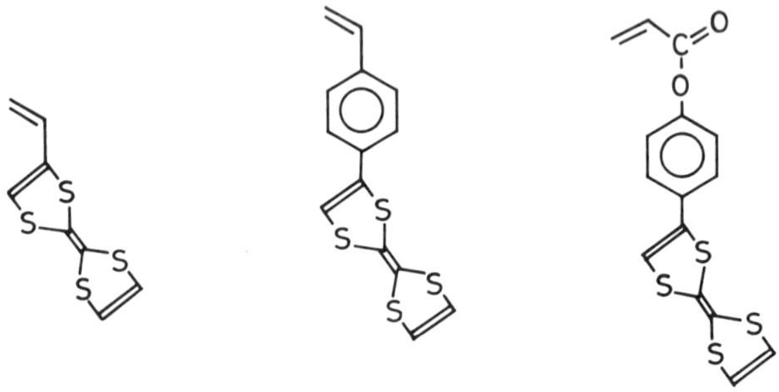

6.3.3.q Synthesis of poly(*p*-phenylene selenide) [149,150]

Interest in chemically doped poly(*p*-phenylene sulfide) which exhibits a high electrical conductivity has led to the synthesis of analogous poly(*p*-phenylene selenide) (PPSe). Poly(*p*-phenylene selenide) is prepared by reacting *p*-dibromobenzene with sodium selenide prepared in situ directly from its elements in *N*,*N*-dimethylformamide for 20 h. It is obtained as a light-yellow powder in 80% yield after removal of low-molecular-weight material with warm tetrahydrofuran. The room-temperature conductivity of the compacted black material of AsF_5-doped PPSe is 10^{-2} to 10^{-3} S cm^{-1} [149].

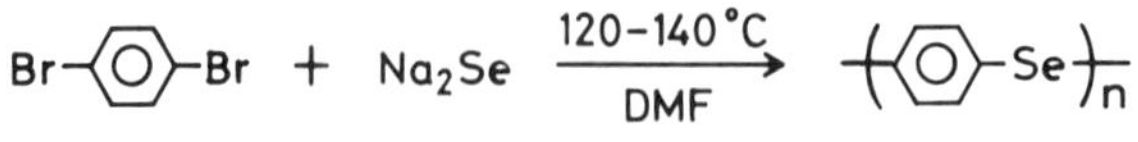

Another route for the synthesis of PPSe is as follows. Diazotization of *p*-bromoaniline followed by reaction with potassium selenocyanate affords *p*-bromophenylselenocyanate, which is converted to bis(*p*-bromophenyl)diselenide on treatment with alkali. The diselenide is converted to PPSe. The reaction of the diselenide with electrolytic copper in refluxing *n*-hexanol-pyridine for 4 h affords cuprous *p*-bromoselenophenoxide, which is polymerized by heating with pyridine at 200°C for 24 h [150].

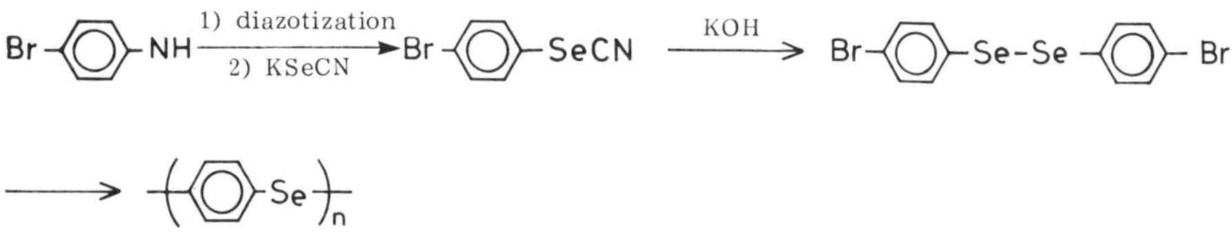

6.3.4 Applications of Organic Polymeric Conductors

Generally, organic polymeric materials have characteristic properties such as light weight, flexibility, and film-forming capability, thus offering advantages in the fabrication of devices. It is expected that electrically conducting polymeric materials will find a broad range of applications as functional materials rather than the substitutes for conventional metals and inorganic semiconductors.

Electrically conducting polymeric materials are classified into two types: electrically conducting polymers and composite materials where conducting fillers are dispersed in insulating polymer binders. The electrically conducting plastics and rubbers that have been in practical use for antistatic coatings, shielding against electromagnetic interference, and interconnection of electronic circuits are mostly composite materials in which silver or carbon black is dispersed in insulating plastics and rubbers. In principle it is possible to replace carbon black with conducting organic materials. It has been reported that the TCNQ anion radical salt shown below forms filament-like structures when relatively small amounts of the salt are dispersed in a polymer binder, and that the polymer film exhibits a relatively high conductivity of about 1 S cm [151].

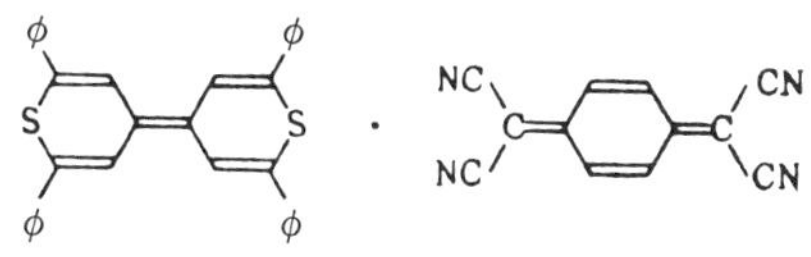

Various potential applications have been suggested for electrically conducting polymers [152,153]. They include primary and secondary batteries, diodes, solar cells, electrode materials for chemical reactions, sensors, and displays. It is well known that a miniature primary battery consisting of poly(2-vinylpyridine)-iodine charge-transfer complex as the cathode and lithium as the anode has been used as a power source for implantable cardiac pacemakers. The characteristic feature of this solid-state cell is that the cathode material is placed in contact with the lithium anode and the solid electrolyte LiI is formed in situ as the discharge product [153,154]. It has been reported that the battery has a high energy density of 120 Wh/kg and is durable for 10 years [153,154].

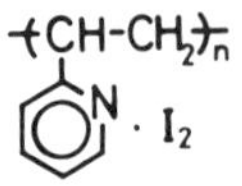

$$2Li + I_2 \longrightarrow 2LiI$$

PVP : I_2 = 1 : 7~24

Solid-state electrochemical cells using other polymeric charge-transfer complexes have also been devised [155].

Lightweight, rechargeable storage batteries using polyacetylene [156,157] and poly(*p*-phenylene) [158] as the electrode material have attracted considerable attention. Electrochemical doping of polyacetylene and poly(*p*-phenylene) occurs through the charging process and the doped polymers return to their initial neutral polymers through the discharge process. It has been reported that 0.5-cm^2 piece of the 6% doped polyacetylene film (1.0 × 0.5 × 0.01 cm; 3 mg) exhibits an open-circuit voltage of 3.7 V and an initial short-circuit current of 25 mA [156]. The overall discharge reaction is

$$[CH^{+0.06}(ClO_4^-)_{0.06}]_x + 0.06xLi \longrightarrow (CH)_x + 0.06xLiClO_4$$

Other polymers are also used as electrode materials [159,160,170].

Photovoltaic devices using polyacetylene and other polymers have been studied [161-163,171]. Polypyrrole [99-101], $(SN)_x$ [164], and TTF-TCNQ charge-transfer complex [165] function as the electrode materials for chemical reactions. Potential applications using properties other than conductivity have been reported. For examples, polymers containing the pendant TTF group can be used as materials for electrochromic display [166], and its cation radical salt with bromine as a counteranion functions as an electron beam resist in microlithography [167]. The photochemical doping process might find some applications [168]. Active studies are now being continued to find various potential applications of organic polymeric conductors [169,172].

6.4 ORGANIC POLYMERIC PHOTOCONDUCTORS

6.4.1 Survey of Organic Photoconductors and Organic Polymeric Photoconductors

Photoconduction is observed when the concentration of charge carriers (holes and electrons) is larger under illumination than in the thermal equilibrium state. In case materials show high conductivity in the dark, photoconduction is not observable in most cases because it is hidden by dark conduction. Therefore, photoconduction is observed for compounds that fall into the class of insulators or poor semiconductors. Although it is not an exaggeration to say that photoconduction is more or less observed for almost all compounds, including saturated polymers such as polyethylene, what is called photoconductors show a high ratio of photocurrent to dark current (e.g., more than two orders of magnitude larger photocurrent relative to dark current). Such a difference in conductivity between conditions of dark and those of illumination leads to applications in electrophotography technology, which is discussed later in this section.

Organic photoconductors are largely classified into low-molecular-weight compounds and polymers. Both low-molecular-weight compounds and polymers can be used as photoreceptor materials in electrophotography. Usually, low-molecular-weight organic photoconductors are used by dispersing in polymeric binders that form films. This blend system is called a molecularly dispersed system. Polymeric photoconductors possess the functions of both photoconductivity and processibility; as a result, they are readily formed into films and the serious problem of compatibility that arises for molecularly dispersed systems can be avoided.

Low-molecular-weight organic photoconductors can be classified into several classes based on their structures. They are (a) polycyclic aromatic hydrocarbons, (b) heterocyclic aromatic compounds, (c) arylamines, (d) amino-substituted triarylmethanes, (e) charge-transfer complexes, and (f) dyes and pigments. Typical examples of each class are (a) anthracene, (b) *N*-isopropylcarbazole, (c) triphenylamine, (d) bis(2-methyl-4-dimethylaminophenyl)phenylmethanes, (e) phenanthrene-pyromellitic dianhydride charge-transfer complex, and (f) cupper phthalocyanine [24f]. Similarly, organic polymeric photoconductors are classified into several classes: (a) linear π-conjugated polymers, (b) planar π-conjugated polymers, (c) polymers containing pendant photoconductive groups, (d) polymers containing photoconductive groups in the polymer main chain, (e) polymeric charge-transfer complexes, (f) biological polymers, and (g) others. Polymers of classes c, d, and e find analogy with low-molecular-weight organic photoconductors. That is, pendant photoconductive groups contained in the polymers of class c are those which are classified for low-molecular-weight organic photoconductors. Polymers of class d include polymers that are known to be thermally stable polymers. Polymeric charge-transfer complexes of class e are the systems where sensitization to visible light is achieved.

A variety of photoconducting polymers as classified above have been synthesized in view of understanding photoconductivity or developing new photoreceptor materials. Among these, polymers of class c have been studied most extensively; however, because of their poor xerographic properties or lack of processibility or productivity, only poly(*N*-vinylcarbazole) (PVC_Z) and its halogenated derivatives have been put into practical use. Examples of structures of some representative photoconducting polymers are given in Figure 9. Compounds 1 and 2 are the linear conjugated polymers of class a; compounds 3 to 24 are the polymers of class c; compounds 25 and 26 are the polymers of class d [219,220]; and compound 27 is an example of the polymeric charge-transfer complexes of class e [221]. An example of the polymers of class b is heat-treated polyacrylonitrile, the definite structure of which is unknown.

Polydiacetylenes of the general structure 1 have been synthesized by the solid-state polymerization of single crystals of diacetylenes via a 1,4-addition reaction [173-176]. The polymerization is achieved by irradiation with UV light or high-energy radiation or simply by annealing the monomer crystals below their melting point [175]. Polydiacetylenes are unique in that they can be obtained as large-dimension single crystals. Poly[2,4-hexadiyne-1,6-diol bis(*p*-toluene sulfonate)] ($R = CH_2SO_3C_6H_4CH_3$ in 1) has been most intensively studied with regard to its photoconductivity [177-179]. It has been found that single crystals of this polymer show a high carrier mobility of $\mu \sim 20$ $m^2/(V \cdot s)$ at low fields [179]. A carrier mobility value of $\mu = 2800$ $cm^2/(V \cdot s)$ has been observed for poly[1,6-bis(*N*-carbazolyl)-2,4-hexadiyne] [179]. The photoconductivity of polyacetylene [180], poly(*p*-xylylidene) [181], and poly(arylene vinylene) [182] has also been studied.

Although anthracene has been known for some time to have photoconductive properties, high-molecular-weight polymers with anthracene groups have not been synthesized until recently. Since anthracene acts as an efficient radical scavenger, free-radical polymerization techniques usually did not give high-molecular-weight polymers; however, 1-(2-anthryl)ethyl methacrylate and 9-(*p*-vinylphenyl) anthracene have been polymerized by free-radical methods to give high-molecular-weight polymers [183,184]. Recently, high-molecular-weight vinyl polymers with anthracene groups have been prepared by anionic polymerization techniques [185-187]. It has been established that the monomer purity is extremely important in attaining high molecular weight for the polymers. Poly(1-vinylpyrene) has also been synthesized [188].

Poly(*N*-vinylcarbazole) (PVC_Z) has received the most attention as a representative example of polymeric photoconductors. A variety of related polymers, such as compounds 9 to 16 in Figure 9, have been synthesized [189-198]. Poly(*N*-vinylcarbazole), which is produced by the polymerization of *N*-vinylcarbazole, is commercially available. The polymerization of *N*-vinylcarbazole is induced by various types

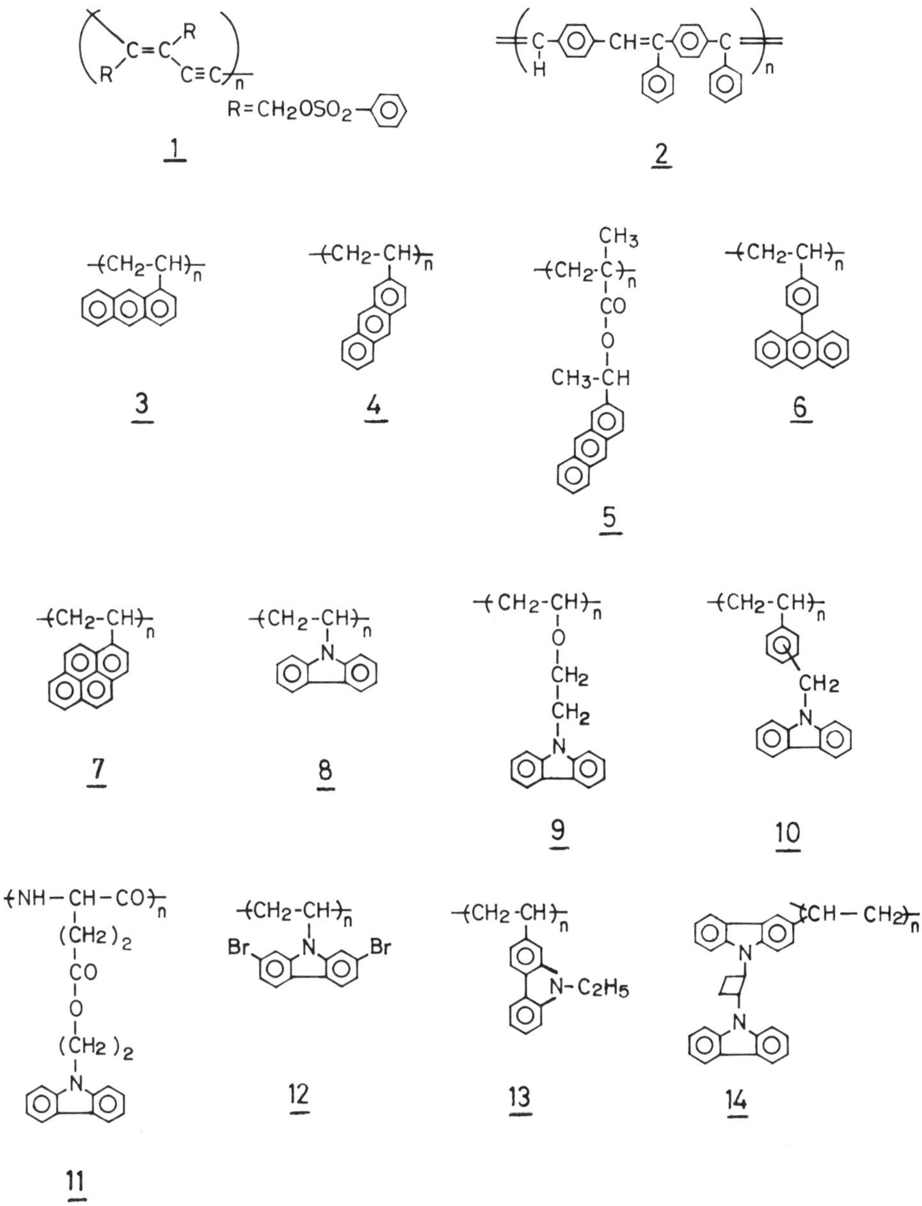

Figure 9. Examples of photoconducting polymers.

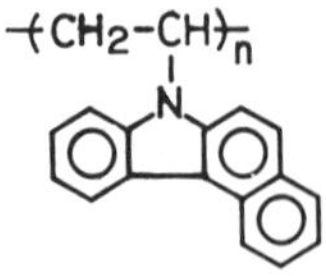

15

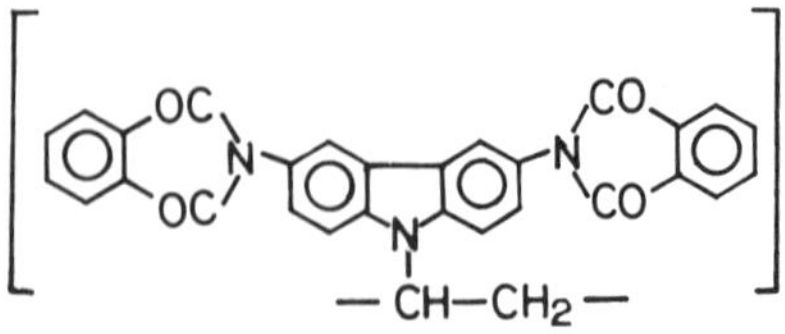

16

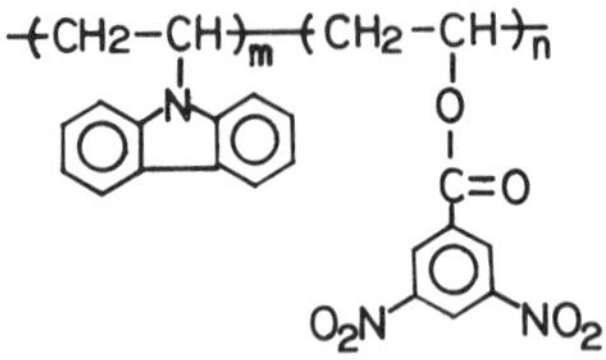

17

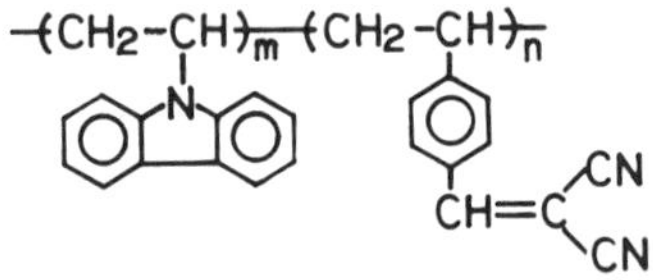

18

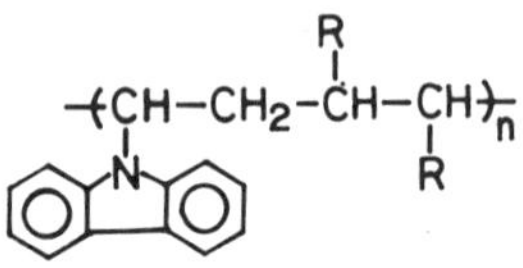

R= CN, CO_2Et

19

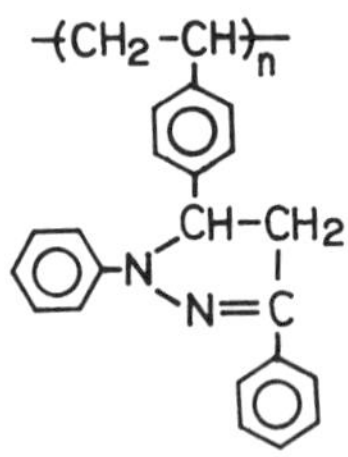

20

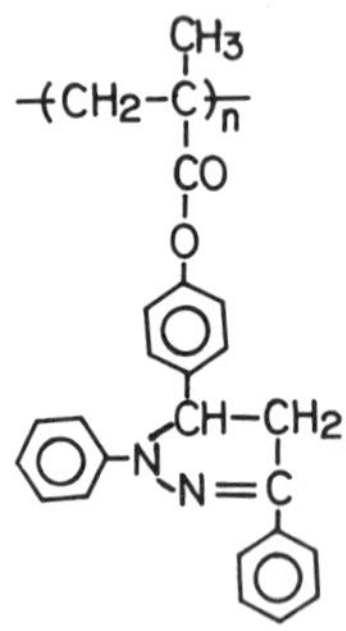

21

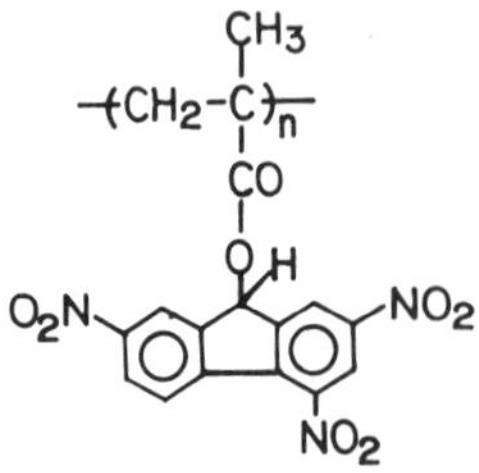

22

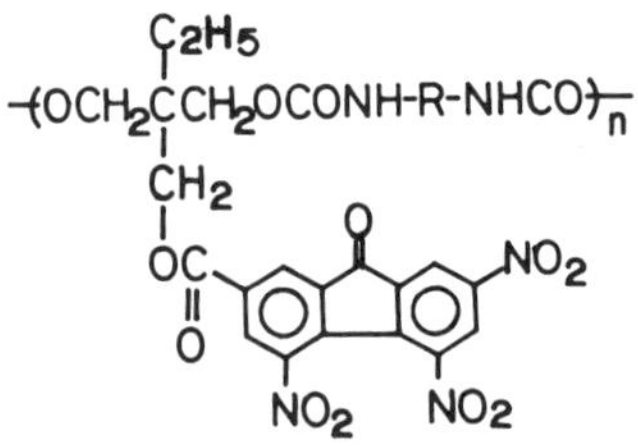

23

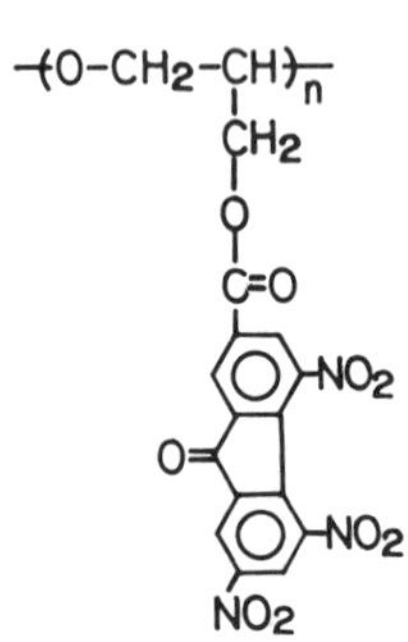

24

Figure 9 (continued)

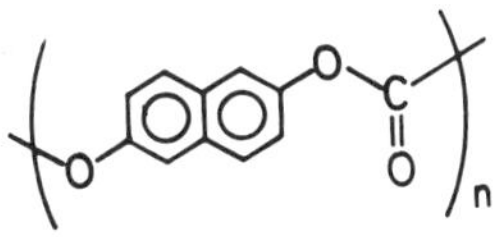

25

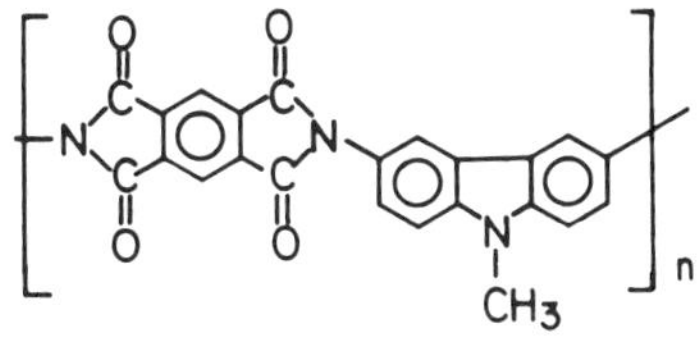

26

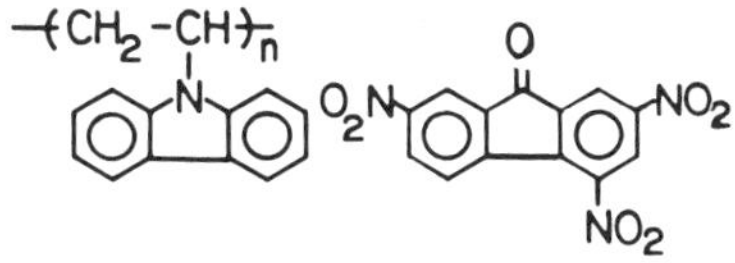

27

Figure 9 (continued)

of initiation except anionic initiation. Not only is it initiated by conventional cationic, free-radical, and Ziegler-Natta-type catalysts but also by photoirradiation and γ-ray irradiation, and thermal and photochemical initiation in the presence of electron acceptors. The last type of polymerization is called charge-transfer polymerization [199].

The photoconductivity of related polymers containing pendant carbazolyl groups has also been examined in relation to their molecular structures [191, 198, 200-202]. Polymers 17 and 18 contain molecules that function as sensitizers by forming ground-state charge-transfer complexes with incorporated PVC_Z [198]. Alternating 1:1 copolymers containing *N*-vinylcarbazole as one component have been synthesized by the free-radical copolymerization of *N*-vinylcarbazole with fumaronitrile [203, 205], diethyl fumarate [204, 205], and maleic anhydride [206]. These serve as model polymers for gaining a deeper insight into the photoconduction phenomenon of PVC_Z. Spectroscopic and photophysical properties of these 1:1 alternating copolymers have been examined [205, 207, 208]. The values of threshold ionization potentials of PVC_Z and the 1:1 alternating copolymers in thin films have been determined by means of UV photoelectron spectroscopy and discussed in relation to photoconductivity [209].

Derivatives of 1,3,5-triphenyl-2-pyrazoline have recently been put into practical use as a photoreceptor material in electrophotography by dispersing them into a polyester binder [210]. Polymers containing a pendant 1,3,5-triphenyl-2-pyrazoline group have been synthesized [211] and their spectroscopic and photophysical properties have been studied [212,213]. Polymers containing a pendant 2,5-diphenyl-1,3-oxazole group have also been synthesized [214].

In contrast to a number of examples of polymers with electron-donating properties, very few examples of electron-accepting polymers have been reported. The lack of examples of electron-accepting polymers is attributed to the fact that free-radical or anionic polymerizations would fail because of free-radical inhibition by nitro aromatics or electron transfer reactions [215]. In view of the fact that 2,4,7-trinitrofluorenone (TNF) acts as an excellent electron transport material [216], polymers containing pendant TNF or its analogous groups (22 to 24 in Figure 9) have been synthesized [215, 217,218].

6.4.2 Syntheses of Organic Polymeric Photoconductors and Their Corresponding Monomers

The synthetic methods for the polymers of classes a, b, and d are described in Section 6.3; therefore, attention is paid here to the polymers of class c.

The methods for the synthesis of photoconducting polymers largely fall into two classifications: (a) incorporation of photoconductive groups into polymers by polymer reactions, and (b) synthesis of monomers containing photoconductive groups followed by their polymerizations. The former method is simple and convenient because it does not involve the process of monomer synthesis and polymerization; however, regular incorporation of photoconductive groups into polymers cannot be achieved. This method has an advantage in that the resulting polymers are expected to retain good film-forming properties similar to those of the starting polymers. Method b enables the synthesis of polymers with regular incorporation of photoconductive groups; however, the resulting polymers do not necessarily have desirable mechanical properties such as film formation and flexibility.

A family of monomers containing pendant photoconductive groups includes vinyl, isopropenyl, ethynyl, and allyl olefins, acrylates and methacrylates, and acrylamides and methacrylamides. Generally, the synthesis of vinyl monomers can be achieved by (a) dehydration of the corresponding alcohols such as 1- and 2-arylethanols, or (b) Wittig reaction of the corresponding aldehydes. Usually, 2-arylethanols are prepared by the reduction of aryl methyl ketones with lithium aluminum hydride or sodium borohydride and by the Grignard reactions with aldehydes. In the acetylation reaction of condensed aromatic hydrocarbons, attention should be paid to the regioselectivity of acetylation. 1-Arylethanols are obtained by reduction of the corresponding carboxylic acids or esters with lithium aluminum hydride. Alcohol dehydration is effected by distillation over $KHSO_4$ or by passing through activated alumina at elevated temperatures. Sodium hydroxide or potassium hydroxide can also be used for alcohol dehydration.

Alternatively, vinyl monomers are prepared by conversion of alcohols to the corresponding chlorides followed by dehydrochlorination. Like vinyl monomers, isopropenyl monomers are synthesized by (a) dehydration of corresponding tertiary alcohols which are prepared by Grignard reactions, such as the reaction of acetone with arylmagnesium bromide, or reactions of aryl methyl ketones with methylmagnesium bromide; and by (b) Wittig reaction with aryl methyl ketones. For the synthesis of vinyl ethers and vinyl esters, trans etherification and trans esterification of the corresponding alcohols or carboxylic esters with vinyl ethers or vinyl esters can be used. Acrylates or methacrylates are readily obtained by condensation of corresponding alcohols with acryloyl chloride or methacryloyl chloride in the presence of triethylamine or pyridine. Similarly, acrylamides or methacrylamides are prepared by the reactions of corresponding primary or secondary amines with acryloyl chloride or methacryloyl chloride in the presence of triethylamine or pyridine.

6.4.2.a Polymer reactions

Synthesis of poly[(N-carbazolyl)methylstyrene] [190]

Poly[(N-carbazolyl)methylstyrene] has been synthesized by the reaction of potassium carbazole with poly(vinylbenzyl chloride). Since poly(vinylbenzyl chloride) is a mixture of isomers (60% meta, 40% para), the resulting polymer is a mixture of meta and para isomers. This polymer can also be synthesized by polymerizing the corresponding monomer [190].

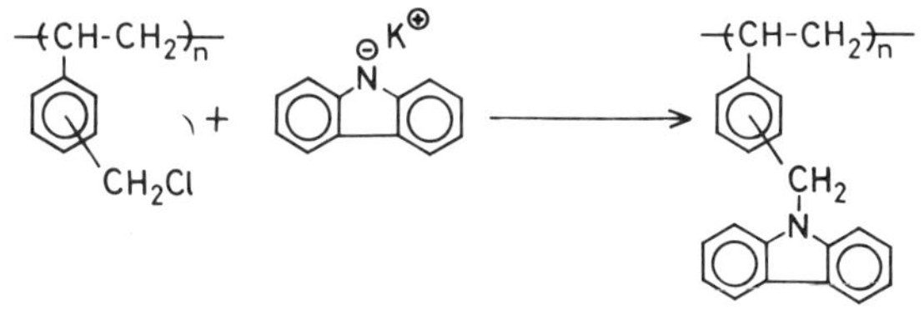

The potassium salt of carbazole is prepared by azeotropic removal of water from an equimolar solution of potassium hydroxide and carbazole in xylene. The solvent xylene is removed and replaced by dimethylformamide and a solution of poly(vinylbenzyl chloride) in dimethylformamide is added. After stirring for 2.5 h, the reaction solution is added dropwise to methanol to yield 97% of the desired poly[*N*-carbazolyl)methylstyrene]. Three reprecipitations from tetrahydrofuran with ethanol give a nearly colorless solid, T_g = 152°C.

Synthesis of poly(N-vinylcarbazole-co-vinyl-3, 5-dinitrobenzoate) [198]

Poly(*N*-vinylcarbazole-co-vinyl-3,5-dinitrobenzoate) has been synthesized by copolymerizing *N*-vinylcarbazole with vinyl acetate followed by hydrolysis and treatment of the resulting copolymer with 3,5-dinitrobenzoyl chloride. The reaction scheme is shown below.

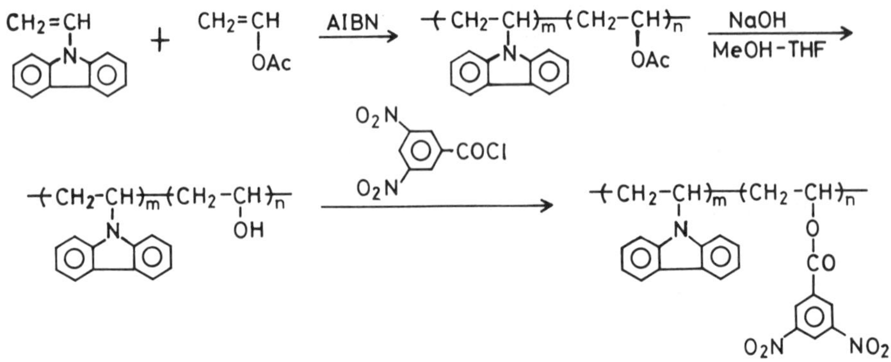

N-Vinylcarbazole (19.3 g, 0.1 mmol) and vinyl acetate (43.0 g, 0.5 mmol) are copolymerized with AIBN (150 mg) initiation in benzene at 60 to 70°C for 20 h. The resulting viscous solution is poured into 600 mL of methanol. The isolated polymer is purified by two sequential precipitations from chloroform with methanol. The copolymer yield is 23% and contains 47.9 mol % *N*-vinylcarbazole.

Poly(*N*-vinylcarbazole-co-vinyl acetate) (5.0 g) is dissolved in 100 mL of tetrahydrofuran, and 25 mL of 1 N methanolic sodium hydroxide is added. The solution is stirred for 3 h at 60°C, precipitated into 600 mL of methanol, and then purified by two additional precipitations from tetrahydrofuran into methanol. Poly(*N*-vinylcarbazole-co-vinyl alcohol) (2.0 g) is dissolved in 50 mL of anhydrous pyridine. 3,5-Dinitrobenzoyl chloride (3.88 g) in 15 mL of anhydrous benzene is added with stirring to the pyridine solution. The resulting orange solution is stirred at room temperature for 4 h and filtered, and then the filtrate is poured into 400 mL of methanol. The isolated polymer is purified by two precipitations from tetrahydrofuran into methanol to give 3.2 g of a yellow powder.

Synthesis of poly(N-vinylcarbazole-3, 6-diphthalimide) [197]

Poly(*N*-vinylcarbazole-3,6-diphthalimide) has been synthesized by the nitration of PVC_Z and then reduction of the resulting poly(3,6-dinitro-*N*-vinylcarbazole) with tin in the presence of concentrated hydrochloric acid, followed by the Schotten-Baumann reaction.

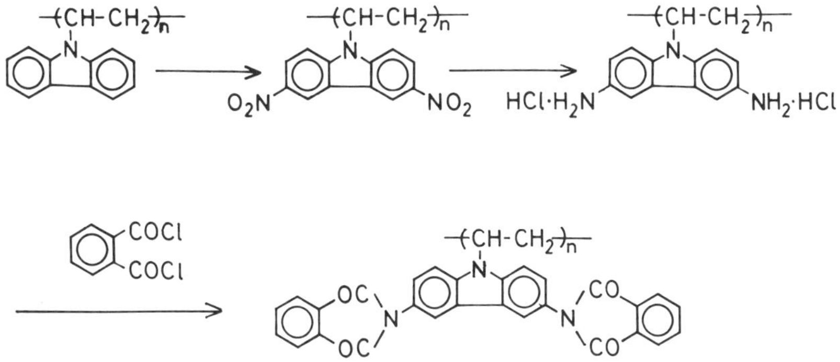

Poly(3, 6-dinitro-N-vinylcarbazole). Poly(*N*-vinylcarbazole) (5.0 g) is dissolved in 100 mL of 1,2-dichloroethane and then 25 mL of glacial acetic acid are added. A nitrating reagent (25 mL of concentrated nitric acid and 25mL of glacial acetic acid) are added dropwise to the solution at room temperature with continuous stirring. A pale yellow compound settles to the bottom of the flask. The total contents are heated at 60°C in a water bath for 2 h with frequent shaking and then left overnight at room temperature. The dark yellow liquid that separates out is poured into 150 mL of cold ethanol, filtered, washed several times with ethanol, and dried in vacuum at 55°C to yield 95% of product.

Poly(N-vinylcarbazole-3, 6-diamine hydrochloride). Poly(3,6-dinitro-*N*-vinylcarbazole) (1.0 g) is mixed with tin (2.0 g) in 25 mL of glacial acetic acid. The reaction mixture is refluxed at 120°C for 2 h and then further refluxed at 120°C for 3 h with the slow addition of 12 mL of concentrated hydrochloric acid. The brown mass that separates is filtered and further refluxed with concentrated hydrochloric acid for 2 h, filtered, washed several times with methanol, and dried in vacuum at 55°C.

Poly(N-vinylcarbazole-3, 6-diphthalimide). Poly(*N*-vinylcarbazole-3,6-diamine hydrochloride) (0.23 g) is mixed with phthaloyl chloride (1.42 g) in a closed vessel and 14 mL of 10% aqueous sodium hydroxide are added to the reaction mixture. The total contents are shaken vigorously for 30 min (exothermic reaction) and finally diluted with water. The dark brown compound is filtered, washed several times with water, then washed with methanol, and dried in vacuum at 100°C to yield, 38% of product.

6.4.2.b Monomer syntheses followed by polymerizations

Synthesis of vinyl- and isopropenylanthracenes: general methods [185,186]

1- and 2-Vinylanthracenes have been synthesized by acetylation in the 1- and 2-positions of anthracene, followed by the reduction of 1- and 2-aryl methyl ketones to the secondary alcohols, conversion to the corresponding chlorides, and dehydrochlorination in a basic medium [185]. Usually, chloroform, methylene chloride, or nitrobenzene are used as solvents to obtain the 1- and 2-aryl methyl ketones. 9-Vinylanthracene has also been prepared by the Wittig reaction of 9-anthraldehyde [185b]. Alternatively, 1-, 2-, and 9-vinylanthracenes have been prepared by dehydration of the corresponding secondary alcohols [186]. 1-, 2-, and 9-Isopropenylanthracenes have been synthesized from the corresponding acetylanthracenes by Wittig reactions [185,186].

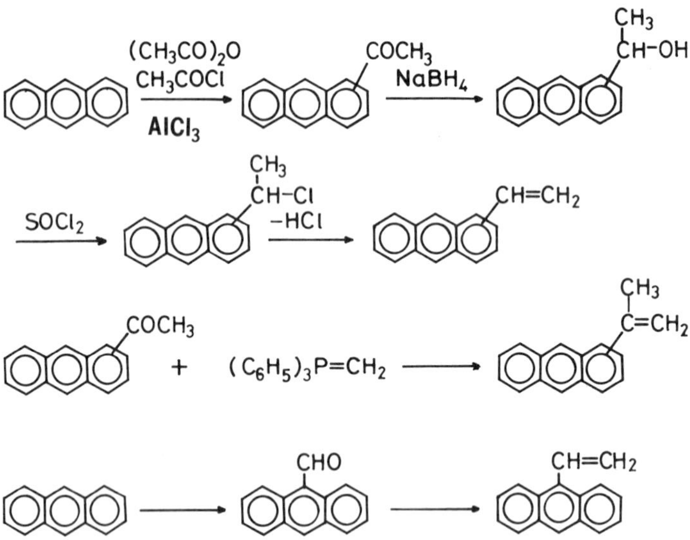

Synthesis of 2-vinylanthracene [185a]

2-Acetylanthracene. In a 3-L flask 150 g (0.84 mol) of anthracene is dispersed in 150 mL of nitrobenzene. The dispersion is cooled to 15°C. In a separate vessel, 255 g (1.9 mol) of $AlCl_3$ is dissolved in

480 mL of nitrobenzene. To this solution (cooled to room temperature) is slowly added 155 mL (1.6 mol) of acetic anhydride. The complex $AlCl_3$-acetic anhydride is cooled and slowly (over at least 1-h period) added to the vigorously stirred anthracene suspension. The reaction mixture is maintained at 15°C for an additional 16 h. The red complex is precipitated by the addition of 1500 mL of cold benzene. The solid is filtered and washed with about 200 mL of cold benzene and then with hexane. It is essential at this step to remove as much nitrobenzene from the complex as possible. The complex is then hydrolyzed in ice-HCl (2 L of broken ice per 200 mL of concentrated HCl) and the greenish product is filtered, washed with water, dried, dissolved in benzene with activated charcoal and is crystallized from benzene-hexane to give 98 g (50%) of light yellow-green crystalline product 2-acetylanthracene, mp 188°C.

1-(2-Anthryl)ethanol. Sodium borohydride ($NaBH_4$) (25 g, 0.53 mol) in 280 mL of H_2O is slowly added to a refluxing suspension of 53 g (0.24 mol) of 2-acetylanthracene in 1800 mL of ethanol. The yellow solid dissolves upon addition of about two-thirds of the borohydride and the solution turns gold-brown. After 2 h of refluxing about 1 L of the solvent is removed by distillation and the product is precipitated with water, washed with dilute HCl and water, dried, and recrystallized (following charcoal treatment) from benzene to give 40 g (75%) of white crystalline 1-(2-anthryl)ethanol.

1-Chloro-1-(2-anthryl)ethane. Thionylchloride ($SOCl_2$) (20 mL) is slowly added to the solution of 39 g of 1-(2-anthryl)ethanol in 500 mL of benzene. The solution is refluxed for 2 h and then about 250 mL of benzene is removed by diltillation. The residual slurry is cooled and poured into an excess of petroleum ether (about 1.2 L). The precipitated solid is washed, dried, and recrystallized from benzene-hexane to give 35 g (83%) of light green powder of 1-chloro-1-(2-anthryl)ethane, mp 170 to 180°C dec.

2-Vinylanthracene. 1-Chloro-1-(2-anthryl)ethane (33 g) is dehydrochlorinated in 300 mL of dimethylformamide containing 20 g of Li_2CO_3 at 130°C. After 4 h the mixture is cooled and 2-vinylanthracene is isolated by precipitation with water. Two recrystallizations from benzene-methanol yield 20 g (60%) of light yellow powder, mp 205 to 210°C. Further purification by column chromatography from alumina using nitrogen-purged solvents and eluents yields white crystalline product, mp 211°C.

Synthesis of 9-vinylanthracene [185b]

To a solution of 180 g of triphenylmethylphosphonium bromide (0.51 mol) in 800 mL of tetrahydrofuran is added under nitrogen 190 mL of a 20% solution of *n*-butyl lithium in hexane. One hour later a warm solution of 103 g of 9-anthraldehyde in 200 mL of tetrahydrofuran is added dropwise and the mixture is refluxed for 1 h. One liter of hexane is added, the precipitate is filtered, and after removal of the solvent the product is purified by column chromatography on basic alumina to yield 50 g (45%) of 9-vinylanthracene, mp 65°C.

Synthesis of poly(1-, 2-, and 9-vinylanthracenes) [185,187]

1- and 2-Vinylanthracenes are polymerized ($M_n > 10^5$) with anionic-addition-type initiators such as living polystyrene and α-methylstyrene tetramer dianion, $\alpha\text{-MeSt}_4^{2-}$. The polymerization reactions are effective at temperatures below -40 to -50°C and optimum results are obtained at -78°C. With an increase in temperature, both the yield and the molecular weight of the polymers tend to decrease. Attempts to initiate polymerization with electron-transfer-type initiators such as sodium naphthalene and sodium biphenyl were unsuccessful or resulted in poor yields and oligomeric products.

Synthesis of 1-(2-anthryl)ethyl methacrylate and poly[1-(2-anthryl)ethyl methacrylate] [183]

1-(2-Anthryl)ethyl methacrylate has been synthesized by condensation of 1-(2-anthryl)ethanol with methacryloyl chloride.

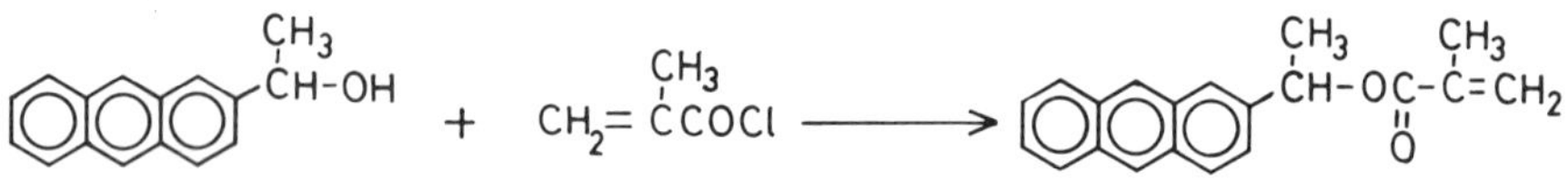

In a typical synthesis, 11.0 g (4.96×10^{-2} mol) of the alcohol is dissolved in 75 mL of dioxane at 70°C in the presence of 7.5 mL of triethylamine. To this solution, 5.5 mL (5.6×10^{-2} mol) of freshly distilled methacryloyl chloride is added and the mixture is heated for 24 h. The product is precipitated in water, filtered, washed, dried, and recrystallized from benzene-alcohol to give 9.3 g (3.2×10^{-2} mol, 65%) of white 1-(2-anthryl)ethyl methacrylate, mp 161°C.

The monomer was polymerized by free-radical methods to give a soluble linear amorphous polymer with molecular weight up to 280,000 MWD = 2.7). Solvents such as benzene, xylene, dimethylformamide, and dioxane, and initiators such as benzoyl peroxide, azobis(isobutyronitrile), and di-*tert*-butyl peroxide, were used for the polymerization.

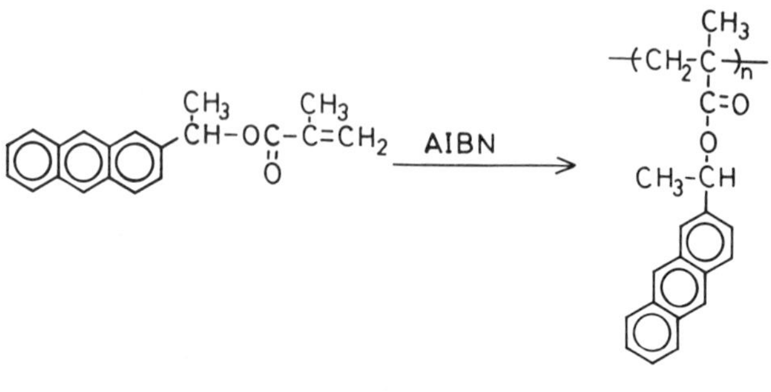

In a typical experiment, 3.0 g of 1-(2-anthryl)ethyl methacrylate is polymerized in 30 mL of benzene in a nitrogen atmosphere using 0.02 g of azobis(isobutyronitrile) at 67°C for 24 h. The system is protected against light. The yield, after reprecipitation from benzene in acetone, is 2.8 g (93%). The polymer is soluble in benzene, xylene, methylene chloride, chloroform, tetrahydrofuran, dimethylformamide, 1,1,2,2-tetrachloroethane, and *o*-dichlorobenzene. No insoluble polymer is formed.

Synthesis of 9-(p-vinylphenyl)anthracene and poly[9-(p-vinylphenyl)anthracene] [184]

9-(*p*-Vinylphenyl)anthracene has been synthesized by the dehydration of 9-(*p*-vinylphenyl)-9-hydroxy-9,10-dihydroanthracene which is obtained by the Grignard reaction of anthrone and *p*-vinylphenylmagnesium halides.

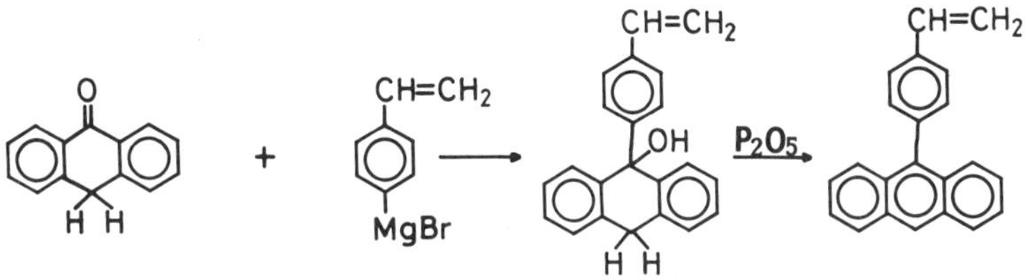

The overall yield of 9-(*p*-vinylphenyl)anthracene is about 20% based on anthrone when 9-(*p*-vinylphenyl)-9-hydroxy-9,10-dihydroanthracene is separated, purified, and then treated with phosphorus pentoxide. When the reaction mixture from the reaction of anthrone with *p*-vinylphenylmagnesium halide is treated directly with phosphorus pentoxide, the yield of 9-(*p*-vinylphenyl)anthracene increases to about 40%.

A high-molecular-weight poly[9-(*p*-vinylphenyl)anthracene] was obtained by free-radical polymerization using di-*tert*-butyl peroxide as an initiator in xylene at 130°C. The molecular weight of the polymer depends on the purity of the monomer, and the polymer yield is affected by the solvent. The polymer (MW = 136,500) has good thermal properties: weight loss at 300°C is 2.2%. The polymer has a glass transition at 270°C and decomposes at 350°C.

Synthesis of 2-vinyl-N-ethylcarbazole [194] *and poly(2-vinyl-N-ethylcarbazole)* [193]

2-Acetylcarbazole (24 g) and aluminum isopropoxide (40 g) are refluxed in xylene (75 mL) for 3 h, during which time the acetone formed is distilled from the reaction mixture. After cooling, the reaction mixture is filtered free of aluminum isopropoxide and poured into water. The xylene layer is separated and the aqueous layer is extracted with ether. The combined extracts are washed with water and dried. Evaporation of the solvent and crystallization from toluene affords 2-vinyl-*N*-ethylcarbazole as pale yellow crystals, 53%, mp 253 to 254°C.

Poly(2-vinyl-*N*-ethylcarbazole) is prepared by conventional free-radical methods. The polymer is precipitated by methanol: M_n = 19.400. In contrast to *N*-vinylcarbazole, 2-vinyl-*N*-ethylcarbazole shows remarkably little reactivity in free-radical homopolymerization.

Synthesis and polymerization of trans-1-(3-vinyl-9-carbazole)-2-(9-carbazolyl) cyclobutane [195]

Methyltriphenylphosphonium bromide (33 g, 0.093 mol) in 100 mL of warm dimethylsulfoxide is added to a solution of methylsulfinyl carbanion which is prepared by the reaction of sodium hydride (0.127 mol as a 55% dispersion in mineral oil) with dimethyl sulfoxide (100 mL). To the resulting reddish-yellow solution which contains the ylide is added *trans*-1-(3-formyl-9-carbazole)cyclobutane (35 g, 0.0845 mol) at room temperature. The mixture is stirred for 1.5 h at room temperature and poured into 1 L of ice water. The precipitated crude product is purified twice on an alumina column with benzene as an eluent to give white fine crystals of *trans*-1-(3-vinyl-9-carbazole)-2-(9-carbazolyl)cyclobutane, mp 52 to 53°C. Yield, 22 g (63%).

Trans-1-(3-vinyl-9-carbazole)-2-(9-carbazolyl)cyclobutane (2.0 g, 4.85 mmol) is polymerized with boron trifluoride etherate (2 μL) in freshly distilled methylene chloride (8 mL) at 72 to 74°C in a dry nitrogen atmosphere. After 6 min the cooling bath is removed and polymerization is continued until stirring by a magnetic stirrer is no longer possible because of the increase in viscosity. Polymerization is then quenched by adding 2 mL of ethanol. The polymer is dissolved in chloroform and precipitated in methanol and then redissolved and precipitated from *n*-hexane to yield 1.73 g (86.5%).

Synthesis of N-carboxyanhydride of γ-(β-N-carbazolylethyl)-L-glutamate (γ-CLG) and poly[γ-(β-N-carbazolylethyl)-L-glutamate] [191]

γ-(β-*N*-Carbazolylethyl)-L-glutamate (γ-CLG) is prepared by the reaction of L-glutamic acid with *N*-(β-hydroxyethyl)carbazole in dioxane in the presence of *p*-toluenesulfonic acid. The *N*-carboxyanhydride of γ-CLG (γ-CLG-NCA) is prepared by reacting γ-CLG suspended in dioxane with phosgene gas until the reaction solution becomes clear.

The γ-CLG-NCA monomer is polymerized by triethylamine in 1,2-dichloroethane at room temperature. The polymer is purified by reprecipitation from 1,2-dichloroethane into methanol.

Synthesis of 1, 3-diphenyl-5-(p-vinylphenyl)-2-pyrazoline [211]

1,3-Diphenyl-5-(*p*-vinylphenyl)-2-pyrazoline has been synthesized by the 1,3-dipolar cycloaddition reaction of *p*-divinylbenzene with *N*-anilinobenzimidoyl chloride in the presence of triethylamine in benzene in 22 to 65% yields.

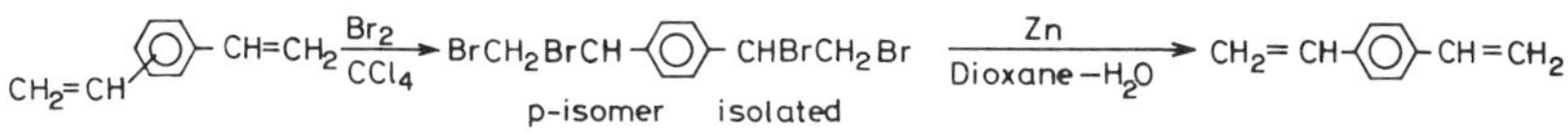

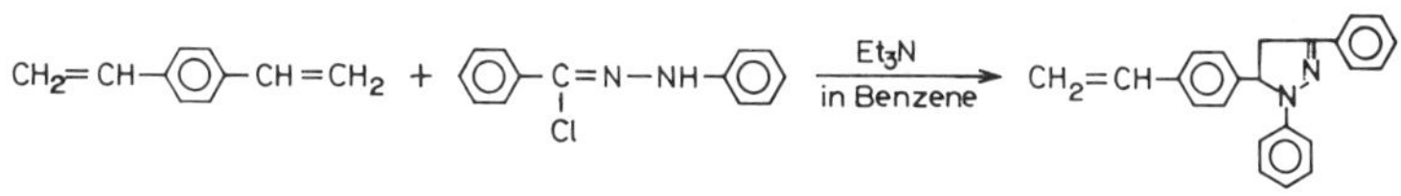

VTPP

p-Divinylbenzene is prepared from commercially available divinylbenzene. Commercially available divinylbenzene, which is a mixture of meta and para isomers, is brominated, and *p*-bis(1,2-dibromoethyl)-benzene is isolated, recrystallized from benzene, and then debrominated with zinc dust in dioxane-water to give *p*-divinylbenzene, which is purified by distillation in vacuo, bp 41°C/0.6 mmHg, mp 28 to 29°C. *N*-anilinobenzimidoyl chloride is prepared by the reaction of phosphorus pentachloride with β-benzoylphenylhydrazine which is prepared from benzoic anhydride on the treatment of phenylhydrazine, mp 129 to 130°C.

p-Divinylbenzene (6.5 g) and *N*-anilinobenzimidoyl chloride (15 g) are dissolved in 300 mL of benzene, and triethylamine (6.6 g) is added dropwise over a period of 1 h to the solution at room temperature. After the reaction solution is allowed to stand for a day, the resulting triethylamine hydrochloride is removed by filtration. The filtrate is evaporated to dryness under reduced pressure. The crude product obtained is chromatographed over neutral alumina with benzene as an eluent, recrystallized three times from methanol and dried in vacuo to give 3.5 g of 1,3-diphenyl-5-(*p*-vinylphenyl)-2-pyrazoline, mp 132 to 133°C.

Synthesis of 1, 3-diphenyl-5-(p-methacryloyloxyphenyl)-2-pyrazoline [211]

1,3-Diphenyl-5-(*p*-methacryloyloxyphenyl)-2-pyrazoline has been synthesized by the reaction of 1,3-diphenyl-5-(*p*-hydroxyphenyl)-2-pyrazoline in the presence of triethylamine in tetrahydrofuran in 55 to 65% yields.

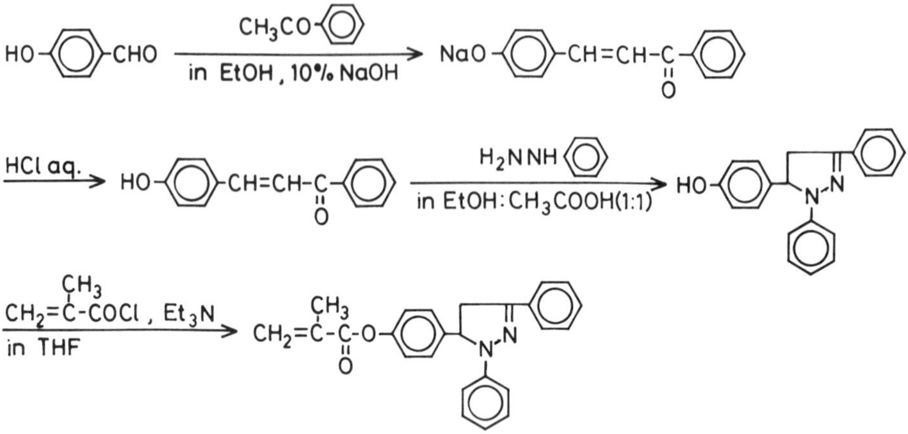

MTPP

1,3-Diphenyl-5-(*p*-hydroxyphenyl)-2-pyrazoline is prepared in ethanol by the acid-catalyzed reaction of phenylhydrazine with *p*-hydroxybenzalacetophenone which is prepared by the alkali-catalyzed reaction of *p*-hydroxybenzaldehyde with acetophenone in ethanol, recrystallized from ethanol-hexane (1:6), mp 143 to 144°C. The reaction of 1,3-diphenyl-5-(*p*-hydroxyphenyl)-2-pyrazoline (37 g) with methacryloyl chloride (25 g) is carried out in anhydrous tetrahydrofuran (300 mL) at room temperature in the presence of triethylamine (30 mL). After removal of the resulting triethylamine hydrochloride, the filtrate is evaporated to dryness under reduced pressure. The crude product obtained is purified by silica gel column chromatography with benzene as an eluent, followed by recrystallization from hexane, and dried in vacuo to give 25 g of 1,3-diphenyl-5-(*p*-methacryloyloxyphenyl)-2-pyrazoline, mp 115 to 116°C.

Synthesis of poly[1,3-diphenyl-5-(p-vinylphenyl)-2-pyrazoline] (PVTPP) and poly[1,3-diphenyl-5-(p-methacryloyloxyphenyl)-2-pyrazoline] (PMTPP) [211]

These monomers have been polymerized free radically using AIBN as an initiator to give the corresponding vinyl polymers. PVTPP and PMTPP are very soluble in tetrahydrofuran, dimethylformamide, soluble in benzene, dioxane, and 1,2-dichloroethane, partly soluble in acetone, but insoluble in hexane, diethyl ether, and alcohol. The number-average molecular weights of PVTPP and PMTPP are in the range 9000 to 19,000 and 10,000 to 25,000, respectively, depending on polymerization conditions. They are thermally stable below 200°C.

Synthesis of 2, 4, 7-trinitro-9-fluorenyl methacrylate and poly[2, 4, 7-trinitro-9-fluorenyl methacrylate] [215]

2,4,7-Trinitro-9-fluorenyl methacrylate has been prepared by the boron trifluoride etherate-catalyzed reaction of 9-diazo-2,4,7-trinitrofluorenone with methacrylic acid in nitromethane.

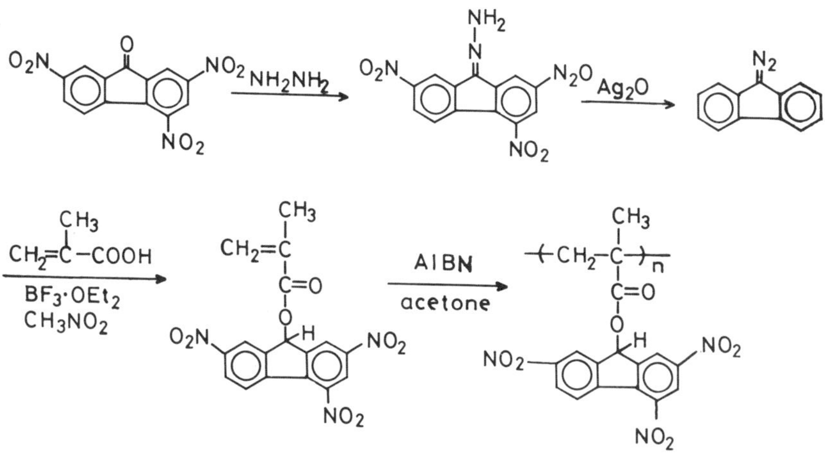

9-Diazo-2,4,7-trinitrofluorenone (100 g, 0.03 mL), which is prepared by the oxidation of 2,4,7-trinitrofluorenone hydrazone with silver oxide, is suspended in 150 mL of nitromethane. To this slurry 20 mL (0.22 mol) of freshly distilled methacrylic acid is added. Next, 4 mL of boron trifluoride etherate is added slowly via a syringe. Gas evolution is noted during addition. After stirring at room temperature for 1 h, a clear red solution results. The nitromethane is then removed on a rotary evaporator. When the reaction material reduces to an oil, the temperature is raised to 50°C and the evaporation is continued until an orange-yellow solid remains (approximately 1 to 2 h). Water (150 mL) is added to the flask and the contents are stirred at room temperature. After about 1 h the suspension is filtered and then dissolved in 500 mL of benzene. This is heated to reflux and filtered hot to remove residual diazo compound. The benzene solution is evaporated on rotary evaporator and the residual oil is allowed to solidify overnight. It is filtered, utilizing a minimum of ether, to effect the transfer. After drying at room temperature under vacuum for 24 h, 6.1 g (61%) of light orange powder (mp 148 to 150°C) is obtained.

The monomer is polymerized readily by standard free-radical techniques to relatively high molecular weight polymers. Acetone and tetrahydrofuran are the best solvents for the polymerization. No polymerization occurs in benzene. The resulting polymer is readily soluble in tetrahydrofuran, dimethylformamide, and dimethylsulfoxide. Brittle films are readily cast from tetrahydrofuran solution.

Synthesis of glycidyl 4, 5, 7-trinitro-9-fluorenone-2-carboxylate and 2'-ethylmethacryl 4, 5, 7-trinitrofluorenone-2-carboxylate [218]

These monomers have been synthesized by coupling 4,5,7-trinitro-fluorenone-2-carboxoyl chloride with glycidol and 2-hydroxyethyl methacrylate, respectively. 4,5,7-Trinitrofluorenone-2-carboxoyl chloride (10.0 g, 0.026 mol) is dissolved in 50 mL of tetrahydrofuran, and triethylamine (2.62 g, 0.026 mol) is added slowly. Glycidol (2.74 g, 0.037 mol), dissolved in 10 mL of tetrahydrofuran, is added drop-wise. The mixture is allowed to stir for 30 to 45 min and then filtered. The amine salt residue is washed with 50 mL of tetrahydrofuran. The combined filtrates are flash evaporated and the black oily residue is taken up in 100 mL of chloroform. Any remaining salt is removed by extracting three times with 100 mL of water. The water layer is back-extracted with 100 mL of chloroform. The chloroform portions are combined and dried over magnesium sulfate. The chloroform is then flash evaporated until about 50 mL remains. At this point this solution is slowly added to about 800 mL of hexane and a light yellow powder results. The sample is redissolved and reprecipitated. After the precipitate is dried under vacuum, 7.9 g, a 63% yield, of glycidyl 4,5,7-trinitro-9-fluorenone-2-carboxylate with a melting range of 65 to 75°C is obtained.

2'-Ethyl methacryloyl-4,5,7-trinitro-9-fluorenone-2-carboxylate is prepared in a similar manner, using 2-hydroxyethyl methacrylate in place of glycidol. Methylene chloride is used instead of chloroform in the workup. The product is a light yellow powder with a melting range of 60 to 70°C; 81.5% yield.

Glycidyl 4,5,7-trinitro-9-fluorenone-2-carboxylate and 2'-(ethyl methacryl)-4,5,7-trinitro-9-fluorenone-2-carboxylate have been polymerized by cationic and free-radical techniques, respectively, by using boron trifluoride etherate and AIBN as initiators.

The epoxide (1.0 g) and methylene chloride (20 mL) are placed in a 100-mL three-necked round-bottomed flask equipped with a magnetic stirrer. A deep-brownish solution forms. Both the reaction vessel and contents and the initiator solution are brought to 0°C. A positive nitrogen pressure is kept on the reaction. A solution of boron trifluoride etherate in methylene chloride (0.2 mL, 1 wt %) is introduced via a syringe through a rubber septum; cloudiness appears immediately. During the course of the 2-h polymerization time, solid deposits are formed on the sides of the flask. The polymerization is quenched by pouring into methanol. Yield, 46%.

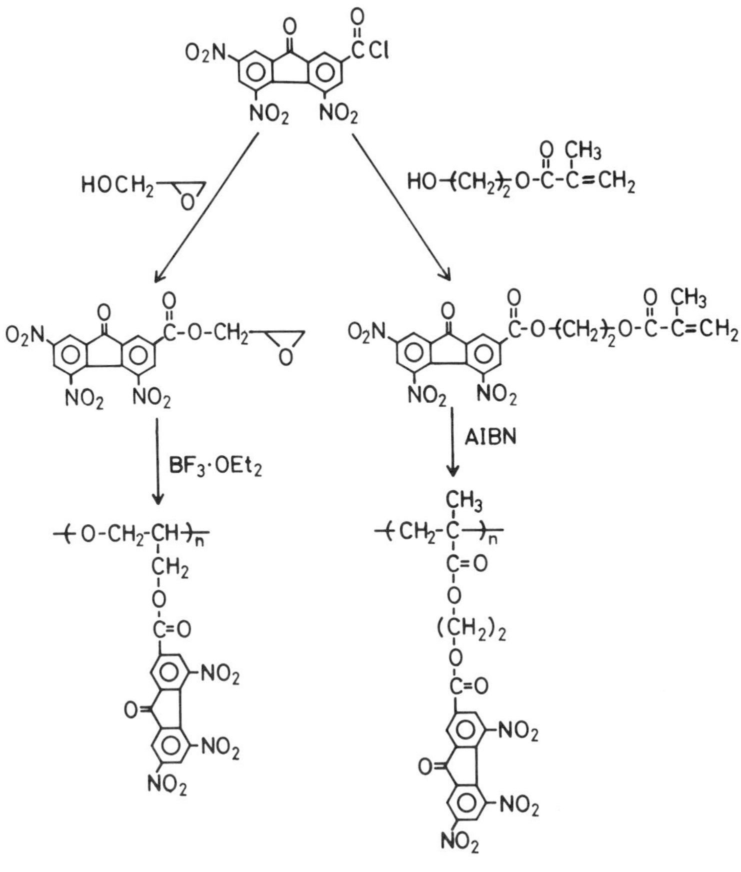

6.4.3 Applications of Organic Photoconductors

Photoconductive materials have been practically used as photoreceptors in electrophotography. Electrophotographic systems have been developed significantly in the area of document copying because of their facile operation, short access time, and low cost. Recently, electrophotographic techniques have been applied in the production of printing plates and in laser printers for use in computer terminal output and intelligent copiers and printers in which the functions of copying, facsimile, and printing are combined. It is expected that electrophotography will replace silver-halide photography in some areas, such as x-ray diagnosis. These new applications of electrophotographic techniques, together with need for higher-performance copiers than the existing ones, require new photoconductive materials that fit each system. Furthermore, photoconductive materials are potentially useful for solar cell applications such as solar energy conversion into electrical energy.

6.4.3.a Electrophotographic systems

Electrophotographic techniques are the image-recording techniques which employ a combination of electrostatic and photoconductive phenomena of materials. Electrophotographic systems are based on Carlson's method, which is termed xerography. As shown in Figure 10, the xerographic system involves a sequence of processes: corona charging, image exposure, toner development, transfer to paper, fixing of toner images, and cleaning of a photoreceptor for repeated cycles. The characteristics of charging and photodischarge are shown in Figure 11.

As is evident from Figure 11, the necessary properties for photoreceptors are as follows: (a) acceptance of electrostatic charge and its retentivity, (b) rapid photodischarge, (c) negligible residual charge at the exposed area, and (d) spectral sensitivity to visible light. Therefore, the materials used as photoreceptors are photoconductive insulators. For the photoreceptor to show rapid photodischarge, materials should have high quantum efficiencies for photogeneration, high drift mobilities, long charge carrier lifetimes, and high absorption coefficients in the visible wavelength region. In addition to these functional properties, materials should meet such requirements as processibility (film-forming), flexibility, durability, and nontoxicity.

6.4.3.b Photoreceptors in practical use

Inorganic photoconductors such as amorphous selenium, zinc oxide, and cadmium sulfide have found practical applications from the early days of research and development. Among these, amorphous selenium,

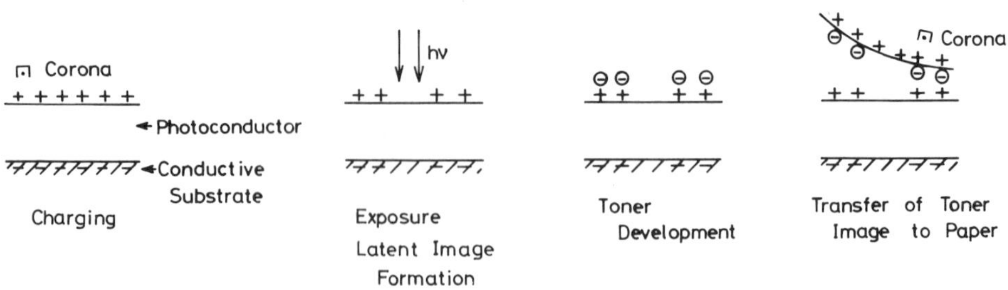

Figure 10. Xerographic processes [153].

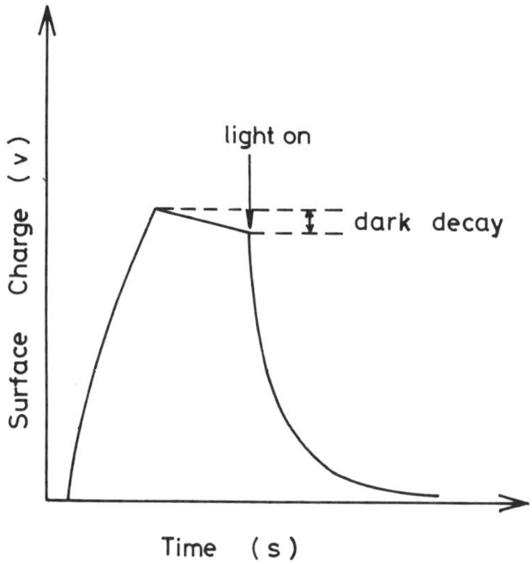

Figure 11. Characteristics of corona charging and photodischarge in xerographic processes.

sensitized by the doping with tellurium or arsenic, has been used extensively. Organic photoconductors also have been put into practical use with good results. Organic photoconductors have potential advantages over inorganic photoconductors because it is easier to modify their molecular structures, they have excellent processibility and spectral sensitization capability, and it is easy to dispose of used photoreceptors.

Among organic photoconductors, poly(*N*-vinylcarbazole) (PVCz) was the first to be put into practical use. Since PVCz has no inherent spectral response to visible light, sensitization is necessary. Generally, spectral sensitization can be achieved by using dye molecules or by forming charge-transfer complexes. The photoreceptor of the PVCz-2,4,7-trinitrofluorenone (TNF) system developed at IBM consists of one molecule of TNF to each monomer unit of PVCz. This ground-state charge-transfer complex absorbs light throughout the visible wavelength region. The photoreceptor is characterized by a high sensitivity approximately equal to that of amorphous selenium for negative corona charging, the absence of residual charge in light decay, and negligible fatigue effects [221]. It is the TNF molecules that conduct electrons [216,221].

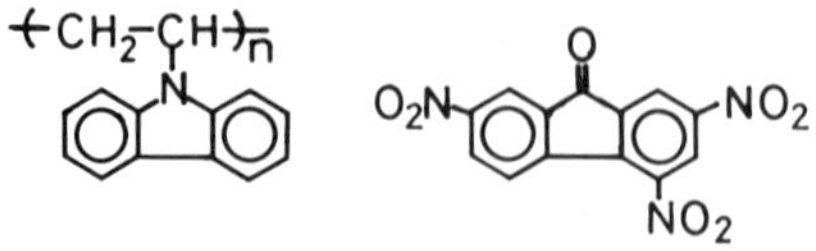

PVCz TNF

The photoreceptor developed at Eastman Kodak, which is called an aggregate photoconductive film, has been reported to show a high sensitivity, a broad spectral response in the 500- to 730-nm wavelength region, and excellent electrical stability characteristics when rapidly recycled. It consists of an aryl-substituted thiapyrylium salt (dye) and a dialkylamino-substituted triphenylmethane dispersed in bisphenol A polycarbonate. A crystalline complex of the polymer and the dye form an aggregate as a separate phase. This causes an expansion of the spectral response of the photoreceptor to a longer-wavelength region. The photoreceptor shows both electron and hole conduction [222].

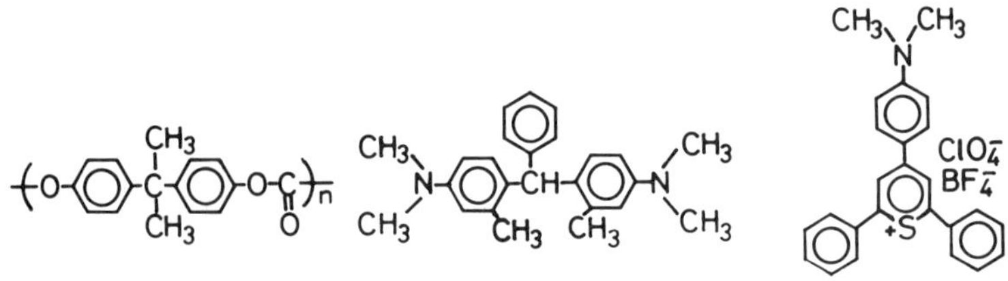

Recently, photoreceptors of layered photoconductor systems have been developed which consists of a charge-carrier generation layer (CGL) and a transport layer (CTL). Usually, CTL is coated over CGL, which is then coated onto a conductive base as shown in Figure 12. Light is absorbed by the CGL through the transparent CTL, and

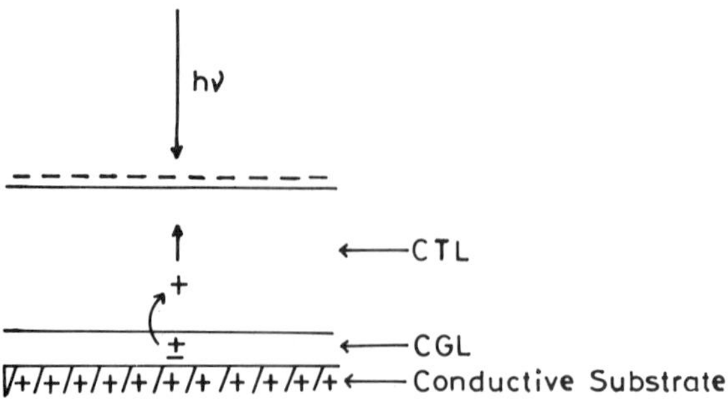

Figure 12. Layered photoreceptor.

Table 4. Materials for a Layered Photoreceptor

	Material for CTL	Material for CGL
(1)	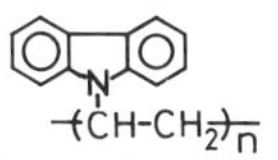	a-Se
(2)	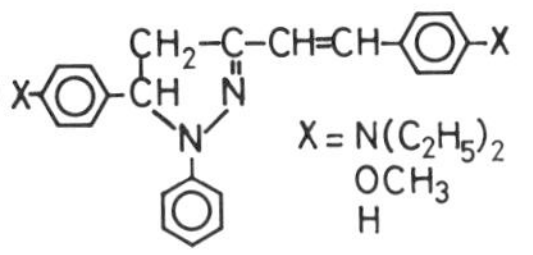	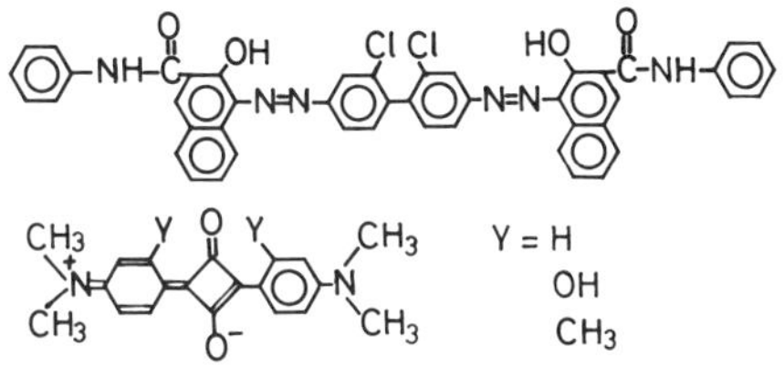
(3)	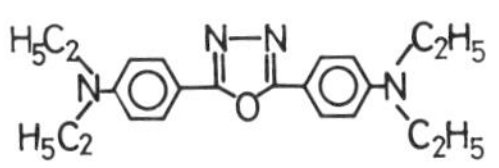	
(4)	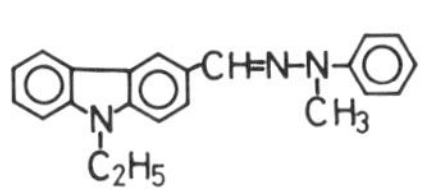	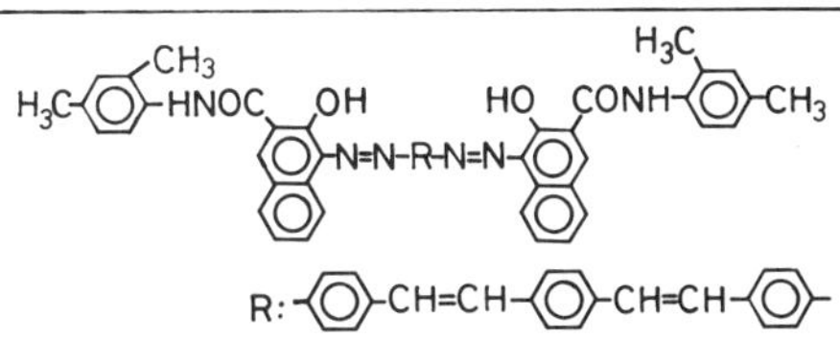

one of the charge carriers generated in the CGL is injected into the CTL and then transported across the CTL to neutralize a surface charge. An important requirement for attaining high sensitivity is a high efficiency of charge carrier injection from CGL to CTL. Several systems have been in practical use (Table 4) [210, 223-225].

In conventional electrophotography, the phenomenon that charge acceptance at the exposed area decreases, which is called the photomemory effect, is undesirable and should be avoided. By contrast, a new type of photoreceptor with a charge acceptance memory can be developed by positively utilizing such a photomemory memory effect. After a single image exposure, the photoreceptor, which functions as an electrostatic duplication master, is repeatedly processed by corona charging, toner development, transfer of toner image to paper, and fixing. This results in a number of copies being obtained by a single exposure. It has recently been suggested that the systems PVCz-TNF-leuco malachite green and PVCz-diazonium salt function as the electrostatic duplication master [226, 227].

REFERENCES

1. A. Pochettino, Atti Acad. Naz. Lincei Cl. Sci. Fis. Mat. Nat. Rend., 15, 355 (1906).
2. D. D. Eley, Nature, 162, 819 (1948).
3. H. Akamatsu and H. Inokuchi, J. Chem. Phys., 18, 810 (1950).
4. H. Akamatsu, H. Inokuchi, and Y. Matsunaga, Nature, 173, 168 (1954).
5. H. Inokuchi, Bull. Chem. Soc. Jpn., 27, 22 (1954).
6. a. R. H. Baughman, J. L. Bredas, R. R. Chance, R. L. Elsenbaumer, and L. W. Shacklette, Chem. Rev., 82, 209 (1982).
   b. A. J. Berlinsky, Contemp. Phys., 17, 331 (1976).
   c. J. Devreese, One-Dimensional Conductors, Plenum Publishing Corporation, New York (1978).
   d. J. T. Devreese, R. P. Evrard, and V. E. Doren, Eds., Highly Conductive One-Dimensional Solids, Plenum Publishing Corporation, New York (1979).
   e. A. A. Diaz, J. M. V. Vallejo, and A. M. Duran, IBM J. Res. Dev., 25, 42 (1981).
   f. E. M. Engler, Chem. Tech., 274 (1976).
   g. A. G. Garito and A. J. Heeger, Acc. Chem. Res., 7, 232 (1974).
   h. E. P. Goodings, Endeavor, 34, 123 (1975); Chem. Soc. Rev., 5, 95 (1976).
   i. W. E. Hatfield, Ed., Molecular Metals, NATO Conference Series VI, Plenum Publishing Corporation, New York (1979).
   j. H. J. Keller, Ed., Low-Dimensional Cooperative Phenomena, NATO Advanced Study Institutes Series, B7, Plenum Publishing Corporation, New York (1974).
   k. H. J. Keller, Ed., Chemistry and Physics of One-Dimensional Metals, NATO Advanced Study Institutes Series, B25, Plenum Publishing Corporation, New York (1976).
   l. M. L. Khidekel and E. I. Zhilyaeva, Synth. Met., 4, 1 (1981).
   m. M. M. Labes, P. Love, and L. F. Nichols, Chem. Rev., 79, 1 (1979).
   n. A. G. MacDiarmid and A. J. Heeger, Synth. Met., 1, 101 (1979-1980).
   o. T. J. Marks, Ann. N.Y. Acad. Sci., 312, 595 (1978).
   p. J. S. Miller and A. J. Epstein, Prog. Inorg. Chem., 20, 1 (1976).
   q. J. S. Miller and A. J. Epstein, Eds., Synthesis and Properties of Low-Dimensional Materials, Ann. N.Y. Acad. Sci., 313 (1978).
   r. M. Narita and C. U. Pittman, Jr., Synthesis, 489 (1976).
   s. L. Pal, G. Gruner, A. Janossy, and J. Solyom, Eds., Organic Conductors and Semiconductors, Springer-Verlag, Berlin (1977).

t. J. H. Perlstein, Angew. Chem. Int. Ed. Engl., 16, 519 (1977).
u. Z. G. Soos, J. Chem. Ed., 55, 546 (1978).
v. G. B. Street and T. C. Clarke, IBM J. Res. Dev., 25, 51 (1981).
w. G. B. Street and R. L. Green, IBM J. Res. Dev., 21, 99 (1977).
x. G. D. Stucky, A. J. Shulz, and J. M. Williams, Ann. Rev. Mater. Sci., 301 (1977).
y. J. B. Torrance, Acc. Chem. Res., 12, 79 (1979).
7. D. S. Acher, R. J. Harder, W. R. Hertler, W. Mahler, L. R. Melby, R. E. Bensen, and W. E. Mochel, J. Am. Chem. Soc., 82, 6408 (1960).
8. R. G. Kepler, P. E. Bierstedt, and R. E. Merrifield, Phys. Rev. Lett., 5, 503 (1960).
9. H. Kobayashi, Bull. Chem. Soc. Jpn., 47, 1346 (1974).
10. L. R. Melby, Can. J. Chem., 43, 1448 (1965).
11. F. Wudl, G. M. Smith, and E. J. Hufnager, Chem. Commun., 1453 (1970).
12. J. P. Ferraris, D. O. Cowan, V. V. Walatka, Jr., and J. H. Perlstein, J. Am. Chem. Soc., 95, 947 (1973).
13. L. B. Coleman, M. J. Dohen, D. J. Sandman, F. G. Yamagishi, A. F. Garito, and A. J. Heeger, Solid State Commun., 12, 1125 (1973).
14. K. Bechgaard, C. S. Jacobsen, K. Mortensen, H. J. Pedersen, and N. Thorup, Solid State Commun., 33, 1119 (1980).
15. D. Jerome, A. Mazaud, M. Ribault, and K. Bechgaard, J. Phys. Paris Lett., 41, L-95 (1980).
16. K. Bechgaard, K. Carneiro, F. B. Rasmussen, M. Olsen, G. Rindorf, C. S. Jacobson, H. J. Pedersen, and J. C. Scott, J. Am. Chem. Soc., 103, 2440 (1981).
17. V. V. Walatka, Jr., M. M. Labes, and J. H. Perlstein, Phys. Rev. Lett., 31, 1139 (1973).
18. R. L. Green, G. B. Street, and L. J. Suter, Phys. Rev. Lett., 35, 577 (1975).
19. a. C. K. Chiang, C. R. Fincher, Jr., Y. W. Park, A. J. Heeger, H. Shirakawa, E. J. Louis, S. C. Gau, and A. J. MacDiarmid, Phys. Rev. Lett., 39, 1098 (1977).
b. H. Shirakawa, E. J. Louis, A. G. MacDiarmid, C. K. Chiang, and A. J. Heeger, Chem. Commun., 578 (1977).
20. a. W. A. Little, Phys. Rev., 134, 1416 (1964).
b. D. Davis, H. Gutfreund, and W. A. Little, Phys. Rev., B13, 4766 (1976).
21. O. H. LeBlanc, J. Chem. Phys., 33, 626 (1960).
22. R. G. Kepler, Phys. Rev., 119, 1226 (1960).
23. A. C. Brown, Phys. Rev., 97, 355 (1955).

24. a. J. H. Dessauer and H. E. Clark, Eds., Xerography and Related Processes, Focal Press, London (1965).
b. J. Mort and D. M. Pai, Eds., Photoconductivity and Related Phenomena, Elsevier Science Publishing Co., Inc., New York (1976).
c. A. V. Patsis and D. A. Seanor, Eds., Photoconductivity in Polymers. An Interdisciplinary Approach, Technomic Publishing Co., Inc., Westport, Conn. (1976).
d. R. C. Penwell, B. N. Ganguly, and T. W. Smith, J. Polym. Sci. Macromol. Rev., 13, 63 (1978).
e. R. M. Schaffert, Electrophotography, Focal Press, London (1974).
f. M. Stolka and D. M. Pai, Adv. Polym. Sci., 29, 1 (1978).
25. H. Akamatsu and H. Inokuchi, J. Chem. Phys., 18, 810 (1950).
26. H. Akamatsu, H. Inokuchi, and Y. Matsunaga, Nature, 173, 168 (1954).
27. Y. Matsunaga, J. Chem. Phys., 42, 2248 (1965).
28. L. R. Melby, R. J. Harder, W. R. Hartler, W. Mahler, R. E. Benson, and W. E. Mochel, J. Am. Chem. Soc., 84, 3374 (1962).
29. J. E. Lane, T. R. Fico, A. Lombardo, and J. S. Blakemore, J. Chem. Phys., 69, 3981 (1978).
30. D. O. Cowan, C. Le Vande, J. Park, and F. Kaufman, Acc. Chem. Res., 6, 1 (1973).
31. E. A. Perez-Albuerne, L. C. Isett, and R. K. Haller, Chem. Commun., 417 (1977).
32. A. N. Bloch, D. O. Cowan, K. Beckgaard, R. E. Pyle, R. H. Banks, and T. O. Poehler, Phys. Rev. Lett., 34, 1561 (1975).
33. E. M. Engler, Chemtech, 274 (1976).
34. M. Narita and C. U. Pittman, Jr., Synthesis, 489 (1976).
35. R. C. Wheland and J. L. Gibson, J. Am. Chem. Soc., 98, 3916 (1976).
36. R. C. Wheland, J. Am. Chem. Soc., 98, 3926 (1976).
37. E. M. Engler, B. A. Scott, S. Etemad, T. Penny, and V. V. Patel, J. Am. Chem. Soc., 99, 5909 (1977).
38. E. M. Engler, R. A. Craven, Y. Tomkiewicz, and B. A. Scott, Chem. Commun., 337 (1976).
39. J. P. Ferraris, D. O. Cowan, V. V. Walatka, Jr., and J. H. Perlstein, J. Am. Chem. Soc., 95, 947 (1973).
40. H. Strzelecka, W. Schoenfelder, and J. Rivory, Mol. Cryst. Liq. Cryst., 52, 307 (1979).
41. R. G. Kepler, P. E. Bierstedt, and R. E. Merrifield, Phys. Rev. Lett., 5, 503 (1960).
42. L. R. Melby, Can. J. Chem., 43, 1448 (1965).
43. K. Kameras, G. Mihaly, G. Gruner, and A. Janossy, Chem. Commun., 974 (1978).

44. P. Kathigamanathan and D. R. Rosseinsky, Chem. Commun., 365 (1982).
45. H. Tanaka, T. Nogami, and H. Mikawa, Chem. Lett., 727 (1982).
46. H. J. Keller, D. Nothe, H. Pritzkow, D. Wehe, and M. Werner, Mol. Cryst. Liq. Cryst., 62, 181 (1980).
47. J. S. Miller and A. J. Epstein, Prog. Inorg. Chem., 20, 1 (1976).
48. C. J. Sehramm, D. R. Stojakovic, B. M. Hoffman, and T. J. Marks, Science, 200, 47 (1978).
49. K. Bechgaard, C. S. Jacobson, K. Mortensen, H. J. Pedersen, and N. Thorup, Solid State Commun., 33, 1119 (1980).
50. D. Jerome, A. Mazaud, M. Ribault, and K. Bechgaard, J. Phys. Paris Lett., 41, L-95 (1980).
51. K. Bechgaard, K. Carneiro, F. B. Rasmussen, M. Olsen, G. Rindorf, C. S. Jacobson, H. J. Pedersen, and J. C. Scott, J. Am. Chem. Soc., 103, 2440 (1981).
52. K. Bechgaard, Mol. Cryst. Liq. Cryst., 79, 1 (1982).
53. H. Shirakawa, T. Itoh, and S. Ikeda, Macromol. Chem., 179, 1565 (1978).
54. G. Wegner, Makromol. Chem., 154, 35 (1972).
55. I. Moritani, T. Nagai, and Y. Shirota, J. Ind. Chem. Jpn. (Kogyo Kagaku Zasshi), 68, 296 (1965).
56. H.-H. Horhold and J. Opfermann, Makromol. Chem., 131, 105 (1970).
57. D. Wohle, Makromol. Chem., 175, 1751 (1974).
58. T. Hayashi and H. Sato, Bull. Chem. Soc. Jpn., 50, 1023 (1977).
59. a. J. H. T. Kho and H. A. Pohl, J. Polym. Sci., A1(7), 139 (1969).
    b. H. A. Pohl and E. H. Engelhardt, J. Phys. Chem., 66, 2085 (1962).
60. R. Knoesel, B. Gebus, J.-P. Roth, and J. Parrod, Bull. Soc. Chim. Fr., 294 (1969). See E. P. Goodings, Chem. Soc. Rev., 5, 95 (1976).
61. a. M. H. Litt and J. W. Summers, J. Polym. Sci., A1(11), 1359 (1973).
    b. J. W. Summers and M. H. Litt, J. Polym. Sci., A1(11), 1379 (1973).
62. W. R. Hertler, J. Org. Chem., 41, 1412 (1976).
63. J. H. Lupinski, K. D. Kopple, and J. J. Hertz, J. Polym. Sci., C16, 1561 (1967).
64. A. Rembaum, A. M. Hermann, F. E. Stewart, and A. Gutmann, J. Phys. Chem., 73, 513 (1969).
65. V. Hadek, H. Noguchi, and A. Rembaum, Macromolecules, 4, 494 (1971).
66. J. Pecherz, W. Ciesielshi, and M. Kryszewski, Macromolecules, 14, 1139 (1981).

67. M. Spiratos, A. Airinei, G. I. Rusa, and A. Ciobanu, Polym. Bull., 6, 243 (1982); Angew. Makromol. Chem., 107, 33 (1982).
68. J. T. D'Sa, V. J. Rao, K. C. Patel, and R. D. Patel, Angew. Makromol. Chem., 79, 133 (1979).
69. K. F. Schoch, Jr., B. R. Kundalkar, and T. J. Marks, J. Am. Chem. Soc., 101, 7071 (1979).
70. O. Schneider and M. Hanack, Angew. Chem., 92, 391 (1980).
71. N. M. Rivera and E. M. Engler, Chem. Commun., 184 (1979).
72. M. Hatano, S. Kambara, and S. Okamoto, J. Polym. Sci., 51, 26 (1961).
73. D. J. Beretz and D. S. Smith, Trans. Faraday Soc., 64, 823 (1968).
74. J. M. Pochan, H. W. Gibson, and F. C. Bailey, J. Polym. Sci. Polym. Lett. Ed., 18, 447 (1980).
75. T. Masuda, N. Sasaki, and T. Higashimura, Macromolecules, 8, 717 (1975).
76. M. Hatano, H. Nomori, and S. Kambara, J. Ind. Chem. Jpn. (Kogyo Kagaku Zasshi), 67, 1608 (1964).
77. P. M. Kuznesof and K. J. Wynne, Chem. Commun., 121 (1980).
78. M. Hanack, K. Mitulla, G. Pawlowski, and L. R. Subramanian, J. Organomet. Chem., 204, 315 (1981).
79. H. B. Brom, Y. Tomkiewicz, A. Aviram, A. Broes, and B. Sunners, Solid State Commun., 35, 135 (1980).
80. A. G. MacDiarmid and A. J. Heeger, Synth. Met., 1, 101 (1979-1980).
81. H. Shirakawa, E. J. Louis, A. G. MacDiarmid, C. K. Chiang, and A. J. Heeger, Chem. Commun., 578 (1977).
82. C. K. Chiang, Y. W. Park, A. J. Heeger, H. Shirakawa, E. J. Louis, and A. G. MacDarmid, Phys. Rev. Lett., 39, 1098 (1977).
83. C. K. Chiang, M. A. Drny, S. C. Gau, A. J. Heeger, E. J. Louis, A. G. MacDiarmid, Y. W. Park, and H. Shirakawa, J. Am. Chem. Soc., 100, 1013 (1978).
84. C. K. Chiang, Y. W. Park, A. J. Heeger, H. Shirakawa, E. J. Louis, and A. G. MacDiarmid, J. Chem. Phys., 69, 5098 (1978).
85. T. C. Clarke, R. H. Geiss, W. D. Gill, P. M. Grant, J. W. Macklin, M. Morawitz, J. F. Rabolt, D. Sayers, and G. B. Street, Chem. Commun., 332 (1979).
86. S. C. Gau, J. Mieliken, A. Pron, A. G. MacDiarmid, and A. J. Heeger, Chem. Commun., 663 (1979).
87. A. Pron, I. Kulszewiez, D. Billaud, and J. Przyluski, Chem. Commun., 783 (1981).
88. T. C. Clarke, R. H. Geiss, J. F. Kwak, and G. B. Street, Chem. Commun., 489 (1978).
89. L. R. Anderson, G. P. Pez, and S. L. Hsu, Chem. Commun., 1066 (1978).

90. H. Selig, A. Pron, M. A. Druy, A. G. MacDiarmid, and A. J. Heeger, Chem. Commun., 1288 (1981).
91. P. J. Nigrey, A. G. MacDiarmid, and A. J. Heeger, Chem. Commun., 594 (1979).
92. D. M. Ivory, G. G. Miller, J. M. Sowa, L. W. Shacklette, R. R. Chance, and R. H. Baughman, J. Chem. Phys., 71, 1506 (1979).
93. L. W. Schacklette, R. R. Chance, D. M. Ivory, G. G. Miller, and R. H. Baughman, Synth. Met. 1, 307 (1980).
94. A. F. Diaz, K. K. Kanazawa, and G. P. Gardini, Chem. Commun., 635 (1979).
95. K. K. Kanazawa, A. F. Diaz, R. H. Geiss, W. D. Gill, J. F. Frank, U. A. Logan, J. F. Rabolt, and G. B. Street, Chem. Commun., 854 (1979).
96. K. K. Kanazawa, A. F. Diaz, W. D. Gill, P. M. Grant, G. B. Street, G. P. Gardini, and J. F. Kwak, Synth. Met., 1, 329 (1979-1980).
97. K. K. Kanazawa, A. F. Diaz, M. T. Krounbi, and G. B. Street, Synth. Met., 4, 119 (1981).
98. M. Salmon, A. F. Diaz, A. J. Logan, M. Krounbi, and J. Bargon, Mol. Cryst. Liq. Cryst., 83, 265 (1982).
99. A. F. Diaz and J. I. Castillo, Chem. Commun., 397 (1980).
100. A. F. Diaz, J. I. Castillo, J. A. Logan, and W.-Y. Lee, J. Electroanal. Chem., 129, 115 (1981).
101. G. B. Street, T. C. Clarke, M. Krounbi, K. Kanazawa, V. Lee, P. Pfluger, J. C. Scott, and G. Weiser, Mol. Cryst. Liq. Cryst., 83, 253 (1982).
102. J. F. Rabolt, T. C. Clarke, K. K. Kanazawa, J. R. Raynolds, and G. B. Street, Chem. Commun., 347 (1980).
103. R. R. Chance, L. W. Shacklette, G. G. Miller, D. M. Ivory, J. M. Sowa, R. L. Eisenbaumer, and R. H. Baughman, Chem. Commun., 348 (1980).
104. T. Yamamoto, K. Sanechika, and A. Yamamoto, J. Polym. Sci. Polym. Lett. Ed., 18, 9 (1980); Bull. Chem. Soc. Jpn., 56, 1497, 1503 (1983).
105. H. W. Gibson, F. C. Bailey, A. J. Epstein, H. Rommelmann, and J. M. Pochan, Chem. Commun., 426 (1980).
106. M. Tanaka, A. Watanabe, H. Fujimoto, and J. Tanaka, Chem. Lett., 907 (1980).
107. G. E. Wnek, J. C. W. Chien, F. E. Karasz, and L. P. Lillya, Polymer, 20, 1441 (1979).
108. D. O. Cowan, J. Park, C. U. Pittman, Y. Sasaki, T. Mukherjee, and N. A. Diamond, J. Am. Chem. Soc., 94, 5110 (1972).
109. C. U. Pittman and B. Surynarayanan, J. Am. Chem. Soc., 96, 7916 (1974).
110. A. M. Hermann and A. Rembaum, J. Polym. Sci., C17, 107 (1967).

111. a. H. Kanega, Y. Shirota, and H. Mikawa, Chem. Commun., 158 (1984).
b. Y. Shirota, T. Kakuta, and H. Mikawa, Makromol. Chem., Rapid Commun., 5, 337 (1984).
c. Y. Shirota, N. Noma, H. Kanega, and H. Mikawa, Chem. Commun., 470 (1984).
112. a. G. Tourillon and F. Garnier, J. Electroanal. Chem., 135, 173 (1982); J. Phys. Chem., 87, 2289 (1983).
b. K. Kaneto, K. Yoshino, and Y. Inuishi, Jpn. J. Appl. Phys., 21, L567 (1982); Chem. Commun., 382 (1983).
c. S. Hotta, T. Hosaka, and W. Shimotsuma, Synth. Met., 6, 317, 319 (1983).
113. G. Natta, G. Mazzanti, and P. Corradini, Rend. Acad. Naz. Lincei, 8, 25, 3 (1958).
114. H. Shirakawa and S. Ikeda, Polym. J., 2, 231 (1971).
115. T. Itoh, H. Shirakawa, and S. Ikeda, J. Polym. Sci. Polym. Chem. Ed., 12, 11 (1974).
116. T. Itoh, H. Shirokawa, and S. Ikeda, J. Polym. Sci. Polym. Chem. Ed., 13, 1943 (1975).
117. T. C. Clarke and G. B. Street, Synth. Met., 1, 119 (1979-1980).
118. H. Shirakawa and S. Ikeda, Synth. Met., 1, 715 (1979-1980).
119. K. Seeger and W. D. Gill, Colloid Polym. Sci., 258, 252 (1980).
120. C. K. Chiang, A. J. Heeger, and A. G. MacDiarmid, Ber. Bunsenges. Phys. Chem., 83, 407 (1979).
121. M. Rolland, M. Aldissi, P. Bernier, M. Cadene, and S. Schue, Nature, 294, 60 (1981).
122. G. Lieser, G. Wegner, W. Muller, and V. Enkelman, Makromol. Chem. Rapid Commun., 1, 621 (1980).
123. G. Wegner, Makromol. Chem. Suppl., 4, 155 (1981).
124. M. Aldissi, C. Linaya, J. Stedz, F. Schue, L. Giral, J. M. Fabre, and M. Rolland, Polymer, 23, 243 (1982).
125. M. Aldissi, F. Schue, L. Giral, and M. Rollan, Polymer, 23, 246 (1982).
126. F. Wang, X. Zhas, Z. Gong, Y. Cao, Q. Yang, and R. Qiàn, Makromol. Chem. Rapid Commun., 3, 929 (1982).
127. A. Bradley and J. P. Hammes, J. Electrochem. Soc., 110, 15 (1963).
128. M. F. Farona, P. Lofgren, and P. S. Woon, Chem. Commun., 246 (1974).
129. P. S. Woon and M. M. Farona, J. Polym. Sci. Polym. Chem. Ed., 12, 1749 (1974).
130. T. Masuda, K. Hasegawa, and T. Higashimura, Macromolecules, 7, 728 (1974).
131. Y. M. Pauskin, T. P. Vishnyakova, S. A. Nisova, A. F. Lunin, O. Y. Omarov, Y. Y. Markov, F. F. Machers, I. A. Golubeva,

L. S. Polak, I. I. Patalakh, V. A. Stychenko, and T. A. Sokolinskaya, J. Polym. Sci., A1(5), 1203 (1967).
132. K. Soga, Y. Kobayashi, S. Ikeda, and S. Kawakami, Chem. Commun. 931 (1980).
133. R. N. McDonald and T. W. Campbell, J. Am. Chem. Soc., 82, 4669 (1960).
134. J. G. Speight, P. Kovacic, and F. D. Koch, J. Macromol. Sci. Rev. Macromol. Chem., C5, 295 (1971).
135. P. Kovacic and H. Kyriakis, J. Am. Chem. Soc., 85, 454 (1963).
136. P. Kovacic and R. M. Lange, J. Org. Chem., 28, 968 (1963).
137. P. Kovacic and J. Oziomek, J. Org. Chem., 29, 100 (1964).
138. T. Yamamoto, Y. Hayashi, and A. Yamamoto, Bull. Chem. Soc. Jpn., 51, 2091 (1978).
139. N. Kitajima, Y. Hakone, and Y. Ono, Chem. Lett., 871 (1982).
140. G. G. Engstrom and P. Kovacic, J. Polym. Sci. Polym. Chem. Ed., 15, 2453 (1977).
141. P. Kovacic and W. B. England, J. Polym. Sci. Polym. Lett. Ed., 19, 359 (1981).
142. L. W. Shacklette, H. Eckhardt, R. R. Chance, G. G. Miller, D. M. Ivory, and R. H. Baughman, J. Chem. Phys., 73, 4098 (1980).
143. J. T. Edmonds, Jr., and H. W. Hill, Jr., U.S. Patent 3,354,129 (1967).
144. L. W. Shacklette, R. L. Eisenbaumer, R. R. Chance, H. Eckhardt, J. E. Frommer, and R. H. Baughman, J. Chem. Phys., 75, 1919 (1981).
145. C. U. Pittman, Jr., Y.-F. Liang, and M. Ueda, Macromolecules, 12, 355, 541 (1979).
146. Y. Ueno, Y. Masuyama, and M. Okawara, Chem. Lett., 603 (1975).
147. D. C. Green and R. W. Allen, Chem. Commun., 832 (1978).
148. M. L. Kaplan, R. C. Haddon, F. Wudl, and E. D. Feit, J. Org. Chem., 43, 4642 (1978).
149. D. J. Sandman, M. Rubner, and M. Samnelsen, Chem. Commun., 1133 (1982).
150. K. Y. Jen, M. V. Lakshmikantham, M. Albeck, M. P. Cava, W. S. Huang, and A. G. MacDiarmid, J. Polym. Sci. Polym. Lett. Ed., 21, 1141 (1983).
151. J. H. Perlstein, J. A. VanAllan, L. C. Isett, and G. A. Reynolds, Ann. N.Y. Acad. Sci., 313, 61 (1978).
152. S. Yoshimura, in Molecular Metals (W. H. Hatfield, Ed.), NATO Conference Series VI, Plenum Press, New York (1979), p. 471.
153. E. P. Goodings, Chem. Soc. Rev., 5, 95 (1976).
154. P. M. Skarstad and B. B. Owens, Proceedings of the Workshop on Lithium Nonaqueous Battery Electrochemistry, P-Vol. 80-7, Electrochemical Society, Inc. (1980), p. 276.

155. F. Gutman, A. M. Hermann, and A. Rembaum, J. Electrochem. Soc., 114, 323 (1967).
156. P. J. Nigrey, D. MacInnes, Jr., D. P. Nairns, A. G. MacDiarmid, and A. J. Heeger, J. Electrochem. Soc., 128, 1651 (1981).
157. C. K. Chiang, Polymer, 22, 1454 (1981).
158. L. W. Shacklette, R. L. Elsenbaumer, R. R. Chance, J. M. Sowa, D. M. Ivory, G. G. Miller, and R. H. Baughman, Chem. Commun., 361 (1982).
159. A. Nishino, I. Tajima, and T. Muranaka, Jap. Pat. 55-99714 (1980).
160. T. Nogami, M. Nawa, and H. Mikawa, Chem. Commun., 1158 (1982).
161. M. Ozaki, D. L. Peebles, B. R. Weinberger, C. K. Chiang, S. C. Gau, A. J. Heeger, and A. G. MacDiarmid, Appl. Phys. Lett., 35, 83 (1979); J. Appl. Phys., 51, 4252 (1980).
162. J. Tsukamoto, H. Ohigashi, K. Matsumura, and A. Takahashi, Synth. Met., 4, 177 (1982).
163. B. R. Weinberger, M. Akhtar, and S. C. Gau, Synth. Met., 4, 187 (1982).
164. R. J. Nowak, W. Kutner, and H. B. Mark, J. Electrochem. Soc., 125, 232 (1978).
165. C. D. Jaeger and A. J. Bard, J. Am. Chem. Soc., 101, 1690 (1979).
166. F. B. Kaufman, A. H. Schroeder, E. M. Engler, and V. V. Patel, Appl. Phys. Lett., 36, 422 (1980).
167. D. C. Hofer, F. B. Kaufman, S. R. Kramer, and A. Aviram, Appl. Phys. Lett., 37, 314 (1980).
168. a. T. C. Clarke, M. T. Krounbi, V. Y. Lee, and G. B. Street, Chem. Commun., 384 (1981).
b. S. Petchumani and F. Willig, Chem. Commun., 809 (1983).
169. K. Kaneto, K. Yoshino, and Y. Inuishi, J. Appl. Phys., 22, 412 (1983).
170. T. Kakuta, Y. Shirota, and H. Mikawa, Chem. Commun., 553 (1985).
171. Y. Shirota, T. Kakuta, H. Kanega, and H. Mikawa, Chem. Commun., 1201 (1985).
172. H. Kanega, Y. Shirota, and H. Mikawa, Kobunshi Ronbunshu, 41, 201 (1984).
173. G. Wegner, Z. Naturforsch., B24, 824 (1969).
174. G. Wegner, Makromol. Chem., 145, 85 (1971); 154, 35 (1972).
175. R. H. Baughman, J. Polymer Sci. Polym. Phys. Ed., 12, 1511 (1974).
176. W. Schermann and G. Wegner, Makromol. Chem., 175, 667 (1974).
177. R. R. Chance and R. H. Baughman, J. Chem. Phys., 64, 3889 (1976).

178. K. J. Donavan and E. G. Wilson, J. Phys., C12, 4857 (1979); Philos. Mag. B44, 9 (1981).
179. L. Sebastian and G. Weiser, Phys. Rev. Lett., 46, 1156 (1981).
180. T. Tani, P. M. Grant, W. D. Gill, G. B. Street, and T. C. Clarke, Solid State Commun., 33, 499 (1980).
181. H.-H. Horhold and J. Opfermann, Makromol. Chem., 131, 105 (1970).
182. H.-H. Horhold and J. Opfermann, Makromol. Chem., 178, 195 (1977).
183. M. Stolka, Macromolecules, 8, 8 (1975).
184. A. Mukoh and H. Morishita, J. Polym. Sci. Polym. Lett. Ed., 18, 35 (1980).
185. a. M. Stolka, J. F. Yanus, and J. M. Pearson, Macromolecules, 9, 710 (1976).
b. M. Stolka, J. F. Yanus, and J. M. Pearson, Macromolecules, 9, 715 (1976).
186. E. G. E. Hawkins, J. Chem. Soc., 3216 (1958).
187. J. F. Yanus, M. Stolka, and J. M. Pearson, Macromolecules, 9, 719 (1976).
188. K. Tanikawa, S. Kusabayashi, and H. Mikawa, J. Polym. Sci., B6, 275 (1968).
189. S. R. Turner and D. M. Pai, Macromolecules, 12, 1 (1979).
190. W. O. Gibson and F. C. Baily, Macromolecules, 9, 688 (1976); 10, 602 (1977).
191. K. Tanikawa, Z. Okuno, T. Iwaoka, and M. Hatano, Makromol. Chem., 178, 1779 (1977).
192. K. Morimoto and H. Mitsuda, J. Polym. Sci., B5, 27 (1967).
193. W. W. Limburg and D. J. Williams, Macromolecules, 6, 787 (1973).
194. D. Hyde, L. J. Kricka, and A. Ledwith, Polymer, 14, 124 (1973).
195. T. Inoue and S. Tazuke, J. Polym. Sci. Polym. Chem. Ed., 19, 2861 (1981).
196. K. Okamoto, M. Yamada, A. Itaya, T. Kimura, and S. Kusabayashi, Macromolecules, 9, 654 (1976).
197. M. Biswas and S. K. Das, J. Polym. Sci. Polym. Lett. Ed., 19, 235 (1981).
198. a. D. M. Chang, S. Gromelski, R. Rupp, and J. E. Mulvaney, J. Polym. Sci. Polym. Chem. Ed., 15, 571 (1977).
b. J. E. Mulvaney and R. A. Brand, Macromolecules, 13, 244 (1980).
199. Y. Shirota and H. Mikawa, J. Macromol. Sci. Rev. Macromol. Chem., C16, 129 (1977-1978).
200. K. Okamoto, A. Itaya, and S. Kusabayashi, J. Polym. Sci. Polym. Chem. Ed., 14, 869 (1976).
201. D. J. Williams, W. W. Limburg, J. M. Pearson, A. O. Goedde, and J. F. Yanus, J. Chem. Phys., 62, 1501 (1975).

202. M. Biswas and S. K. Das, J. Polym. Sci. Polym. Lett. Ed., 20, 333 (1982).
203. Y. Shirota, A. Matsumoto, and H. Mikawa, Polym. J., 3, 643 (1972).
204. M. Yoshimura, Y. Shirota, and H. Mikawa, J. Polym. Sci., B11, 457 (1973).
205. Y. Shirota, M. Yoshimura, A. Matsumoto, and H. Mikawa, Macromolecules, 7, 4 (1974).
206. Y. Shirota, K. Takemura, H. Mikawa, T. Kawamura, and K. Matsuzaki, Makromol. Chem. Rapid Commun., 3, 913 (1982).
207. M. Yokoyama, T. Tamamura, M. Atsumi, M. Yoshimura, Y. Shirota, and H. Mikawa, Macromolecules, 8, 101 (1975).
208. Y. Shirota, Y. Matsumoto, T. Aukuda and H. Mikawa, Unpublished results.
209. Y. Shirota, Y. Matsumoto, H. Mikawa, K. Seki, and H. Inokuchi, Chem. Phys. Lett., 97, 57 (1983).
210. P. J. Metz, R. B. Champ, L. S. Chang, C. Chiou, G. S. Keller, L. E. Liclican, R. R. Neiman, M. D. Shattuck, and W. J. Weiche, Photogr. Sci. Eng., 21, 73 (1977).
211. F. Iinuma, H. Mikawa, and Y. Shirota, Polym. J., 13, 641 (1981).
212. F. Iinuma, H. Mikawa, and Y. Shirota, Macromolecules, 14, 1747 (1981).
213. F. Iinuma, H. Mikawa, and Y. Shirota, Mol. Cryst. Liq. Cryst., 73, 309 (1981).
214. F. Iinuma, Y. Shirota, and H. Mikawa, Polym. Prepr. Jpn., 30(2), 445 (1981).
215. S. R. Turner and C. AuClair, Macromolecules, 9, 868 (1976).
216. W. D. Gill, Amorphous Liq. Semicond., 2, 901 (1974).
217. S. Tazuke and H. Nagahara, J. Polym. Sci. Polym. Lett. Ed., 16, 525 (1978).
218. S. R. Turner, Macromolecules, 13, 782 (1980).
219. T. Nakayama, K. Demura, K. Kamisako, and H. Sasabe, Reports on Progress in Polymer Physics in Japan, 23, 433 (1980).
220. M. Biswas and S. K. Das, Eur. Polym. J., 17, 1245 (1981).
221. R. M. Schafert, IBM J. Res. Dev., 75 (1971).
222. P. M. Borsenberger, A. Chowdry, D. C. Hoesterey, and W. Mey, J. Appl. Phys., 49, 5555 (1978).
223. Y. Hayashi, Nat. Tech. Rep., 22, 818 (1976).
224. E. G. Schlosser, J. Appl. Photogr. Eng., 4, 118 (1978).
225. T. Kazami, T. Enomoto, and K. Sakai, 41st Annual Meeting of Jpn. Chem. Soc., Preprint I, 370 (1980).
226. E. Inoue, I. Shimizu, and Y. Nishio, Photogr. Sci. Eng., 22, 194 (1978).
227. a. J. Hanna and E. Inoue, Photogr. Sci. Eng., 25, 209 (1981).
b. Y. Nishio and E. Inoue, Photogr. Sci. Eng., 25, 35 (1981).

7

Crown Ether Polymers as Functional Polymers in Analytical Chemistry

Keiichi Kimura and Toshiyuki Shono

Osaka University, Osaka, Japan

7.1 INTRODUCTION

The extensive development in the synthesis of monomers and polymers bearing functional groups has stimulated research efforts in various fields of chemistry. Particularly in the field of organic synthesis, an enormous number of functional polymers have been synthesized, mainly for polymeric reagents and catalysts, because of the easy retrieval of the polymers and the easy purification of the resulting product after the reactions.

In the field of analytical chemistry, applications of functional polymers have been concentrated primarily on separation chemistry. A large number of chelating polymers and ion exchangers have been developed for the separation and concentration of metal ions. However, these functional polymers were generally deficient in selectivity on binding metal ions. High selectivity in the separation and the concentration has been required in recent analytical chemistry.

Macrocyclic polyether, which is often referred to as crown ether, is one of the most sensational compounds. Crown ethers can selectively complex alkali and alkaline earth metal ions, and sometimes the other metal ions, by ion-dipole interaction. Taking advantage of the cation-binding selectivity, they have been widely used in analytical chemistry in the separation and the analyses of metal ions. More recently, functional polymers containing crown ether moiety have just begun to be used in this field for the purpose of gaining excellent cation selectivities, immobilizing crown ether moiety, and alleviating the toxicity of crown ether.

This chapter deals with the synthesis and applications of the crown ether polymers which have a great potential as specific functional polymers in analytical chemistry. The goal is to encourage the practical use of the crown ether polymers in this area and stimulate further work to solve the remaining analytical and separation problems using these polymers. Hence in some cases, experimental procedures covering the synthesis and analytical use of the crown ether polymers are described in detail. In the terminology, polymers sometimes include oligomers such as dimers.

7.2 CATION-BINDING PROPERTIES OF CROWN ETHER DERIVATIVES

A number of books [1-4] have been devoted to the detailed description of the cation-binding properties and the recent progress of crown ether derivatives. Since the pioneering work of Pedersen [5,6] on crown ethers, one of the emphases on crown ether design has been directed on the selective binding of metal ions, notably alkali and alkaline earth metal ions as well as alkyl ammonium ions. One of the most important things in their analytical application is the cation-binding selectivity of the crown ethers, which depend primarily on (a) the relative size of ion and crown ether cavity; (b) the type, number, and placement of heteroatoms (binding sites); (c) (d) the electrical charge of ions; and (e) solvent interaction with crown ether. Pedersen carried out solvent extractions of alkali metal picrates with various oxygen-containing crown ethers which he had synthesized for the purpose of rapid screening of their cation-binding abilities. In this case the cation-binding selectivities for the alkali metal ions were explained in terms of the size-fit concept, the first factor governing the complexation described above. For example, the 14-crown-4 derivative is selective for Li^+, the 15-crown-5 for Na^+, and the 18-crown-6 derivative for K^+. The relative size of ion and crown ether cavity is undoubtedly a significant factor governing the cation selectivity of crown ethers, although factors (a) through (e) all affect the concerted selectivity.

Macrobicyclic or macrotricyclic polyethers, which are called cryptands, have appeared [7] as more powerful cation-binding ligands than crown ethers. Since cryptand molecules bear a three-dimensional cavity, the relative size of the ion and the cavity become more important factors in the cation selectivity of cryptands than for the crown ethers. Noncyclic crown ethers, such as polyoxyethylene derivatives, are normally equipped neither with powerful cation-binding ability nor with high cation selectivity, unlike the "cyclic" crown ethers. Incorporation of a rigid structure unit like quinoline rings into the basic polyoxyethylenes sometimes improves the cation-binding ability and selectivity, as substantiated by Tummler et al. [8]. Crown ethers form mostly 1:1 crown ether-to-cation complexes with various cations. However, some crown ethers possess the interesting potential to form 2:1 crown ether-to-cation complexes in which a metal ion is sandwiched by two crown ether rings. Although crown ethers tend to form the 1:1 complexes even with metal ions which are larger than the crown ether cavity, the existence of a large excess of crown ether molecules to cations permits the formation of the 2:1 complexes.

Polymeric crown ethers have attractive cation-binding abilities compared to their monomeric analogs. Notably, polymers bearing crown ether moiety in the side chain, such as vinyl-type crown ether polymers [9-17], bind certain metal ions more strongly. The cation is sandwiched intramolecularly by two adjacent crown ether rings in the polymer molecule. In other words, the poly(crown ether)s can easily form 2:1 crown ether ring-to-cation complexes with particular metal ions which are a little larger than the cavity of the crown ether ring. Therefore, the poly(crown ether)s often exhibit excellent cation-binding selectivities, which are very different from those of the corresponding monomeric analogs. For instance, poly(benzo-15-crown-5)s 1 (n = 1) [9], 3 (n = 1; R = H, CH_3) [10-12], and 5 (n = 1; R = H, CH_3) [13-15] are selective for K^+ in binding alkali metal ions, and poly(benzo-18-crown-6)s 1 (n = 2), 3 (n = 2; R = H, CH_3), and 5 (n = 2; R = H, CH_3) for Cs^+. Poly(benzo-12-crown-4) 3 (n = 0; R = H) is expected to bind Na^+ selectively, but it does not complex very strongly with any alkali metal ion, probably because of the low basicity of the aromatic oxygen atoms—therefore, the drastic decrease in the number of the binding sites from 4 to 2 [16]. The removal of the benzene ring from benzo-12-crown-4 augments the cation-binding ability markedly. Thus poly(12-crown-4) 6 (n = 1; R = H, CH_3) can bind Na^+ strongly and selectively [17].

The dimeric model compounds, so-called bis(crown ether)s, behave similar to the poly(crown ether)s on binding metal ions. They form sandwich-type 2:1 crown ether ring-to-cation complexes [11,17-22]. An approach to see if the bis(crown ether) derivatives [or the poly(crown ether)s] really form the intramolecular 2:1 complexes, as shown schematically in Figure 1, is the synthesis of stereoisomeric bis(crown ether)s such as 10 (n = 1, 2; cis, trans) [23,24]. Optical spectra of alkali metal picrate tetrahydrofuran (THF) solutions were measured in the presence of varying amounts of the bis(crown ether)s to get some information about their complexing properties with alkali metal ions. Without crown ethers, the optical spectrum of potassium picrate in THF exhibits an absorption maximum at 357 nm, indicating that tight ion pairs of crown ether-potassium complexes are involved. It is well known that the maximum of the main absorption band of the picrate THF solution shifts to higher wavelengths with increasing distance of the ion pairs [25]. Also, the formation of the 2:1 crown ether ring-to-cation complexes generally gives loose ion pairs, while the 1:1 complexes remain as tight pairs.

Addition of only an equimolar of *cis*-bis(crown ether) 10 (n = 1) to the potassium picrate THF solution leads to a pronounced bathochromic shift in the main absorption from 357 to 381 nm (Figure 2), suggesting that the *cis*-bis(crown ether) easily complex K^+ with 2:1 crown ether ring-to-cation ratio owing to the favorable cis configuration. On the other hand, only a slight shift in the spectrum from 357 to 365 nm was observed even on addition of a large excess of *trans*-bis(crown ether) 10 (n = 1). This result is quite similar to that observed by the addition of the corresponding monomeric analog, benzo-15-crown-5. This means that due to the unfavorable configuration for the 2:1 complex formation, *trans*-bis(crown ether) 10 (n = 1) can form only 1:1 crown ether ring-to-cation complexes. One of the

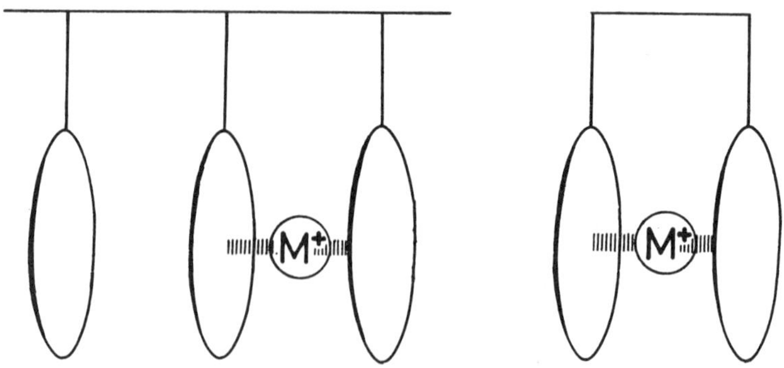

Figure 1. Formation of 2:1 crown ether ring to metal ion complexes by poly- and bis(crown ether)s.

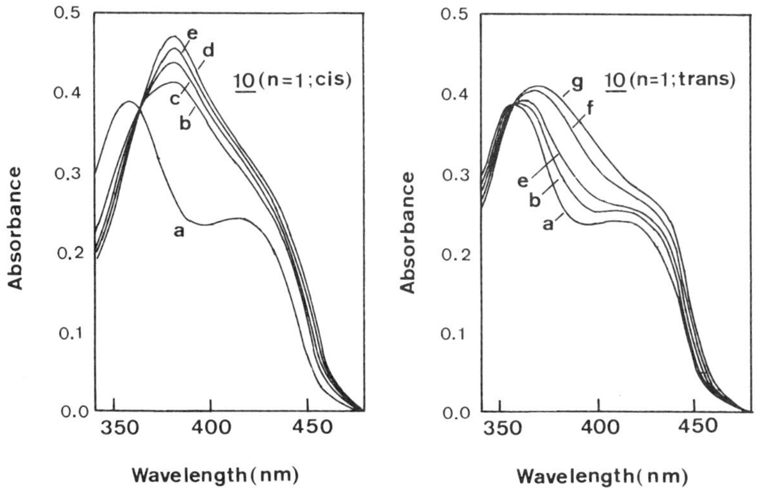

Figure 2. Optical spectra of potassium picrate solution in THF in the presence of various amount of *cis*- and *trans*-bis(crown ether)s 10 (n = 1). [Potassium picrate], 2.5×10^{-5} M, [crown ether unit]/[potassium picrate]. a, 0; b, 2; c, 3; d, 5; e, 10; f, 30; g, 50. (From Ref. 24.)

two crown ether rings in the bis(crown ether) molecule takes part in complexing a metal ion and the other is forced to be distant far from the complexing crown ether ring. Similar spectral changes to that in the *cis*-bis(crown ether) systems were observed in flexible bis(crown ether)s such as 8 and 9 (n = 1), in which two crown ether rings are simply connected by a short aliphatic chain [18,24]. These observations present direct evidence that bis(crown ether) derivatives easily form the sandwich-type 2:1 crown ether ring-to-cation complexes with particular metal ions intramolecularly. It is likely that the length and rigidity of the chain connecting two crown ether rings affects the cation-binding ability in bis(crown ether)s. Too-long or too-short chain length depresses the formation of the intramolecular 2:1 complexes [17-19]. In addition, the replacement of an aliphatic chain by a more flexible oxyethylene chain promotes the formation of the 2:1 complex [18].

Tacticity (or stereoregularity) of poly(crown ether)s is expected to affect the 2:1 complex formation, cation-binding ability, and selectivity. Compared with syndiotactic and atactic poly(crown ether)s 5 (R = CH_3), the isotactic polymer can bind more strongly those metal ions that are large enough to be sandwiched by the crown ether rings [13,15], although differences in the cation-binding ability are not as significant as observed for the *cis*- and *trans*-bis(crown ether) systems.

In condensation-type crown ether polymers, where the crown ether moiety is contained chiefly in the main chain, any improvement or change of cation-binding ability and selectivity over the corresponding monomeric analogs can hardly be obtained. This behavior is unlike the vinyl-type crown ether polymers. The cation-binding ability may sometimes be influenced by substituent effects and the participation of the neighboring heteroatoms [26]. An exception is the epoxy polymer of diazacrown ether, which forms a 2:1 crown ether ring to cation complex with Ba^{2+} intramolecularly [48]. Some cooperative effect such as observed in the vinyl-type crown ether polymers might be also expected for crown ether-modified resins in which the monocyclic crown ether moieties are immobilized close to one another on an organic or inorganic support by a covalent bonding. In the case of the vinyl-type cryptand polymers synthesized [27,28], there does not seem to be any polymer effect such as the cooperative action of two adjacent cryptand moieties.

7.3 SYNTHESIS OF CROWN ETHER POLYMERS

Polymers containing crown ether moieties consist of vinyl-type crown ether polymers, condensation-type polymers, and immobilized crown ethers. The vinyl-type crown ether polymers may include their dimeric model compounds, bis(crown ether)s, in a broader sense.

7.3.1 Vinyl-Type Crown Ether Polymers

Vinyl-type crown ether polymers, so-called poly(crown ether)s, possess attractive properties compared to their monomeric crown ethers, as described in Section 7.2. First, the poly(crown ether)s as well as the bis(crown ether)s are capable of binding particular metal ions strongly by the cooperative action of two adjacent crown ether rings. Also, some water-soluble poly(crown ether)s such as 1 (n = 2) forms polymeric micelles in water, resulting in the binding of organic solutes to the poly(crown ether) micelle [29-32]. Furthermore, insoluble polysalt complexes are readily formed between the poly(crown ether)s and polyacids or their salts [14,33,34].

Several poly(benzocrown ether)s have been synthesized by polymerization of benzocrown ethers through the vinyl group. Compared to aliphatic crown ethers not containing a benzene ring in the structural unit of the crown ether ring, the benzocrown ethers are easily modified on the benzene skeleton either by direct introduction of functional groups to the benzocrown ethers or by cyclization reaction of oligo(oxyethylene)s with catechol derivatives containing an appropriate substituent.

Poly(vinylbenzocrown ether)s 1 (n = 1, 2) were synthesized first by Smid et al. [9,35], as illustrated in Scheme 1. The parent crown

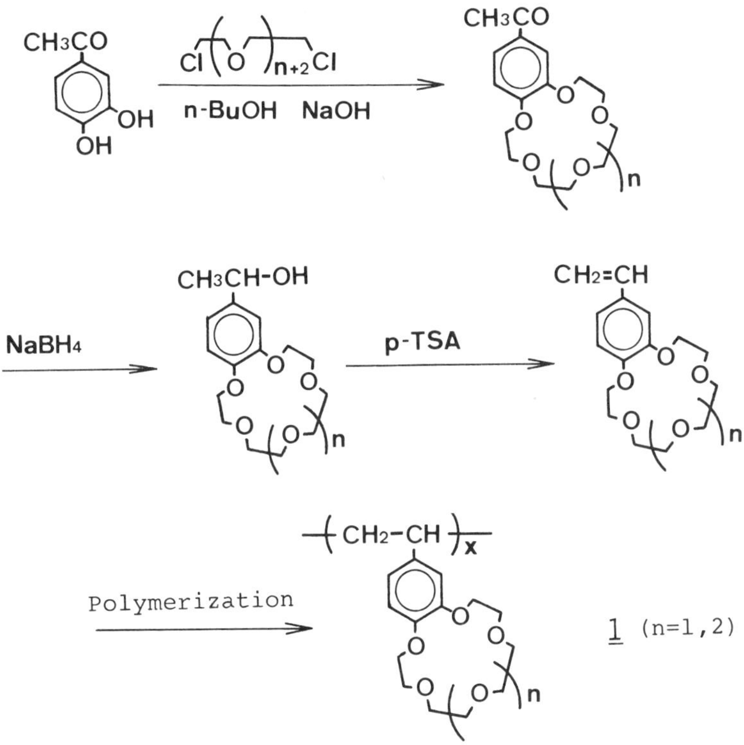

Scheme 1

ether, 4'-acetobenzocrown ether, was prepared in a moderate yield by the cyclization reaction of 3,4-dihydroxyacetophenone with an appropriate α,ω-dichlorinated oligo(oxyethylene) in butanol-water in the presence of NaOH. The yield in the cyclization reaction can be improved by replacing the dichloride, NaOH, and the solvent with the ditosylate, CsF (or KF), and acetonitrile, respectively [36]. The $NaBH_4$ reduction of the acetobenzocrown ethers in ethanol affords their 4'-(1"-hydroxyethyl) derivatives quantitatively. The hydroxyethylbenzocrown ethers can also be obtained by reaction of 4'-formylbenzocrown ethers with CH_3MgI in ethyl ether-benzene [35]. The resulting secondary alcohol derivatives are then dehydrated to yield 4'-vinylbenzocrown ethers, the typical procedure being as follows.

4'-Vinylbenzo-18-crown-6 [35]

A trace of *p*-toluenesulfonic acid monohydrate was added to 8.0 g of 4'-(1"-hydroxyethyl)benzo-18-crown-6 in 350 mL of benzene. The mixture was refluxed with removal of water for 14 h, using a Dean-Stark apparatus. The reaction mixture was then cooled to room temperature and 5 drops of pyridine added. The benzene was evaporated to yield an oil that crystallized upon standing after 15 min. The solid product was dissolved in 100 mL of chloroform, extracted four times with 100 mL of water, and then the chloroform solution dried over Na_2SO_4. The residue obtained after evaporating was recrystallized from petroleum ether (1 g of the product to 75 mL of the solvent) to give 6.2 g (80%) of pure 4'-vinylbenzo-18-crown-6, mp 59 to 61°C. The synthesis of 4'-vinylbenzo-15-crown-5 can be performed in a similar way, mp 43 to 44°C.

These crown ether monomers can easily be polymerized to high-molecular-weight polymers using either radical or anionic initiator. The radical polymerization was carried out for 24 h at 70°C, a benzene solution of the monomer and α,α'-azobisisobutyronitrile (AIBN) in a degassed sealed tube. A few copolymers with different ratios of the vinylcrown ether to styrene can also be obtained. The anionic polymerization was run in THF at 80°C using the sodium tetramer of α-methylstyrene as the initiator. The poly(crown ether)s obtained by the living polymerization are essentially the same in the tacticity, molecular weight, and solubility as those by radical polymerization. The poly(vinylcrown ether)s do not seem to possess any unusual regularity in the polymer chain and seems to be mostly atactic polymer. The number-average molecular weight of the poly(crown ether)s ranges from 3×10^4 to 1.2×10^5. An abnormal thing in their solubility is that poly(benzo-18-crown-6) 1 (n = 2) is water soluble (8.1 g of polymer per 100 mL of water) unlike poly(vinylbenzo-15-crown-5) 1 (n = 1), which forms a polymeric micelle in water. A styrene-type cryptand monomer can be synthesized via a macrocyclic diamide and easily polymerized to polycryptand 2 [27].

Acrylamide-type poly(crown ether)s 3 (n = 1, 2; R = H, CH_3) have been synthesized by Kimura et al. [10,16], and more recently by Yagi et al. [12]. As compared to the preceding 4'-vinylbenzocrown ethers, these 4'-acryloylaminobenzocrown ethers are easier to synthesize. The synthesis of the crown ether monomers involves direct substitution of nitro group at the 4'-position of the benzene ring of the parent benzocrown ethers, as shown in Scheme 2. The nitration of the benzocrown ethers was accomplished at room temperature with nitric acid-acetic acid in chloroform [37]. The acryloylaminobenzocrown ethers were obtained by the hydrogenation of the nitro derivatives with palladium on carbon [37], followed by the condensation reaction with acrylic anhydride. The aminobenzocrown ethers, which are labile to

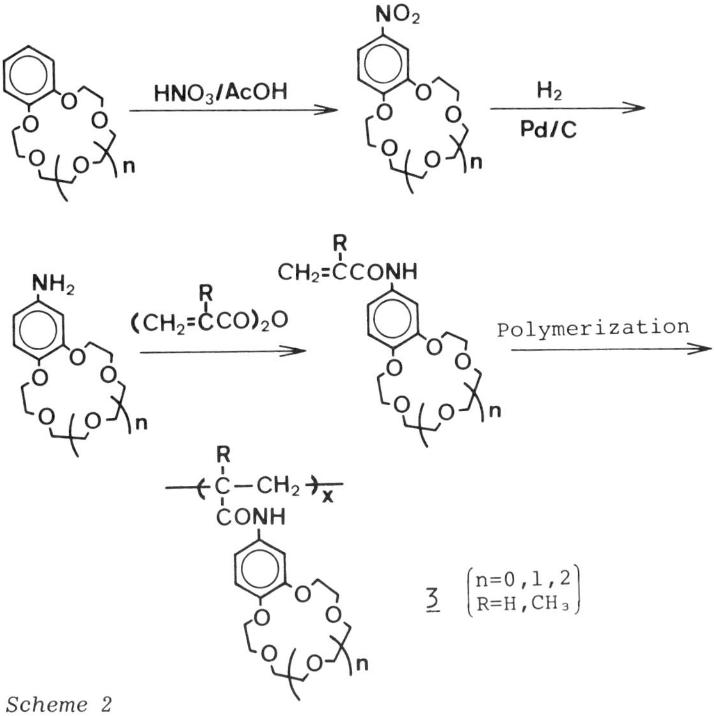

Scheme 2

air oxidation, were subjected to the subsequent reaction under nitrogen atmosphere without isolation.

4'-Acryloylaminobenzo-18-crown-6 [10]

A solution of 7.1 g of 4'-nitrobenzo-18-crown-6 dissolved in 60 mL of dioxane was stirred at room temperature in a glass autoclave under a hydrogen pressure of 6×10^5 psi in the presence of 2.8 g of 5% palladium on carbon, until the hydrogen absorption of the reaction mixture ceased. After the reaction was completed, the catalyst was filtered off rapidly, and then the filtrate cooled in an ice bath. To the aminobenzocrown ether in dioxane 2.5 g of acrylic anhydride in dioxane was added dropwise. The reaction mixture was allowed to stir overnight at room temperature. After the reaction the bulk of the solvent was evaporated off, keeping the temperature below 30°C. The residue was again dissolved in dioxane and decolorized with activated charcoal powder. The dioxane solution was freeze-dried to yield 5.8 g (76%) of 4'-acryloylaminobenzo-18-crown-6, mp 113 to 114°C. Other acryloylaminobenzocrown ethers and the methacryloylamino derivatives were made similarly. The methacryloylamino derivatives were easier to work up

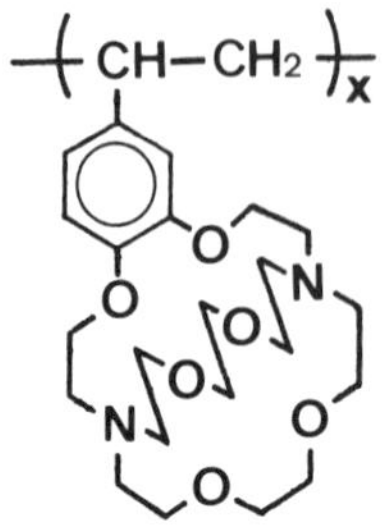

2

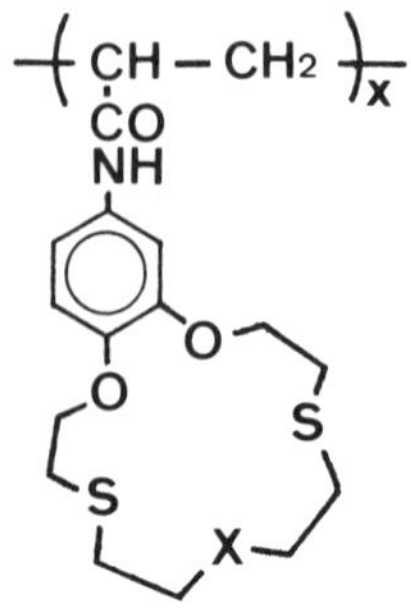

4 (X= -S-, -S$(CH_2)_3$S-)

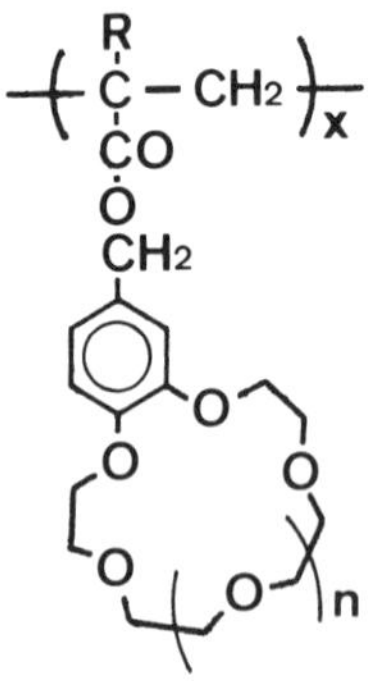

5 (n=1,2; R=H,CH_3)

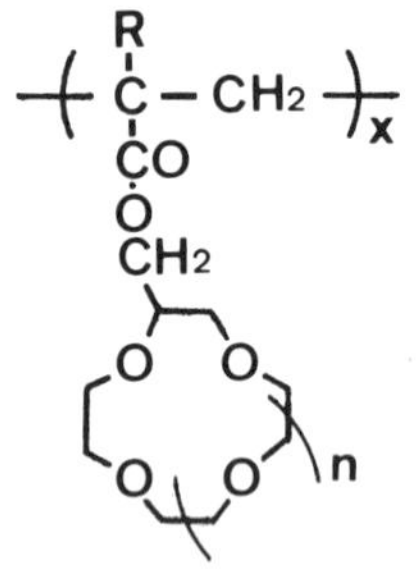

6 (n=1,2,3; R=H,CH_3)

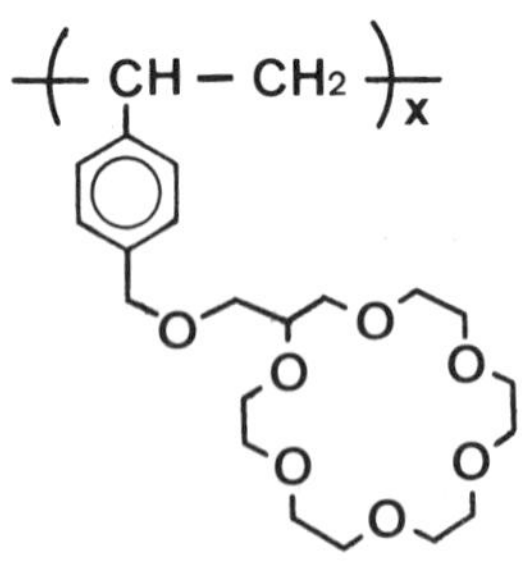

7

than the acryloylamino ones. These can be purified by recrystallization from heptane—mp: monomer of 3 (n = 1; R = H) [10], 137 to 138°C; monomer of 3 (n = 0; R = H) [16], 105 to 106°C; monomer of 3 (n = 2; R = H) [127], 113 to 114°C; monomer of 3 (n = 1; R = CH_3), 145 to 146°C; monomer of 3 (n = 2; R = CH_3), 94 to 95°C.

The acrylamide- or methacrylamide-type crown ether monomers were polymerized radically in dimethylformamide (DMF) at 60°C to yield poly(crown ether)s 3. The polymers were reprecipitated twice from chloroform in ethyl ether to remove unreacted crown ether monomers. The hydrogen-transfer polymerization of the acryloylaminobenzocrown ethers 3 (n = 1, 2; R = H) initiated by potassium *t*-butoxide did not give the expected novel poly(crown ether)s with polyamide structure in the main chain [38]. The polyacrylamide-type poly(thiacrown ether)s 4 [X = -S-, $-S(CH_2)_3S-$] can be synthesized [94] in a manner similar to poly(crown ether)s 3.

Polyacrylate type of poly(crown ether)s 5 (n = 1, 2; R = H, CH_3) are also easy to prepare [13, 14]. The acryloyl- or methacryloyloxymethylbenzocrown ethers were obtained by triethylamine-assisted condensation of acrylic or methacrylic chloride with the corresponding 4'-hydroxymethylbenzocrown ethers, which were obtained by $NaBH_4$ reduction of the formyl derivatives. These crown ether monomers can be polymerized radically in benzene or toluene to yield poly(crown ether)s 5 with a molecular weight of about 1×10^5.

The foregoing vinyl-type poly(crown ether)s, even the polymers obtained by anionic polymerization, are atactic. Unsuccessful attempts have been made to polymerize the methacryloyloxymethylbenzocrown ethers anionically in toluene and THF using butyllithium as the initiator [13]. Tactic poly(crown ether)s 5 (n = 1, 2; R = CH_3) were synthesized successfully by the polymer of tactic poly(methacrylic acid) with 4'-chloromethylbenzocrown ethers in DMF in the presence of dicyclohexylamine. The parent poly(methacrylic acid)s with various tacticities were obtained either by anionic polymerization of diphenylmethyl methacrylate, followed by treating with hydrogen bromide, or by anionic polymerization of trimethylsilyl methacrylate, followed by hydrolysis with NaOH aqueous solution [39]. In this polymer reaction, three- or fourfold excess of the chloromethylbenzocrown ether and the amine were required for the quantitative conversion to the corresponding poly(crown ether)s 5 (n = 1, 2; R = CH_3). This procedure allows the synthesis of poly(crown ether)s with the same tacticity as the respective poly(methacrylic acid) used. In this procedure, it is possible to synthesize the poly(crown ether)s with extremely high molecular weight, whereas the polymerization of the vinyl crown ether monomers did not give the poly(crown ether)s with a molecular weight of more than 1.2×10^5. The experimental procedure for the polymer reaction is described below.

Tactic poly(methacryloyloxymethylbenzo-15-crown-5)
5 (n = 1; R = CH_3) [13]

To a solution of a mixture of 0.25 g of tactic poly(methacrylic acid) and a 3.68 g of chloromethylbenzo-15-crown-5 in 40 mL of DMF was added dropwise 2.30 g of dicyclohexylmethylamine over a period of 6 h at 60°C with continuous stirring. The reaction was allowed to continue for 5 days. After the solvent was replaced with a small quantity of chloroform, the tactic poly(crown ether) was isolated by pouring the chloroform solution into a large excess of methanol. The conversion of the methacrylic acid unit to the crown methacrylate unit was almost quantitative, more than 96%, which was evaluated by pH titration. The ^{1}H NMR spectral data of the poly(crown ether)s obtained showed that no change in the tacticity occurred during the polymer reaction.

The vinyl monomers and polymers appended with benzocrown ether moiety may be easy to synthesize, but they have a distinct disadvantage concerning cation-binding ability. The benzene ring incorporated in the crown ether ring decreases the basicity of the heteroatoms (binding sites), especially the aromatic ones, due to the electron delocalization, which in turn lowers the cation-binding ability compared to the corresponding aliphatic crown ethers. For example, benzo-12-crown-4 possesses much lower cation-binding ability than 12-crown-4, so in the poly(benzo-12-crown-4) 3 (n = 0; R = H) [16] there is little enhanced binding ability, unlike in the poly(crown ether)s of benzo-15-crown-5 and benzo-18-crown-6. Acrylate-type aliphatic crown ethers [17] were synthesized by reaction of acryloyl chloride with hydroxymethyl-12-crown-4 [40] in the presence of triethylamine. The monomers were polymerized to yield poly(12-crown-4) 6 (n = 1; R = H). Poly(crown ether)s bearing 15-crown-5 and 18-crown-6 moieties 6 (n = 2, 3) can be obtained as well. Another poly(18-crown-6) 7 was also accessible from the crown ether monomer, which is given by reaction of chloromethylstyrene with the hydroxymethyl-18-crown-6 [41].

7.3.2 Bis(crown ether)s

Bis(crown ether)s, the dimeric model compounds of the poly(crown ether)s, exhibit similar cation-binding properties. The length and the rigidity of an aliphatic (or oxyethylene) chain which connects two crown ether moieties in a bis(crown ether) molecule generally affects the cation-binding abilities of the various types of bis(crown ether)s prepared so far. Two similar types of bis(crown ether)s 8 and 9 have been synthesized by the condensation reactions of aliphatic diols or oligo(oxyethylene) with 4'-carboxybenzocrown ethers [18], and of dicarboxylic acids with 4'-hydroxymethylbenzocrown ethers [19], respectively.

Bis(benzo-15-crown-5) 9 *(n = 1; m = 5)* [19]

To a mixture of 1.6 g of 4'-hydroxymethylbenzo-15-crown-5 and 0.6 g of triethylamine in 10 mL of chloroform was dropwise added 0.5 g of pimelic chloride, while stirring and cooling. The reaction mixture was then allowed to reflux for 5 h. After the reaction, the mixture was extracted with 5% $NaHCO_3$ aqueous solution, the chloroform layer washed with water, and then the mixture dried over $MgSO_4$. After the filtration and solvent evaporation, an oil was left, which crystallized upon standing. The crude product was recrystallized from ethanol to yield pure bis(crown ether) 9 (n = 1; m = 5) (87%), mp 94 to 95°C.

Rigid bis(crown ether) stereoisomers, *cis*- and *trans*-bis(crown ether)s 10 (n = 1, 2), can be synthesized by esterification of potassium salts of maleic and fumaric acids with 4'-chlormethylbenzocrown ethers in acetonitrile [23, 24]. Diamide-type bis(crown ether)s 11 are also obtained from appropriate dicarboxylic chlorides and the 4'-aminobenzocrown ether. More recently, bis(crown ether)s containing oligo(oxy-

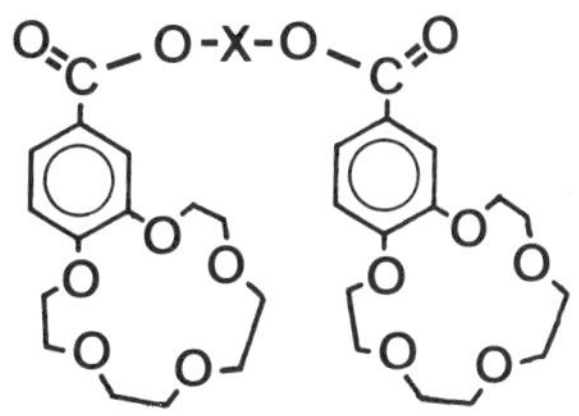

8

(X= $-(CH_2)_m-$, m=2,5,8; $-(CH_2CH_2O)_jCH_2CH_2-$, j=0,1)

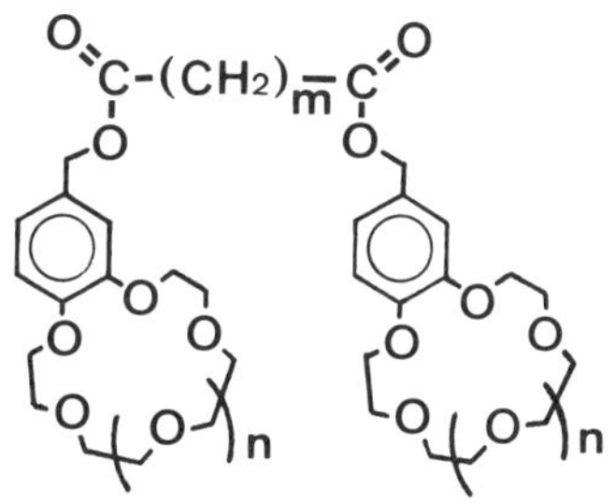

9

(n=1,2; m=1,3,5,7)

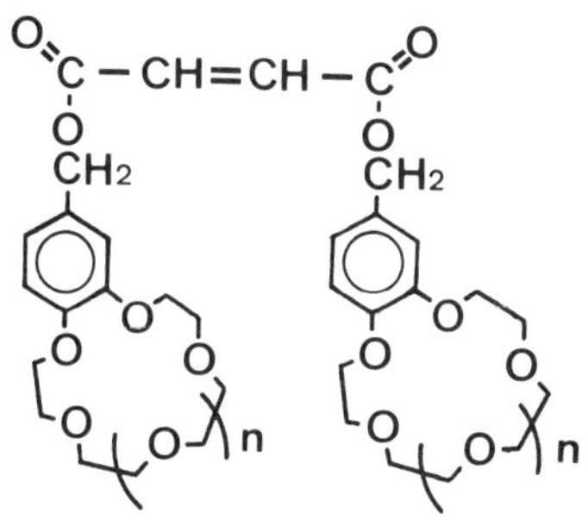

10 (n=1,2; *cis*, *trans*)

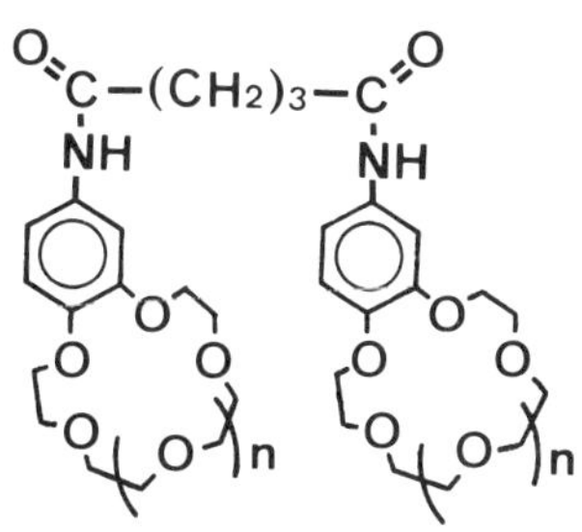

11 (n=0,1,2)

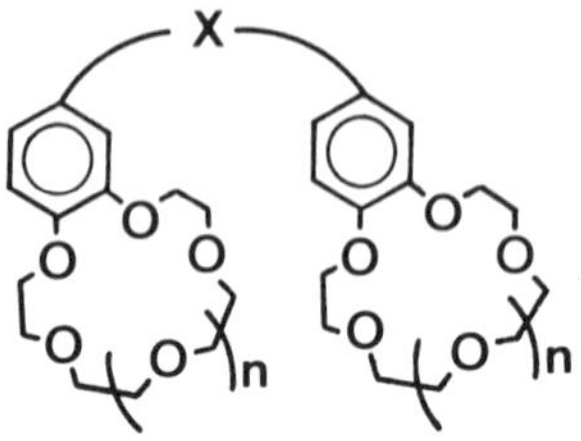

12

(n=1,2;
X= $-(OCH_2CH_2)_mCH_2CH_2O-$,
m=1-4, X=
$-CH_2(OCH_2CH_2)_jOCH_2-$
j=0-3)

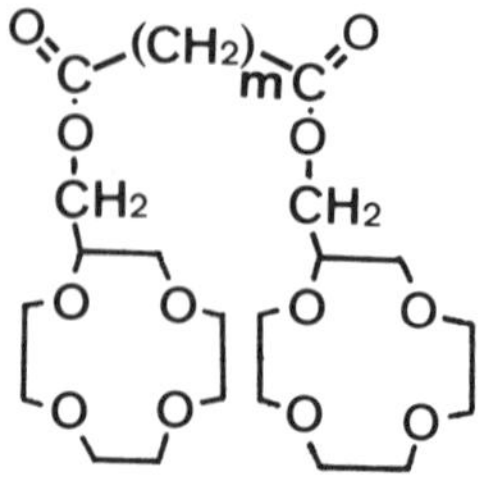

13

(m=1-3)

ethylene)s in the main chain 12 have been prepared by Williamson synthesis [20, 21].

Various bis(crown ether)s not containing benzocrown ethers but aliphatic crown ethers such as 13 [17] and 14 [22] have also been synthesized. They both are highly sodium-selective bis(crown ether)s because of the favorable formation of intramolecular 2:1 crown ether ring to sodium complexes. Malonate bis(crown ether)s incorporating lipophilic chains at the α-position, 15 (n = 1), were developed for the neutral carriers of sodium-selective electrodes [43]. These bis(crown ether)s can be prepared by transesterification of the diethyl malonate derivatives with hydroxymethylcrown ethers.

Dodecylmethylmalonate bis(12-crown-4)
15 ($n = 1$; $R_1 = CH_3$; $R_2 = C_{12}H_{25}$) [42]

To a mixture of 2.2 g of hydroxymethyl-12-crown-4 and 0.02 g of potassium metal dissolved in 2 mL of ethanol was added 0.94 g of

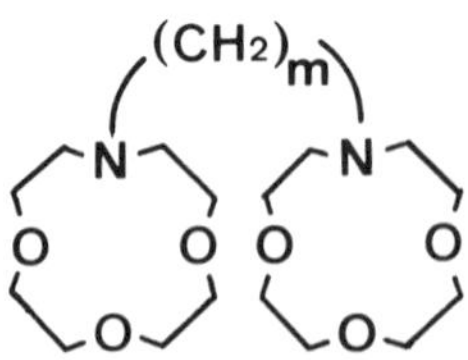

14

(m=2,3)

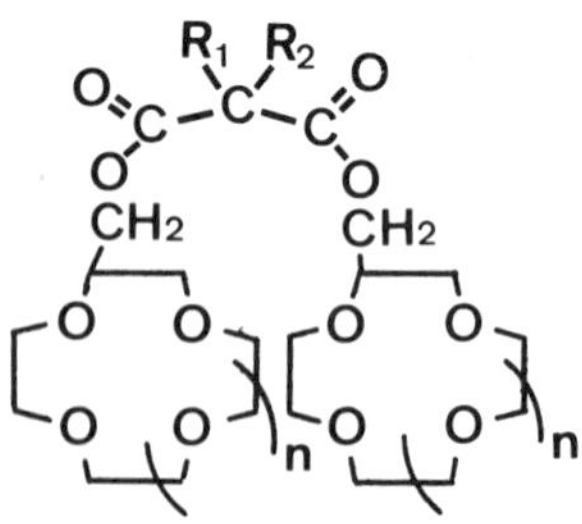

15

(n=1-3;
$R_1=R_2=C_2H_5$;
$R_1=CH_3$, $R_2=C_{12}H_{25}$)

diethyl dodecylmethylmalonate. The solution was heated to 80°C to evaporate the solvent. The reaction mixture was further heated at the same temperature over a period of 24 h. Silica gel column chromatography of the crude product yielded a viscous oil (36%).

7.3.3 Condensation-Type Crown Ether Polymers

Condensation polymerization of bifunctional crown ethers or benzocrown ethers may be a rather easy way to incorporate crown ether moiety into polymers. The first report of the condensation polymerization of crown ethers concerns the synthesis of polyamide-type crown ether polymers 16 by Feigenbaum and Michael [44]. Dibenzo-18-crown-6 was first nitrated to the 4,4'-dinitro derivatives, which can be separated to the cis- and trans isomers by fractional crystallization, and then hydrogenation of the dinitrobenzo-18-crown-6 gives the diamino derivatives. The condensation polymerization of these diaminobenzocrown ethers with iso- or terephthaloyl chloride in hexamethylphosphoramide (HMPA) yields polyamide 16. The consistently best results are obtained with the *trans*-diamino derivative and isophthaloyl chloride. The resulting crown ether polymer, which is rather insoluble in the good solvents for the conventional polyamides, gave fairly tough films by casting from hexafluoroisopropanol. The same type of dibenzo-18-crown-6 polymer was also prepared by Shchori and Jagur-Grodzinski [45].

Polyamide-type dibenzo-18-crown-6 polymer 16 [45]

Into a 250-mL reaction flask, 8.3 g of isophthaloyl chloride was distilled under vacuum. To the reaction flask was added 15.8 g of diaminodibenzo-18-crown-6 in 200 mL of HMPA. The reaction mixture was slowly stirred under an argon atmosphere in an ice bath. After 30 min all isophthaloyl chloride crystals disappeared. After another hour, the solution became extremely viscous. After dilution of the solution with *N*-methylpyrrolidone, the polymer was precipitated by pouring into vigorously stirred methanol. The product was filtered, placed in a Soxhlet, and the residual solvents were extracted with acetone.

Epoxy polymers containing diazacrown ethers 17 and cryptands, and polyesters and polyamides incorporating diazacrown ethers, have been synthesized by Gramain et al. [46-49]. Polyureas from diisocyanate and diaza-18-crown-6 were prepared by Mathias and Al-Jumah [50].

Extensive synthesis of condensation-type crown ether polymers for stationary phases of liquid chromatography has been reported by Blasius et al. [51-62]. Various types of crown ether and cryptand polymers have been synthesized; the typical ones made were polymers

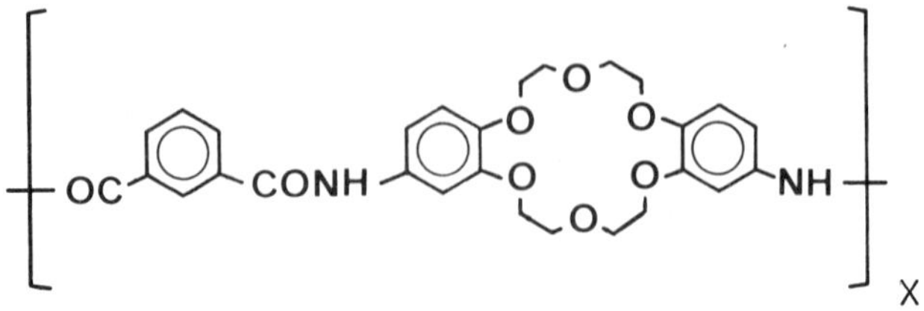

16

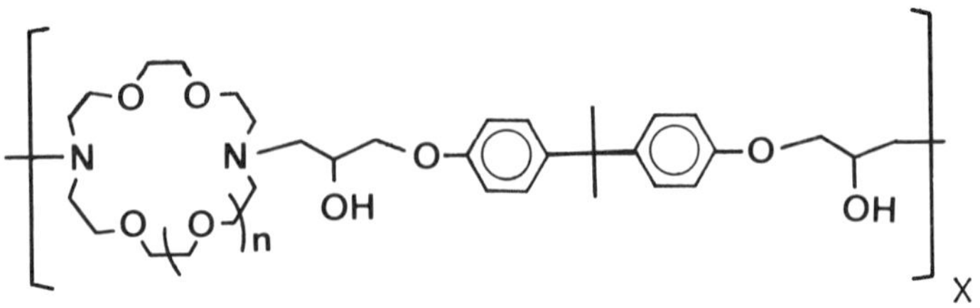

17 (n=0,1)

18 to 20. The phenol formaldehyde resins containing crown ethers were obtained by condensation polymerization of formaldehyde with dibenzocrown ethers in formic acid, or with monobenzocrown ethers in a mixture of formic and sulfuric acids, using cross-linking agents such as phenol, resorcinol, and xylol. Some crown ethers have been immobilized on cross-linked polystyrene by the reaction of hydroxymethylbenzocrown ethers with the chloromethylated resin. Cryptands were employed instead of crown ethers in some cases. These crown ether and cryptand polymers were characterized in detail by means of pyrolysis-gas chromatography and pyrolysis-mass spectroscopy [62].

Phenol-formaldehyde resin of dibenzo-18-crown-6 19 (m = n = 0) [51]

In 1 L of boiling formic acid was dissolved 180 g of dibenzo-18-crown-6. To the formic acid solution was added 60 g of formaldehyde in 500 mL of hot formic acid. The mixture was allowed to heat for 5 h, and then poured into 5 L of cold water. Dark brown resin was obtained, extracted with methanol for 48 h, and dried at 100°C under vacuum. The yield of the crown ether polymer amounted to 100 g.

Phenol-formaldehyde resin of cryptand[2.2.2B] 20 [55]

In 60 mL of 15% sulfuric acid was dissolved 5 g of dihydrochloride salt of cryptand[2.2.2B]. To the solution was added 3.5 mL of 35% formalde-

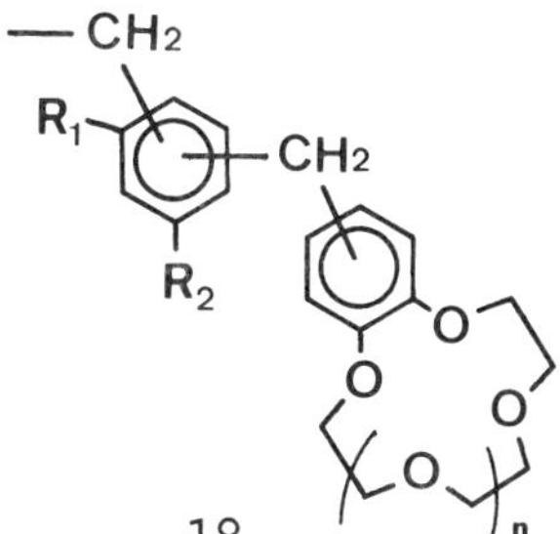

18

(n=2,3; R_1,R_2= H,OH,CH_3)

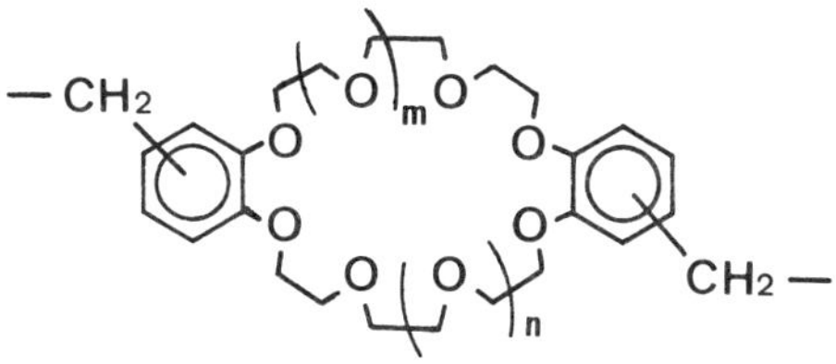

19

(m,n=0,1,2)

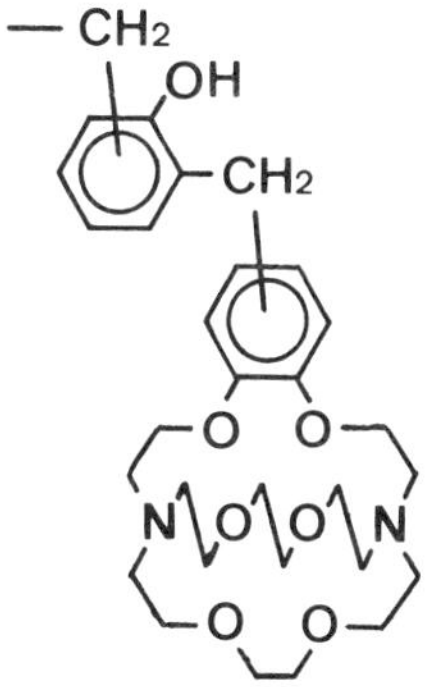

20

hyde aqueous solution, and then the mixture was heated at 90 to 110°C for 48 h, for the formation of hydroxymethylether and acetal groups. After cooling, 2 g of phenol (10% excess) was added and stirred at 100°C. After 24 h an additional 25 mL of formaldehyde solution was added and again stirred at the same temperature. The reaction mixture was neutrallized by 10% NaOH aqueous solution, the water being distilled off. The residual dark resin was crushed in a powder mill and treated with 50 mL of 3% Na_2CO_3 aqueous solution. The resin was purified by 24-h extraction in a Soxhlet with water and methanol (75% of yield).

7.3.4 Immobilized Crown Ethers on Organic and Inorganic Supports

The easiest way to gain crown ether polymers may be immobilization of crown ethers on appropriate polymer matrices. This can be achieved by the reaction of crown ethers carrying a functional group with func-

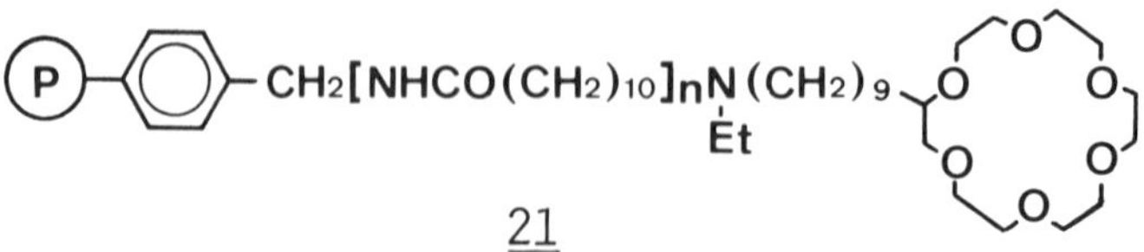

21
(n=0,1,2)

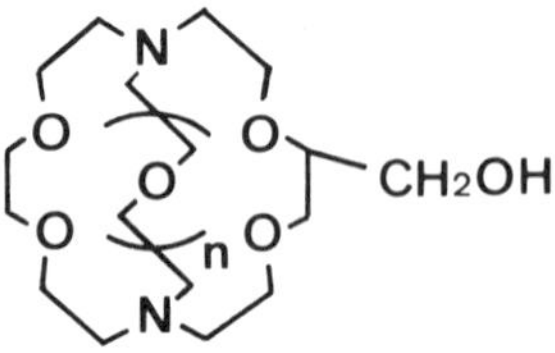

22 (n=1,2)

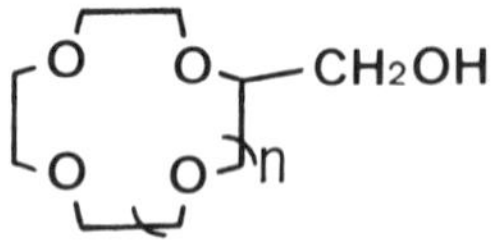

23 (n=1-3)

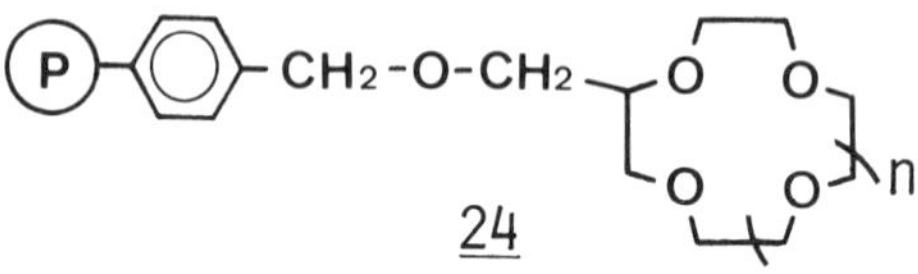

24
(n=1-3)

tionalized polymer matrices like chloromethylated polystyrene [63-67, 70,71], silica gel [61,74,75], polypeptide [76], and polyethyleneimine [126]. Cross-linked polystyrene is the most often used polymer matrix for immobilization of functional groups. Crown ethers and cryptands have been attached to the polymer matrix mainly for the use as phase-transfer catalysts [63-65].

To prepare crown ether polymer 21 (n = 0, 1, 2), crown ethers are bound to cross-linked polystyrene by a long spacer chain, which ensures enough distance between the polymeric matrix and the crown ether moiety. This results in greater accessibility of reagents to the catalytic centers in phase-transfer catalyst use. The reaction of ω-

(ethylamino)nonyl-18-crown-6 with chloromethyl polystyrene or polystyrenes appended with an appropriate ω-bromo-substituted linear chain in DMF in the presence of $NaHCO_3$ gives 18-crown-6-immobilized polystyrenes with various spacer chains. Cryptand[2.2.2] can be replaced with the 18-crown-6.

Since hydroxy group-containing crown ethers and cryptands can easily be bound to chloromethylated polystyrene under Williamson conditions, several kinds of hydroxymethyl-substituted crown ethers and cryptands, such as 22 and 23, have been synthesized [66-70] and some convenient synthetic methods have also been developed [40,71]. A typical synthetic procedure of crown ether-immobilized polystyrene 24 is described below.

18-Crown-6-immobilized polystyrene 24 *(n = 3)* [71]

Hydroxymethyl-18-crown-6 (0.76 g, 2.58 mmol) dissolved in 40 mL of THF was stirred with sodium hydride (0.19 g, 4.1 mmol) for 1 h under a nitrogen atmosphere at room temperature. Chloromethylated cross-linked polystyrene (1.00 g, 1.76 mmol of chlorine, 1% divinylbenzene, microporous, 20 to 40 mesh) was then added to the solution and the mixture refluxed without stirring for 24 h. The product mixture was cooled to room temperature and 5 mL of methanol added. The resin was filtered, washed successively with THF (5 × 10 mL), THF-water (1:1, 3 × 10 mL), water (2 × 20 mL), and THF (10 mL), and then dried under reduced pressure for 12 h, yielding 1.40 g of crown ether polymer 24 (n = 3). For quantitative replacement of chlorine by the crown ether, the theoretical expected weight of the product is 1.46 g. Similarly, 4'-hydroxymethylbenzocrown ethers can be attached to the polystyrene [72].

There is an interesting way to immobilize benzocrown ether moiety on cross-linked polystyrene, such as shown in Scheme 3. Polystyrene was modified first by catechol, and thereafter the catechol was reacted

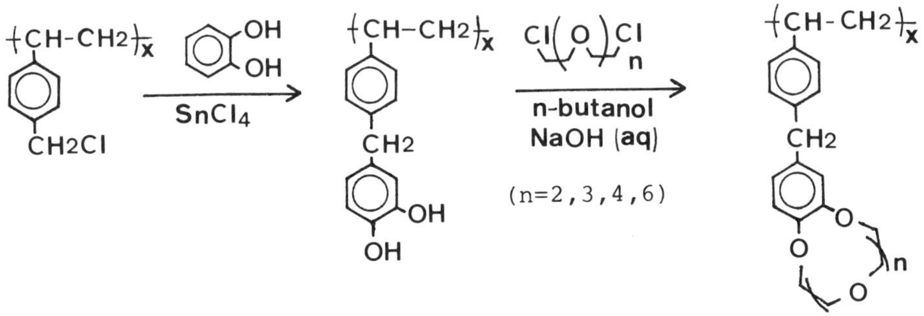

Scheme 3

with an appropriate oligo(oxyethylene) dichloride in the presence of NaOH to give benzocrown ether-immobilized polystyrenes. The disadvantage in this method is that unreacted catechol and uncyclized benzoglyme are contained in the resulting crown ether polymers as well.

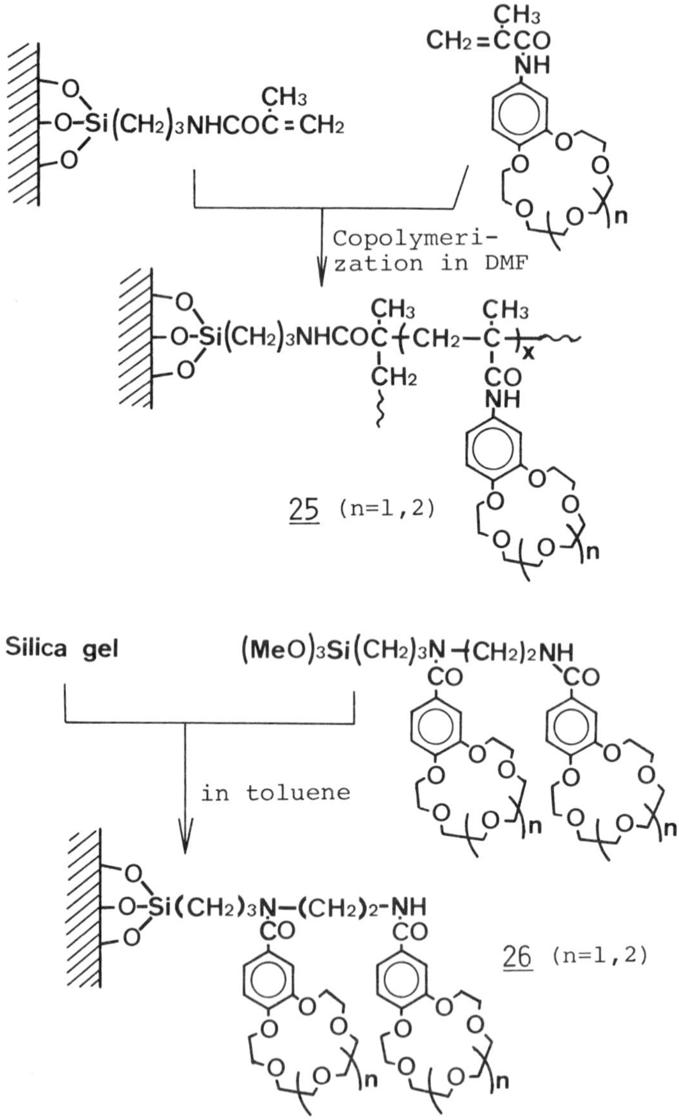

25 (n=1,2)

26 (n=1,2)

Scheme 4

Some crown ethers have been immobilized on silica gel by covalent bonding [61,74,75]. Poly- and bis(crown ether)s were attached to silica gel, as depicted in Scheme 4. Poly(crown ether)-modified silica 25 was prepared by incorporating vinyl groups on silica gel first, followed by radical copolymerization of the vinyl-modified silica with crown ether vinyl monomers. Bis(crown ether)-modified silica 26 was synthesized by reaction of silica gel with silane-coupling reagents containing bis(crown ether), which are derived from *N*-(2-aminoethyl)-3-aminopropyltrimethoxysilane and 4'-chlorocarbonylbenzocrown ethers.

Poly(benzo-15-crown-5)-modified silica 25 (n = 1) [74]

Silica gel (for HPLC, 10 μm) was activated by refluxing in concentrated hydrochloric acid for 4 h, followed by washing with water until the washing was neutral, and then dried under vacuum at 80°C. Onto the activated silica gel (13.4 g) a toluene solution of 3-aminopropyltriethoxysilane (10.0 g, 0.045 mol) was placed and then stirred for 30 min. The mixture was heated at 80°C to evaporate the solvent gradually, yielding aminopropyl-modified silica. To a suspension of the aminopropyl-modified silica in dry methylenechloride was dropwise added methacrylic anhydride (6.61 g, 0.05 mol) with cooling. The reaction mixture was stirred for 12 h at room temperature. After the reaction the vinyl-modified silica was washed successively with methylenechloride, chloroform, and methanol and then dried overnight under vacuum at 80°C. A glass tube containing a mixture of the vinyl-modified silica (6.97 g), 4'-methacryloylaminobenzo-15-crown-5 (2.50 g, 7.11 mmol), and AIBN (0.04 g) in 20 mL of DMF was degassed by a freeze-and-thaw method and then sealed. The polymerization was carried out by shaking the sealed tube in an incubator at 70°C for 21 h. After the polymerization the poly(benzo-15-crown-5)-modified silica was washed successively with DMF, chloroform, and methanol and then dried in a similar way to the previous modified silicas. The capacity of the benzocrown ether-modified silica ranged from 0.1 to 0.3 mmol of crown ether monomer unit per gram of the modified silica. Monomeric benzocrown ethers can also be attached readily to silica gel by a covalent bonding [61,74].

The other possibility to immobilize crown ethers is polymer reactions of functional crown ethers with polyacrylamide [89] and poly(glycidyl methacrylate) [90] to yield 27 and 28, which also bring cross-linking to the resulting crown ether polymers.

Crown ether-immobilized resins which possess chiral center in the crown ether ring, the spacer chain, or the backbone have been synthesized, aiming at their use for chromatographic optical resolution [77-79]. Either on cross-linked polystyrene [77] or on silica gel [78] were covalently bonded chiral bis(dinaphthyl)-22-crown-6 derivatives 29, which were prepared by sophisticated synthetic methods. The

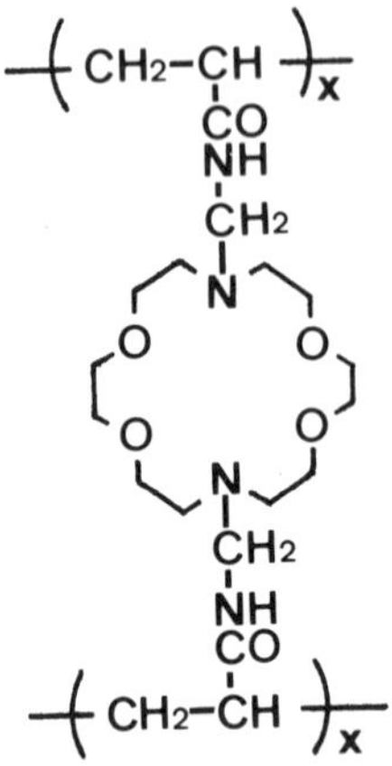

27

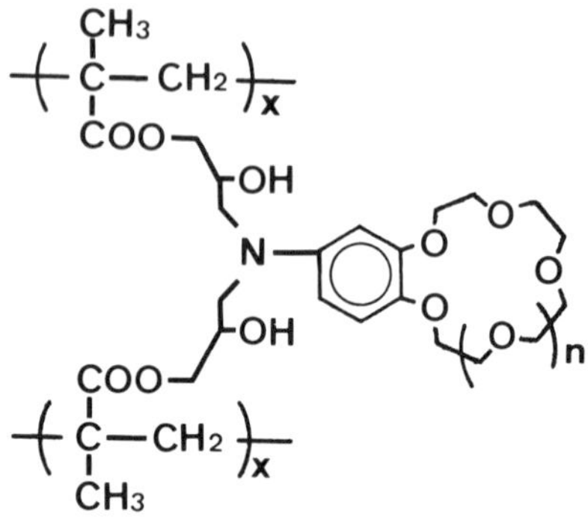

28 (n=1,2)

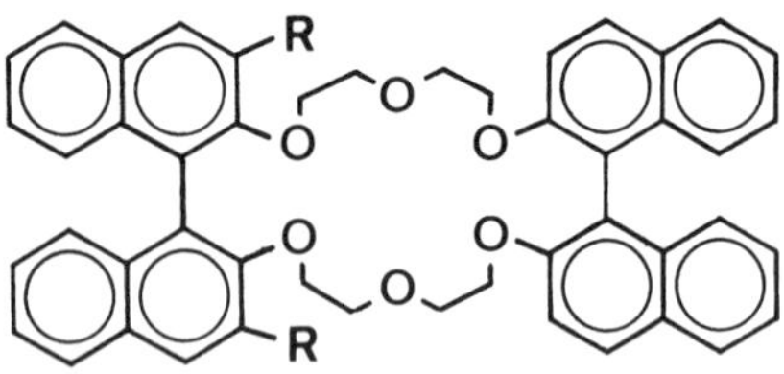

29 (R=H,CH_3)

immobilization of the chiral crown ethers was carried out by the reaction of the hydroxymethyl-substituted crown ether with chloromethylated polystyrene, or of the tetra(dimethylmethoxysilyl)-substituted crown ether with silica gel. There are some other crown ether-immobilized polymers which contains a chiral center somewhere except in the crown ether ring, such as 30 and 31 [79].

Crown ether polymer 30 has chirality on the backbone of polypropyleneimine, which can be obtained by ring-opening polymerization of 4-methyl-2-oxazoline, followed by the removal of the formyl group [80]. The reaction of 4'-chlorocarbonylbenzo-18-crown-6 with the chiral backbone polymer in the presence of imidazole affords 30. In chiral crown ether polymer 31, the chiral center is incorporated into the spacer chain between the crown ether moiety and the polymer backbone, an amino acid derivative being used as the chiral source. The chiral crown ether polymer was prepared by successive condensation reactions of (S)-alanine derivative and 4'-carboxybenzo-18-crown-6 with chloromethyl polystyrene using dicyclohexylcarbodiimide, in a similar way to Merrifield polypeptide synthesis. The copolymers of chiral crown ether monomer 32 [81] with appropriate comonomers may be alternatives to the preceding chiral crown ether polymers.

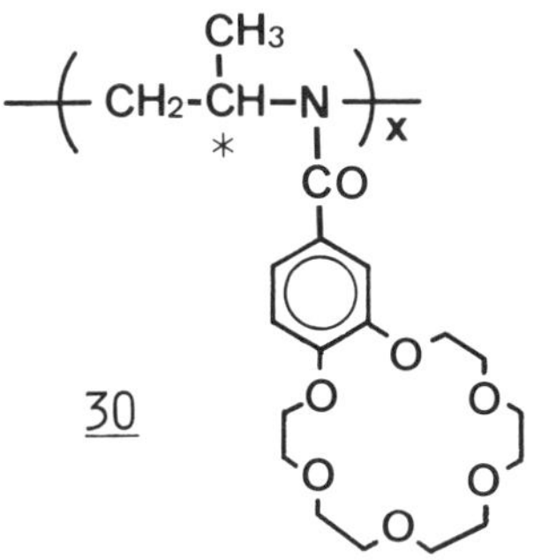

30

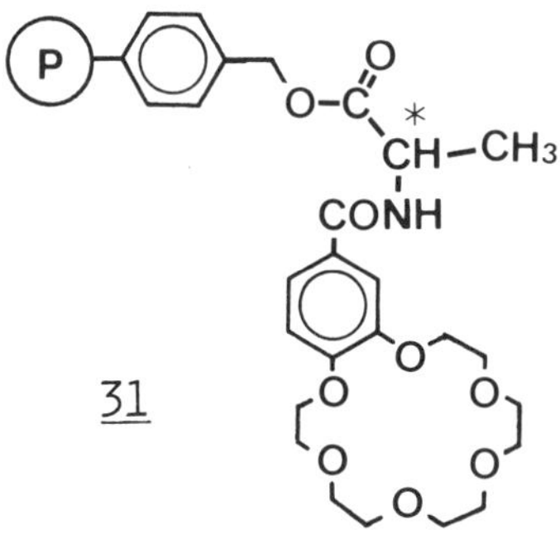

31

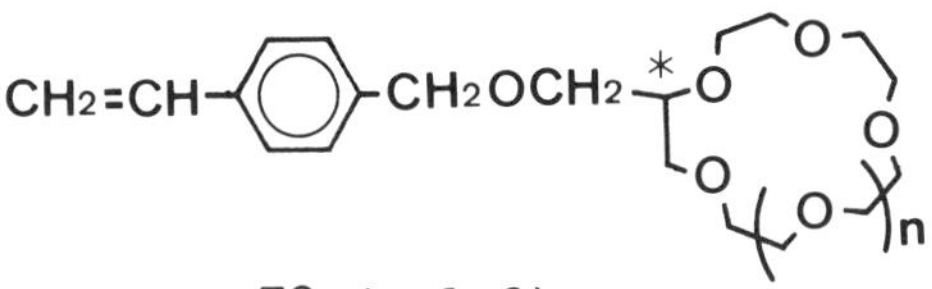

32 (n=1,2)

Polyketone-type macrocycles, which tend to exhibit great affinity for transition-metal and uranyl ions [82,83], unlike the conventional crown ethers, have been bound to polystyrene (non-cross-linked) [84,85]. The synthesis of 33 and 34 was achieved by polymer reactions of appropriate polyketone-type macrocycles with chloromethyl polystyrene in the presence of sodium hydride. Some other analogous macrocycles might be immobilized as well.

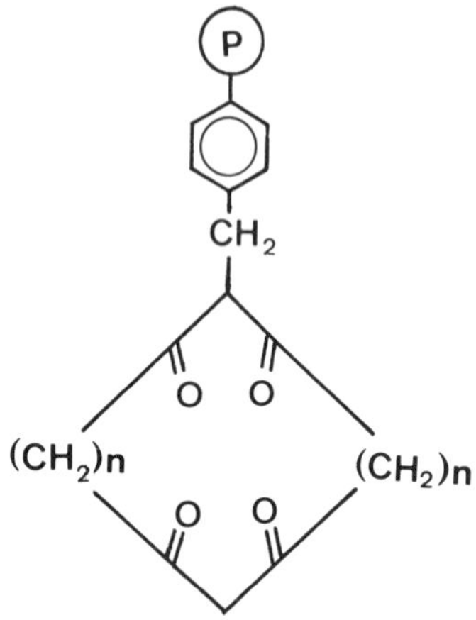

33 (n=7,8,10)

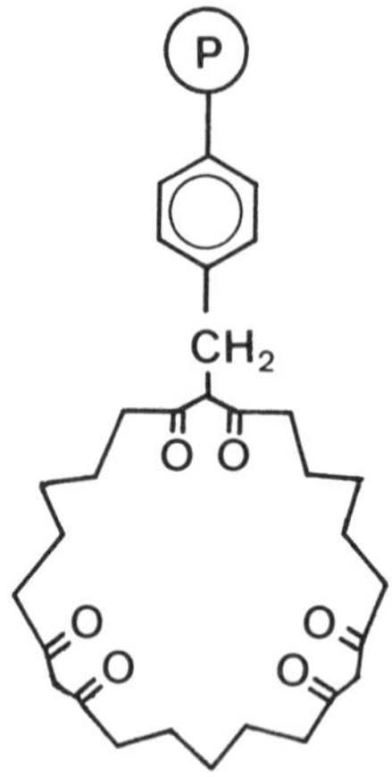

34

7.3.5 Immobilized Noncyclic Crown Ethers

Noncyclic polyethers such as poly(oxyethylene) sometimes act as if they were the corresponding cyclic analogs, and exhibit cation-binding abilities similar to those of crown ether derivatives. Therefore, they are often called noncyclic crown ethers. Poly(oxyethylene)-immobilized polystyrene [86] can easily be obtained by sodium hydride-assisted Williamson synthesis of appropriate poly(oxyethylene)s and chloromethyl polystyrene resins, as shown in Scheme 5. Polymeric pseudocrown

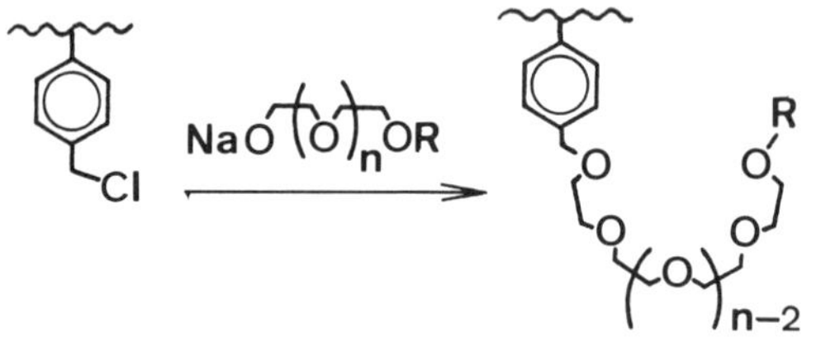

35 (n=2-9, R=H, CH_3)

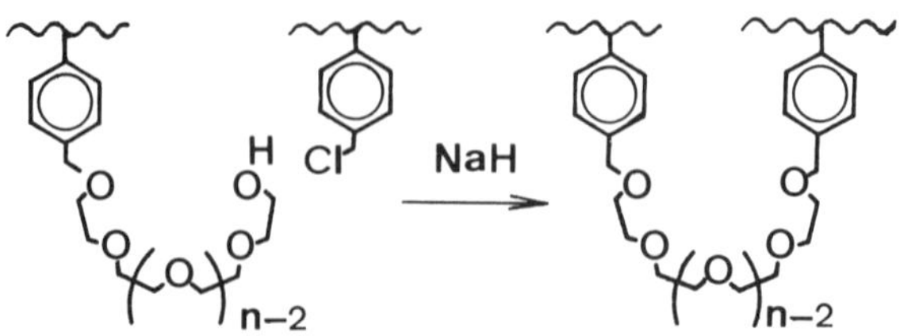

36 (n=2-9)

Scheme 5

ethers synthesized by Warshawsky et al. [87] are essentially the same as noncyclic crown ether polymer 36.

Polymeric pseudocrown ether 36 [87]

Chloromethylated polystyrene resin (Amberlite XE-305, 100 g) was placed in a three-necked 3-L flask and immersed in a thermally controlled oil bath. One liter of distilled dioxane, predried over anhydrous sodium sulfate, was charged, followed by 1 mol of an appropriate polyoxyethylene. Sixty grams of 50% sodium hydride in oil was washed rapidly with dry ether and added slowly to the reaction mixture. The temperature was allowed to rise spontaneously to 100°C. Alternatively, the disodium salt of polyoxyethylene was prepared first, and then the polymer was added. After 24 to 36 h, the reaction mixture was cooled to room temperature and methanol added slowly to destroy excess sodium hydride. The polymer was filtered and washed successively with distilled water, methanol, water, and ethanol, and then dried at 50°C under vacuum to constant weight. The polymer was purified by washing in a Soxhlet extractor for 24 h.

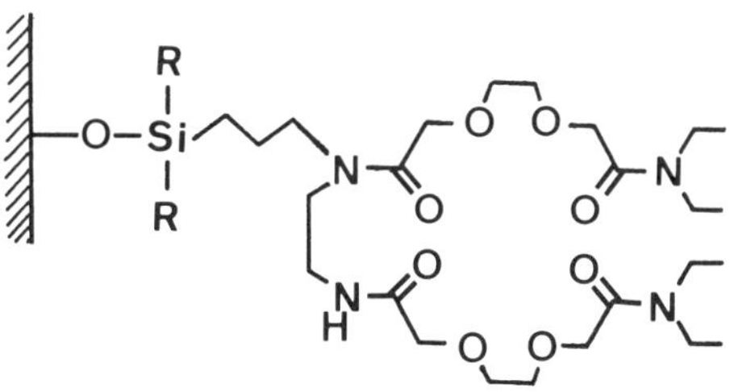

37

The immobilization of 3,6-dioxaoctanedioic acid diamide derivatives has also been attempted [88], cross-linked polystyrene, silica gel, and porous glass being employed as the supports. One of the noncyclic crown ether-immobilized silicas is 37.

7.4 APPLICATIONS OF CROWN ETHER POLYMERS IN ANALYTICAL CHEMISTRY

The main analytical applications of the crown ether polymers lie in chromatographic separation and concentration of various ions, solvent extraction, and ion-selective electrode, and the others in membrane transport of ions and derivatization prior to gas or liquid chromatography.

7.4.1 Liquid Chromatography

Insoluble crown ether polymers, such as the condensation-type ones and the immobilized crown ethers on organic or inorganic supports, are most likely to be applied for the stationary phases in liquid chromatography.

7.4.1.a Condensation-type crown ether polymers and immobilized crown ethers on polymeric supports

Blasius and co-workers have reported extensive investigations of chromatographic behavior of insoluble crown ether and cryptand polymers [51-62]. Some of the crown ether polymers are polymers 18 to 20. The cation-binding abilities of the crown ether polymers were examined by batch extraction of alkali and alkaline earth metal salts in methanol, in order to obtain some preliminary information about their chromatographic behavior for various metal ions. The ability depends remarkably on the type of metal ion, its counterion, and solvent. In general, the polymers containing benzo-15-crown-5 or dibenzo-18-crown-6 bind K^+ effectively in the group of alkali metal ions. The dibenzo-24-crown-8 polymers prefer Cs^+, while Rb^+ is favored by the dibenzo-21-crown-7 polymers. Li^+ does not seem to be bound very strongly by any of these crown ether polymers. For alkaline earth metal ions, Ba^{2+} is favored by all these crown ether polymers. Although the anion dependence of the cation-binding ability for the crown ether polymers varies slightly from one polymer to another, it seems to follow approximately the polarizability sequence

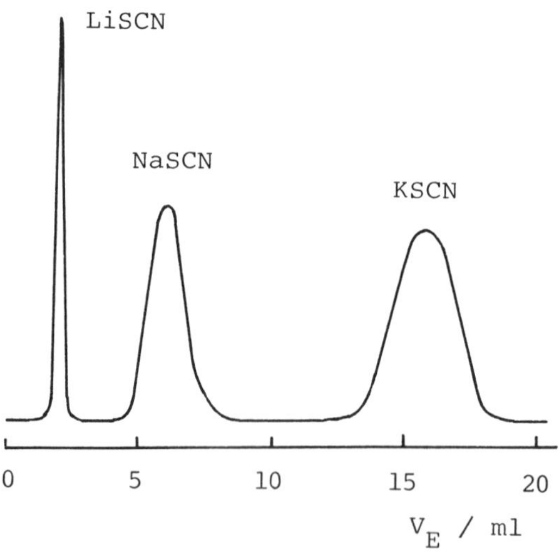

Figure 3. Chromatographic separation of alkali metal thiocyanates on a column of crown ether polymer 19 (m = 1, n = 0). LiSCN, 0.065 mg; NaSCN, 0.48 mg; KSCN, 1.77 mg; mobile phase, H_2O. (From Ref. 56.)

of the anions. For example, the order of the anion dependence for the dibenzo-18-crown-6-containing polymer is $SCN^- > I^- > Br^- > Cl^- > F^- > OH^-$ in methanol. When methanol is replaced with water, the cation-binding ability decreases drastically as anticipated.

Crown ether polymers have been investigated for use as the stationary phase of liquid chromatography. Demonstrated in Figure 3 is a typical chromatogram, in which the thiocyanates of Li^+, Na^+, and K^+ are separated by dibenzo-21-crown-7 polymer 19 (m = 1, n = 0). Chromatographic separations of other metal ions, such as alkaline earth metal, transition metal, and rare earth metal, were also attempted using the crown ether polymers. The separation of various alkylammonium chlorides is also possible using dibenzo-18-crown-6 polymer 19 (m = n = 0), as illustrated in Figure 4. In this chromatography, the primary ammonium ions were eluted more slowly, reflecting the high stability of the primary ammonium ion - dibenzo-18-crown-6 complex. It is interesting that unlike the conventional cation exchangers, these crown ether polymers are capable of separating anions as well as various

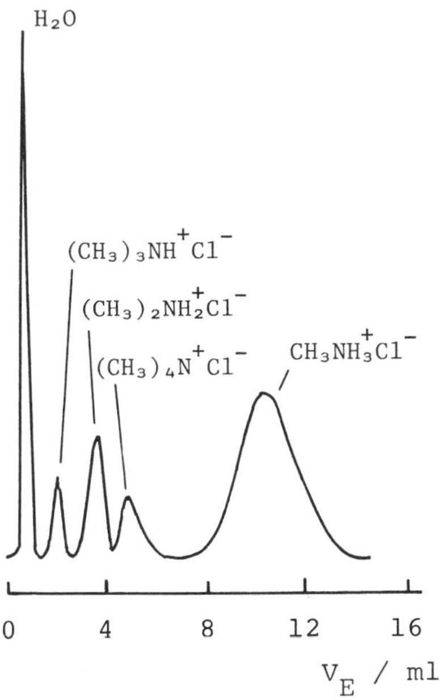

Figure 4. Separation of alkylammonium chlorides on a column of crown ether polymer 19 (m = n = 0) by elution of methanol. CH_3NH_3Cl, 0.094 mg; $(CH_3)_3NHCl$, 0.048 mg; $(CH_3)_2NH_2Cl$, 0.049 mg; $(CH_2)_4NCl$, 0.081 mg. (From Ref. 57.)

cations. This is attributed to the marked anion dependence of the cation-binding ability of the "neutral" ion exchangers. Depicted in Figure 5 is a chromatogram of a mixture of potassium halides and thiocyanate on a column of crown ether polymer 19 (m = 1, n = 0). The interconversion of mixtures of two salts was found to be feasible. For example, when a mixture of NH_4SCN and RbCl was injected on a column of benzo-15-crown-5 polymer 18 (n = 2; R_1 = H, R_2 = OH) and then eluted with water, only two peaks of NH_4Cl and RbSCN were observed on the chromatogram. This means complete interconversion from a mixture of NH_4SCN and RbCl to that of NH_4Cl and RbSCN. This chromatographic technique may be used for preparing valuable salts which are not commercially available.

Another interesting application of the crown ether polymers is chromatographic separation of neutral organic compounds. A chromatogram of furan, thiophene, and pyrrole on a column of dibenzo-30-crown-10 polymer 19 (m = n = 2) is shown in Figure 6. It appears that sulfur- and nitrogen-containing compounds interact with the crown ether polymer more strongly than oxygen-containing ones, although the binding force between the organic compounds and the crown ether polymers has not been explained. Chromatographic separation of urea-thiourea mixtures and halobenzene mixtures (Figure 7) are also possible on the crown ether polymers. Another application

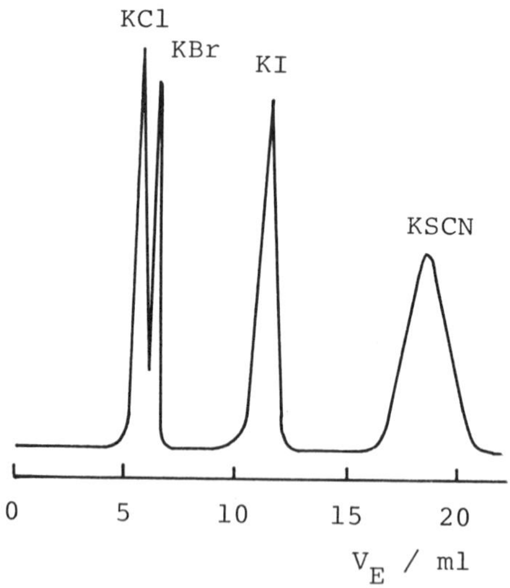

Figure 5. Chromatographic separation of anions on a column of crown ether polymer 19 (m = 1, n = 0). KCl, 0.42 mg; KBr, 0.60 mg; KI, 1.60 mg; KSCN, 1.77 mg; mobile phase, H_2O. (From Ref. 56.)

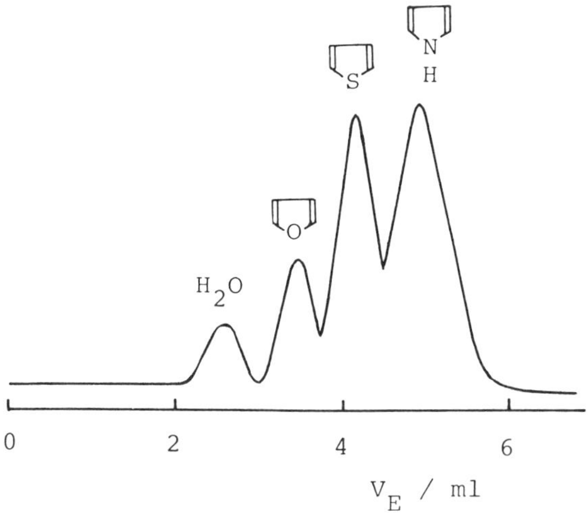

Figure 6. Chromatographic separation of neutral organic compounds on a column of crown ether polymer 19 (m = n = 2). Furan, 0.2 mg; thiophene, 0.2 mg; pyrrole, 0.2 mg; mobile phase, MeOH. (From Ref. 58.)

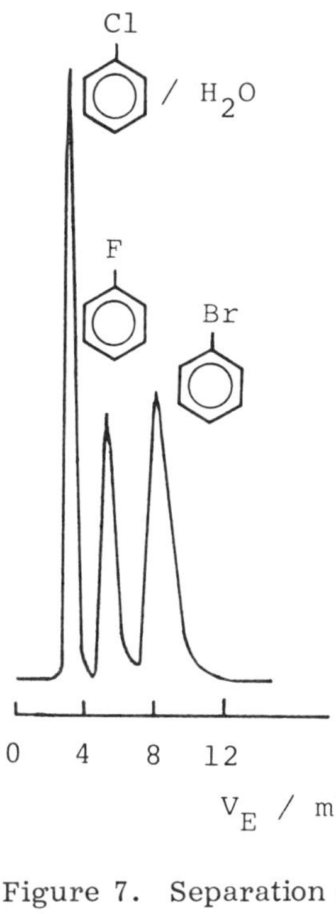

Figure 7. Separation of halobenzene derivatives on a column of crown ether polymer 19 (m = n = 0) by elution with methanol. Fluorobenzene, 1.0 mg; chlorobenzene, 2.0 mg; bromobenzene, 1.0 mg. (From Ref. 60.)

is the determination of hydration water of salts by liquid chromatography using these crown ether polymers. When $CaCl_2 \cdot H_2O$ was eluted with methanol on a column of dibenzo-21-crown-7 19 (m = 1, n = 0), $CaCl_2$ and the water of hydration were separated completely. The linear calibration plot afforded the determination of the water of hydration in the salt.

Condensation-type crown ether polymers 17 were grafted on cross-linked polystyrene to yield a stationary phase of liquid chromatography, which was applied to the separation of alkali and alkaline earth metal ions [49]. Condensation-type crown ether polymer 16 was also tested for use as the stationary phase in liquid chromatography [91]. A moderate separation of alkali metal chlorides was attained on a column packed with this crown ether polymer and eluted with water (Figure 8). The immobilized crown ethers on polyacrylamide 27, poly(glycidyl methacrylate) 28, and cellulose [92] may be good candidates for the stationary phases of liquid chromatography.

7.4.1.b Immobilized crown ethers on silica gel

Crown ether polymers, which consist of the condensation-type polymers, the crown ether-immobilized polystyrene, and the cross-linked crown

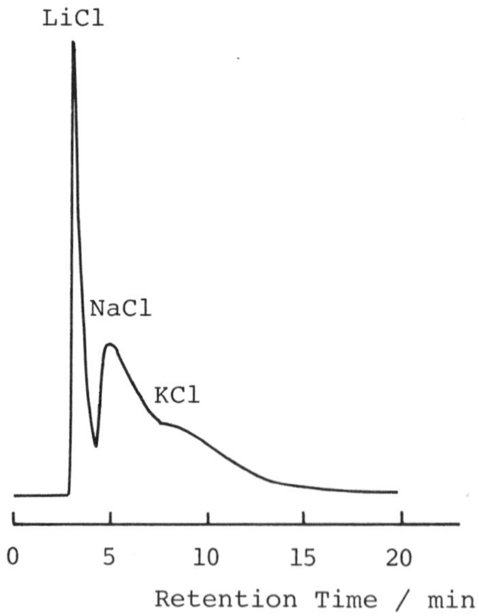

Figure 8. Chromatogram of a mixed solution of alkali metal chlorides on a column of crown ether polymer 16. Mobile phase, H_2O, flow rate, 0.44 mL/min. (From Ref. 91.)

ether polymers, may not be suited for high-performance liquid chromatography (HPLC), where pressure-resistant stationary phases are required. Inorganic materials containing hard core such as silica gel seem to be preferable as the packing supports of HPLC. Poly(crown ether)-modified silicas 25 (n = 1, 2) has been synthesized specifically for HPLC [74, 75]. The liquid chromatography studies of alkali and alkaline earth metal ions were conducted using the columns packed with the crown ether modified silicas by a conventional slurry packing method. The poly(crown ether) silicas bind alkali metal salts, even the halides, so strongly that methanol cannot elute out K^+ and Rb^+ salts from the columns. It was found that water containing a small quantity of methanol is generally a good solvent for the chromatography of the alkali metal salts. Typical chromatographic separations of alkali metal halides by the poly(benzo-15-crown-5)- and poly(benzo-18-crown-6)-modified silicas are depicted in Figures 9 and 10, respectively. The retention sequence of alkali metal chlorides on 25 (n = 1) is $Li^+ < Na^+ < Cs^+ < Rb^+ < K^+$, and that of 25 (n = 2) $Li^+ < Na^+ < Rb^+ < K^+ < Cs^+$. As mentioned in the preceding sections, the poly(crown ether)s with the crown ether moiety on the side chain possess much better cation-binding properties compared to the corresponding monomeric analogs. Thus, by the formation of 2:1 crown ether ring to cation complexes, poly(benzo-15-crown-5) and poly(benzo-18-crown-6)

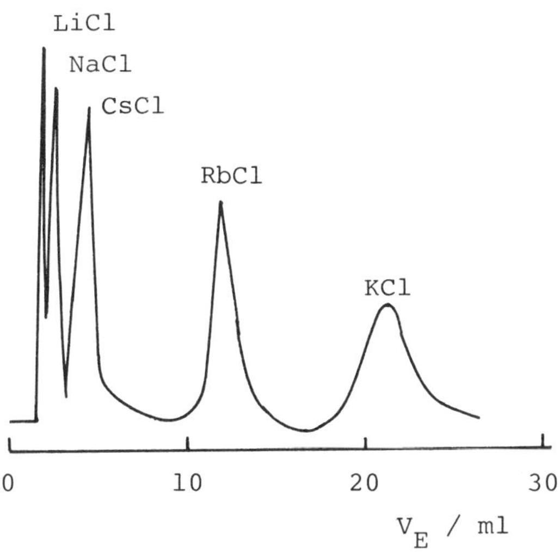

Figure 9. Separation of alkali metal chlorides on a column of poly(crown ether)-modified silica 25 (n = 1) by elution with water-methanol (90/10). LiCl, 6.4 μg; NaCl, 9.4 μg; KCl, 17.2 μg; RbCl, 17.2 μg; CsCl, 26.9 μg. (From Ref. 74.)

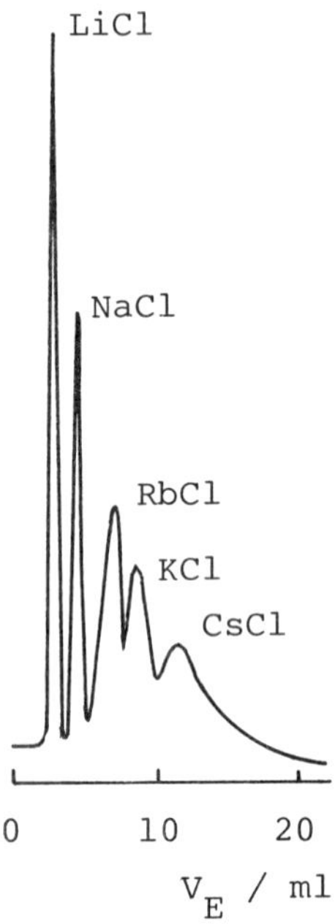

Figure 10. Separation of alkali metal chlorides on a column of poly(crown ether)-modified silica 25 (n = 2). LiCl, 0.80 mg; NaCl, 0.12 mg; KCl, 0.15 mg; RbCl, 0.24 mg; CsCl, 0.34 mg; mobile phase, H_2O-MeOH 90/10).

exhibit pronounced selectivities for K^+ and Cs^+, respectively. The cation-binding properties of the poly(crown ether)s seem to be reflected in the chromatographic behavior of these poly(crown ether)-modified silicas. Poly(12-crown-4)-modified silica, in which poly(12-crown-4) 6 (n = 1; R = CH_3) is immobilized on silica gel in a manner similar to 25, was also attempted for its use as a stationary phase of ion chromatography [93] (Figure 11). Since the modified silica interacts with Na^+ strongly, as expected from the cation-binding tendency of the poly(12-crown-4) [17], the metal ion is eluted more slowly on a column of modified silica. The poly(crown ether) silicas were also tested for their usefulness in liquid chromatography of alkaline earth metal ions (Figure 12). Moreover, chromatographic separation of various anions is feasible with the poly(crown ether)-modified silicas, in analogy with Blasius's crown ether polymers. The capacity factors of potassium halides, which are of course dependent on the composition of mobile phases (Figure 13), increase in the order $Cl^- < Br^- < I^-$. Bis(crown

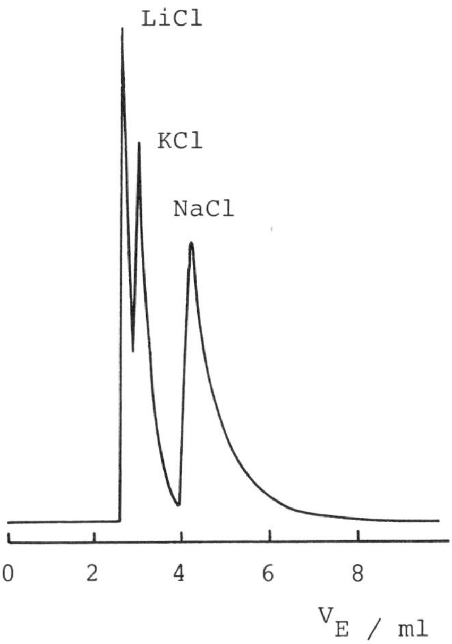

Figure 11. Separation of alkali metal chlorides on a column of poly(12-crown-4)-modified silica. LiCl, 0.08 mg; NaCl, 0.23 mg; KCl, 0.15 mg; mobile phase, H_2O-MeOH (50/50).

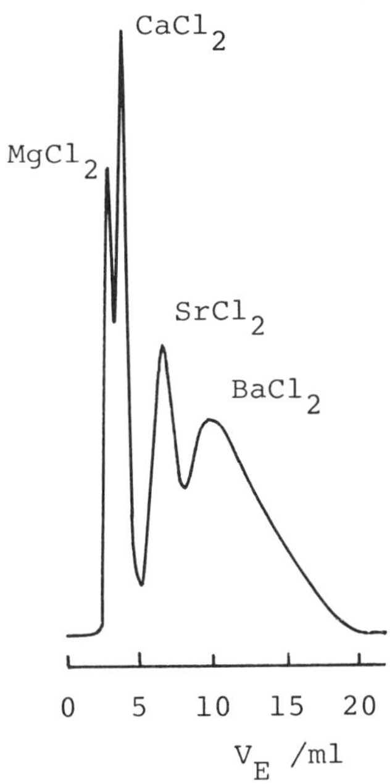

Figure 12. Separation of alkaline earth metal chlorides on a column of poly(crown ether)-modified silica 25 (n = 2). $MgCl_2$, 0.10 mg; $CaCl_2$, 0.15 mg; $SrCl_2$, 0.27 mg; $BaCl_2$, 0.63 mg; mobile phase, H_2O/MeOH (90/10).

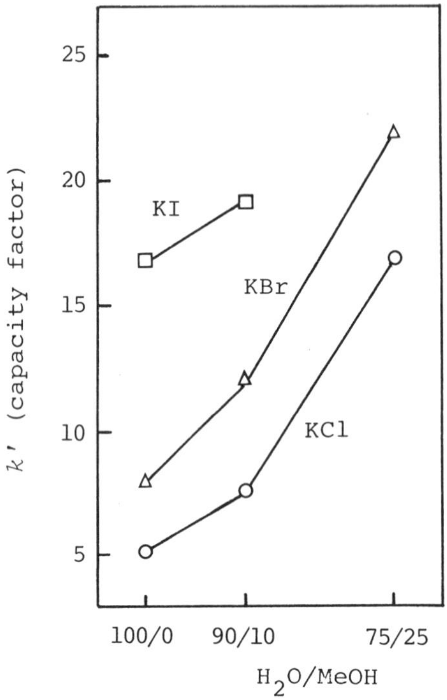

Figure 13. Dependence of solvent composition on capacity factor of potassium halides on a column of poly(crown ether)-modified silica 25 (n = 1).

ether)-modified silicas 26 are almost the same in the chromatographic behavior for alkali and alkaline earth metal ions as the corresponding poly(crown ether)-modified silicas. Furthermore, the silica gel similarly modified by poly(thiocrown ether) 4, which possesses a great affinity for Ag^+ [94], is a promising candidate of the stationary phase for the chromatographic separation of heavy metals such as Ag, Hg, Cd, and the other transition metals.

Monomeric benzo-15-crown-5 [but not poly- or bis(crown ether) derivative] was immobilized covalently on silica gel and used as the stationary phases of ion chromatography [61,74]. In the liquid chromatography of alkali metal chlorides using the benzo-15-crown-5 silica [61], the elution with pure methanol allowed good separation among Li^+, Na^+, Cs^+, and K^+ (or Rb^+). The cation-binding ability of the benzo-15-crown-5 silica is, however, not as high as that of the poly(benzo-15-crown-5) silica. Moreover, the separation between

K^+ and Rb^+ is incomplete on the monomeric crown ether silica, unlike the polymeric crown ether silica, although they resemble each other in the retention sequence of the alkali metal ions.

Chromatographic behavior of silica gel coated by polyamide-type crown ether polymer 16 has also been investigated [95]. In the liquid chromatography of alkali metal salts using the crown ether-coated silica, the metal ions can hardly be separated, unlike another stationary phase containing the same type of crown ether polymer [91], whereas good separation among various anions was observed (Figure 14). In other words, this crown ether-coated silica permits the separation of anions but not of cations. This phenomenon has never been observed for the crown ether polymers or the immobilized crown ethers. A column packed with this crown ether-coated silica was attempted for its usefulness on chromatographic analysis of various anions in seawater.

7.4.1.c Immobilized noncyclic crown ethers

Poly(oxyethylene)s with various chain lengths were immobilized on chloromethylated cross-linked polystyrene, aiming at their chromatographic application [86]. The noncyclic crown ether polymers cannot

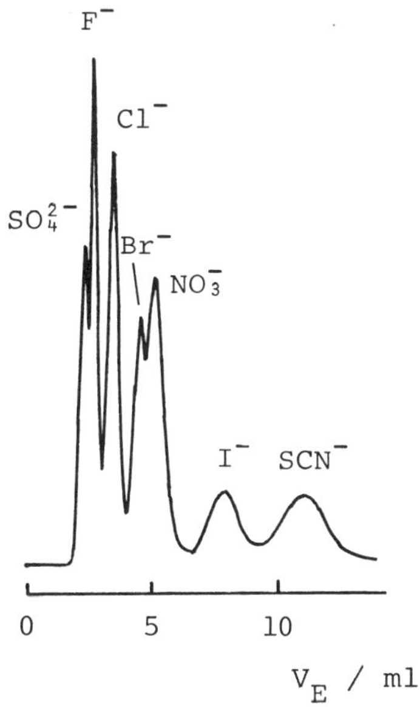

Figure 14. Chromatogram of a mixture of anions on a column of silica coated by polyamide-type crown ether polymer 16. Mobile phase, H_2O; volume injected, 20 μL; salt concentration, 0.1 mol/L. (From Ref. 95.)

interact with alkali metal ions as strongly as the "cyclic" crown ether polymers. However, good separation of thiocyanates of Li^+, Na^+, and K^+ was attained in the liquid chromatography using a column of noncyclic crown ether polymer 35 (n = 10) and water-acetone (10/90) mobile phase, as shown in Figure 15. Polymeric pseudocrown ether 36 could also be used as the stationary phases of liquid chromatography [87].

The immobilized 3,6-dioxaoctanedioic acid diamide derivatives on cross-linked polystyrene, NH_2-modified silica gel, or glass (37) were applied for chromatographic separation of alkali and alkaline earth metal ions [88]. Elution with water or water-alcohol mixtures, only poor separation of the chlorides salts was obtained in liquid chromatography on immobilized noncyclic crown ethers. However, using a technique of solvent gradient of a water-alcohol system, good separation of the iodide salts was achieved, as demonstrated in Figure 16.

7.4.1.d Optical resolution

Crown ethers are known to form stable complexes with ammonium ions, especially primary ones. Some crown ethers incorporating chiral centers in the crown ether cavity can form complementary complexes

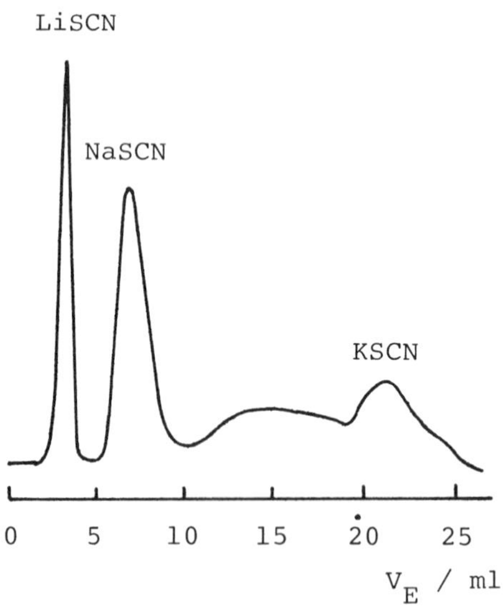

Figure 15. Chromatographic separation of alkali metal thiocyanates on a column of noncyclic crown ether polymer 35 (n = 9). Sample, LiSCN/NaSCN/KSCN (1.4/1/1.6); mobile phase, H_2O-acetone (10/90). (From Ref. 86.)

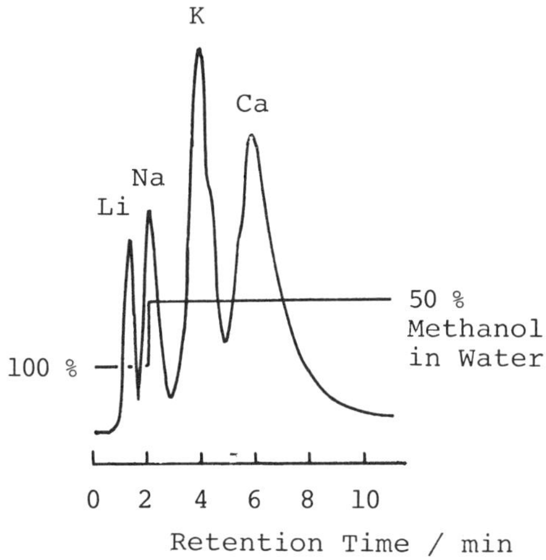

Figure 16. Separation of lithium, sodium, potassium, and calcium iodides on noncyclic crown ether polymer 37 by using a solvent gradient technique. Mobile phase, H_2O-MeOH; flow rate, 1.4 mL/min; sample, 20 μL, 4.1×10^{-3} M LiI, 4.6×10^{-3} M NaI, 1.3×10^{-2} M KI, 9.2×10^{-3} M CaI_2. (From Ref. 88.)

with ammonium salts of chiral amines and amino acids (or the esters) [96]. That is, the chiral crown ethers possess the potential of chiral recognition, therefore optical resolution of the amino compounds. The feasibility of resolving chromatographically racemic mixtures of amines and amino acid derivatives was demonstrated using the immobilized chiral crown ethers by Cram and co-workers [77, 78]. Bis(dinaphthyl)-22-crown-6 derivatives 29 were bound covalently to silica gel and cross-linked polystyrene to yield the stationary phases for the chromatographic optical resolution. Excellent optical resolution of phenylglycine perchlorate or methyl phenylalaninate hydrochloride was obtained with the (R,R)-bis(dinaphthyl)-22-crown-6 silica; a typical chromatogram is shown in Figure 17. In this case the stable complex formation of (S)-methyl phenylalaninate hydrochloride with the (R,R)-crown ether resulted in the slower elution of the (S)-isomer compared to the (R)-isomer.

Some attempts were also made to resolve racemic mixtures of threonine, aspartic acid, histidine, tryptophan, phenylalanine, and valine by the liquid chromatography using phenol-formaldehyde resin of dibenzo-18-crown-6 19 (m = n = 0), which does not contain any chirality in the chemical structure [53]. A slight resolution of the

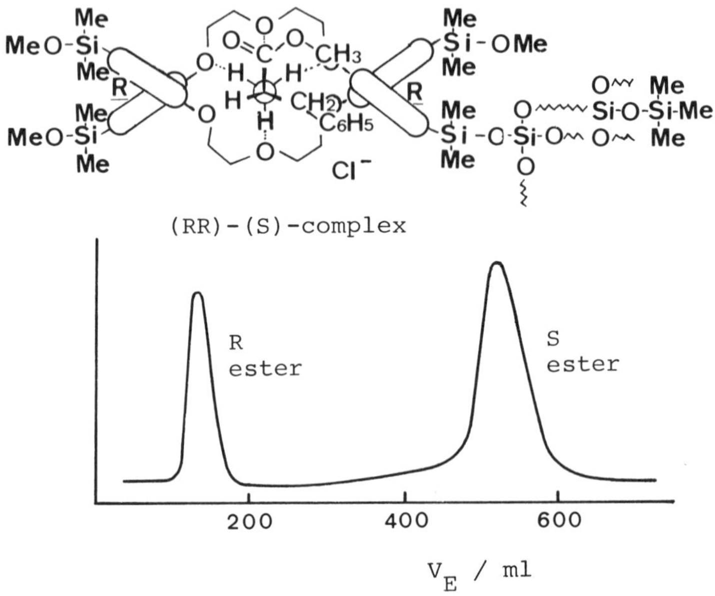

Figure 17. Chromatographic optical resolution of methyl phenylalaninate hydrochloride salt by (R,R)-bis(dinaphthyl)-22-crown-6-bound silica. (From Ref. 77.)

racemates was observed (Figure 18), which might suggest that some chirality was produced in the crown ether polymer during the preparation.

7.4.1.e Isotope separation

Isotope separation is a major target in analytical application of crown ethers. The single-stage separation factor on the isotope separation is generally very small. For example, the factor of $^{44}Ca/^{40}Ca$ is 1.0010 ± 0.0002 using dicyclohexyl- or dibenzo-18-crown-6 [97]. However, this single-state separation factor of $^{44}Ca/^{40}Ca$ is very good compared with those for the conventional ion-exchanger method (1.00011). Separation of calcium isotope ($^{48}Ca/^{40}Ca$ and $^{44}Ca/^{40}Ca$) has been examined by column chromatography using a commercially available cryptand polymer [98]. The chromatography was carried out in a thermostated column of 60 cm length and 1.2 cm i.d. packed with 19 g of cryptand polymer ([2.2.2B] polymer, E. Merck). In each run 40 mg of calcium of neutral isotopic composition was applied to the column as $CaCl_2$ dissolved in 1 mL of methanol-chloroform-water (70:30 v/v + 1.65 vol % water) and then eluted with this solvent (ca. 0.5 mL/min). In Figure 19 is depicted the dependence of the isotopic ratio R for $^{48}Ca/^{40}Ca$ on the amount of substrate eluted, $\Delta m/m$, showing

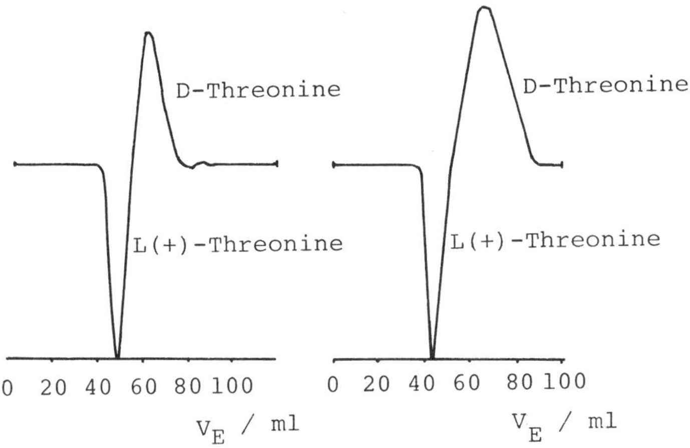

Figure 18. Chromatographic resolution of DL-threonine on a column of crown ether polymer 19 (m = n = 0). Left, 40 μL of DL-threonine; right, 40 μL of DL-threonine + 40 μL of D-threonine; mobile phase, MeOH. (From Ref. 53.)

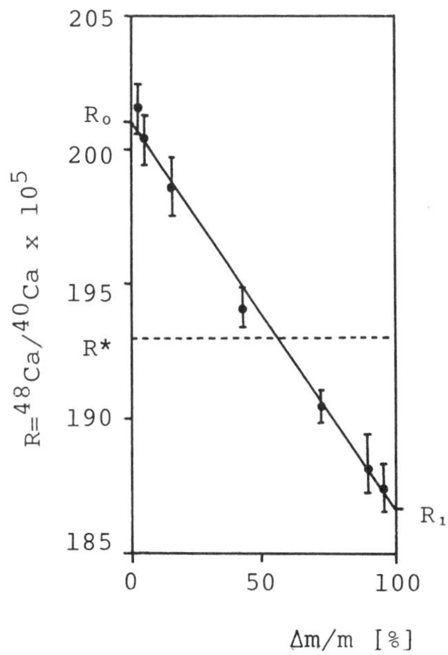

Figure 19. Calcium isotope separation as function of the eluted amount of substance $\Delta m/m$ at 20°C. R^*, initial isotopic ratio; R_0 and R_1, extrapolated isotope ratio at $\Delta m/m$ = 0 or 1. (From Ref. 98.)

that this cryptand polymer realizes the enrichment of calcium isotope in the solution. In fact, 210 mg of calcium having a natural isotope abundance produced 30 mg of calcium, in which ^{48}Ca was enriched by 3.3%. The other crown ether polymers of this type may also be applicable to the isotope separation of alkali and alkaline earth metals.

7.4.2 Solvent Extraction

Crown ethers can generally lipophilize alkali and alkaline earth metal ions by complexation and extract them from aqueous to organic layers, together with their counteranions. Therefore, they have been extensively utilized as effective extraction agents. Some crown ether polymers have also been attempted for their usefulness as extraction agents, taking advantage of their high extractability and selectivity.

7.4.2.a Efficient extraction by poly- and bis(crown ether)s

The solvent extraction of alkali and alkaline earth and heavy metal picrates had been investigated with two types of poly(crown ether)s 1 and 3, and bis(crown ether) 11, using chloroform or dichloromethane as the organic solvent. The extraction data [9] for poly(crown ether)s 1 (n = 1,2) are summarized in Table 1, together with those for their corresponding monomeric analogs. Poly(benzo-15-crown-5) 1 (n = 1) has a much greater extraction capacity than the monomeric analog; the difference is remarkable in the case of K^+, Rb^+, and Cs^+. For benzo-18-crown-6 systems, the poly(crown ether) again extracts alkali metal picrates more efficiently than does its monomeric analog, but the extraction capacity difference is not so remarkable as that observed for benzo-15-crown-5 systems. In this case, the extraction capacity for Cs^+ is improved drastically by poly(benzo-18-crown-6) 1 (n = 2). The main reason for the higher extraction capacity of the poly(crown ether)s is that they can intramolecularly form lipophilic 2:1 crown ether ring to cation complexes with certain ions which are slightly larger than the respective crown ether cavity. The lipophilicity of the polymer backbone may also contribute to the higher extraction capacity of the poly(crown ether)s. Polyacrylamide-type poly(crown ether)s 3 are essentially the same in the extraction capacity tendency as 1. Thus the poly(benzo-15-crown-5) derivative is selective for K^+ or Rb^+ in the extraction of alkali metal picrates, and the poly(benzo-18-crown-6) derivative for Cs^+ or K^+. Poly(benzo-21-crown-7) 3 (n = 3; R = H) does not exhibit as enhanced extraction capacity for alkali and alkaline earth metal ions as that observed for the formation of the 2:1 crown ether ring-to-cation complexes by two adjacent crown ether rings, because benzo-21-crown-7 ring is too large to form the 2:1 complexes with the metal ions [127]. However, the extraction of relatively large organic ions such as guanidinium ion by the poly(benzo-

Table 1. Solvent Extraction of Picrate Salts by Poly(Crown Ether)s 1 as Compared to Their Monomeric Analogs[a]

	Picrate salt extracted (%)			
Cation	1 (n = 1)	Monomer of 1 (n = 1)	1 (n = 2)	Monomer of 1 (n = 2)
Li^+	5.3	< 1	3.1	< 1
Na^+	13.1	2.3	11.2	4.2
K^+	68.8	8.1	72.2	55.5
Rb^+	67.8	4.7	68.0	30.7
Cs^+	38.2	1.6	82.2	20.0
NH_4^+	6.0	< 1	19.3	3.8
Ba^{2+}	14.7	< 1	17.5	2.1

[a][Picric acid]: 7×10^{-5} M, [crown ether unit]: 3.5×10^{-4} M, [metal hydroxide]: 0.01 M, $CHCl_3$-H_2O system.
Source: Ref. 9.

21-crown-7) is possible. Poly(benzo-12-crown-4) 3 (n = 0; R = H) was also applied to the solvent extraction of alkali metal picrates [16], but it was found to be less efficiently than the poly(benzo-15-crown-5) and poly(benzo-18-crown-6) derivatives. This may be due to the poor cation-binding ability. Poly(12-crown-4) 6 (n = 1; R = H) can be used instead of the poly(benzo-12-crown-4), being capable of extracting effectively alkali metal ions, especially Na^+ [17].

Bis(crown ether) derivatives, in which two crown ether rings are connected at the end of a short aliphatic chain, exhibit similar cation-binding properties [18, 19] and extraction capacities [11] to those of the corresponding poly(crown ether)s. However, *trans*-bis (crown ether)s 10 (n = 1, 2) cannot extract alkali metal picrates as effectively as the other bis(crown ether)s and its cis isomer [23, 24]. A comparison of the extraction capacity of *cis*- and *trans*-bis(benzo-15-crown-5) 10 (n = 1) is presented in Figure 20. The marked difference in the extractability for K^+ and Rb^+ suggested that in the *trans*-bis(crown ether) the two crown ether rings are forced to be far from each other, making it difficult to form the intramolecular 2:1 crown ether ring to cation complex. The cis isomer, however, can form the 2:1 complex easily because of the favorable configuration. The different cation-

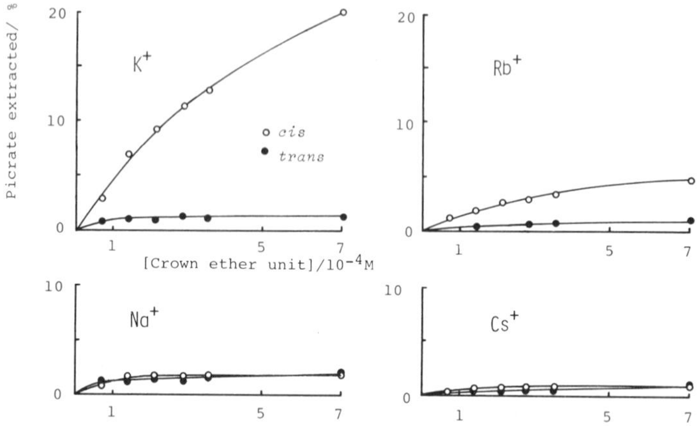

Figure 20. Extraction of alkali metal picrates by *cis*- and *trans*-bis(crown ether)s 10 (n = 1). Chloroform-water solvent system, [MOH] = 0.01 M, picric acid = 7×10^{-5} M in H_2O, at 25°C.

binding properties of the *cis*- and *trans*-bis(crown ether)s were also observed spectrophotometrically as described in detail in Section 7.2 (Figure 2) [24]. The configuration of poly(crown ether)s, that is, their tacticities, should affect their extractabilities. In the solvent extraction of alkali metal picrates with isotactic, syndiotactic, and atactic poly(crown ether)s 5 (n = 1, 2; R = CH_3), the isotactic poly(benzo-15-crown-5) extracted K^+ and Rb^+ more favorably than did the syndiotactic one, and the isotactic poly(benzo-18-crown-6) extracted Cs^+ and Rb^+. The effect of the tacticity upon the extraction capacity of the poly(crown ether)s is, however, not as drastic as observed in the *cis*- and *trans*-bis(crown ether) systems.

Poly(benzocrown ether)s 3 (n = 1, 2; R = H) have been employed as extraction agents for alkaline earth metal ions [100] and some heavy metal ions such as Ag^+ and Tl^+ [101]. In alkaline earth metal extraction, Ba^{2+} and Sr^{2+} were extracted effectively by both poly(crown ether)s 3 (n = 1, 2; R = H). Poly(benzo-15-crown-5) 3 (n = 1; R = H) is also an excellent extraction agent for Tl^+ and Ag^+. The intramolecular formation of the 2:1 crown ether ring-to-cation complexes bind these ions strongly, especially Tl^+, which resembles K^+ in ionic size. The poly(benzo-15-crown-5) derivative would be applied to the removal and the concentration of highly toxic Tl^+ in aqueous samples.

Poly(thiocrown ether)s 4 and their bis(crown ether) derivatives, in which some oxygen atoms are replaced with sulfur in poly- and bis(crown ether)s 3 and 11, are powerful extraction agents for Ag^+

[94]. The partial replacement of the oxygen atoms with sulfur causes the poly(crown ether) to lose affinity for alkali and alkaline earth metals drastically, so these metal ions are ineffectively extracted.

7.4.2.b Extraction equilibrium constant

Attempts have been made to evaluate the extraction equilibrium constants (K_e) for the extraction systems of the poly- and bis(crown ether)s using chloroform as the organic solvent. The extraction equilibrium between aqueous solution of metal ions (M^+), picrate anion (A^-), and a chloroform solution of crown ether (Cr) can be defined for the 1:1 crown ether ring to cation complex by the following equations:

$$M^+_{aq} + A^-_{aq} + Cr_{org} \xrightleftharpoons{Ke} MCA_{org}$$

$$K_e = \frac{[MCrA]_{org}}{[M^+]_{aq}[A]_{aq}[Cr]_{org}} \tag{1}$$

Since chloroform is used as the organic solvent, the dissociation of the ion pair MCrA is negligible in the organic layer [25], and the concentration of the uncomplexed cation in the chloroform layer is extremely low compared with that of the complex. The concentration of crown ether in the aqueous layer is also very low. The following assumptions can then be made:

$$\frac{[MCrA]_{org}}{[M^+]_{aq}} = D \tag{2}$$

$$[M^+]_{aq} = [A^-]_{aq} = M° - \underset{\sim}{A} \tag{3}$$

$$[Cr]_{org} = Cr° - \underset{\sim}{A} \tag{4}$$

where D, M°, and Cr° denote the distribution ratio of metal picrates between organic and aqueous layers, the initial concentrations of metal picrates, and crown ethers [for the monomer unit in case of poly- and bis(crown ether)s], respectively. $\underset{\sim}{A}$ refers to the concentration of picrates transferred to the organic layer, which can be determined spectrophotometrically. Thus Equation 1 can be rewritten

$$D = K_e(M° - \underset{\sim}{A})(Cr° - \underset{\sim}{A}) \tag{5}$$

If these assumptions are reasonable, the plots of -log D against $-\log(M° - \underset{\sim}{A})(Cr° - \underset{\sim}{A})$ should give a straight line with a slope of 1, and log K_e can be obtained from the intercept of the straight line. If 2:1 complexes of crown ether ring and metal ion are mainly formed, the following equation can be applied:

$$D = K_e (M^\circ - \underset{\sim}{A})(Cr^\circ - 2\underset{\sim}{A})^2 \qquad (6)$$

In the case of the intramolecular 2:1 complex formation by the poly- and bis(crown ether)s, two adjacent crown ether rings can be considered as a single entity, and then the following equation 7 can be used instead of Equation 6:

$$D = K_e(M^\circ - \underset{\sim}{A})(Cr^\circ - 2\underset{\sim}{A}) \qquad (7)$$

The stoichiometry of the complexes can be determined conductometrically and spectrophotometrically. A graphical determination of the extraction equilibrium constant in the alkali metal extraction systems by the monomeric analogs of polyacrylamide-type poly(crown ether) 3 (n = 1, 2; R = H) is shown in Figure 21, and in Table 2 are summarized the log

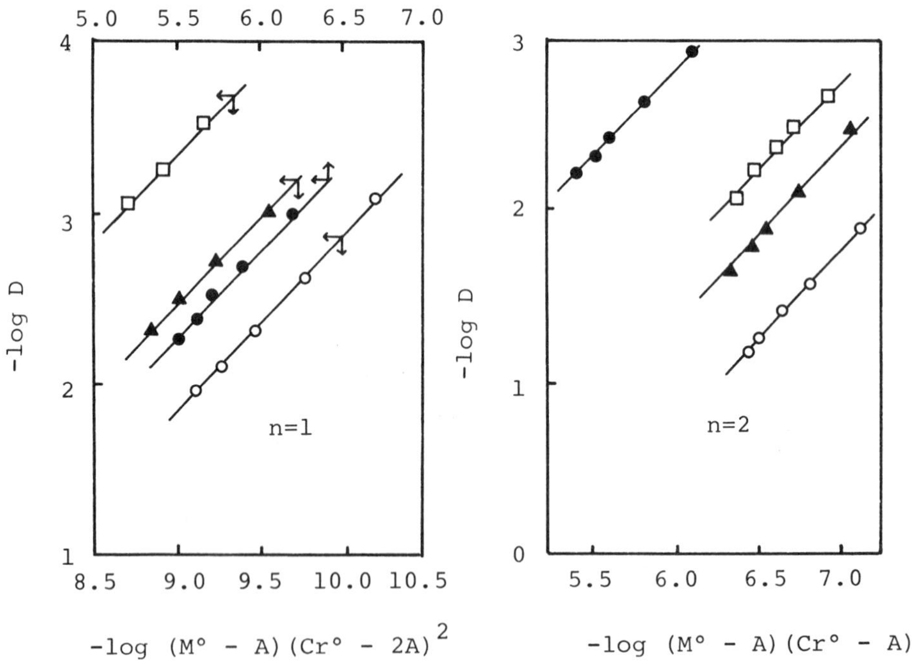

Figure 21. Graphical determination of extraction equilibrium constant for the monomeric analog system of poly(crown ether) 3 (n = 1, 2; R = H). Na^+ (●), K^+ (○), Rb^+ (▲), Cs^+ (□). (From Ref. 11.)

Table 2. Extraction Equilibrium Constants (K_e) in Extraction Systems of Poly- and Bis(crown ether)s 3 and 11

	log K_e[a]			
Crown ether	Na^+	K^+	Rb^+	Cs^+
3 (n = 1)	3.80	7.75	7.20	5.44
Monomer of 3 (n = 1)	3.10	(7.10)	(6.54)	(5.59)
11 (n = 1)	3.46	6.63	6.13	4.85
3	3.61	5.82	5.59	5.62
Monomer of 3 (n = 2)	3.16	5.22	4.64	4.24
11 (n = 2)	3.41	5.46	5.35	6.36

[a]The units for K_e are L^2/mol^2 (L^3/mol^3 for the values in parentheses at 25°C.
Source: Ref. 11.

K_e values for the extraction systems of poly(crown ether)s 3 (n = 1, 2; R = H), their monomeric analogs, and bis(crown ether)s 11 (n = 1, 2).

7.4.2.c Other crown ether polymers and their practical use

Soluble condensation-type crown ether polymers 17 have been utilized as extraction agents in water-chloroform solvent system [46,47]. The crown ether polymers strongly complex alkali (Na^+, K^+, and Rb^+) and alkaline earth (Ca^{2+}) metal ions, the complex stoichiometry is 1:1 crown ether ring to cation. The crown ether polymers, especially the polymers with larger crown ether rings show high extraction capacities for transition-metal ions such as Cu^{2+}, Cd^{2+}, and Pb^{2+}, chiefly because of the coordination ability of the nitrogen atoms in the crown ether cavity. Some experiments have been performed with mixtures of alkali and alkaline earth metal ions, to evaluate the possible use of the crown ether polymers for separation purposes. However, the extraction selectivity of the polymers does not seem to be very high.

Soluble functional polystyrenes on which polyketone-type macrocycles were immobilized covalently (33 and 34) have been employed for the selective concentration of transition-metal and uranyl ions [84,85]. Particularly, hexaketone-bound polystyrene 34 has been tested for the concentration of uranyl ion in seawater using solvent extraction method. The typical procedure is as follows. Hexaketone-bound polystyrene 34 (0.1 mmol equivalent of macrocycle) dissolved in 30 mL of chloroform was stirred magnetically with 3.5 L of seawater sample. The chloroform layer was separated after 18 h and the seawater was extracted with chloroform (5 mL × 2) to recover any dissolved polymer and/or polymer-uranyl complexes. From the combined chloroform solution, uranyl ion was back-extracted with 1 M hydrochloric acid solution (5 mL). The resulting aqueous solution contained 36% of uranyl ion in the original seawater. Repeated extraction and back-extraction (three times) of the uranyl ion from 15 L (3 × 5 L) of seawater accumulated 13.8 μg in the recovered polymer. Little deterioration of the hexaketone-bound polystyrene was observed on the repeated employment. Analogous hexaketone-bound polystyrene 34, in which the polystyrene backbone is partially sulfonated and therefore more hydrophilic, may be more effective in the extraction of uranyl ion from seawater.

7.4.3 Ion-Selective Electrode

Ion-selective electrodes have been given considerable attention as one of the most conventional apparatus for the determination of ion concentration in solutions. Crown ethers can be applied to the neutral carrier-type ion-selective electrodes. Attempts have been made to use some poly- and bis(crown ether) derivatives as the neutral carrier.

7.4.3.a Selectivity coefficient

The selectivity of a measured ion over interfering ions is the most important factor for ion-selective electrodes. This is generally represented as potentiometric selectivity coefficient k_{IJ}^{pot} (the degree of selectivity of a measuring ion I over an interfering ion J). The selectivity coefficient is defined by the Nicolsky (modified Nernst) equation.

$$E = \text{constant} + S\left[\log a_I + k_{IJ}^{pot}\, a_J^{(z_I/z_J)}\right]$$

where

E = experimentally observed potential
S = Nernst constant
a_I = activity of the measuring ion, I
a_J = activity of the interfering ion, J

z_I = charge of I ion
z_J = charge of J ion

In analytical applications, the selectivity for a measuring ion should be as high as possible, so that the electrode potential has a Nernstian dependence on the ion activity over as wide a range of activity as possible. To fulfill this condition, the selectivity coefficient k_{IJ}^{pot} should be very "small." The various equilibria with respect to a measuring ion (I^+), its counterion (X^-), crown ether (S), and the complex (IS^+) are present between the aqueous layer and the liquid membrane, as shown in Figure 22. This is also the case in the interfering ion (J^+) system. It is generally accepted that in the liquid membrane-type electrode k_{IJ}^{pot} is adequately approximated by the equilibrium constant K_{IJ} for the following ion-exchange reaction, that is, the ratio of the extraction equilibrium constants for I^+ and J^+, K_J/K_I.

$$(IS)^* + J^+ \underset{}{\overset{K_{IJ}}{\rightleftharpoons}} (JS)^* + I^+$$

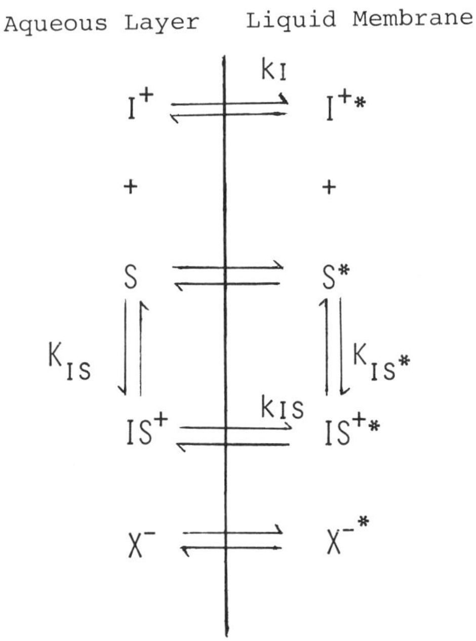

Figure 22. Various equilibria between the aqueous layer and the liquid membrane.

(The asterisk denotes species in the liquid membrane). In other words, the electrode exhibits high selectivity for the cation, which can easily be extracted by the crown ether employed. Furthermore, the selectivity coefficient k_{IJ}^{pot} can be rewritten as follows:

$$k_{IJ}^{pot} = \frac{K_J}{K_I} = \frac{K_{JS}^{*}k_J}{K_{IS}^{*}k_I} = \frac{K_{JS}k_{JS}}{K_{IS}k_{IS}}$$

If the distribution constant of the crown ether-cation complex is independent of the kind of the cation ($k_{IS} = k_{JS}$), the selectivity coefficient equals approximately the ratio of the stability constants of the crown ether complexes in the aqueous layer ($k_{IJ}^{pot} \doteqdot K_{JS}/K_{IS}$). Valinomycin, which is often utilized as a neutral carrier of K^+-selective electrode, wraps around certain metal ions completely on its complex formation. Therefore, the selectivity coefficient k_{KJ}^{pot} for the K^+-selective valinomycin electrode are in good agreement with the ratio of the complex stability constants for K^+ and the interfering ion J^+, as can be seen in Figure 23 [128]. In the case of complex formation by crown ethers, the solvation effect of the cation complexes formed is not negligible, so that k_{IJ}^{pot} does not necessarily equal the ratio of the complex stability constants. Thus the selectivity of the electrodes is governed more or less by the polarity of organic solvent and the charge of metal ion as well as the cation-binding ability of crown ether.

7.4.3.b Prerequisites for neutral carriers of ion-selective electrode

The response mechanism on the neutral carrier-type liquid membrane electrodes can be interpreted in terms of a type of metal ion transport across a liquid membrane. The ion transport by crown ethers consists of the following three processes: (a) the complex formation of crown ether with metal ion, (b) the dissolution and diffusion of the crown ether-metal complex in the liquid membrane, and (c) the release of the metal ion from the complex. Three of the processes must be in balance so that the metal ion is allowed to move across the liquid membrane effectively. For process (a) the crown ether should bind the metal ion strongly and selectively. To fulfill process (b) smoothly, the crown ether is given enough lipophilicity, but the crown ether should not be too big to diffuse freely. For process (c) the decomplexation rate of the crown ether complex must be facile so that the complexed metal ion is readily liberated. Listed in Table 3 are the rates of complexation and decomplexation for some crown ethers, cryptand, and valinomycin complexes with Na^+ and K^+. The complexation rate does not vary markedly from one complexation system to

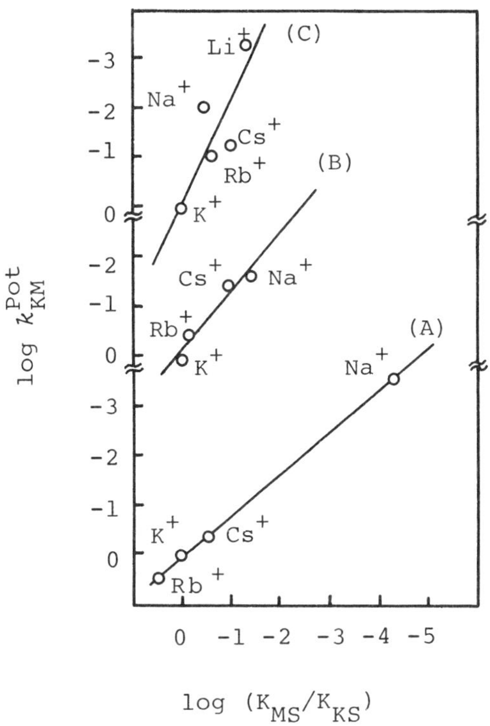

Figure 23. Correlation between selectivity coefficients and complex stability constants for electrodes based on neutral carriers. A, valinomycin; B, nonactin; C, dicyclohexyl-18-crown-6. (From Ref. 128.)

another, but it should be noted that the decomplexation rates for the cryptand systems are tremendously slow compared to those for the other systems. These results may suggest that rigid neutral carriers such as cryptand are not suited for ion-selective electrodes.

7.4.3.c Bis(crown ether)-based electrodes and their practical use

Monocyclic crown ethers have been utilized as the neutral carriers of K^+-selective electrodes [102-106]. An electrode body is incorporated in a liquid membrane in which a small quantity of crown ether is dissolved in water-immiscible solvents such as nitrobenzene and diphenyl ether, or a polymeric membrane such as poly(vinyl chloride) (PVC) and polysiloxane membranes containing the neutral carrier

Table 3. Rate Constants of Complexation and Decomplexation for Neutral Carrier Complexes

Neutral carrier	Cation	Solvent	Complexation ($L \cdot mol^{-1} \cdot S^{-1}$)	Decomplexation (S^{-1})
Valinomycin	Na^+	MeOH	1.4×10^7	2×10^6
	K^+	MeOH	4×10^7	1.3×10^3
Cryptand[2.2.2]	Na^+	H_2O	2×10^5	27
	K^+	H_2O	7.5×10^6	38
18-Crown-6	Na^+	H_2O	2.2×10^8	3.4×10^7
	K^+	H_2O	4.3×10^8	3.7×10^6
Dibenzo-18-crown-6	Na^+	MeOH	1.6×10^7	1.3×10^5
	K^+	MeOH	6×10^8	1.6×10^4

Source: Ref. 2.

together with an appropriate plasticizer. The liquid membrane is sometimes hard to handle, so at present the polymeric membrane is employed preferentially. Most of the K^+-selective electrodes based on the monocyclic crown ethers possess k_{KNa}^{Pot} (selectivity coefficient of k^+ over Na^+) of about 10^{-2}. A rather good k_{KNa}^{pot} value is 2.2×10^{-3} for the electrode based on dimethyldibenzo-30-crown-10 derivative [105], but even the k_{KNa}^{pot} value is far from comparable to about 10^{-4}, the value for the valinomycin-based electrode, which is one of the most excellent K^+-selective electrodes and now commercially available. If the crown ether concentration in the liquid or polymeric membrane is raised extremely, to something like 1 M, the selectivity coefficient for the K^+-selective electrodes may be improved [107].

However, the crown ether concentration in the membrane should normally be as low as possible to prevent electrode deterioration caused by the dissolution of the crown ether into the aqueous sample layer. Thus it appears that the electrodes based on monocyclic crown ethers are not very ion selective. Bis(crown ether) derivatives, which are highly ion selective because of the easy formation of the intramolecular sandwich-type complexes, have so far proved to be excellent neutral carriers for the ion-selective electrodes of various alkali metal ions [19,43,108-113,117].

Bis(crown ether)s 9 (n = 1; m = 1, 3, 5, 7) have been utilized as the neutral carrier of PVC membrane K^+-selective electrode, taking advantage of their high selectivity for K^+ [19,109]. The typical procedure for the preparation of the K^+-sensitive PVC membrane and the electrode construction is described below. PVC (100 mg), crown ether (1-10 mg), and plasticizer such as *o*-nitrophenyl octyl ether (NPOE), dipentyl phthalate (DPP), and dibutyl phthalate (DBP) (250 mg) were dissolved in THF (3 mL). The solution was poured into a petri dish of 30 mm diameter. The solvent (THF) evaporated slowly at room temperature and a film of 0.15 mm thickness was obtained. A disk of 4 mm diameter was cut out from the PVC membrane and incorporated into an electrode body (Orion Model 92). The diameter for the exposed membrane area was about 2 mm. After injection of 1×10^{-3} M KCl solution as the internal solution, the electrode was conditioned overnight soaking in a potassium salt solution. The external reference was a standard calomel electrode with a 0.1 M ammonium nitrate electrolyte bridge. The composition of the electrochemical cell is

$$Ag\cdot AgCl \mid 0.001\ M\ KCl \mid PVC\ membrane \mid measured\ solution \mid 0.1\ M\ NH_4NO_3 \mid KCl(sat.) \mid Hg_2Cl_2(s), Hg$$

The electrode properties of the K^+-selective electrodes based on bis(crown ether)s 9 (n = 1; n = 1, 3, 5, 7) are summarized in Table 4, together with the corresponding monocyclic crown ether and valinomycin. The bis(crown ether)-based electrodes are superior to the monocyclic crown ether electrode in the selectivity coefficient k^{pot}_{KM}, especially in that over Na^+, k^{pot}_{KNa}. It is noteworthy that the K^+ electrode based on bis(crown ether) 9 (n = 1; m = 5) possesses an excellent k^{pot}_{KNa} value of 4×10^{-4}, NPOE being used as the plasticizer of the PVC membrane. This k^{pot}_{KNa} value is comparable to the k^{pot}_{KNa} value for the valinomycin electrode of 2×10^{-4}. Moreover, Rb^+ does not interfere as much with the determination of K^+, while it does seriously in the valinomycin electrode, the k^{pot}_{KRb} value of which is not lower than unity. The length of the chain connecting two crown ether moieties in the bis(crown ether)s affects their cation-binding ability [18], which in turn changes the selectivity coefficient for the bis(crown ether)-based electrodes. The electrode selectivity also seems to be governed partially by the type of the plasticizer in the membrane. Bis(crown ether) 11 (n = 1) was also tested for its usefulness as a neutral carrier of K^+-selective electrode [19], but the electrode did not work well because of the poor solubility in the plasticizers.

Since the bis(benzo-15-crown-5) derivatives possess great affinity for Tl^+ as well as for K^+, they can be used as the neutral carrier of Tl^+-selective electrode [110]. Naturally, the Tl^+-selective electrode shows relatively poor preference of Tl^+ over K^+ and Rb^+ [$k^{pot}_{TlK} = 3 \times 10^{-1}$, $k^{pot}_{TlRb} = 2 \times 10^{-1}$ for the bis(crown ether) 9 (n = 1; m = 5)-NPOE electrode system], but the selectivity coefficients of Tl^+ over

Table 4. Properties of K^+-Selective PVC Membrane Electrodes Based on Bis(crown ether)s 9 (n = 1), and 11 (n = 1), Poly(crown ether)s 3 (n = 1) as Compared to Their Monomeric Analog and Valinomycin

Crown ether	Maximal slope (mV/decade)	Range pK^+	k_{KM}^{pot}			
			Na^+	Rb^+	Cs^+	NH_4^+
9 (n = 1; m = 1)	57	5-1	2×10^{-3} (2×10^{-3})[a]	1×10^{-1}	7×10^{-3}	1×10^{-2}
9 (n = 1; m = 3)	57	5-1	1×10^{-3} (1×10^{-3})[a]	1×10^{-1}	5×10^{-3}	8×10^{-3}
9 (n = 1; m = 5)	57	5-1	6×10^{-4} (7×10^{-4})[a]	1×10^{-1}	3×10^{-3}	7×10^{-3}
9 (n = 1; m = 7)	57	5-1	2×10^{-3} (1×10^{-3})[a]	1×10^{-1}	7×10^{-3}	1×10^{-2}
11 (n = 1)	54	5-1	2×10^{-3} (2×10^{-3})[a]	1×10^{-1}	1×10^{-2}	1×10^{-2}
3 (n = 1; R = H)	46	5.5-3	3×10^{-4} [b]	2×10^{-1} [b]	2×10^{-3} [b]	2×10^{-3} [b]
Copolymer of 3 (n = 1; R = H)	50	5.5-2.5	5×10^{-4} [b]	8×10^{-2} [b]	3×10^{-3} [b]	4×10^{-3} [b]
Monomer of 3 (n = 1; R = H)	58	5.5-1	6×10^{-2} (7×10^{-2})[a]	4×10^{-1}	1×10^{-1}	6×10^{-2}
	58[c]	5.5-1[c]	3×10^{-2} [c]	2×10^{-1} [c]	1×10^{-1} [c]	7×10^{-2} [c]
Valinomycin[d]	60	5-2	2×10^{-4}	3	5×10^{-1}	2×10^{-2}

[a]Measured by taking a constant background of sodium and varying the potassium concentration.
[b]Apparent selectivity coefficient [19].
[c]Liquid membrane.
[d]From Ref. 106.

the other alkali metal, alkaline earth metal, and heavy metal ions such as Ag^+, Pb^{2+}, and As^{3+} are very small, the interference on Tl^+ determination being negligible, unlike the Tl^+ electrodes based on thallium salts [130].

Bis(benzo-18-crown-6)s 9 (n = 2; m = 3, 5, 7) are Cs^+-selective neutral carriers [108], forming 2:1 crown ether ring to cation complexes intramolecularly. Although the Cs^+ selectivity of the bis(benzo-18-crown-6)-based electrodes may not be remarkably good [k_{CsM}^{pot} for the bis(crown ether) 9 (n = 2; m = 5)-NPOE electrode system, Na^+: 9×10^{-4}; NH_4^+: 9×10^{-3}; K^+: 8×10^{-2}; Rb^+: 8×10^{-2}], the selectivity is drastically improved compared to that of the corresponding monocyclic crown ether electrode.

Recently, the high K^+ and Cs^+ selectivities of the bis(crown ether)-based electrodes have been confirmed, again, using the analogous type of bis(benzocrown ether) derivatives 12 (n = 1, 2; X = $-CH_2(OCH_2CH_2)_mOCH_2-$, m = 1, 2, 3) [117].

From CPK molecular model considerations, bis(benzo-12-crown-4) derivatives of this type were expected to be Na^+ selective, but unfortunately this is not the case with bis(benzo-12-crown-4)-based electrodes, probably because of their poor Na^+-binding ability. Some of bis(12-crown-4) derivatives [13 and 15 (n = 1)], which do not contain any benzene ring in the crown ether ring, exhibit outstanding Na^+ selectivity [17]. Hence, the properties of the Na^+-selective electrodes based on the bis(12-crown-4)s have been elucidated using PVC membranes containing them [43]. The selectivity coefficients of the Na^+ electrodes listed in Table 5 show that the Na^+ selectivity of the bis(12-crown-4) derivatives are markedly dependent on the length of the chain connecting the two crown ether moieties, as has been observed slightly in the electrodes of bis(benzo-15-crown-5) and bis(benzo-18-crown-6) derivatives. The electrode of bis(12-crown-4) derived from malonic acid [13 (m = 1)] is relatively Na^+ selective, which reflects the easy formation of the intramolecular 2:1 crown ether ring to Na^+ complex. On the other hand, the Na^+ selectivity is rather poor for the bis(12-crown-4) 13 (m = 3) electrode, in which the two crown ether rings are connected by a glutaric ester linkage, being too distant from each other to cooperate on the formation of the 2:1 complex.

More striking are the excellent Na^+-selective electrodes based on α,α'-disubstituted malonate type of bis(12-crown-4)s 15 (n = 1). One of the advantages of bis(12-crown-4)s 15 (n = 1) over 13 (n = 1) as the neutral carrier of Na^+-selective electrodes is the improved Na^+ selectivity, which is attributed mainly to the back strain of the two special substituents at the α-position. The other is the high lipophilicity caused by the incorporation of two aliphatic chains, which in turn brings about outstanding electrode properties, such as stable EMF reading, high responsiveness, and long electrode life. Bis(12-

Table 5. Properties of Na^+-selective PVC Membrane Electrodes Based on Bis(12-crown-4) Derivatives 13 and 15 (n = 1)

Crown ether	Maximal slope (mV/decade)	Range pNa	k_{NaM}^{pot} [a]				
			Li^+	K^+	Rb^+	Cs^+	NH_4^+
13 (m = 1)	40	4-0	—	4×10^{-2}	4×10^{-2}	4×10^{-2}	2×10^{-2}
13 (m = 2)	55	4-1	—	6×10^{-2}	2×10^{-2}	4×10^{-2}	3×10^{-2}
13 (m = 3)	55	4-1	—	8×10^{-1}	7×10^{-1}	7×10^{-1}	3×10^{-1}
15 (n = 1; $R_1 = R_2 = C_2H_5$)	52	4.5-1	1×10^{-3}	2×10^{-2}	2×10^{-2}	2×10^{-2}	2×10^{-2}
15 (n = 1; $R_1 = CH_3$, $R_2 = C_{12}H_{25}$)	53	4-1	1×10^{-3}	9×10^{-3}	4×10^{-3}	1×10^{-2}	1×10^{-3}
12-Crown-4	52	4-1	—	6×10^{-1}	2	1	8×10^{-1}

[a] The k_{NaM}^{pot} values for alkaline earth metal ions are below 2×10^{-4}.

Source: Ref. 43.

crown-4) 13 (m = 1) is slightly water soluble, which causes the properties of the Na^+-selective electrode to deteriorate. For example, the Na^+ electrode of 13 (m = 1) shows a slope of only 40 mV/decade in the calibration plots, although the slope should theoretically be around 59 mV for the monovalent metal ion at 25°C.

The most prominent of these bis(12-crown-4) derivatives as the neutral carrier of Na^+-selective electrodes is dodecylmethylmalonic ester type, 15 (n = 1; $R_1 = CH_3$, $R_2 = C_{12}H_{25}$). The selectivity coefficient of Na^+ over K^+, k^{pot}_{NaK}, is 9×10^{-3}, compared to the bis(crown ether) electrode, which reach about 1×10^{-3}, the k^{pot}_{NaK} value for the conventional Na^+-selective glass electrode [114]. There is, however, a great advantage of the crown ether-based PVC membrane electrode over the glass electrode. For example, the glass electrode normally requires special treatment at regular intervals to avoid the deterioration of the electrochemical behavior, while the PVC membrane electrode does not suffer from such contamination.

Attempts have been made practically to use the K^+- and Na^+-selective electrodes based on bis(benzo-15-crown-5) 9 (n = 1; m = 5) and bis(12-crown-4) 15 (n = 1; $R_1 = CH_3$, $R_2 = C_{12}H_{25}$), respectively [112,113]. The coated-wire ion-selective electrodes (CWISE), which have been developed by Cattrall and Freiser [115], were employed for this purpose, taking advantage of its simple construction and potential as a microelectrode. The construction of the bis(crown ether)-based CWISE is shown schematically in Figure 24. A silver wire of 0.4 mm o.d. was carefully soldered to a shielded cable and incorporated in a glass tube of 8 mm o.d. with epoxy resin adhesive. The junction was covered with sealing tape to prevent its being wetted by sample solutions. The exposed silver wire of 5 mm length was cleaned by rinsing in diluted nitric acid, followed by washing carefully with distilled water and drying with acetone. The exposed area of the silver wire was then immersed in a coating solution 10 times, the solvent (THF) being evaporated with a heat gun each time. The typical coating solution includes 10 mg of bis(crown ether), 100 mg of PVC, 200 mg of NPOE, and 3 mL of THF. The resulting PVC membrane coated on the silver wire has 3.2 wt % of crown ether content and about 0.5 mm of thickness. Finally, the junction between the PVC membrane and the taped area was covered with poly(cyanoacrylate) adhesive to adjust the length of the exposed membrane area to about 3 mm. The electrode response of the K^+ and Na^+ CWISEs is fast enough for practical use, the response time (t_{95}) being 10 s in incremental methods such as Grans's plot method [129], as depicted in Figure 25. Actually, the concentration of K^+ and Na^+ in human blood and urine have been determined successfully using bis(crown ether)-based CWISEs. For comparison, the metal ion determination was also carried out by flame photometry. The correlation of the results of K^+ and Na^+ measurements

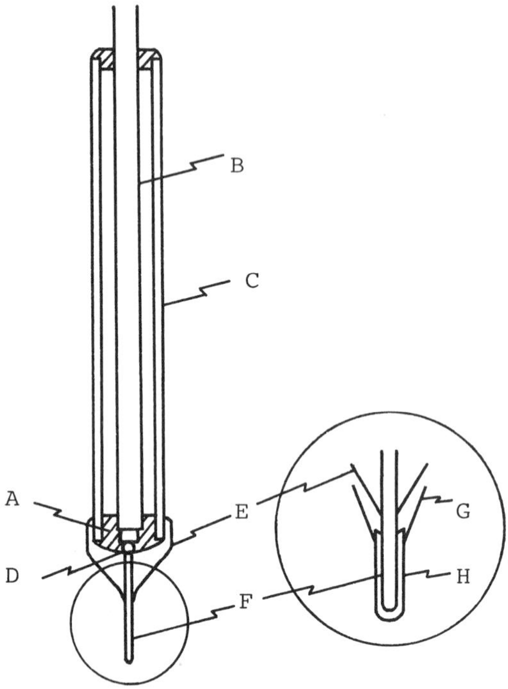

Figure 24. Construction of coated-wire electrode. A, Epoxy resin; B, shielded cable; C, 8-mm-o.d. glass tube; D, soldering; E, seal tape; F, 0.4-mm-o.d. silver wire; G, cyanoacrylate resin; H, PVC matrix. (From Ref. 112.)

in urine samples between the ion-selective methods and flame photometry shows that both results are in good agreement (Figure 26), verifying that the CWISEs can be put to practical use.

More recently, bis(15-crown-5) 15 (n = 2; $R_1 = CH_3$, $R_2 = C_{12}H_{25}$) was utilized instead of bis(benzo-15-crown-5) 9 (n = 1; m = 5) as the neutral carrier of the K^+-selective electrode [116]. The selectivity coefficients k_{KNa}^{pot} of the K^+-CWISE based on the dodecylmethylmalonic ester-type bis(15-crown-5) are 2×10^{-4}, 2×10^{-1}, 2×10^{-2}, and 3×10^{-4} for Na^+, Rb^+, Cs^+, and NH_4^+, respectively. The selectivity over Na^+ is still slightly better than that for the bis(benzo-15-crown-5) electrode. Most outstanding is the highly stable EMF reading and the long life in the K^+-selective electrode based on dodecylmethylmalonic ester-type bis(15-crown-5) 15 (n = 2; $R_1 = CH_3$, $R_2 = C_{12}H_{25}$) (Figure 27). The slope of the calibration plots for the dodecylmethyl-bis(15-

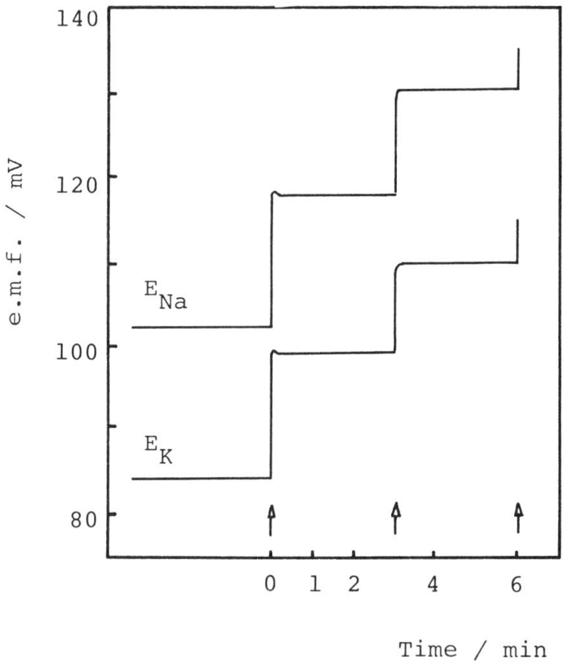

Figure 25. EMF-time profile of Na- and K-coated wire ion-selective electrodes for artificial urine. Arrows indicate the injection of the standard solution. Sample solution, 1.29×10^{-2} M Na^+, 4.25×10^{-3} M K^+, 7.2×10^{-4} M Ca^{2+}, 5.0×10^{-4} M Mg^{2+}, 3.0×10^{-5} M $(NH_4)_2HPO_4$, 2.0×10^{-3} M urea, $V_0 = 10$ mL. Standard solution, 1.30 M Na^+, 4.30×10^{-1} M K^+, 0.1 mL each time. (From Ref. 112.)

crown-5) electrode hardly changes even after 200 measurements, whereas that for the bis(benzo-15-crown-5) electrode decreased significantly after 60 measurements. Thus the dodecylmethyl-bis(15-crown-5) electrode would be a real substitute for valinomycin as the neutral carrier of K^+-selective electrodes.

Crown ether and cryptand polymers, such as the vinyl-type 3 (n = 1; R = H), its copolymer with *N*-phenylacrylamide [19], and the condensation-type [2B.2B.2] polymer [118] similar to 19, were also evaluated for K^+-selective PVC membrane electrodes. They were blended with PVC in a similar way to the bis(crown ether) electrode systems. The K^+-selective electrodes based on the crown ether (or cryptand) polymers were found to be of little practical use, due to the poor responsiveness.

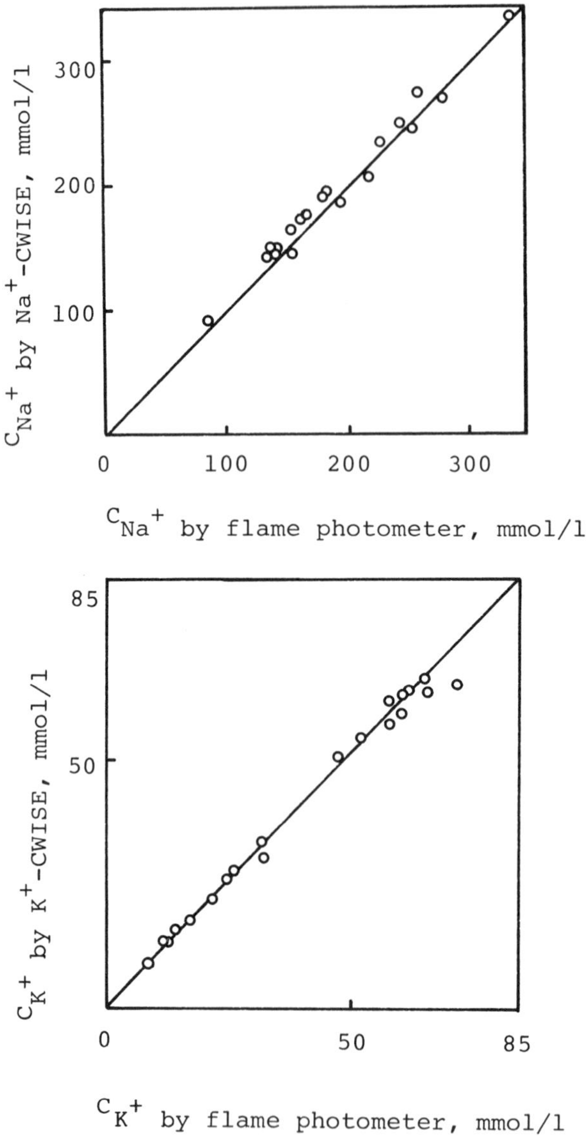

Figure 26. Correlation of results of Na^+ and K^+ measurements in urine obtained by flame photometry and potentiometry using coated-wire ion-selective electrodes based on bis(crown ether)s 9 (n = 1, m = 5) and 15 (n = 1; $R_1 = CH_3$, $R_2 = C_{12}H_{25}$).

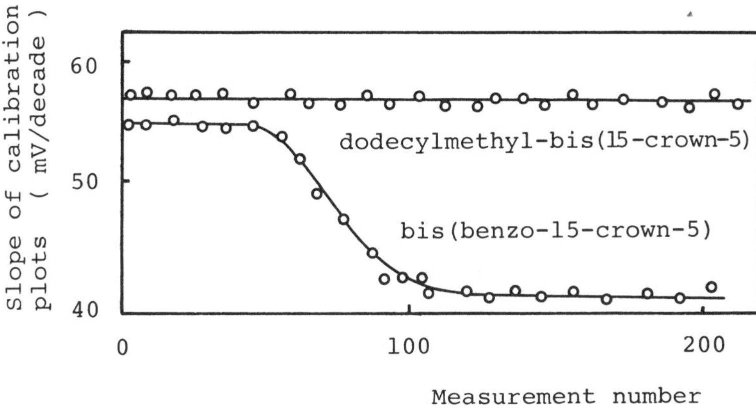

Figure 27. Comparison of electrode life between K^+-selective electrodes based on 9 (n = 1, m = 5) and 15 (n = 2; $R_1 = CH_3$, $R_2 = C_{12}H_{25}$).

7.4.4 Ion Transport and Reverse Osmosis

Ion transport and reverse osmosis across a membrane containing functional substances may be useful separation techniques. The ion transport across a liquid membrane is known to be accelerated by macrocyclic antibiotics and crown ether derivatives.

Poly(crown ether)s 1 have also been utilized as the carrier in liquid membrane [25]. The transport of sodium and potassium picrates was carried out across liquid membranes of chloroform solutions containing poly(crown ether) 1, their copolymers with styrene, and the monomeric analog. The results with the crown ether copolymer are depicted in Figure 28. The salt migration across the liquid membrane is controlled by the diffusion of the ion-pair complex in the chloroform layer, the extraction equilibrium across the interphases being rapidly established. Hence the salt migration rate for the bulky poly(crown ether) is slower than for the corresponding monomeric analog. The selectivity ratio of Na^+/K^+ on the ion transport for the poly(crown ether) was not as high as expected from the extraction data [9].

Polymeric solid membranes for ion transport and reverse osmosis is another application of the crown ether polymers. Only a few reports concerning the crown ether polymer membrane have appeared so far. The condensation-type dibenzo-18-crown-6 polymer 16 was cast from DMF containing NaSCN or hexafluoroisopropanol (HFP) to yield mechanically strong film [45]. The polymeric blends of the polyamide-type crown ether polymer with polyvinylpyrrolidone (PVP) also gave tough

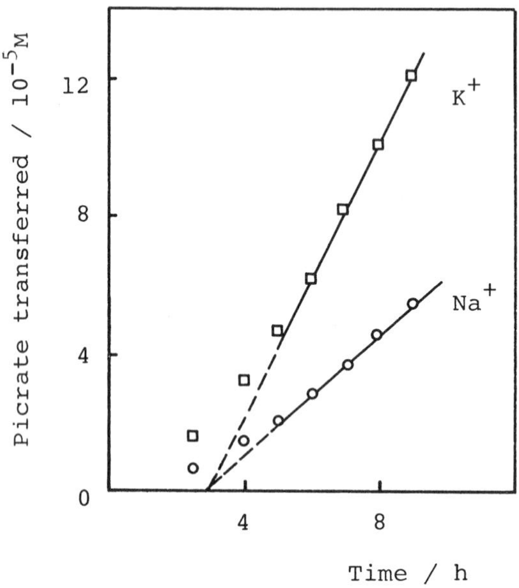

Figure 28. Concentration of picrate salt transferred as a function of time on membrane transport of alkali metal picrates across a chloroform solution of a copolymer of poly(crown ether) 1 (n = 2). Initial picrate concentration: K^+, 2.0×10^{-3} M; Na^+, 2.0×10^{-2} M. (From Ref. 25.)

film. The added PVP does not change the mechanical properties of the crown ether polymer film. Also, the polymeric blends of crown ether polymer 16 with PVP form so-called polymeric alloys, a fact supported by the following observations. Although PVP itself is slightly water soluble, in the polymeric blend it is hardly extracted by water, which indicates an intimate mixing of the two polymers. The thermochemical analyses and the differential scanning calorimetry of the polymeric blend showed only one transition in place of those of the two compounds, which again supports the homogeneity of the polymeric blend. The polyamide-type crown ether polymer and its polymeric alloys with PVP are much more permeable to water than to alkali and alkaline earth metal salts, although the properties and the permeabilities of the films depend on how they are prepared and conditioned. Therefore, the films can be used as the polymeric membranes for reverse osmosis. The reverse osmotic behavior of the membrane of crown ether polymer 16 was compared with those of the conventional membranes such as commercially available aromatic polyamide and cellulose acetate, which are often used in reverse osmosis water desalination units. The water permeabilities of the crown ether polymer membranes are at least one order of magnitude higher than those of the conventional aromatic polyamides, while their salt rejection is comparable (more than 95%). The membrane-osmotic properties of polyamide-type crown ether polymers are also superior to those of cellulose acetate membranes.

It appears that it is difficult to form mechanically strong films using vinyl-type poly(crown ether)s. One of the ways to obtain tough films of polymers anchoring crown ether moiety is through the polymer reaction of functional crown ethers with the conventional polymers, for which are easy to form tough polymeric films. Benzocrown ether moiety has been introduced partially on poly(vinyl alcohol) by the reaction of the polymer with 4'-formylbenzo-15-crown-5. The partially crown ether-modified poly(vinyl alcohol) (substitution of less than 10%) forms tough films [119]. This membrane may be promising for the salt rejection by reverse osmosis.

Polysalt complexes containing poly(crown ether) may be good candidates as polymeric membranes for ion transport and reverse osmosis. The formation of soluble or insoluble complexes on mixing two polyions containing opposite charges is a phenomenon of high polymers. In many cases the electrostatic interaction between polycations and polyanions forces the polysalt complexes to coprecipitate.

In the presence of alkali metal salts, poly(benzo-18-crown-6) 1 (n = 2) in water forms the precipitate of the polysalt complex quantitatively by adding polyanion species such as sodium carboxymethycellulose [33], poly(styrene sulfonate) [14], poly(acrylic acid) [14], and polynucleotides. For the carboxymethylcellulose system the precipitation occurs on addition of KCl to a mixture of the poly(crown ether) and the polyanion in water (Figure 29). The maximal formation of the polysalt complex was attained at the point of electroneutrality, that is, when equimolar of KCl and the polyanion (the concentration of the monomer unit) was added to the poly(crown ether) aqueous solution. The addition of a high concentration of salts causes counterion shielding, therefore redissolution of the precipitate. Similar precipitates to the K^+-mediated polysalt complex is also formed with CsCl, but not with NaCl, due to the poor Na^+-binding ability of the poly(crown ether). The poly(benzo-18-crown-6) can be precipitated with polyacid in the absence of the salts as well. For example, addition of an equimolar poly(acrylic acid) to an 1×10^{-4} M poly(crown ether) aqueous solution in the presence of 0.01 N HCl leads to precipitation of about 90% of the poly(crown ether). Apparently, the complex formation results from hydrogen bonding between the crown ether moiety and the nonionized carboxyl group. Polymeric films consisting of the poly(crown ether)-poly(acrylic acid) complex can be formed at the interphase of a toluene solution of the poly(crown ether) in contact

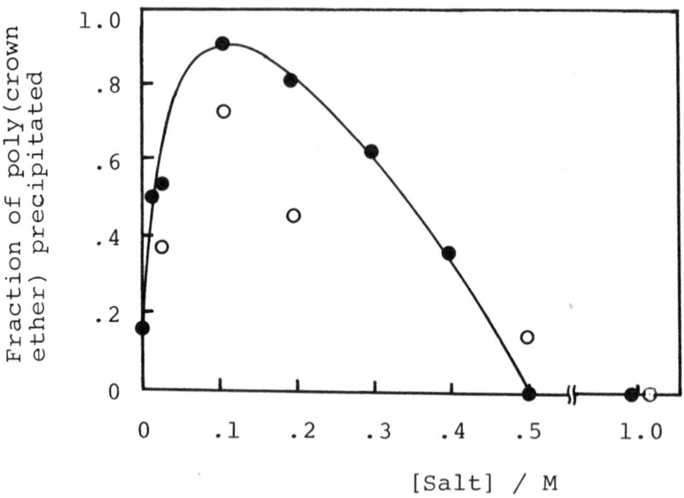

Figure 29. Precipitation of polysalt complex between poly(crown ether) 1 (n = 2) and carboxymethylcellulose as a function of salt concentration. ●, KCl; ○, CsCl. (From Ref. 33.)

with 0.01 N HCl aqueous solution of the polyacid [99]. The film at the interphase can be removed only as a fiber. The poly(crown ether) and the polyacid dissolve in dioxane independently, but the dried fiber of the polysalt complex does not. However, the problem with the application of the polysalt complexes to the polymeric membrane is how to prepare the membranes mechanically strong enough to be of practical use. Similarly, the use of the polysalt complexes as microcapsules may also be of interest [99].

7.4.5 Catalyst of Derivatization for Chromatography

The chemical derivatization for chromatographic analyses is one of the ways to solve problems concerning volatility in gas chromatography and the detecting sensitivity in both gas and liquid chromatography. Crown ethers have been utilized as phase-transfer catalysts of the derivatization prior to chromatography [120-122]. They are crown ether-catalyzed derivatizations of various fatty acids and phenols to the phenacyl esters for UV detection [120], to the coumarin esters for fluorescence detection in HPLC [122], and to the pentafluorobenzyl esters for electron-capture detection in gas chromatography. Highly efficient derivatization is possible using the crown ether catalysts. There are, however, some problems such as the interference of volatile crown ethers on chromatograms, toxicity, and the retrieval of the expensive crown ethers. The low-molecular-weight crown ethers employed for the derivatization are likely to interfere in the gas chromatogram because of their volatility, unless tedious removal of the crown ethers from the reaction mixture is performed before injection into the gas chromatograph. Poly(crown ether) 3 (n = 1; R = CH_3) has been used for the derivatization catalyst in the gas chromatographic determination of lower fatty acids [123], taking advantage of its nonvolatility and also its low toxicity compared to the monomeric crown ethers. The poly(crown ether)-catalyzed reaction of lower fatty acids with *p*-bromophenacyl bromide in acetonitrile in the presence of large excess of K_2CO_3 allows quantitative derivatization within 30 min, at room temperature (Figure 30). In the absence of the poly(crown ether), the yield of the esters was low, below 40% in an hour.

Figure 31 shows a typical gas chromatogram of a mixture of lower fatty acids that have been successfully derivatized in the presence of the poly(crown ether). The nonvolatile poly(crown ether) never interferes the chromatogram, whereas monomeric crown ethers such as dicyclohexyl-18-crown-6 and 18-crown-6, which are often used as the derivatization catalysts, seems to do so severely (Figure 32). The calibration plots for each acid showed linearity in the range of about 10 nmol to 10 pmol, the lowest detection limit of which (14 pmol for *n*-valeric acid, 28 pmol for acetic acid) is close to that of flame ionization detector (FID). Recently, a process of preconcentration

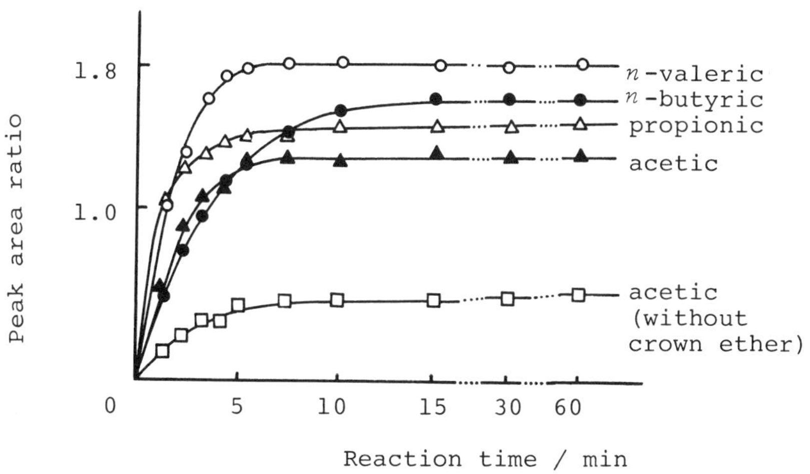

Figure 30. Time dependence of peak area ratio of the derivatized fatty acids on gas chromatogram. (From Ref. 123.)

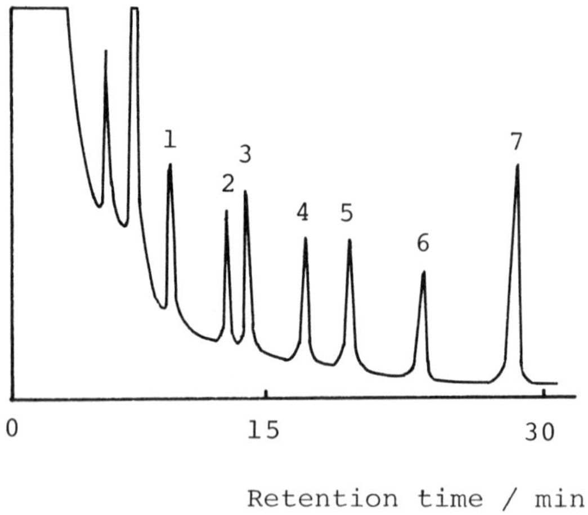

Figure 31. Typical gas chromatogram of fatty acid esters obtained by poly(crown ether)-catalyzed derivatization. 1, Acetic; 2, propionic; 3, *iso*-butyric; 4, *n*-butyric; 5, *iso*-valeric; 6, *n*-valeric; 7, *iso*-caproic (internal standard). (From Ref. 123.)

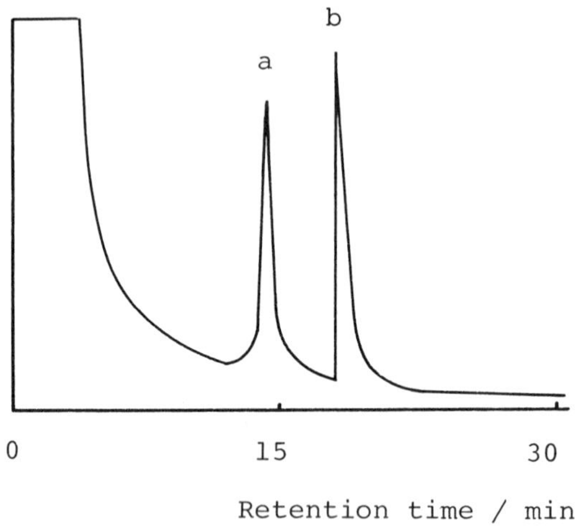

Figure 32. Gas chromatogram of monomeric crown ethers under the identical gas chromatographic condition with that in Figure 31. a, Dicyclohexyl-18-crown-6; b, 18-crown-6. (From Ref. 123.)

of lower fatty acids in gas samples has been combined with subsequent poly(crown ether)-catalyzed derivatization of the acids [124]. That is, attempts have been made to determine successfully very low level of fatty acids in gaseous samples by the preconcentration of them on a column of strontium hydroxide-coated glass beads first, followed by the in situ derivatization of the fatty acid strontium salts into their *p*-bromophenacyl or the pentafluorobenzyl esters in the presence of poly(crown ether) 3 (n = 1; R = CH_3). The typical procedure for the in situ derivatization of lower fatty acids in artificial samples is as follows. Traces of lower fatty acids were preconcentrated by passing 20 L of a sample gas, at a flow rate of 1 L/min, through a glass column (70 × 15 mm i.d.) packed with 2 g of strontium hydroxide-coated glass beads (1%, 20 to 30 mesh). To all of the glass beads was then added 1 mL of an acetonitrile solution containing 2×10^{-2} M *p*-bromophenacyl bromide, 1.4×10^{-3} M (for crown ether unit) poly(crown ether), and 1.2×10^{-3} M *p*-bromophenacyl *iso*-caproate (the internal standard). The mixture was allowed to reflux with vigorous stirring for 30 min. An aliquot (2 μL) of the supernatant solution was then injected into a gas chromatograph without further treatment. As the alternative derivatizing reagent, pentafluorobenzyl bromide can be employed for a high-sensitivity electron-capture detector (ECD).

Excellent separation of the pentafluorobenzyl esters of C_2 to C_6 acids was again obtained, as shown in Figure 33. The calibration plots gave straight lines over a wide concentration range, the lowest detection limit of which was at the nanogram level for the *p*-bromophenacyl bromide system and at the picogram level for the pentafluorobenzyl bromide system (Figure 34).

Table 6 represents the results of the in situ derivatization of some lower fatty acids followed by their gas chromatographic determination. The calculated and found amounts of fatty acid in a 20-L polyethylene bag are in good agreement within acceptable negative errors of only a few percent, which is due mainly to the experimental errors in preparing the sample gases. For 25 ng, the lowest amount of *n*-valeric acid contained in 20 L of the sample gas (Table 6), the acid concentration is 0.27 ppb, which is still lower than its threshold value for perception (0.62 ppb). Thus for all practical purposes, this in situ poly(crown ether)-catalyzed derivatization technique for the lower fatty acids combined with their preconcentration can be utilized.

The retrieval of crown ethers from the reaction mixture and the reuse of them are also important on using them as the derivatization catalyst. Insoluble crown ethers or crown ether-immobilized resins may be employed for this purpose. In this case the insoluble catalyst can readily be removed from the reaction mixture by simple filtration, which causes no contamination of the reaction mixture by the crown

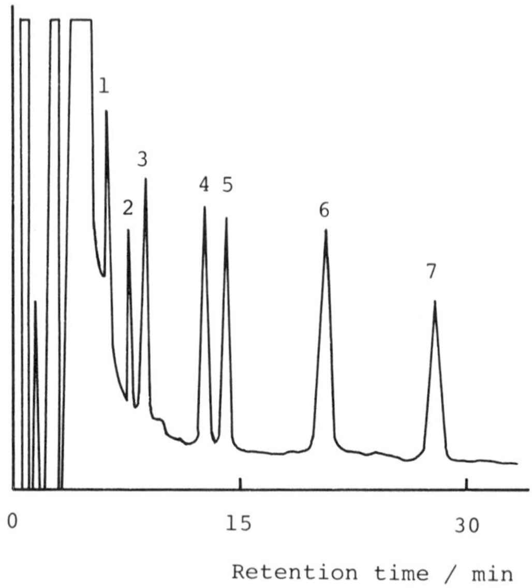

Figure 33. Gas chromatogram of fatty acid pentafluorobenzyl esters (ECD system). 1, Acetic; 2, propionic; 3, *iso*-butyric; 4, *n*-butyric; 5, *iso*-valeric; 6, *n*-valeric; 7, *iso*-caproic (internal standard).

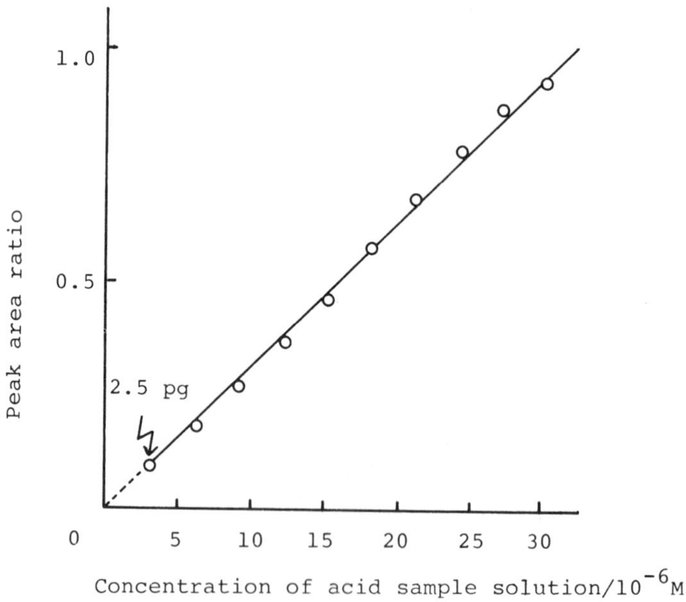

Figure 34. Calibration plots for ECD gas chromatography of *n*-valeric acid pentafluorobenzyl ester obtained by poly(crown ether)-catalyzed derivatization (at the lowest concentration range). (From Ref. 124.)

Table 6. Determination of Lower Fatty Acids in Artificial Sample Gases via the Poly(crown ether)-Catalyzed in situ Derivatization

Fatty acid	Calculated amount (ng)	Found amount (ng)	Error (%)
Acetic	72.0	71.1	-1.3
	36.0	34.9	-3.1
	14.4	13.6	-5.6
Propionic	88.8	86.9	-2.1
	44.4	42.0	-5.4
	17.8	16.1	-9.6
n-Butyric	106	105	-0.9
	52.8	49.2	-6.8
	21.1	19.3	-8.5
n-Valeric	122	119	-2.5
	61.2	58.3	-4.7
	24.5	22.9	-6.5

Source: Ref. 124.

ethers and thus no interference on the gas chromatogram. The catalyst can then be recycled several times without a drastic decrease in catalytic activity. However, the catalytic activities of the insoluble crown ether resins are generally not superior to those of conventional crown ethers because of the heterogeneity of the reaction, the low mobility of crown ether moiety in the resin, and the steric barrier of the polymeric support. To overcome this disadvantage of the insoluble crown ether resins [125], a long spacer chain was incorporated between the crown ether moiety and the polymeric support [64,65]. This idea is based on the assumption that by allowing the crown ether moiety to protrude into the solution, a higher degree of solvation could be achieved, thus providing easier collision with the reaction substances. A continuous increase in the catalytic activity was observed upon lengthening the spacer chain. Finally, the insoluble crown ether resin incorporating a rather long spacer chain exhibited catalytic activity similar to that of the corresponding "nonimmobilized" crown ethers. Thus this type of insoluble crown ether resin might be applied in the derivatization catalyst for the chromatographic analyses without any problems.

7.5 CONCLUSION

A great number of crown ether polymers, such as vinyl-type ones, condensation-type ones, and immobilized crown ethers on polymeric supports, have been synthesized, in an attempt to realize the many advantages of these polymers over the corresponding low-molecular-weight crown ethers: for example, the immobilization of crown ethers, the excellent binding ability and selectivity for metal ions caused by some polymer effects, and the low toxicity on handling derived from their nonvolatility and poor solubility. The insoluble crown ether polymers and immobilized crown ethers on supports have been used extensively as the stationary phases of liquid chromatography, taking advantage of their insolubility. Chromatographic separation of various metal ions and some neutral organic compounds, and chromatographic optical resolution, have been achieved successfully. The crown ether polymers can also be utilized in isotope separation. Some crown ether-modified silicas containing vinyl-type poly(crown ether)s exhibit higher cation selectivity than those for the corresponding monomeric analogs in their chromatographic application. Excellent cation selectivities were again attained in the application of poly- and bis(crown ether)s to extraction agents and the neutral carriers of ion-selective electrodes. In particular, some bis(crown ether)s have proved to be outstandingly excellent neutral carriers for K^+- and Na^+-selective electrodes. The formation of polymeric films for ion transport and reverse osmosis is allowed only for the crown ether polymers, not for low-molecular-weight crown ethers. Furthermore, the crown ether polymers are very useful as the catalyst of derivatization for chromatography because of their noninterference on the chromatogram, easy retrieval, and reuse.

Thus the crown ether polymers are very useful functional polymers in the field of analytical chemistry. Since investigations have recently begun regarding their analytical application, numerous published studies can be expected to appear in the future.

REFERENCES

1. J. M. Lehn, Structure and Bonding, Vol. 16, Springer-Verlag, Berlin (1973), p. 1.
2. R. M. Izatt and J. J. Christensen, Eds., Synthetic Multidentate Macrocyclic Compounds, Academic Press, Inc., New York (1978).
3. R. M. Izatt and J. J. Christensen, Eds., Progress in Macrocyclic Chemistry, Vols. 1 and 2, John Wiley & Sons, Inc., New York (1979, 1981).

4. F. Vogtle, Ed., Topics in Current Chemistry, Vols. 98 and 101, Springer-Verlag, Berlin (1981, 1982).
5. C. J. Pedersen, J. Am. Chem. Soc., 89, 7017 (1967).
6. C. J. Pedersen, Fed. Proc., 27, 1305 (1968).
7. J. M. Lehn, Pure Appl. Chem., 49, 857 (1977).
8. B. Tummler, G. Maass, F. Vogtle, H. Sieger, U. Heimann, and E. Weber, J. Am. Chem. Soc., 101, 2588 (1979).
9. S. Kopolow, T. E. Hogen Esch, and J. Smid, Macromolecules, 6, 133 (1973).
10. K. Kimura, T. Maeda, and T. Shono, Anal. Lett., 11(A10), 821 (1978).
11. K. Kumura, T. Maeda, and T. Shono, Talanta, 26, 945 (1979).
12. K. Yagi, J. A. Ruiz, and M. C. Sanchez, Makromol. Chem., Rapid Commun., 1, 263 (1980).
13. K. Kimura, H. Tamura, T. Maeda, and T. Shono, Polym. Bull., 1, 403 (1979).
14. A. J. Varma, T. Majewicz, and J. Smid, J. Polym. Sci. Polym. Chem. Ed., 17, 1573 (1979).
15. K. Kimura, T. Maeda, and T. Shono, Makromol. Chem., 182, 1579 (1981).
16. T. Maeda, K. Kimura, and T. Shono, Bull. Chem. Soc. Jpn., 55, 3506 (1982).
17. T. Maeda, M. Ouchi, K. Kimura, and T. Shono, Chem. Lett., 1573 (1981).
18. M. Bourgoin, K. H. Wong, J. Y. Yui, and J. Smid., J. Am. Chem. Soc., 97, 3462 (1975).
19. K. Kimura, T. Maeda, H. Tamura, and T. Shono, J. Electroanal. Chem., 95, 91 (1979).
20. F. Wada, Y. Wada, T. Goto, K. Kikukawa, and T. Matsuda, Chem. Lett., 1189 (1980).
21. K. H. Wong and H. L. Ng, J. Coord. Chem., 11, 49 (1981).
22. M. J. Calverley and J. Dale, J. Chem. Soc. Chem. Commun., 684 (1981).
23. K. Kimura, H. Tamura, T. Tsuchida, and T. Shono, Chem. Lett., 611 (1979).
24. K. Kimura, T. Tsuchida, T. Maeda, and T. Shono, Talanta, 27, 801 (1980).
25. K. H. Wong, K. Yagi, and J. Smid, J. Membr. Biol., 18, 379 (1974).
26. P. Gramain, M. Kleiber, and Y. Frere, Polymer, 21, 915 (1980).
27. G. Manecke and P. Reuter, Makromol. Chem., 182, 1973 (1981).
28. G. Manecke and P. Reuter, J. Mol. Catal., 13, 355 (1981).
29. L. H. Wong and J. Smid, J. Am. Chem. Soc., 99, 5637 (1977).
30. S. C. Shah and J. Smid, J. Am. Chem. Soc., 100, 1426 (1978).

31. K. Kimura and J. Smid, Makromol. Chem. Rapid Commun., 2, 235 (1981).
32. K. Kimura, K. H. Wong, and J. Smid, Polym. Bull., 6, 395 (1982).
33. A. J. Varma and J. Smid, J. Polym. Sci. Polym. Chem. Ed., 15, 1189 (1977).
34. G. D. Jaycox and J. Smid, Makromol. Chem. Rapid Commun., 2, 299 (1981).
35. J. Smid, B. El Haj, T. Majewicz, A. Nonni, and R. Sinta, Org. Prep. Proced. Int., 8, 193 (1976).
36. D. N. Reinhoudt, F. de Jong, and H. P. M. Tomassen, Tetrahedron Lett., 2067 (1979).
37. R. Ungaro, B. El Haj, and J. Smid, J. Am. Chem. Soc., 98, 5198 (1976).
38. K. Kimura and T. Shono, unpublished data.
39. N. N. Aylward, J. Polym. Sci., A1(3), 319 (1970).
40. I. Ikeda, S. Yamamura, Y. Nakatsuji, and M. Okahara, J. Org. Chem., 45, 5355 (1980).
41. G. Manecke and A. Kramer, Makromol. Chem., 182, 3017 (1981).
42. I. Ikeda, T. Katayama, M. Okahara, and T. Shono, Tetrahedron Lett., 22, 3615 (1981).
43. T. Shono, M. Okahara, I. Ikeda, K. Kimura, and H. Tamura, J. Electroanal. Chem., 132, 99 (1982).
44. W. M. Feigenbaum and R. H. Michel, J. Polym. Sci., A1(9), 817 (1971).
45. E. Shchori and J. Jagur-Grodzinski, J. Appl. Polym. Sci., 20, 773, 1665 (1976).
46. S. Bormann, J. Brossas, E. Franta, P. Gramain, M. Kirch, and J. M. Lehn, Tetrahedron, 31, 2791 (1975).
47. P. Gramain and Y. Frere, Macromolecules, 12, 1038 (1979).
48. P. Gramain and Y. Frere, Polymer, 21, 921 (1980).
49. P. Gramain and Y. Frere, Ind. Eng. Chem. Prod. Res. Dev., 20, 524 (1981).
50. L. J. Mathias and K. Al-Jumah, J. Polym. Sci. Polym. Chem. Ed., 18, 2911 (1980).
51. E. Blasius, W. Adrian, K.-P. Janzen, and G. Klautke, J. Chromatogr., 96, 89 (1974).
52. E. Blasius and K.-P. Janzen, Chem. Ing. Tech., 47, 594 (1975).
53. E. Blasius, K.-P. Janzen, and G. Klautke, Fresenius Z. Anal. Chem., 277, 374 (1975).
54. E. Blasius and P. G. Maurer, J. Chromatogr., 125, 511 (1976).
55. E. Blasius and P. G. Maurer, Makromol. Chem., 178, 649 (1977).
56. E. Blasius, K.-P. Janzen, W. Adrian, G. Klautke, R. Lorscheider, P.-G. Maurer, V. B. Nguyen, T. Nguyen-Tien, G. Scholten, and J. Stockemer, Fresenius Z. Anal. Chem., 284, 337 (1977).
57. E. Blasius, K.-P. Janzen, and W. Neumann, Mikrochim. Acta, 337 (1977).

58. E. Blasius, K.-P. Janzen, H. Luxenburger, V. B. Nguyen, H. Klotz, and J. Stockemer, J. Chromatogr., 167, 307 (1978).
59. E. Blasius, K.-P. Janzen, M. Keller, H. Lander, T. Nguyen-Tien, and G. Scholten, Talanta, 27, 107 (1980).
60. E. Blasius, K.-P. Janzen, W. Adrian, W. Klein, H. Klotz, H. Luxenburger, E. Mernke, V. B. Nguyen, T. Nguyen-Tien, R. Rausch, J. Stockemer, and A. Toussaint, Talanta, 27, 127 (1980).
61. E. Blasius, K.-P. Janzen, W. Klein, H. Klotz, V. B. Nguyen, T. Nguyen-Tien, R. Pfeiffer, G. Scholten, H. Simon, H. Stockemer, and A. Toussaint, J. Chromatogr., 201, 127 (1980).
62. E. Blasius and H. Lander, Fresenius Z. Anal. Chem., 303, 177, 272 (1980).
63. M. Cinouini, S. Colonna, H. Molinari, and F. Montanari, J. Chem. Soc. Chem. Commun., 394 (1976).
64. H. Molinari and F. Montanari, J. Chem. Soc. Chem. Commun., 639 (1977).
65. F. Montanari and P. Tundo, J. Org. Chem., 46, 2125 (1981).
66. M. Tomoi, O. Abe, M. Ikeda, K. Kihara, and H. Kakiuchi, Tetrahedron Lett., 3031 (1978).
67. M. Tomoi, K. Kihara, and H. Kakiuchi, Tetrahedron Lett., 3485 (1979).
68. F. Montanari, Tetrahedron Lett., 5055 (1979).
69. B. Czech, Tetrahedron Lett., 4197 (1980).
70. F. Montanari and P. Tundo, J. Org. Chem., 47, 1298 (1982).
71. K. Fukunishi, B. Czech, and S. L. Regen, J. Org. Chem., 46, 1218 (1981).
72. R. Sinta and J. Smid, J. Am. Chem. Soc., 103, 6962 (1981).
73. A. Warshawsky and N. Kahana, J. Am. Chem. Soc., 104, 2663 (1982).
74. M. Nakajima, K. Kimura, and T. Shono, Anal. Chem., 55, 463 (1983).
75. M. Nakajima, K. Kimura, and T. Shono, Bull. Chem. Soc. Jpn., 56, 3052 (1983).
76. J. Anzai, A. Ueno, Y. Suzuki, and T. Osa, Makromol. Chem. Rapid Commun., 3, 55 (1982).
77. L. R. Sousa, G. D. Y. Sogah, D. H. Hoffman, and D. J. Cram, J. Am. Chem. Soc., 100, 4569 (1978).
78. G. D. Y. Sogah and D. J. Cram, J. Am. Chem. Soc., 101, 3035 (1979).
79. K. Kimura and T. Shono, unpublished data.
80. T. Saegusa, S. Kobayashi, and M. Ishiguro, Macromolecules, 7, 958 (1974).
81. G. Manecke and H.-J. Winter, Makromol. Chem. Rapid Commun., 2, 569 (1981).
82. A. H. Alberts and D. J. Cram, J. Am. Chem. Soc., 101, 3545 (1979).

83. I. Tabushi, J. Kobuke, and T. Nishiya, Tetrahedron Lett., 3515 (1979).
84. Y. Ito and T. Saegusa, J. Macromol. Sci. Chem., A13, 503 (1979).
85. I. Tabushi, Y. Kobuke, and T. Nishiya, Nature, 280, 665 (1979).
86. H. Fujita, S. Yanagida, and M. Okahara, Anal. Chem., 52, 869 (1980).
87. A. Warshawsky, R. Kalir, A. Deshe, H. Berkovitz, and A. Patchornik, J. Am. Chem. Soc., 101, 4249 (1979).
88. P. Grossmann and W. Simon, J. Chromatogr., 235, 351 (1982).
89. P. Kutchukov, A. Ricard, and C. Quivoron, Eur. Polym. J., 16, 753 (1980).
90. K. Yagi and M. C. Sanchez, Makromol. Chem. Rapid Commun., 2, 311 (1981).
91. M. Igawa, I. Ito, and M. Tanaka, Bunseki Kagaku, 29, 580 (1980).
92. M. G. Djamali, P. Burba, and K. H. Lieser, Angew. Makromol. Chem., 92, 145 (1980).
93. M. Nakajima, K. Kimura, E. Hayata, and T. Shono, J. Liquid Chromatogr., 7, 2115 (1984).
94. M. Oue, A. Ishigaki, Y. Matsui, T. Maeda, K. Kimura, and T. Shono, Chem. Lett., 275 (1982).
95. M. Igawa, K. Saito, J. Tsukamoto, and M. Tanaka, Anal. Chem., 53, 1942 (1981).
96. D. J. Cram, R. C. Helgeson, L. R. Sousa, J. M. Timko, M. Newcomb, P. Moreau, F. De Jong, G. W. Gokel, D. H. Hoffman, L. A. Domeier, S. C. Peacock, K. Madan, and L. Kaplan, Pure Appl. Chem., 41, 327 (1975).
97. B. E. Jepson and R. DeWitt, J. Inorg. Nucl. Chem., 38, 1175 (1976).
98. K. G. Heuman and H.-P. Schiefer, Angew. Chem. Int. Ed. Engl., 19, 406 (1980).
99. G. D. Jaycox, R. Sinta, and J. Smid, J. Polym. Sci. Polym. Chem. Ed., 20, 1629 (1982).
100. T. Maeda, K. Kimura, and T. Shono, Fresenius Z. Anal. Chem., 313, 407 (1982).
101. T. Maeda, K. Kimura, and T. Shono, Fresenius Z. Anal. Chem., 298, 363 (1979).
102. G. A. Rechnitz and E. Eyal, Anal. Chem., 44, 370 (1972).
103. O. Ryba, E. Knizakova, and J. Petranek, Collect. Czech. Chem. Commun., 38, 497 (1973).
104. O. Ryba and J. Petranek, J. Electroanal. Chem., 44, 425 (1973).
105. J. Petranek and O. Ryba, Anal. Chim. Acta, 72, 375 (1974).
106. M. Mascini and F. Pallozzi, Anal. Chim. Acta, 73, 375 (1974).
107. M. Yamauchi, A. Jyo, and N. Ishibashi, Anal. Chim. Acta, 136, 399 (1982).

108. K. Kimura, H. Tamura, and T. Shono, J. Electroanal. Chem., 105, 335 (1979).
109. H. Tamura, K. Kimura, and T. Shono, Bull. Chem. Soc. Jpn., 53, 547 (1980).
110. H. Tamura, K. Kimura, and T. Shono, J. Electroanal. Chem., 115, 115 (1980).
111. H. Tamura, K. Kimura, and T. Shono, Nippon Kagaku Kaishi, 1648 (1980).
112. H. Tamura, K. Kimura, and T. Shono, Anal. Chem., 54, 1224 (1982).
113. H. Tamura, K. Kumami, K. Kimura, and T. Shono, Mikrochim. Acta, 287 (1983).
114. M. Guggi, M. Oehme, E. Pretsch, and W. Simon, Helv. Chim. Acta, 59, 2417 (1976).
115. R. W. Cattrall and H. Freiser, Anal. Chem., 43, 1905 (1971).
116. K. Kimura, A. Ishikawa, H. Tamura, and T. Shono, J. Chem. Soc., Perkin Trans. 2, 447 (1984).
117. K. W. Fung and K. H. Wong, J. Electroanal. Chem., 111, 359 (1980).
118. J. Gajowski, B. Rieckemann, and F. Umland, Fresenius Z. Anal. Chem., 309, 343 (1981).
119. Y. Makino, J. Komiyama, and T. Iijima, Polym. Reprints, Soc. Polym. Sc. Jpn, 30(2), 358 (1981).
120. H. D. Durst, M. Milano, E. J. Kikta, S. A. Connelly, and E. Grushka, Anal. Chem., 47, 1797 (1975).
121. B. Davis, Anal. Chem., 49, 832 (1977).
122. S. Lam and E. Grushka, J. Chromatogr., 158, 207 (1978).
123. K. Kimura, M. Sawada, and T. Shono, Anal. Lett., 12(A10), 1095 (1979).
124. K. Kimura, M. Sawada, and T. Shono, J. Chromatogr., 240, 361 (1982).
125. J. M. Brown and J. A. Jenkins, J. Chem. Soc. Chem. Commun., 458 (1976).
126. J. Anzai, Y. Sakata, A. Ueno, and T. Osa, Makromol. Chem. Rapid Commun., 3, 399 (1982).
127. K. Kimura, S. Kitazawa, T. Maeda, and T. Shono, Fresenius Z. Anal. Chem., 313, 132 (1982).
128. W. E. Morf, D. Ammann, E. Pretsch, and W. Simon, Pure Appl. Chem., 36, 421 (1973).
129. G. Gran, Analyst, 77, 661 (1952).
130. W. Szczepaniak and K. Ren, Anal. Chim. Acta, 82, 37 (1976).

8

Biomedical Polymers and Polycarboxylic Acid Polymer Drugs

Raphael M. Ottenbrite

Virginia Commonwealth University
Richmond, Virginia

Kiichi Takemoto and Mikiji Miyata

Osaka University
Osaka, Japan

8.1 INTRODUCTION

The chemistry of biomedical polymers has received much attention in recent years; interest in utilizing both natural and synthetic polymers in the biomedical field is increasing rapidly.

In this chapter, emphasis is focused particularly on the biomedical polymers that have been recently developed, especially the experimental procedures for their preparation and utilization. The content of this chapter is classified as follows:

1. Biomedical polymeric materials that have little contact with blood and are used primarily as supplementary material, such as contact lenses, dental polymers, adsorbable sutures, and artificial skin.
2. Materials that are in contact with blood but only for a short period, such as artificial red blood cells, artificial liver, artificial kidney, and artificial lung membranes.
3. Materials that are in contact with blood for a long time or permanently, such as heart valves and shunts. The artificial heart may be included in the latter category.

Several reviews have been published [1-6] in these areas and should be referred to for further information.

8.2 CONTACT LENSES

Presently about 10 million people worldwide wear plastic contact lenses. The optical advantage of contact lenses over the usual spectacle lenses is considerable. The latter alter rays of light before they reach the eye, whereas contact lenses make the optical correction on the surface of the eye. Contact lenses simulate the normal eye by directing the rays to the retina.

Contact lens development was very rapid following World War II with the availability of poly(methyl methacrylate). The pioneering work of Wichterle and Lim in the late 1950s using acrylic-type hydrogels opened the way for soft contact lenses, which have completely revolutionized the industry [7-11].

The materials used for contact lenses should be (a) optically transparent, (b) permeable to oxygen, (c) physiologically inert, and (d) mechanically stable. It appears to be relatively easy to find polymeric materials that have good transmittance of light and a refractive index close to that of the cornea: 1.37. Major problems arise when we look for material that is permeable to atmospheric oxygen, because the cornea needs oxygen for the metabolic processes of its living cells. The lenses should be physiologically inert and afford minimal disturbance to the cornea and surrounding eye tissues. This implies that the lenses should be finished absolutely free of impurities.

Two types of the polymers are currently being used as materials for contact lenses: thermoplastics and hydrogels. The thermoplastic materials are generally hard and hydrophobic; the hydrogels are soft and hydrophilic.

Most hard contact lenses are made from poly(methyl methacrylate), which is the mainstay material because of its optical properties, toughness, and physiological inertness. Most hard lenses are cross-linked with ethylene glycol dimethacrylate comonomer to improve their strength, hardness, and scratch resistance. Bulk polymerization is the preferred method for preparation, although there seems to be no problem of solvent removal, and the polymer can be processed immediately after curing. The most serious drawbacks of hard contact lenses are their hydrophobic property and resultant low permeability to oxygen.

Soft contact lenses are usually manufactured from acrylic hydrogels, which are derived from slightly cross-linked poly(2-hydroxyethyl methacrylate). The most common cross-linking agent is ethylene glycol dimethacrylate. In addition, the second component, vinyl-pyrrolidone, is frequently used as the material for hydrogel contact lenses. Cross-linked or grafted copolymers of vinylpyrrolidone and 2-hydroxyethyl methacrylate find widespread application. The fundamental property of a hydrogel is its equilibrium water content, by which the permeation of oxygen through the hydrogel is substantially facilitated.

8.2.1 Preparation of Cross-linked Hydrophilic Polymers [12]

Hydrogels containing 20 to 97% of an aqueous liquid and a cross-linked hydrophilic polymer can be prepared from esters of acrylic acid and methacrylic acid with alcohols having hydrophilic groups. The resulting hydrogels act as semipermeable membranes, permitting the passage of water and certain dissolved materials. Instead of the esters of acrylic and methacrylic acids, many of their derivatives or monomers having polymerizable vinyl groups can be used. Thus an aqueous solution containing 20% poly(vinyl alcohol), 2% adipic acid, and 5% HCl is filled in molds for contact lenses or pessaries and heated to 50 to 80°C until a clear soft hydrogel is formed; the molded article is then washed in water and stored in an aqueous solution of 3 to 5% hydroboric acid, which diffuses into the gel. To produce a diaphragm, 100 equivalents of poly(vinyl alcohol) is reacted with 1 equivalent of methacrylic acid anhydride; the partly esterified poly(vinyl alcohol) is then dissolved in water to obtain a 25% aqueous solution and polymerized in the presence of potassium persulfate and sodium thiosulfate in a flat mold. After removal of the water-soluble materials by washing, a clear diaphragm is obtained. To prepare a surgical implant, an aqueous 15% solution of poly(acrylic acid) is mixed with an equimolar amount of a 20% aqueous solution of poly(vinyl alcohol) and HCl is added to obtain a concentration of 5%. The mixture is poured into a mold of the desired shape and heated for several hours until a self-supporting gel-like transparent mass is formed. It is washed with water and made isotonic by immersion in a physiological saline solution. To produce ion exchangers, acrylamide (20 g), dimethyl (0.1 g) are dissolved in water (60 mL) and polymerized in the presence of potassium persulfate and sodium thiosulfate; the block of hydrogel thus obtained is cut into small pieces for use as ion exchangers.

8.3 DENTAL POLYMERS

Most people develop caries of the teeth. In some diseases the teeth are lost, and as they do not regenerate naturally, artificial materials must be used as a replacement. In addition, synthetic polymers play an important role in curing caries and can also serve as effective materials for the prevention of caries.

To develop materials for dental uses, the following properties are necessary: (a) chemical stability, (b) mechanical strength, (c) biocompatibility, and (d) good adhesive property to teeth. Since dental materials used as fillings remain in the mouth for the long time, they should not be degraded by the action of saliva, microorganisms, or food. Furthermore, they should not wear out under strong repetitive occlusal stresses. The polymerization of monomers must proceed in the mouth under biologically mild temperature in an excess amount of water on and in the tissues. The heat of polymerization must be low to avoid damage to the surrounding tissues. It should also be noted that shrinkage must be limited because any space between the teeth and the filling materials forms an ideal place for further caries to develop.

Methyl methacrylate is a common monomer used for dental filling materials. Composite resins containing a large amount of filler and methacrylate monomers have been widely used as restorative materials for hard tooth tissue. The polymerization initiator should function at body temperature. Benzoyl peroxide-*N*,*N*-dimethyl-*p*-toluidine initiator system was first developed, but it was found that the amine causes color change in the resulting polymers. More recently, *p*-toluenesulfonic acid, trimethylbarbituric acid ester, or tri-*n*-butylborane derivatives have been found to be suitable as well. Photoinitiator systems are now becoming more practical for use [13,14].

Although poly(methyl methacrylate) has a wide application as a dental polymer, it has a serious defect in that it does not adhere substantially to tooth surfaces. As is well known, polymer dental fillings often fall out. Furthermore, the poor interface contributes significantly to secondary caries because of the percolation of microorganisms, liquids, and other materials into the marginal areas of the restoration.

An effective way to overcome this drawback is to develop adhesive dental filling materials. One approach is an acid etching of enamel to effect a mechanical interlock. Another is the application of functional monomers that could chelate with calcium ions or couple with dentine collagen. Recently, it was found that the monomers having both hydrophilic and hydrophobic groups have good adhesion property for tooth tissue:

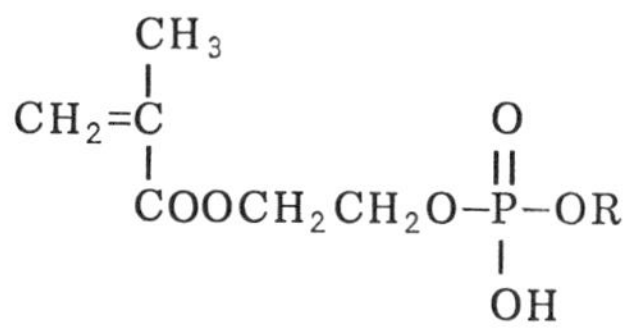

R: $-CH_2CH_2Br$, $-C_6H_5$

(1)

8.3.1 Preparation of 2-Methacryloyloxyethyl-2'-bromoethylphosphoric Acid (1, R = $-CH_2CH_2Br$) [15,16]

Phosphorus trichloride (210 g) in dry carbon tetrachloride is treated dropwise with 124 g of bromohydrin and refluxed for 30 min to give 61% (116.7 g) 2-bromoethyl phosphoryl dichloride. This is further treated dropwise with 59.8 g of 2-hydroxyethyl methacrylate and 54 g of triethylamine in chloroform at -20°C. Then 44 g of triethylamine and 13.5 g of water are added to give 26% 2-methacryloyloxyethyl-2'-bromoethylphosphoric acid.

8.3.2 Preparation of Dental Adhesive Composition [17]

Powdered poly(methyl methacrylate) (98 parts) containing 2 parts of benzoyl peroxide is mixed with 95 parts of methyl methacrylate containing 5 parts of 2-methacryloyloxyethyl phenylphosphoric acid (1, R = $-C_6H_5$) and 2.5 parts of *p*-toluenesulfinic acid. This mixture is applied to an ivory surface and allowed to harden at 37°C for 72 h to give an adhesive bond strength of 160 kg/cm^2.

8.4 ADSORBABLE SUTURES

Special fibers are required for suturing body parts such as skin, muscle, and blood vessels together after an operation or an accidental cut. Many sutures are required to be degradable and absorbable since they remain in the body after the operation. Poly(glycolic acid) was designed as an artificial surgical suture which combines many of the factors necessary for a suitable biodegradable implant. Sutures are considered to be the first successful example of made-to-order biomedical polymers [18,19].

The characteristics required for an acceptable biodegradable suture are as follows: formability, adequate initial strength and dimensional stability, controlled rate of strength loss, complete absorbability, low order of toxicity of both implant and degradation products, and sterilizability. A combination of these unique properties is necessary to produce a useful biodegradable system.

8.4.1 Collagen

Collagen, the major connective tissue protein, has found wide application as a surgical suture in the form of catgut as well as purified reconstituted fiber. It fulfills the basic requirements described above. This material serves as a model for the development of other biodegradable materials. It is not difficult to understand why some of the early work in synthetic biodegradable polymer development involved polymers containing peptide linkages.

8.4.2 Polyamino Acids

Studies of polyglutamates and poly(glutamic acid) revealed that these substances were not sufficiently strong for use as absorbable surgical sutures. In general, poly(glutamic acid) is a relatively weak material that loses strength and solubilizes within a few days after implantation. On the other hand, its corresponding methyl and benzyl esters have appreciable long-term strength in vivo, but are not sufficiently absorbable.

8.4.3 Polyester

The simplest polyester is polyglycolide or poly(glycolic acid). These can be conveniently prepared from a glycolide melt through a salt-catalyzed ring-opening mechanism. Braided sutures prepared from polyglycolide were reported to exhibit high strength, excellent handling properties, minimal tissue reactivity, and an absorption profile similar to that of catgut.

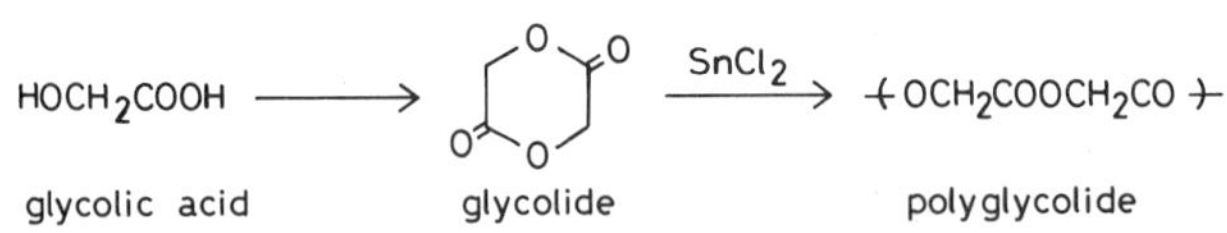

8.4.4 Preparation of a Suture [20]

Glycolic acid (400 parts) is heated to 200°C for 4 h. When the pot temperature reaches 185°C, the pressure of the system is reduced from atmospheric to 15 mm, so that the water of condensation and (or) esterification is removed by distillation. The residue is cooled and pulverized to give 280 parts of a powder which is added in small increments to a suitable pyrolysis chamber maintained at 250 to 285°C and

< 15 mm. The distillate is collected and dissolved in a minimum amount of hot ethyl acetate. After decolorizing and purifying with activated carbon, the product is crystallized to provide 160 parts of glycolide, mp 82.5 to 4°C. A glass tube having a bore of 0.3 in. and containing a magnetic steel ball 0.15625 in. in diameter is charged with a mixture of 3 parts glycolide composition, 0.1% ether solution of 0.04 part $SnCl_2 \cdot 2H_2O$, and 0.0166 part lauryl alcohol. The tube is evacuated, purged with argon gas, evacuated again to < 1 mm, and the top is sealed. The tube is placed in vertical position in a closed glass chamber through which dimethyl phthalate is refluxed at 222°C. At periodic intervals after melting, the viscosity of the reaction mixture is measured by raising the steel ball by means of a magnet and measuring the rate of the fall in seconds per inch. Ninety minutes after the melt is first achieved, the ball drop time is 550 s/in. or 7200 poise and after 120 min, the ball drop time is 580 s/in. or 7600 poise. The rate of fall of the ball is a significant tool in determining the viscosity of the polymeric material and is related to the molecular weight of the polymerized glycolide composition. Equations are presented to show that a direct proportion existed between the measured ball drop time and the conventional melt viscosity units. The melt viscosity in poise is equal to the product of the ball drop time in seconds and 13.13. Additional polymers are similarly prepared using hexyl alcohol, cyclohexanol, diethylene glycol, trimethylol ethane, or pentaerythritol as the alcohol.

8.5 ARTIFICIAL SKIN

The most important function of the skin is simply as a bag to contain our body parts and fluids. The skin protects these parts from drying out and constitutes our first line of defense against microbial invasion. Both excretory and secretory glands contained in this complex organ are used to regulate the body's fats, water, and electrolytes. Our skin is also the organ for the reception of thermal, tactile, and painful stimuli.

Artificial skin made of synthetic polymers should satisfy the following requirements:

1. It should control the amount of loss of body fluids, protein salt, or heat from the wound surface.
2. It should protect against a variety of infections.
3. It should promote the formation of granulation and epithelium.

Biomaterials such as autografts, cadaver skin, or pig skin find a wide application to current treatment for fluid resuscitation and prompt covering of wounds. In addition to these materials, collagen, the main component of the skin, takes advantage from the viewpoint of biocompatibility. Natural collagen can be regenerated enzymatically and chemically in the form of a membrane, sponge, or unwoven cloth [21, 22].

Although the application of many synthetic polymers, such as nylon and Dacron, were studied, a single material did not satisfy the number of requisites for artificial skin. Consequently, multiphase polymers of laminate type have been found to be applicable for this use. Laminated polymers made from polyurethane-collagen or silicone-nylon system have also been explored.

More recently, Yannas's group has developed a successful artificial skin for practical use. The skin in question consists of a bilayer polymeric membrane made of silicone and biodegradable protein layers. Assuming that the assimilation of the artificial skin with the natural skin, the use of epidermal cells of patients themselves may afford a promising way to prepare a well-designed artificial skin in the future.

8.5.1 Example of the Artificial Skin [23]

Represented in Figure 1 is a schematic standard version of the noncellular bilayer polymeric membrane. The moisture flux through the layers ranges from 1 to 10 mg/cm^2 per hour. Fibroblasts and endothelial cells migrate into the bottom layer, while epithelial cell sheets advance from the wound edge in the plane of the membrane between the two layers. The top layer is a conventional medical-grade silicone which undergoes cross-linking at room temperature following exposure to ambient moisture. The bottom layer is a cross-linked network of collagen and chondroitin 6-sulfate a GAG. The collagen-GAG layer undergoes biodegradation at a controlled rate and is replaced by neodermal tissue, while the silicone layer is spontaneously ejected following formation of a confluent neoepidermal layer under it.

8.6 ARTIFICIAL RED BLOOD CELLS

Several attempts have been made to prepare artificial red blood cells by making use of microencapsulation technique. The ideal is to be able to make blood transfusions independent of blood type and without any infection [24].

Microcapsules made of collodion, nylon, and dextran stearate have been developed. However, none of these materials survived for a long time in the circulation system after injection into the blood. In addition, it is necessary for the artificial red blood cells to be less than 0.5 μm in size with high deformability in order to pass freely through capillary blood vessels.

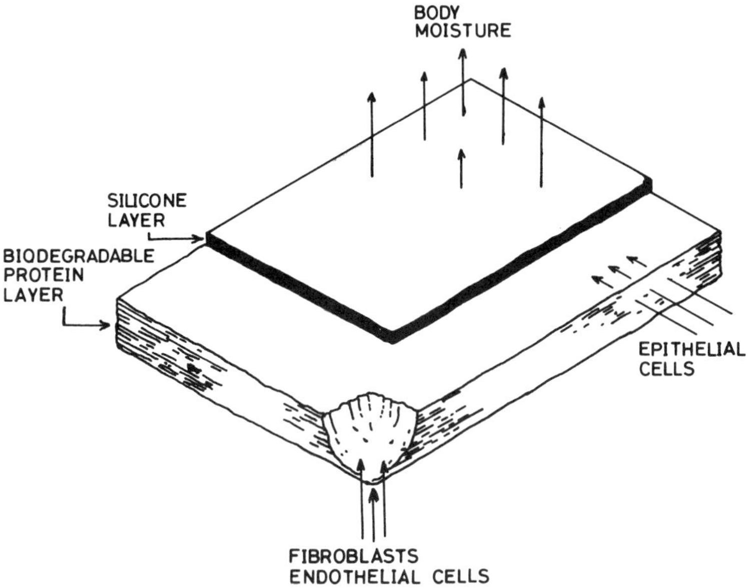

Figure 1. Schematic representation of a standard version of noncellular bilayer polymeric membrane. (From Ref. 23.)

In recent years, promising artificial red blood cells have been prepared using an interfacial polycondensation technique. These cells are hemoglobin-loaded microcapsules that are negatively charged and are similar in size and deformability to those of natural red blood cells. It is expected that these synthetic cells are practical enough to use with respect to the absorbability of oxygen, and enzymatic activity and fluidity.

8.6.1 Preparation of Hemolysate-Loaded Poly(N^{α},N^{ε}-L-lysinediylterephthaloyl) (PPL) Capsules [25]

Sheep hemolysate, the concentration of which has been adjusted to about one-third of that in the sheep cells, is dispersed as very fine droplets into a hydrophobic phase by means of electrocapillary emulsification using the apparatus shown in Figure 2. The hemolysate contains 0.4 M L-lysine and 0.63 M sodium carbonate. The hydrophobic phase consists of 3 volumes of cyclohexane and 1 volume of chloroform, into which terephthaloyl dichloride, tetraethylammonium chloride, and sorbitan trioleate are dissolved. Concentrations of 0.04 M, 1.5×10^{-4} M, and 5% are used for the acid dichloride, quartenary salt, and the emul-

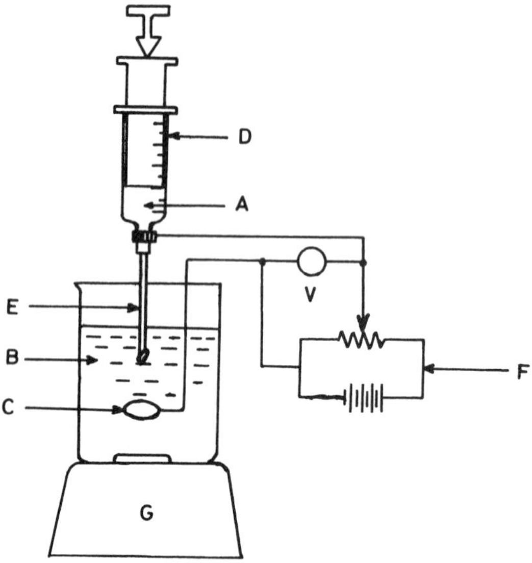

Figure 2. Schematic diagram of the apparatus for electrocapillary emulsification. A, aqueous phase; B, oil phase; C, platinum wire; D, syringe; E, needle; F, direct-current supply; G, magnetic stirrer. (From Ref. 25.)

sifier, respectively. The quarternary ammonium salt serves to give electroconductivity to the oil phase.

Into 50 mL of the hydrophobic phase (B) in a beaker surrounded by ice (see Figure 2) is added 1 mL of the aqueous phase (A) from the syringe (D) driven by a motor. The syringe needle (E) also serves as the anode. The addition is made at a constant rate of 0.042 mL/min under the influence of an applied potential of 850 V between the needle and a platinum wire as cathode (C) immersed in the hydrophobic phase. The distance between the electrodes is about 5 mm and they are connected to a variable direct-current supply (F). The hydrophobic phase is gently stirred during the addition of the aqueous phase by a magnetic stirrer (G). Since the applied potential is above the critical value for spontaneous emulsification, a shower of very fine hemolysate droplets is formed.

On the surface of each droplet, polycondensation takes place between L-lysine and terephthaloyl dichloride to form PPL membrane. Hydrogen chloride formed in the reaction as the by-product is immediately neutralized with alkali in the aqueous phase. PPL capsules thus formed are centrifuged at 1500 rpm (330 × g) for 10 min. The separated capsules are dispersed into water with the aid of polyoxyethylene sorbitan monolaurate, a dispersing agent. Then the aqueous dispersion is centrifuged consecutively at 1000, 2000, and 3000 rpm (150, 590, and 1300 × g) to collect the capsule fraction with a mean diameter of about 400 nm, which is the sediment in the last centrifugation. After washing with isotonic phosphate buffer (pH 7.4) the capsules are redispersed in the same buffer. For oxygen absorption measurement an isotonic solution containing 0.55% glucose and 0.88% NaCl (pH 5.9) is used instead of the phosphate buffer because the presence of glucose is found indispensable to ensure reversible binding of oxygen to hemoglobin, and oxygen absorption is insensitive to changes in pH of the medium around pH 7.

The PPL capsules, which should have a mean diameter of about 400 nm, are washed with the phosphate buffer on the centrifuge and treated with 2.5% glutaraldehyde for 2 h to cross-link the terminal amino groups of the PPL molecules constituting the capsules. After washing with phosphate buffer, the glutaraldehyde-treated capsules are dehydrated by dipping them successively into a series of ethanol-water mixtures the ethanol concentration of which ranges from 50 to 100%. The dehydrated capsules are then placed on an aluminum sample holder to allow the solvent to evaporate into the air.

8.7 ARTIFICIAL LIVER

Activated charcoal has excellent adsorbability for many organic molecules. It can, therefore, be used as an artificial liver for adsorption-type impurities, with the main purpose of removing metabolic materials that the body excretes. It is particularly effective for the removal of poisons, drugs, and drug metabolites introduced in toxic excess in the body [26, 27].

Direct use of activated charcoal as artificial liver resulted in failure because of the release of activated charcoal microparticles which have no blood compatibility. These caused a blockade of capillary blood cells by thrombocytopenia. To overcome these drawbacks, encapsulated activated charcoal has been developed. For example, coconut and pitch-base spherical activated charcoal particles were coated with gelatin, collodion, or poly(2-hydroxyethyl methacrylate). This pitch-base spherical encapsulated activated charcoal is a promising adsorbent for metabolites and poisons and is being readied for clinical trials.

8.7.1 Encapsulation of Pitch Base Spherical Activated Charcoal with Poly-HEMA for Direct Hemoperfusion [27]

Pitch-base activated charcoal (PAC) spheres (750 g) which were boiled in distilled water and dried (particle diameter, 0.8 nm) are immersed into 1 kg of 95% ethanol solution containing 0.1% poly-HEMA ([η] = 1.5, in DMF at 25°C) for 10 min. The particles are separated on a stainless steel mesh from the ethanol solution, and dried at 80°C for 20 h. They are then washed with 3 L of boiling distilled water in three portions, each for 30 min. The procedure is repeated using a 1% polyHEMA solution. The water washings are carried out 12 times. The weight increase in PAC during this double coating is 9 g.

8.8 ARTIFICIAL KIDNEY

In 1960, dialysis began to be used in the medical field as a substitute for the human kidney. The blood of patients suffering from renal poisoning is purified by passing it through cellulosic membranes against a dialysate. Subsequent development of small semipermeable hollow fibers provided the basis for the design of small, efficient commercial dialyzers. The blood to be purified is passed through the inside of the hollow fibers, while the dialysate is distributed over the outside. Metabolic wastes, such as urea, uric acid, and creatine, rapidly diffuse from the blood through the membrane and into the dialysate.

In addition to the dialysis, blood ultrafiltration affords an effective method for separating middle-molecular-weight components in blood.

The requisites for the membrane for the purpose of artificial kidney are that:

1. The membrane have an appropriate channel diameter for dialysis and ultrafiltration.
2. The essential components of blood not be adsorbed or converted.
3. Blood clotting should not occur.
4. Convenient sterilization must be possible and toxic components should not solute out.

The first material used for the artificial kidney in the cure of an acute renal insufficiency was cellulose. This material still finds extensive use in the cure of chronic conditions. Other synthetic materials have also been found to be efficient membranes for this purpose [28-32].

Regenerated cuprammonium cellulose, called cuprophane, has the greatest usage in the world. It has good mechanical strength, which enables the manufacture of thinner cellulose membranes. The thickness of the membrane is 6 μm, which is 20 μm less than that of the original membrane. This results in improved filtration rates and permeability of middle-molecular-weight components.

Regenerated cellulose derived from cellulose acetate is produced by hydrolyzing or deacetylating the cellulose acetate fibers after the melt-spinning of cellulose acetate containing a plasticizer such as sulforane or polyether. During the process of deacetylation, the plasticizer leaches out of the membrane. By this method, the permeability of the membrane can be regulated.

Simple membranes made of poly(methyl methacrylate) cannot satisfy solute permeability and the mechanical strength requirements simultaneously. However, suitable membranes can be produced for blood dialysis with excellent properties when a mixture of isotactic and syndiotactic poly(methyl methacrylate) is used. The original membrane performs well for blood ultrafiltration, but the modified membrane is applicable to the blood dialysis.

Other synthetic membranes made of polyacrylonitrile or copolymer of ethylene with vinyl alcohol are also being used commercially.

8.8.1 Hollow Fibers from Copper Oxide-Ammonia-Cellulose Spin Baths [33]

Hollow fibers for blood dialysis are manufactured from regenerated cuprammonium cellulose by extrusion through the outer circle of an orifice, simultaneously adding an organic solvent through an inner concentric circle, stretching the fiber in a gaseous atmosphere, and coagulating in an NaOH bath. The resulting hollow fiber has a better dialysis rate, no pin holes, smoother inner and outer surfaces, a skin-free surface, higher tenacity, and a better network of pores with a diameter of 200 Å compared to a conventionally manufactured rayon hollow fiber.

8.8.2 Preparation of Another Hollow Fiber [34]

A conventionally prepared cuprammonium cellulose spinning solution contains cellulose (10.0 wt %), ammonium (7.0 wt %), and copper (3.6 wt %) with a viscosity of 2000 poise. This is extruded through an outer ring-shaped orifice, 5 mm in diameter, of the duplicate style at a delivery rate of 20 mL/min. At the same time, perchloroethylene is extruded through a central core orifice, 1 mm in diameter, from the same duplicate nozzle at a delivery rate of 5 mL/min, so as to fill the core space of the ring-shaped liquid string.

This sheath-core composite string is allowed to fall under gravity for a vertical length of 300 nm in an open-air atmosphere. The outside diameter of this composite liquid string is reduced to 600 μm under the influence of gravity stretch. The liquid string is then introduced vertically into a coagulation bath comprised of an aqueous NaOH solution (11 wt %) kept at 25°C and extending horizontally for a length of 8 m. The traveling speed of the filament through the coagulation bath is set to 100 m/min.

The coagulated hollow-fiber filament is taken out on a revolving hank frame and kept there for about 2 h. The hank is suspended from a bar and rinsed with fresh water by means of a shower. The hank is then dipped into a sulfuric acid bath and rinsed with fresh water. The hank is then suspended from a moving frame and passed through a hot-air tunnel drier at 130°C. The hank is mechanically cut into proper lengths to remove the core liquid solvent and kept in a room under quiet conditions at normal temperature for several hours.

8.9 ARTIFICIAL LUNG MEMBRANES

The artificial lung is a prosthetic device presently used in an extracorporeal blood circuit for partial or total gas exchange during cardiac surgery. Besides ability for gas exchange the requirements for the materials and the apparatus for an artificial lung are that the treatment and management are as simple as possible, that complete sterilization can be performed, and that damage to the blood during use is negligible. Therefore, it is recognized that disposable materials are most desired.

The artificial lungs used now are classified principally into three types:

1. *The bubble oxygenator.* Gas exchange is carried out by bubbling oxygen gas in a finely distributed form through the blood. The structure of the apparatus as well as the treatment is simple and the materials are disposable. However, blood foams when stirred with oxygen, causing severe problems, thus making it necessary to provide defoaming equipment.
2. *The film-type oxygenator.* In this case the blood comes in contact with oxygen gas as it passes over an array of films, a process that allows for gas exchange. In comparison with the bubble oxygenator, the damage to the blood during the process is much lower; however, this oxygenator is not disposable and the cleaning, sterilization, and reconstruction are inconvenient.
3. *The membrane oxygenator.* Using polymeric membranes with high gas permeability, gas exchange takes place in the blood. This gentle process minimizes the progressive destruction of blood cells and denaturation of plasma proteins compared with other types of oxygenators.

The membrane used for the artificial lung was first made of Teflon, because of its availability in thin sheets to produce large areas and its fair blood compatibility. During the early 1960s, thin membranes of silica-filled silicone rubber were developed. Silicone rubber has the highest specific gas permeability, and pure filler-free silicone rubber has been shown to minimize thrombosis and platelet sequestration [35]. Presently, membrane lungs are also being prepared from porous polypropylene, Teflon, and silicone-cellulose, which have extremely high gas permeabilities.

8.9.1 Thin Defect-Free Silicone Rubber Films and Membranes [36]

Silicone rubber consisting of a methylvinyl polysiloxane gum compounded with a fumed silica filler was finely cut and swelled overnight in a toluene solvent and was then dispersed to form a 20% silicone rubber dispersion. The mixture was filtered to remove undispersed silicone rubber and fibrous strands and then centrifuged for 20 h at 10,000 g to remove undispersed silicone filler aggregates. The supernatant was treated with 2% 2,4-dichlorobenzoyl peroxide, based on the weight of the rubber. The dispersion was then cast under a doctor blade onto a moving aluminum foil carrier. The resulting film was dried at 75°C for 6 min to evaporate the solvent and was then cured in air for 6 min at 180°C. The cured film, 5 mil in thickness, was removed from the carrier and was observed by light microscopy to have a highly uniform pinhole-free appearance. Tensile strength, burst strength, and elongation measurements made on several different areas of the film varied by ±5%, indicating a highly uniform film.

8.9.2 Preparation of a Hollow Filament [37]

Polypropylene, having $[\eta] = 1.40$ in tetralin at 135°C, was spun from a nozzle of bridge type having a diameter of 4 mm at a spinning temperature of 230°C, at a take-up velocity of 610 m/min and at a draft of 445. The unstretched hollow filaments thus obtained were wound on a metal reel and subjected to heat treatment at 140°C for 30 min in a hot-air box. Sets of filaments were stretched at stretching ratios of 50, 100, 200, 300, and 400% of the original length at room temperature and subjected to a second heat treatment at 145°C for 30 min in a hot-air drier while maintaining the filaments at the stretched state. The surface area obtained for these porous hollow filaments was measured by the gas adsorption method. The gas permeability of the hollow filaments was measured by profusing the outside of the filaments with nitrogen gas at 1 atm and measuring the flow of gas from the interior portion of the hollow filaments. It was found that the distribution curves of the pore radius of the hollow filaments with a stretching ratio of 50 to 200% have at least one maximum point between 200 and 1200 Å and that the hollow filaments with this stretching ratio have much higher gas permeability than those with higher stretching ratios.

8.10 BLOOD-COMPATIBLE POLYMERS

Blood generally coagulates causing blood clotting when it encounters foreign solid surfaces. This phenomenon is assumed to begin with initial platelet adhesion to the surfaces, followed by platelet aggregation, which leads to the formation of thrombus masses. Therefore, the biomaterials used in contact with blood should be free from causing blood clotting.

In cases where polymeric materials are used in the form of blood bags for temporary storage or for circulation of blood outside of the body, blood clotting can be inhibited by adding calcium citrate in the former case or heparin in the latter case. Both of these techniques work well for short-term contact with blood components. However, when it is desired to keep the biomaterials in the body in contact with blood for longer periods, as in the case of artificial hearts or blood vessels, the development of completely antithrombogenic materials is essential [38-42].

A guide to obtaining the optimal material design for such purposes has not been clearly defined. However, the polymeric antithrombic materials studied can be classified as follows:

1. Materials with weak interaction with blood
   a. Unreactive surfaces
   b. Negatively charged surfaces
   c. Heterogeneous structures
   d. Hydrogels
2. Materials that inhibit the formation of blood clotting
   a. Heparinization
   b. Self-washing action
3. Material from the living body itself

Research and development of synthetic polymers with anticlotting properties have produced many suitable materials, such as heparinized inorganic and organic polymers, silicone elastomers, segmented polyurethane, silicone-polyurethane copolymers, hydrogels, special carbon black, microfibers, and biolized polymers. In recent years, however, it has been pointed out that both hydrophilic and hydrophobic properties are essential for realizing antithrombogeneity. An example is avcothane, a block or graft copolymer of hydrophilic polyetherurethane with hydrophobic polydimethylsiloxane, which has excellent antithrombogeneity and is already commercially available. Block copolymers of hydrophilic polyether with hydrophobic urethane-urea, developed by Lyman, is another example of this type. It was found that copolymers with both hydrophilic and hydrophobic properties show better properties than those of homopolymers with combined hydrophilic and hydrophobic properties. This indicates the importance of hydrophilic and hydrophobic fine structures at the molecular aggregation level.

Okano and coworkers have prepared random, block, and graft copolymers of hydrophilic 2-hydroxyethylmethacrylate with hydrophobic styrene. It was found that the block copolymer has a typical microphase-separated structure, and such a surface shows excellent antithrombogeneity compared with that of the homopolymer or the random copolymers. The relationship between fine molecular structure of the surfaces and the antithrombogeneity is gradually being elucidated.

8.10.1 Preparation of Heparinized Hydrophilic Polymer [43]

Commercial graft copolymer composed of vinyl chloride, ethylene, and vinyl acetate (Graftmer R_3) is reacted with sodium *N*,*N*-diethyl dithiocarbamate in DMF solution at 50 to 60°C for about 2 h to introduce dithiocarbamate groups, effecting photo-induced graft copolymerization. Water and methanol are added to precipitate the polymer. After washing with water and methanol several times, the polymer is vacuum dried.

The graft copolymerization of methoxypolyethyleneglycol methacrylate (SM) monomer and *N*,*N*-dimethylaminoethyl methacrylate (DAEM) monomer and Graftmer R_3, containing dithiocarbamate groups, is carried out in cyclohexanone by UV irradiation. This solution is added to methanol and the precipitate soaked in methanol for 2 to 3 days, to extract unreacted monomers, and then vacuum dried at room temperature. The *N*,*N*-dimethylaminoethyl groups of the graft copolymer are quarternized in DMF at 50 to 60°C for 1 to 2 h by ethyl bromide; the cationic polymer forms ionic bonds with heparin.

Catheters, films, and tubes of the cationic polymer are prepared by solvent casting from a mixture of cyclohexanone-DMF containing 2 to 3% of the cationic polymer. After casting, the polymer is dried at 70°C for about 5 h; the remaining solvent was extracted by soaking in methanol at 50°C for 1 day. These materials were finally heparinized at 80°C by soaking in a 2% aqueous sodium heparin solution for 1 to 3 days.

The polymer composition is determined by elementary analysis, the amount of nitrogen atom was measured by titrating bromide, and heparin content as well as degree of water absorption are measured gravimetrically.

8.10.2 Preparation of Segmented Polyetherurethanes [38]

The segmented polyether-urethane system in question consists of a series of block copolyether-urethane elastomers based on methylene bis(4-phenylisocyanate) and either polyoxypropylene (PPG) or polyoxytetramethylene glycol (PTG). Prepolymers formed by these reagents are chain extended with various diamines to form the final elastomer with the following generalized structure:

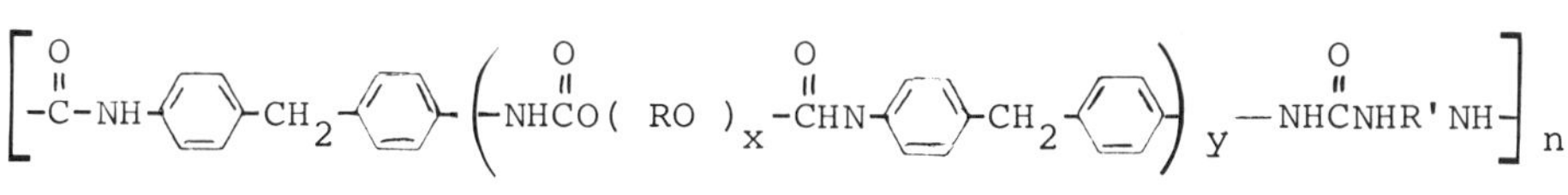

The following procedure for the synthesis of small batches of elastomer (3-1000-1-E) is typical. A solution of PPG (100.5 g, 0.1 mol), methylene bis(4-phenylisocyanate) (50 g, 0.2 mol), and DMSO (300 mL) is heated rapidly to 100 to 110°C and maintained at that temperature for 3 h with continuous stirring under nitrogen. The solution is then cooled to room temperature and ethylenediamine (6 g in 200 mL of DMSO) added. The resulting viscous solution is stirred for 30 min to ensure complete reaction and is then poured into a large volume of water to precipitate to the polymer thus formed. The polymer is then washed in several changes of water, chopped in a Waring Blendor, filtered, and extracted with ethanol for 48 h at 40°C and dried in an air oven at 80°C. With wholly aromatic diamines, the chain-extending reaction is considerably slower and heat must be used to achieve finite polymerization rates.

Inherent viscosities of 0.5% solutions in DMF can be used as a relative estimate of molecular weight. The molecular weight distribution is determined by gel permeation chromatography. Stress-strain curves are obtained for films approximately 4 mil thick with an Instron Universal tester.

8.10.3 Preparation of Avcothane: Block Copolymers of Polyurethanes and Polydimethylsiloxane [44, 45]

Block copolymers were prepared from polydimethylsiloxane with hydroxy or acetoxy end groups and poly(ether-urethanes) or poly(ester-urethanes). The poly(ether-urethanes) were prepared from polypropylene glycol with either methylene bis(4-phenylisocyanate) or 2,4- or 2,6-tolylene diisocyanate. The poly(ester-urethanes) were prepared from poly(ethylene terephthalate) and 1,5-naphthalene diisocyanate. Thus, 70 parts poly(ether-urethane), prepared from poly(propylene glycol) and methylene bis(4-phenylisocyanate) are heated with 30 parts of acetate-protected polydimethylsiloxane, 70 parts THF, and 90 parts dioxane for 6 to 10 h at 40 to 50°C. These polymers are used for the storage and handling of blood.

To determine the degree of blood compatibility of the block copolymers thus formed, solutions are coated onto the walls of 16-mm-id test tubes and allowed to dry at room temperature for 24 to 48 h, during which time all the acetic acid liberated from the acetate-end-blocked polydimethylsiloxane evaporated. Film thicknesses on the walls of the test tubes are all about 0.003 in. About 3 to 5 mL of freshly drawn blood is dispensed into the coated test tubes and the time to coagulate at 37°C determined. For purposes of comparison, freshly drawn blood of the same donor is dispensed into uncoated test tubes and test tubes coated with either the poly(ether-urethane) homopolymer or the polydimethylsiloxane homopolymer.

8.11 ARTIFICIAL HEART

In recent years the fabrication of a number of implantable cardiovascular devices using synthetic polymers has received much attention. Rigid polymers have a variety of applications in the fabrication of implantable blood pump housings, as blood-contacting components used in valves, and as connectors to couple biologic tissue to the blood pump. A variety of rigid polymers are now available; however, each has certain disadvantages [46].

The first study on the fabrication of implantable cardiovascular devices was begun by Kolff and Akutsu in 1958. In parallel with the active research in the United States, Atsumi and his group developed new devices independently in Japan. Of all the systems concerned with the development of an artificial heart, the most important one is the blood pump. It is recognized that the development of special polymeric materials suitable for the blood pump is essential. The important properties that these polymeric materials should possess are durability, processability, and antithrombogenity. Studies to produce more desirable polymeric materials are still going on.

The materials now available can be classified in two groups. The first group are the synthetic polymers that have smooth surfaces, such as silicone rubber, polyurethane (Biomer and its copolymer, Avcothane) and poly(vinyl chloride) paste; the second group are the biological and modified biological materials.

Several elastomeric materials have been evaluated for the fabrication of prosthetic blood pumps. One is a type of segmented polyurethane, the chemical structure of which is similar to that of Biomer, although the mechanical properties are apparently inferior. The others, Avcothanes, are said to be copolymers of polyurethane and silicone rubber. This material is suitable for intravascular use and has been used in the fabrication of the intra-aortic balloon pump (IABP). The fabrication techniques are more difficult than those for Biomer [46]. Atsumi, Imachi, and their groups, in Japan, have produced a new blood pump material by covering a poly(vinyl chloride) pump body with Avcothane so that properties such as durability and processability were greatly improved.

8.11.1 Fabrication of a Blood Pump [49]

A blood pump is a sac-type pump whose sac has a 60- or 90-mL volume. The maximum pump output is between 5 and 7 min. The fabrication procedure for the Avcothane-coated poly(vinyl chloride) pump is shown in Figure 3. Two male molds for the sac and housing are made of metal or epoxy resin. They are heated at 200°C in an oven and dipped into poly(vinyl chloride) paste solution for 30 to 60 s. They are heated

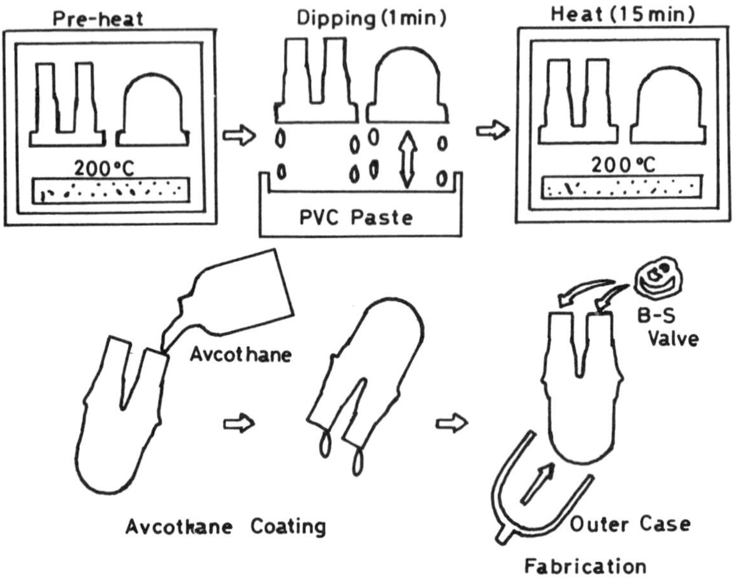

Figure 3. Fabrication procedure for Avcothane-coated poly(vinyl chloride) paste pump. (From Ref. 49.)

again at the same temperature for 10 to 15 min. The poly(vinyl chloride) solution on the mold changes to a gel and forms a transparent film on the mold. The film thickness can be controlled by changing the dipping time and heat treatment. Usually, the sac and housing are formed in one dipping. After the molds are removed, the sac and housing are bonded by heat. The pump is filled with Avcothane prepolymer and the excess prepolymer is removed by turning the pump upside down. The pump is dried at 25°C for 40 to 60 min. This coating process is repeated three times. To evaporate the solvent from the polymer completely, the pump is dried for half a day after the final coating. Two Björk-Shiley (B-S) valves, 18 mm in orifice diameter, are inserted in the pump and tied with nylon from the outside. An outer case made of polycarbonate is then bonded to the pump. All of these processes are carried out in a class 100 clean room. By this method, two blood pumps can be made within 2 days. The polyurethane pump made by Toray Inc. has almost the same design as the Avcothane-poly(vinyl chloride) pump.

8.11.2 Artificial Heart Cannula and Artificial Chest Wall [49]

Four types of cannulae 20 cm long were developed to connect a pump to the circulatory system. The inner diameter of these cannulae is

15 mm at the pump side and 12 mm at the blood vessel side. The cannulae for the left and right atria have six side holes, with 4 mm diameter at the blood vessel side end. At the end of the cannulae for the descending aorta and pulmonary artery, expanded polytetrafluoroethylene vascular grafts are sutured to descending aorta and pulmonary artery by side-to-end anastomosis. All these cannulae are made of poly(vinyl chloride) and their inner surfaces coated with Avcothane.

For the artificial heart described, the blood pump is set into the chest wall and the four cannulae penetrate the chest wall through the intercostal space. With this technique, infection at the interface between tissue and cannulae is a serious problem for long-term animal survival. Subsequently, an artificial chest wall, made of poly(vinyl chloride) and its surface covered with polyester velour, was developed to close the chest completely and to prevent infection.

8.11.3 Evaluation of Thrombogenicity [50]

The thrombogenicity of the pump material was evaluated both by visual inspection and by scanning electron microscopic inspection. After the experimental animal has been sacrificed, the pumps are taken out and rinsed gently with saline. Macroscopic photographs are taken of the pump. Then the pumps are filled with 1.25% phosphate-buffered glutaraldehyde, pH 7.4, for a few hours to fix any blood elements on the inner surface of the sac, rinsed again with saline, and rinsed finally with distilled water. Two specimens for scanning electron microscopic observation are obtained from each pump, one from the sac center, where sac movement amplitude is largest, and one from the sac lower part, where the sac movement amplitude is small.

8.12 BIOLOGICALLY ACTIVE POLYCARBOXYLIC ACID POLYMERS

A large number of naturally occurring polyanionic polymers are known to possess innate physiological properties. These natural macromolecules include polysaccharides, glycoproteins, and polynucleotides. In recent years, comparison of the physiological properties of several synthetic polyanions with those of naturally occurring polyanions has become of interest to investigators. Perhaps the most widely investigated polymers are the carboxylate polyanions, in particular, the cyclocopolymer pyran, which was first synthesized by Butler in 1951 [51]. Although the structure-reactivity relationship remains to be determined for these synthetic polyanions, they have been shown to possess a wide spectrum of biological activity [52].

Several synthetic polyanions have been found to elicit significant effects on the immune system [53]. They have been shown to induce the production of interferon [54]; to modify reticuloendothelial function [55]; and to invoke immunoadjuvant [56], antiviral [57], and antitumor activity [58]. The antineoplastic properties of these polyanionic polymers have largely been attributed to their ability to activate macrophages [59]. The prolonged protective action of these synthetic polyanions, when given prior to a viral infection, has interesting clinical potential. Consequently, a variety of pathophysiologists have become interested in the fundamental role of polyanions in controlling host resistance.

Studies indicate that the most important feature of the carboxylic acid polymers affecting interferon induction, antiviral activity, and tumor growth inhibition is a high density of carboxylate groups attached to a long-chain polymeric backbone [52]. To effect these biological activities, the molecular weight per chain should be greater than 1000, and the carboxyl groups can be in either alternate or adjacent positions. Most of the synthetic polyanionic polymers studied have been stable and not readily biodegradable; this may account for their prolonged activity as well as some of the adverse toxic effects observed.

8.12.1 Carboxylic Acid Homopolymers

Several homopolymers with polycarboxylic acid groups have been studied with regard to biological activity. These include polyacrylic acid (PAA), polymethacrylic acid (PMAA), polyethacrylic acid (PEA), poly(maleic acid) (PMA), and poly(itaconic acid) (PIA). These polymers have shown activity both as interferon inducers, and against several viruses, including Semliki Forest virus, sarcoma 180, vaccinia virus, and herpes simplex II.

The most extensively studied homopolymer for biological drug application is poly(acrylic acid) (PAA). The structure of this polymer engenders a polymer with a carboxylic acid to carbon atom density attached to the chain in a 1:2 ratio:

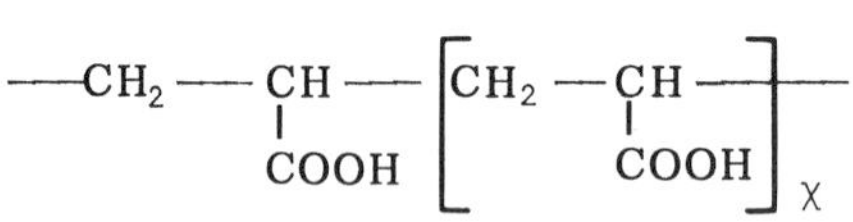

PAA was reported to give complete protection against Semliki Forest virus [60], a 55% reduction in pox count for vaccina virus infection [61], and a 32% increase in survival time for animals with sarcoma 180 [62]. Although PAA is a poor interferon inducer [63], it is capable of effecting a 25% survivor rate against lethal doses of Mengo virus [64]. The antiviral activity of a number of different molecular weight fractions of poly(acrylic acid) with narrow molecular weight distribution have been carried out by Muck et al. [65]. Molecular weights of less than 5000 were ineffective against encephalomyocarditis virus and Colombia SK virus. The optimum efficacy of poly(acrylic acid) against the viruses evaluated was observed for samples with molecular weight between 6000 and 15,000, as molecular weight samples greater than 25,000 are exceedingly toxic at the dosage levels required.

Muck et al. also found that the tacticity of the polymer may also play an important role in biological activity. Based on interferon induction and the ability to provide protection against picornavirus infection in vivo, isotactic poly(acrylic acid) was reported to be more effective than the atactic polymer [65]. Isotactic poly(acrylic acid) samples, with narrow molecular weight distribution, were found to be more active than isotactic poly(acrylic acid) samples, with broad molecular weight distribution. The heterotactic polymer is even less effective and biologically does not exhibit this molecular weight relationship.

8.12.1.a Preparation of poly(acrylic acid)

A solution of ethyl acrylate (9 g, 0.1 mol) in 100 mL of benzene containing 0.2 g of AIBN is purged with N_2 and heated at 60°C for 3 h. The polymer is precipitated by adding the solution to 500 mL of petroleum ether. Reprecipitation from toluene/petroleum ether affords 5.8 g, 64% of polymer, which is hydrolyzed with 0.2 N NaOH at 60°C for 8 h.

8.12.1.b Preparation of Isotactic poly(acrylic acid)

Isopropyl acrylate (34 g, 0.33 mol) in 500 mL of toluene is purged with N_2 and cooled to -78°C. Added to this solution is 12 mL of 2.5 M phenylmagnesium bromide in ethyl ether. The -78°C temperature is maintained for 24 h. The reaction mixture is concentrated to 150 mL, filtered, and the polymer precipitated in petroleum ether. Reprecipitation from toluene-petroleum ether affords 19 g (56%) of 95% isotactic polymer. Hydrolysis to the poly(acrylic acid) is carried out in 0.2 N NaOH at 60°C for 8 h.

Although poly(methacrylic acid) is only slightly different structurally from poly(acrylic acid), their biological effects are different:

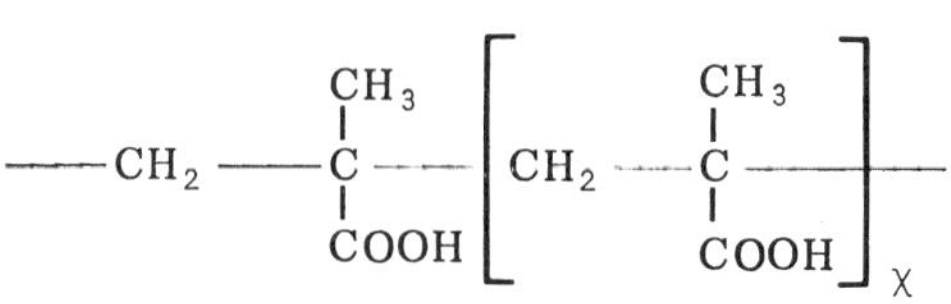

The biological analysis of poly(methacrylic acid) (PMAA) indicates that it is much less effective than PAA both against viruses and as an interferon inducer. When administered 24 h before infection, PMAA exhibits significant inhibition of Sindbis virus and vesicular stomatitus virus. PMAA's activity appears to be related to RNA metabolism since it inhibits Sindbis-RNA but not polio and similar viruses. Much et al. [65] reported that poly(methacrylic acid) molecular weight fractions were generally much less effective as antiviral agents than are similar poly(acrylic acid) fractions. Neither poly(acrylic acid) nor polymethacrylic acid changes the thermolysis profile of erythrocyte ghost cells, indicating very little or no cell membrane interaction [66].

8.12.1.c Preparation of syndiotactic poly(methacrylic acid)

A solution of methyl methacrylate (10 g, 0.1 M) in 100 mL of benzene containing 0.1 g of AIBN is purged with N_2 and heated at 60°C for 3 h. The polymer is precipitated by adding the reaction mixture to rapidly stirred petroleum ether (500 mL). Reprecipitation from toluene petroleum ether yields 6.5 g, 65% of polymer. This material is hydrolyzed with 0.2 M NaOH at 60°C for 8 h. The tacticity is determined by ^{1}H NMR triads (60% syndiotactic).

8.12.1.d Preparation of isotactic poly(methacrylic acid)

Methyl methacrylate 20 g, 0.2 M in 250 mL of toluene is purged with N_2 and cooled to -78°C. Added to this solution is 90 mL of 0.75 M butyllithium in benzene. After 0.5 h, the reaction is filtered, warmed, and the polymer recovered by precipitation into petroleum ether. Reprecipitation from toluene-petroleum ether affords 12 g, 60% of polymer. The acid is formed by heating the polymer in 0.2 M NaOH for 8 h at 60°C (85% isotactic).

Poly(ethacrylic acid) was recently prepared by Tirrell [66]. This material exhibits antitumor activity against Lewis lung less than pyran but much more effectively than PAA and PMAA. Poly(ethacrylic acid) elicits macrophages with higher 5'-nucleotidase levels than the strong macrophage activator pyran and the macrophage stimulant thioglycollate [67]. Structurally, this polymer should be more hydrophobic than PAA, as the carboxylate-to-carbon ratio is 1:4 with an ethyl group pendent to the chain. Tirrell [68] has found that poly(ethacrylic acid) produces a significant effect on the thermolysis profile of erythrocyte ghost cells, indicating that there may be absorption of the polymer by the cell membrane. He also prepared samples with varying amounts of isotacticity in the polymer. These were synthesized with butyl lithium as the anionic initiator at various temperatures to obtain different tacticities. Interestingly, the absorption appears to be inversely proportional to the isotacticity of the polymer. These tactic poly(ethacrylic acids) have not been evaluated biologically.

The poly(maleic acid) homopolymer is ineffective against Lewis lung cells but highly active against Friend leukemia virus [69]. Most interestingly, the supernatant fluids of the peritoneal exudate cells from mice pretreated with poly(maleic acid) transfer protection against

Friend leukemia [70]. It induces phagocytic activity but does not depress the reticuloendothelial response experienced with pyran; nor is the reticuloendothelial response molecular weight dependent, as determined by five different fractions ranging from 1000 to 10,000 Mw. The structure of the polymer indicates a high density of carboxyl group to carbon along the polymer chain in the ratio of 1:1.

Although maleic anhydride does not readily undergo homopolymerization, several methods of polymerization have been reported [71]; these include neat polymerization at 120°C with successive additions of benzoyl peroxide over a 1-h period. This material is usually tan in color and has considerable branching and a very broad polydispersity. Solution polymerization has been effected by ^{60}Co radiation to give a white, linear, and narrower polydispersed material. An excellent compilation of maleic anhydride polymers and copolymers has been made by Culbertson and Trivedi [71].

8.12.1.e Preparation of Poly(maleic acid) [71]

Maleic anhydride is purified by distillation under reduced pressure and then by sublimation. Benzoyl peroxide (wt %) is added to 19.6 g (0.2 mol) of purified maleic anhydride and heated to 85 to 90°C to form a molten mixture. After 10 to 20 min the temperature rises to 140 to 150°C. When cooled to 110°C, additional initiator is added in three successive increments of 1 wt% at 30-min intervals. The final mixture is dissolved in acetone (100 mL) and precipitated into 500 mL of dry ethyl ether. Precipitation is repeated to yield a tan polymer, 12 g (62%), approximately 25,000 MW. This polymer is hydrolyzed in 0.1 N NaOH solution by heating at 60°C for 6 h. The pH of the solution is adjusted to 1.5 with 6 N HCl and the solution freeze dried. The residue is extracted with acetone and the polymer recovered by precipitation in ethyl ether.

Poly(itaconic acid) has not been extensively evaluated for biological use. It has been reported to increase the life span of animals inflicted with ascitic sarcoma 180 by 45% [72]. The structure of this compound is different in that one of the carboxyl groups is pendent to the polymer chain. The carboxyl group-to-polymer backbone carbon ratio is 2:1. Itaconic acid, similar to maleic anhydride, does not readily homopolymerize in organic solvents. The most common way to prepare the homopolymer is to heat an aqueous 0.5 N HCI mixture of monomer in the presence of ammonium persulfite for several hours. As polymerization occurs, the monomer, which is not very soluble in water, slowly dissolves.

8.12.1.f Preparation of poly(itaconic acid) [72]

Itaconic acid (22.4 g, 0.2 mol) and 2 wt % ammonium persulfite is added to 100 mL of 0.5 N HCl and heated at 70°C for 6 h. During this period the insoluble itaconic acid slowly disappears and polymer, which is water soluble, is formed. The solution is cooled to 0°C for 12 h and filtered while cold to remove unreacted monomer. The polymer is recovered by freeze drying; 7.8 g, 35% yield.

8.12.2 Maleic Anhydride Copolymers

Maleic anhydride is a unique comonomer because it does not readily undergo homopolymerization but does form copolymers without difficulty. It is also unique in that the copolymers that form in the presence of radical initiators are in a 1:1 ratio and in an alternating sequence [73-78]. It appears that the electronic structure of this monomer makes it an excellent electron acceptor and consequently, forms a complex with other π systems. During polymerization the complex is added to the chain in such a manner that a 1:1 alternating copolymer is obtained.

The copolymer composition is generally independent of initial monomer ratios. El'Saied et al. [74], in their study of the dependence of copolymer composition on the mole ratio of the monomer feed of the maleic anhydride (MA) and acrylic acid (AA), found that in the presence of an excess of MA, a consistent 1:1 copolymer composition is obtained. When an excess of acrylic acid is used, two polymers are obtained, one a 1:1 copolymer of MA-AA and the other a homopolymer of poly(acrylic acid). It was postulated that copolymerization takes place through an "activated" maleic anhydride complex. It was further postulated that maleic anhydride is not added by either monomer in the "ground" state. In further investigations, El'Saied et al. [75] modified the polarity of maleic anhydride by the addition of an inert electron donor, naphthalene (NPH). When NPH is introduced into the copolymerization mixture of MA and AA, a greater amount of MA is incorporated into the polymer chain, but no NPH. It appears that in the presence of NPH, two maleic anhydride complexes are formed, NA-AA and MA-MA-NPH. Both complexes "activate" MA to form a polymer.

Earlier Kimbrough [76] reported that dihydropyran (DHP) and maleic anhydride form a 1:1 alternating copolymer. A yellow color is observed when MA and DHP are mixed. However, color is no longer apparent after the polymerization is complete. Since charge-transfer complexes are known to produce electronic absorption in the visible and near-ultraviolet spectrum, it was postulated that the yellow color is due to a complex and it may be the precursor to polymerization.

Complexation constants have been obtained for a number of maleic anhydride and monomeric electron donors. Because of the 1:1 alternating nature of the copolymerization products resulting from mixtures of MA with those monomers that exhibit charge-transfer characteristics, it was proposed that the complex is an active intermediate in the polymerization mechanism. For example, MA and dihydropyran give a perfect 1:1 copolymer [75,76]. The alternating sequence appears to be a result of the complex reacting with the propagating free radical.

Many workers have analyzed several of these systems and postulated that the involvement of a charge-transfer complex could explain the alternating copolymer of most monomers with maleic anhydride. For example, maleic anhydride and furan, which had been assumed to undergo the Diel-Alder reaction so forcefully that polymerization is not probable, were found by Butler and Joyce [77] to produce a 1:1 copolymer under specific conditions. These researchers reported that free-radical polymerizations of maleic anhydride and furan not only occur readily, but are maximized when the monomer mole fraction is 0.5. These results are consistent with the participation of a charge-transfer complex.

Since the copolymers of maleic anhydride yield polymers with a consistent 1:1 alternating structure, they are being synthesized for biological evaluation. Their consistency in structure allows for the easy alteration of many characteristics of the polymer chain, such as polarity, lipophilicity, carboxylic acid group density, rigidity, and the specific addition of other functional groups.

8.12.3 Pyran Copolymer

One of the most widely biologically investigated synthetic polyanionic copolymers is pyran copolymer. It is prepared from divinyl ether (DIVE) and maleic anhydride (MA) and is also referred to in the literature by the acronyms DIVEMA and MVE. The term "pyran" is based on the premise that these comonomers cyclopolymerize to yield a six-membered "pyran" ether ring. This polymer was first prepared by Butler [51], who proposed a cyclopolymerization mechanism to describe the formation of a six-membered pyran system in the polymer chain. This polymer not only is formed by a cyclopolymerization process, but the MA and DIVE monomers are incorporated sequentially in a 2:1 ratio, respectively.

The mechanism proposed involves the initial formation of a 1:1 complex of divinyl ether and maleic anhydride. A random radical reaction of these comonomers in this polymerization is inconceivable based on the analytical data [78,79]. The proposed geometry of the complex, coupled with the kinetic data, indicates that rapid ring closure occurs. Furthermore, the orientation of the comonomers through a charge-transfer complex is also compatible with the kinetic data and with the unique cyclic structure obtained. The involvement of a charge-transfer complex with the propagating free-radical chain offers the best explanation as to how the linear polymer with the concomitant cyclic units is obtained. A detailed report on the nature of the charge-transfer complex formation and the cyclopolymerization has been given by Butler et al. [77,80].

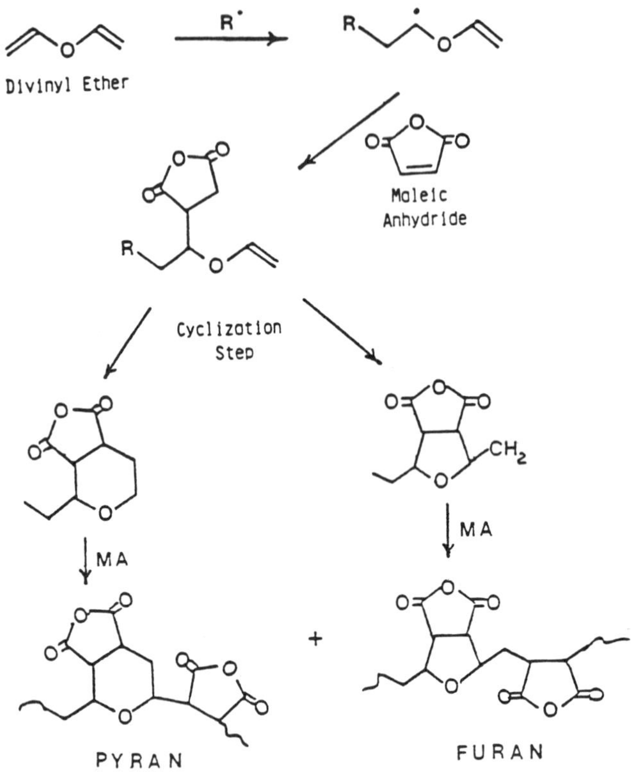

Figure 4. Proposed mechanism for the cyclopolymerization of pyran.

The formation of a six-membered pyran ring in the propagating chain can be obtained by the mechanism proposed in Figure 4. The radical attack on the maleic anhydride-divinyl ether 1:1 CT-complex bonds these two monomers and generates a 2° radical which could intramolecularly attack the β-carbon of the remaining divinyl ether π bond forming the six-membered ether ring. The resulting radical could react with another MA molecule to evolve the polymer chain with a 2:1 ratio of the monomers and a tetrahydropyran ring. Because of this proposed mechanism for the formation of the six-membered ether ring, this copolymer has become known as pyran in much of the literature.

Although this pyran structure has been widely accepted and analysis confirms the 2:1 copolymer ratio, the ring size of the copolymer has been questioned. Recent reports indicate that most cyclopolymerization processes preferentially form five-membered rings instead of earlier proposed six-membered rings in radical cyclization reactions [81-87]. This has been supported by ESR, NMR, and model spectra of the cyclopolymer obtained in a mixed solvent of acetone-carbon disulfide, which gives broad ^{1}H NMR peaks, suggesting that the structure is not uniform. It was estimated that the polymer contained 90% six-membered rings and 10% five-membered rings [86]. The ^{13}C NMR spectrum of the polymer prepared in chloroform indicated a symmetrical bicyclic unit attached by a junction and a monocyclic anhydride unit attached trans along the chain [85]. This indicated that the five-membered ring is the dominant structure of this material. A more recent ^{13}C NMR study, by Freeman and Breslow, of a pyran sample prepared by photoinitiation in acetone-tetrahydrofuran, indicates a very complex system [87]. It was concluded from these ^{13}C NMR evaluations that the ratio of five- to six-membered rings in this polymer is about 50:50.

Other structural complications are prevalent in the determination of the structure of pyran. For example, the rate of homopolymerization of vinyl ether is about one-tenth of its copolymerization rate with maleic anhydride. Consequently, the probability of the incorporation of small segments of divinyl ether homopolymer in the polymer chain is quite high. This would result in the formation of pendent double bonds to the chain, which could polymerize to form branched and cross-linked polymer chains. Therefore, simpler copolymers prepared from maleic anhydride are being used for biological evaluation. These are quantitatively and qualitatively more reproducible, and it has been indicated that they are better therapeutic agents [88,89].

8.12.3.a Preparation of pyran copolymer

Divinyl ether (19.6 g, 0.2 M) and 39.2 g, 0.4 M of maleic anhydride are dissolved in 250 mL of acetone and 1 wt % of benzoyl peroxide is added. The solution is purged with N_2 for 15 min and the flask sealed and heat at 70°C for 8 h. The solution is concentrated and added dropwise into 600 mL of dry ethyl ether through a dispension tube. The polymer is collected and reprecipitated with acetone-ethyl ether, for a 37 g 63% yield. The polymer is hydrolyzed by heating in 0.2 N NaOH for 8 h.

8.12.4 Other Copolymers of Maleic Anhydride

A number of other maleic anhydride copolymers have been evaluated for antitumor activity [90,91]. Several of these polymers are as or more effective than pyran against Lewis lung carcinoma and Erhlich ascites [92]. The more effective polymers seem to be those that have a high degree of hydrophobic character.

Ottenbrite et al. [93,94] have reported two new polycarboxylic acid polymers that after administration produce animal survivors of the Lewis carcinoma challenge drugs. The first polymer, poly(maleic anhydride-co-cyclohexyl-1,3-dioxepin) (MA-CDA), gave a 69% increase in animal survival time and three of five animals survived 90 days after a subcutaneous injection of 2×10^4 Lewis lung tumor cells followed by 100 mg/kg of drug administered 3 days later. The other drug, poly(4-methyl-2-pentenonyl-co-maleic anhydride) (MP-MA), gave a 63% increase in animal survival time and two of four animals survived after a similar Lewis lung challenge and drug administration. Cell counts and cell differentials performed on peripheral blood leukocytes and peritoneal macrophages from MA-CDA- and MP-MA-treated mice corresponded more closely to those of normal mice than to mice treated with the peritoneal macrophage activators pyran and *Corynebacterium parvum* [92]. Furthermore, the lysosomal peroxidase activity of the peritoneal macrophages after MA-CDA and MP-MA administration remained comparable to that of normal resident macrophages, while an increase in peroxidase-positive macrophages was observed after pyran and *C. parvum* administration. These and other data suggest that these two drugs, MA-CAA and MA-MP, activate in situ peritoneal macrophages and prolong survival against transplanted Lewis lung carcinoma [93-96].

8.12.4.a Copolymerization of 4-methyl-2-pentenone and maleic anhydride [97]

To a 100-mL polymerization flask containing 9.8 g (0.1 mol) of maleic anhydride, 11.4 mL (0.1 mol) of mesityl oxide and 25 mL of acetone is added 1 g of AIBN. The solution is purged with nitrogen and the flask sealed and heated at 80°C for 2 h. The flask is cooled and an additional 1 g of AIBN is added and then heated at 80°C for 2 h. This sequence is repeated two more times and the reaction is kept at 80°C for 24 h. The solution is added dropwise to 200 mL of rapidly stirred diethyl ether. The precipitate is recovered by filtration, dissolved in 25 mL of acetone, and reprecipitated in 150 mL of diethyl ether. The polymer is recovered, 6.0 g (31%), and hydrolyzed in 2 M NaOH.

8.12.4.b Copolymerization of maleic anhydride and cyclohexyl-1,3-dioxepin [97]

Maleic anhydride (9.8 g, 0.2 mol), 16.8 g of 1,6-dioxospiro(6,6)dodec-3-one, and 10 mL of acetone are placed into a 100 mL flask with 0.4 g of AIBN. The solution is purged with nitorgen for 10 min, and the flask sealed and heated at 80°C for 7 h. The reaction mixture is cooled and added dropwise to 75 mL of diethyl ether. The polymer is collected as a white powder by filtration and then dissolved in 10 mL of acetone and reprecipitated in diethyl ether; yield 7.1 g, 53%.

8.12.5 Other Copolymers with Carboxylic Acid Groups

Hodnett et al. have prepared five copolymers of acrylic acid and isobutyl vinyl ether [62]. Since the acrylic acid polymerizes readily by free-radical initiation, the concentration of isobutyl vinyl ether has to be kept high to obtain appreciable amounts of this monomer within the polymer chain. These polymers were evaluated against Sarcoma 180 ascites tumor of mice. Maximum antiviral effectiveness was observed for those polymers, with the least number of carboxylic acid groups ionized, at pH 7. Animal toxicity of these polymers, however, increased with decreased carboxylic acid group content.

Ottenbrite investigated the 1:1 copolymer of itaconic acid and styrene. Although this material is extremely toxic, it was effective against Lewis lung carcinoma both in vitro and in vivo [98]. At dose levels of 100 mg/kg, three of five mice died within 2 days. However, one of the surviving animals lived 62 days and the other lived, without tumor, for more than 90 days. This polymer system is also being investigated by Takatsuka and Konda in Japan [99].

8.12.5.a Copolymerization of itaconic acid and styrene [100]

Polymerization is carried out in a 500-mL pressure bottle charged with purified itaconic acid (30 g, 0.23 mol), styrene (20 g, 0.29 mol), benzoyl peroxide (0.5 g, 1 wt % WM) and freshly distilled dioxane (200 g). The contents of the bottle are bubbled with N_2 for 10 min, sealed, and heated in an oil bath at 70°C for 24 h. The itaconic acid is only partially dissolved initially but goes into solution as polymerization proceeds. The polymer product is soluble in dioxane and imparts a slight yellow color to the viscous solution. The small white particles that appear suspended in the solution are removed by filtration. The polymer is precipitated by aspirating the reaction solution with N_2 into 400 mL of rapidly stirred hexane. The gel-like material is collected on a sintered glass filter and air dried overnight. The polymer is purified by dissolving in 0.2 N NaOH and adjusting the pH to 12. Undissolved material is removed by filtration through a sintered glass filter. The polymer is recovered as a precipitate by adjusting the pH to 3 with concentrated HCl. The yield of polymer is 41.8 g, 83.6%. This polymerization yields 20 g of 10,000 to 30,000 molecular weight polymer and very little 1000 to 10,000 MW material. To obtain a less toxic, lower-molecular-weight product, the polymerization can be repeated with the solvent volume increased to 400 mL and the initiator to 2 g. After 48 h at 70°C, 36.2 g of product, 71.4% conversion, is obtained. The monomer composition of the polymer is determined by ^{1}H and ^{13}C NMR, FT-IR, and elemental analysis.

8.12.5.b Separation of polymers into discrete molecular weight fractions

The polymer products are fractionated by an Amicon Ultrafiltration apparatus using appropriate pore size filters. A 1% solution is prepared by dissolving 30 g of polymer in a 1-L flask with dilute NaOH, readjusting the pH to 7, and diluting this solution to 3 L. The solution is introduced into an Amicon filtration cell equipped with the appropriate ultrafiltration membrane. The PM-30 membrane is used to retain the polymer with > 30,000 MW. The effluent is subsequently subjected to a PM-10 membrane which is used to retain the polymer with > 10,000 MW. The fraction collected from the cell at this stage is in the range 10,000 to 30,000 MW. The UM2 membrane is then used to filter the effluent from the PM-10 membrane to remove the polymer that is < 1000 MW. This gives a second fraction with molecular weight ranging between 1000 and 10,000. To recover each molecular weight fraction from solution, the polymer is reprecipitated by adding an acid solution (HCl) and adjusting the pH to 2. The precipitated polymer is collected by filtration and air dried on a sintered glass filter. To remove the inorganic salts, each polymer fraction is stirred in 2 L of distilled deionized water at 50°C for 24 h and filtered. This process is repeated three times. The polymer is dried in a vacuum oven at 5 mm Hg and 60°C for 48 h. The samples were then analyzed by FT-IR, ^{1}H NMR, ^{13}C NMR, and elemental analysis.

REFERENCES

1. R. L. Kronenthal, Z. Oser, and E. Martin, Eds., Polymers in Medicine and Surgery, Plenum Publishing Corporation, New York (1975).
2. H. P. Gregor, Ed., Biomedical Applications of Polymers, Plenum Publishing Corporation, New York (1975).
3. L. G. Donaruma and O. Vogl, Eds., Polymeric Drugs, Academic Press, Inc., New York (1978).
4. E. P. Goldberg and A. Nakajima, Biomedical Polymers, Academic Press, Inc., New York (1980).
5. K. Takemoto and I. Tabushi, Biomedical and Pharmaceutical Polymers, Kodansha Ltd., Tokyo (1979) (in Japanese).
6. Y. Imanishi, Y. Sakurai, M. Seno, and K. Takemoto, Eds., Biomaterials and Human Body, Kodansha Ltd., Tokyo (1981).
7. J. Andrade, Ed., Hydrogels for Medical and Related Applications, ACS Symp. Ser., 31, 81 (1976).

8. W. Timmer, Chem. Techn., 175 (1979).
9. O. Wichterle in Encyclopedia of Polymer Science and Technology, Wiley-Interscience (H. F. Mark, N. G. Gaylord, and N. Bikales, Eds.), New York, 15, 273 (1972).
10. M. F. Refojo, in Polymers in Medicine and Surgery, (R. L. Kronenthal, Z. Oser, and E. Martin, Eds.), Plenum Publishing Corporation, New York (1975), p. 313.
11. O. Wichterle and D. Lim, Nature, 185, 117 (1960).
12. O. Wichterle and D. Lim, U.S. Pat. 3,220,960 (1965).
13. E. P. Goldberg and A. Nakajima, Biomedical Polymers, Academic Press, Inc., New York (1980), p. 85.
14. N. Nakabayashi, E. Masuhara, E. Mochida, and . Ohmori, J. Biomed. Mater. Res., 12, 149 (1978).
15. Y. Kadoma, N. Nakabayashi, E. Masuhara, and J. Yamauchi, Kobunshi Ronbunshu, 35, 423 (1978).
16. J. Yamauchi, E. Masuhara, N. Nakabayashi, and Y. Kadoma, Jpn. Kokai Tokkyo Koho, 79-63, 025; Chem. Abst. 88, 7867 (1979).
17. a. J. Yamauchi, E. Masuhara, N. Nakabayashi, and K. Shibatani, Ger. Pat. 2 711 234.
    b. J. Yamauchi, N. Nakabayashi, and E. Masuhara, Polym. Prepr. Am. Chem. Soc. Div. Polym. Chem., 20(1), 594 (1979).
18. R. L. Kronenthal, in Polymers in Medicine and Surgery (R. L. Kronenthal, Z. Oser, and E. Martin, Eds.), Plenum Publishing Corporation, New York (1975), p. 119.
19. E. J. Frazza and E. E. Schmitt, J. Biomed. Mater. Biomed. Mater. Res. Symp., 1, 43 (1971).
20. E. E. Schmitt, M. Epstein, and R. A. Polistina, U.S. Pat. 3,442,871; Chem. Abstr., 71, 13569 (1969).
21. Y. Sakurai, Chem. Rev., 21, 243 (1978).
22. C. W. Hall and J. J. Ghidoni, in Polymers in Medicine and Surgery (R. L. Kronenthal, Z. Oser, and E. Martin, Eds.), Plenum Publishing Corporation, New York (1975), p. 167.
23. I. V. Yannas, J. F. Burke, D. P. Orgill, and E. M. Skrabut, Science, 215, 174 (1982).
24. T. Kondo, M. Arakawa, and B. Tamamushi, in Microencapsulation (J. R. Nixon, Ed.), Marcell Dekker, Inc., New York (1976), p. 163.
25. M. Arakawa and T. Kondo, Can. J. Physiol. Pharmacol., 58, 183 (1980).
26. T. Kondo, Chem. Today, 7, 57 (1982) (in Japanese).
27. N. Nakabayashi, E. Masuhara, S. Nakagawa, and S. Koshikawa, Kobunshi Ronbunshu, 36, 279 (1979).
28. Y. Kawashima and H. Takano, in Chemistry of Biomaterials (N. Nakabayashi et al., Eds.), Gakkai-Shuppan Center, Tokyo (1978), p. 11.

29. W. M. Zapol and J. Ketteringham, in Polymers in Medicine and Surgery (R. L. Kronenthal, Z. Oser, and E. Martin, Eds.), Plenum Publishing Corporation, New York (1975), p. 287.
30. N. Nakabayashi, in Biomeducal Polymers (m. Seno, Ed.), Kyoritsu Shuppan Co. Ltd., Tokyo (1978), p. 25.
31. Special issue on the hybridization between polymer chemistry and medical science, High Polym. Jpn., 30 (1981).
32. N. Nakabayashi, J. Soc. Fiber Sci. Techn. Jpn. 36, P-490 (1980).
33. M. Tsuge, S. Uematsu, Y. Ono, S. Eiga, and M. Makita, Ger. Pat. 2 328 853 (1973).
34. M. Tsuge, S. Uematsu, Y. Ono, S. Eiga, and M. Makita, U.S. Pat. 3,888,771 (1973).
35. W. M. Zapol and J. Ketteringham, in Polymers in Medicine and Surgery (R. L. Kronenthal, Z. Oser, and E. Martin, Eds.), Plenum Publishing Corporation, New York (1975), p. 287.
36. T. Kolobow, U.S. Pat. 3,819,772 (1974).
37. K. Kamada, S. Minami, and K. Yoshida, U.S. Pat. 4,055,696 (1975).
38. S. D. Bruck, Chem. Tech., 240 (1977).
39. K. Kondo, High Polym. Jpn., 24, 593 (1975).
40. H. Tanzawa, in Biomedical Polymers (M. Seno, Ed.), Kyoritsu Shuppan Co. Ltd., Tokyo (1978), p. 62.
41. Y. Ikada, A. Nakajima, T. Takamatsu, and H. Matsumoto, in Biomaterials and Human Body (Y. Imanishi, Y. Sakurai, M. Seno, and K. Takemoto, Eds.), Kodansha Ltd., Tokyo (1981), p. 202.
42. a. M. Okano and K. Kataoka, Kagaku No Ryoiki, Special Issue 135, 57 (1982).
    b. M. Okano, Y. Sakurai, and I. Shinohara, Kagaku To Kogyo (Tokyo) (Chem. Chem. Ind. Jpn.), 32, 901 (1979).
43. H. Tanzawa, Y. Mori, N. Harumiya, H. Miyama, M. Hori, N. Ohshima, and Y. Idezuki, Trans. Am. Soc. Artif. Intern. Organs, 19, 188 (1973).
44. E. Nyilas, U.S. Pat. 3,562,352 (1971).
45. E. Nyilas, Ger. Pat. 1 944 969 (1970).
46. W. S. Pierce, in Polymers in Medicine and Surgery (R. L. Kronenthal, Z. Oser, and E. Martin, Eds.), Plenum Publishing Corporation, New York (1975), p. 263.
47. T. Kajiyama and A. Takahara, High Polym. Jpn. 30, 511 (1981).
48. H. Imachi and K. Atsumi, in Chemistry of Biomaterials (N. Nakabayashi et al., Eds.), Gakkai-Shuppan Center, Tokyo (1978), p. 9.
49. K. Imachi, I. Fujimasa, H. Miyake, N. Takido, M. Nakajima, K. Motomura, A. Kouno, T. Ono, and K. Atsumi, Artif. Organs Suppl., 5, 423 (1981).

50. K. Imachi, I. Fujimasa, H. Miyake, N. Takido, M. Nakajima, A. Kouno, T. Ono, Y. Mori, S. Nagaoka, S. Kawase, T. Kikuchi, and K. Atsumi, Trans. Am. Soc. Artif. Intern. Organs, 27, 118 (1981).
51. G. B. Butler, J. Polym. Sci., 48, 279 (1960).
52. L. G. Donaruma, R. M. Ottenbrite, and O. Vogl, Eds., Anionic Polymer Drugs, John Wiley & Sons, Inc., New York (1980).
53. R. M. Ottenbrite, W. Regelson, A. Kaplan, R. Carchman, P. Morahan, and A. Munson, in Polymer Drugs (L. G. Donaruma and O. Vogl, Eds.), John Wiley & Sons, Inc., New York (1978).
54. T. C. Merigan and W. Regelson, N. Engl. J. Med., 277, 1283 (1967).
55. A. E. Munson, W. Regelson, W. Lawrence, and W. R. Wooles, J. Reticuloendothel. Soc., 7, 375 (1970).
56. L. G. Baird and A. M. Kaplan, Cell Immunol. 20, 167 (1975).
57. P. S. Morahan, W. Regelson, and A. E. Munson, Antimicrob. Agents Chemother., 16 (1972).
58. a. P. S. Morahan, A. M. Kaplan, Int. J. Cancer, 82 (1976).
b. S. J. Morh, M. A. Chirigos, F. S. Fuhrman, and J. W. Pryor, Cancer Res., 3750 (1975).
59. P. S. Morahan and A. M. Kaplan, Int. J. Cancer, 17 (1976).
60. A. Billeau, J. Desmyter, and P. DeSomer, J. Virol., 5, 321 (1970).
61. P. DeSomer, E. Declercq, A. Billeau, E. Schonne, and M. Claes, J. Virol., 2, 878 (1968).
62. E. M. Hodnett, J. Amirmoazzami, and J. Tien Hai Tai, J. Med. Chem., 21, 652 (1978).
63. T. C. Merigan and M. S. Finkelstein, Virology, 35, 363 (1968).
64. A. Billeau, J. Muyembe, and P. DeSomer, Nature, 232, 183 (1971).
65. K. F. Muck, H. Rolly, and K. Burg, Makromol. Chem., 178, 2773 (1977).
66. D. A. Tirrell, private communication
67. R. M. Ottenbrite, K. Kuus, and A. Kaplan, Polym. Prepr. Am. Chem. Soc. Div. Polym. Chem., 24, 25 (1983).
68. D. A. Tirrell, IUPAC 28th Macromol. Symp. Proc. (1982), p. 372.
69. R. M. Ottenbrite, E. Goodell, and A. E. Munson, Polymer 18, 461 (1977).
70. A. E. Munson, J. M. Veager, S. E. Loveless, and R. M. Ottenbrite, J. Reticuloendothel. Soc., 18, 406 (1975).
71. B. M. Culbertson and B. C. Trivedi, Eds., Maleic Anhydride, Plenum Publishing Corporation, New York (1982).
72. S. LeBras, Rubber Chem. Technol., 19, 313 (1946).
73. C. S. Marvel and T. H. Shepherd, J. Org. Chem., 24, 599 (1959).
74. A. A. El'Saied, Dokl. Akad. Nauk. SSSR, 177, 380 (1967).
75. A. A. El'Saied, Polym. Sci. USSR, 11, 314 (1969).

76. R. D. Kimbrough, J. Polym. Sci., B2, 85 (1964).
77. G. B. Butler and K. C. Joyce, J. Polym. Sci. Polym. Chem. Ed., 22, 45 (1968).
78. G. B. Butler in Anionic Polymer Drugs (L. G. Donaruma, R. M. Ottenbrite, and O. Vogl, Eds.), John Wiley & Sons, Inc., New York (1980), p. 49.
79. G. B. Butler, J. T. Badgett, and M. Sharabash, J. Macromol. Sci. Chem., A4, 51 (1970).
80. J. M. Burton, G. B. Butler, and E. C. Chapman, J. Polym. Sci., A3, 501 (1965).
81. G. B. Butler, in Polyelectrolytes and Their Applications (A. Rembaum and E. Seligny, Eds.), D. Reidel Publishing Co., Dordrecht, The Netherlands (1975), p. 97.
82. R. M. Ottenbrite and D. Shillady, in Polymeric Amines and Ammonium Salts (E. Geothals, Ed.), Pergamon Press, New York (1980), p. 143.
83. D. S. Breslow, Nature, 246, 160 (1973).
84. R. J. Samuels, Polymer, 18, 452 (1976).
85. Y. C. Chu and G. B. Butler, J. Polym. Sci. Polym. Chem. Ed., 17, 859 (1979).
86. T. Kunitake and S. Tsukino, J. Polym. Sci. Polym. Chem. Ed., 17, 877 (1979).
87. W. J. Freeman and D. S. Breslow, in Biological Activities of Polymers (C. E. Carraher and C. G. Gebelein, Eds.), ACS Symp. Ser., 186, 243 (1982).
88. R. M. Ottenbrite, K. Kuus, and A. Kaplan, Fed. Proc., 41, 3364 (1982).
89. R. M. Ottenbrite, IUPAC Macromol. Symp. Proc. (1982), p. 366.
90. R. M. Ottenbrite, and G. B. Butler, in Antitumor and Interferon Inducing Agents: Synthesis and Properties (R. M. Ottenbrite and G. B. Butler, Eds.), Marcel Dekker, Inc., New York (1984), p. 247.
91. R. M. Ottenbrite, in Anionic Polymer Drugs, Vol. I, (G. Donaruma, R. M. Ottenbrite, and O. Vogl, Eds.), John Wiley & Sons, Inc., New York (1980), p. 21.
92. R. M. Ottenbrite, K. Kuus, and A. M. Kaplan, Polym. Sci. Technol., 23, 1 (1983).
93. K. Kuus, R. M. Ottenbrite, and A. M. Kaplan, J. Biol. Response Modif., 4, 46 (1985).
94. R. M. Ottenbrite, K. R. Kuus, and A. M. Kaplan, in Polymers as Biomaterials (S. W. Shalaby, A. S. Hoftman, B. D. Ratner, and T. A. Horbett, Eds.), Plenum Publishing Corporation, New York, (1984), p. 317.
95. R. M. Ottenbrite and A. M. Kaplan, Ann. N.Y. Acad. Sci., 446, 160 (1985).

96. A. M. Kaplan, K. Kuus, and R. M. Ottenbrite, Ann. N.Y. Acad. Sci., A.E., 446, 169 (1985).
97. B. M. Culbertson and A. E. Aulabaugh, in Cyclopolymerization and Polymers with Chain-Ring Structures (G. Butler and J. Kresta, Eds.), ACS Publication 195, American Chemical Society, Washington, D.C. (1982).
98. R. M. Ottenbrite, in Bioactive Polymers (C. Carraher and C. Gebelien, Eds.), Plenum Publishing Corporation, New York, in press.
99. R. Takatsuka and S. Konda, Proc. 1984 Int. Chem. Cong. Pacific Basin Soc., Honolulu, (1984).
100. R. G. Fordyce, and G. E. Ham, J. Am. Chem. Soc., 69, 695 (1947).

9

The Immobilization of Enzymes

Yoshiaki Inaki

Osaka University, Osaka, Japan

9.1 INTRODUCTION

Immobilized enzymes can serve as industrial catalysts and as highly specific reagents for special-purpose analytical procedures. Industrially, continuous large-scale processes can be carried out in immobilized-enzyme reactors. Immobilized enzymes in conjunction with a detector have led to the development of highly specific electrode systems and similar analytical monitoring devices. Immobilized enzymes are also being explored for clinical application in the form of extracorporeal shunts or microcapsules.

The methods most commonly used for the fixation of enzymes onto solid supports can be grouped in four main classes (Figure 1):

1. Adsorption on inert supports or ion-exchange resins
2. Entrappment by occlusion within cross-linked gels or by encapsulation within microcapsules such as hollow fibers, liposomes, and fibers

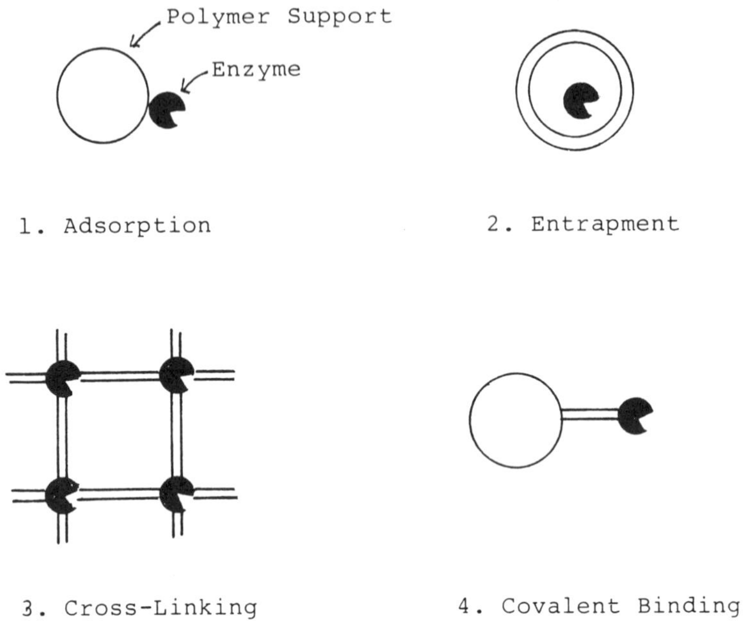

Figure 1. Immobilization of enzymes.

3. Cross-linking by bi- or multifunctional reagents, often following adsorption or entrappment within a structure of defined geometry
4. Covalent binding to polymeric supports, via functional groups nonessential for the biological activity of the protein

The covalent-binding approach has been by far the most widely investigated. A large number of techniques for covalent coupling of enzymes to water-insoluble as well as to water-soluble polymeric carriers of a variety of chemical and physical characteristics have been described in recent literature [1-31].

This chapter is rather strongly oriented toward methods based on the covalent fixation of enzymes. Other approaches are referred to in several comprehensive reviews.

9.2 IMMOBILIZATION OF ENZYMES BY COVALENT BONDING

The type of functional groups on the protein, through which the covalent bond with the polymer support is to be formed, should be nonessential for the catalytic activity of the enzyme. Moreover, binding reactions that can be carried out under relatively mild conditions and in essentially aqueous media are preferred. In selecting the appropriate coupling reaction and functionalized polymer to be employed for the immobilization of a given enzyme, all the available information on the amino acid composition, amino acids involved in the active site, effects of specific chemical modifications on activity, protection of the active-site region by specific chemical agents or inhibitors, and three-dimensional structure of the enzyme should be considered.

The protein functional groups that can be utilized in principle for the covalent binding of enzymes to polymeric supports include:

1. Amino groups: the ε-amino groups of lysine and the α-NH_2 groups of the N termini of the polypeptide chains
2. Carboxyl groups: the β- and γ-carboxyl groups of aspartic and glutamic acid, respectively, and terminal α-carboxyls
3. The phenol rings on tyrosine
4. The mercapto groups on cysteine
5. The hydroxyl groups on serine, threonine, and tyrosine
6. The imidazole groups on histidine
7. The indole groups on tryptophane

Numerous natural and synthetic polymers have been explored as potential supports for enzyme immobilization. This has led to the development of a wide variety of techniques for (a) the grafting of specific functional groups onto preformed polymers, (b) the synthesis of tailor-made polymers and copolymers devised to fulfill particular needs, and (c) the use of "parent carrier polymers," which by consecutive chemical manipulations can be transformed into the chemical species best suited for a given task. The functional groups of the polymer support that can be utilized for the covalent binding of enzymes include:

1. Hydroxyl groups of polysaccharides, poly(vinyl alcohol), poly(hydroxyethyl methacrylate), and inorganic glass
2. Carboxyl and related groups of poly(acrylic acid), poly(glutamic acid), maleic anhydride copolymers, poly(acryl amide), and carboxymethylcellulose
3. Amino groups of aminoethylated polysaccharides and silica gel, arylaminated polysaccharides silicagel and synthetic polymers, and poly(*p*-aminostyrene)
4. Aldehyde and acetal groups of the polymers reacted with glutaraldehyde, and polysaccharides reacted with periodate
5. Mercapto groups on modified polymers
6. Halogen groups of the polymers treated with cyanuric chloride, bromoacetyl bromide, and iodo alkyl alcohol, and polymers containing 3-fluoro-4,6-dinitrophenyl groups
7. Amide groups of polypeptides and polyamides such as nylon

Various types of coupling reaction have been reported:

1. Cyanogen bromide method
2. Cyclic *trans*-2,3-carbonate method
3. Curtius azide method
4. Acid anhydride method
5. Coupling reagent method
6. Four-component condensation reaction
7. Carbamylation, and thiocarbamylation reactions
8. Diazotization reaction
9. Schiff base formation reaction
10. Glutaraldehyde method
11. Thiol-disulfide interchange reaction
12. 3-Fluoro-4,6-dinitrophenyl method
13. *s*-Triazinyl method
14. Imidoester reaction

With the use of these coupling reactions, covalent bonds are formed between functional groups of enzymes and polymer supports to give the covalent-immobilized enzymes. This section deals with the same methods that have been classified according to polymer support and functional group according to the majority of the published covalent-immobilization procedures.

9.3 POLYMERS CONTAINING HYDROXYL GROUPS

Polymers containing hydroxyl groups have been widely used as support polymers for enzyme immobilization. These polymers include poly-

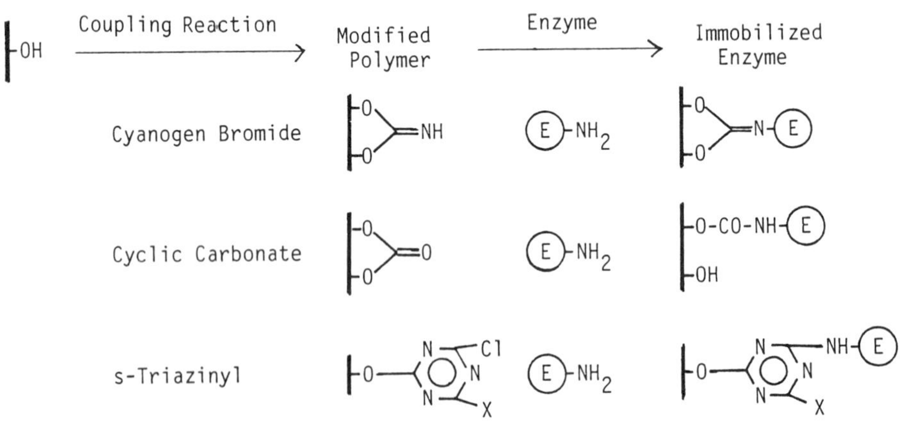

Figure 2

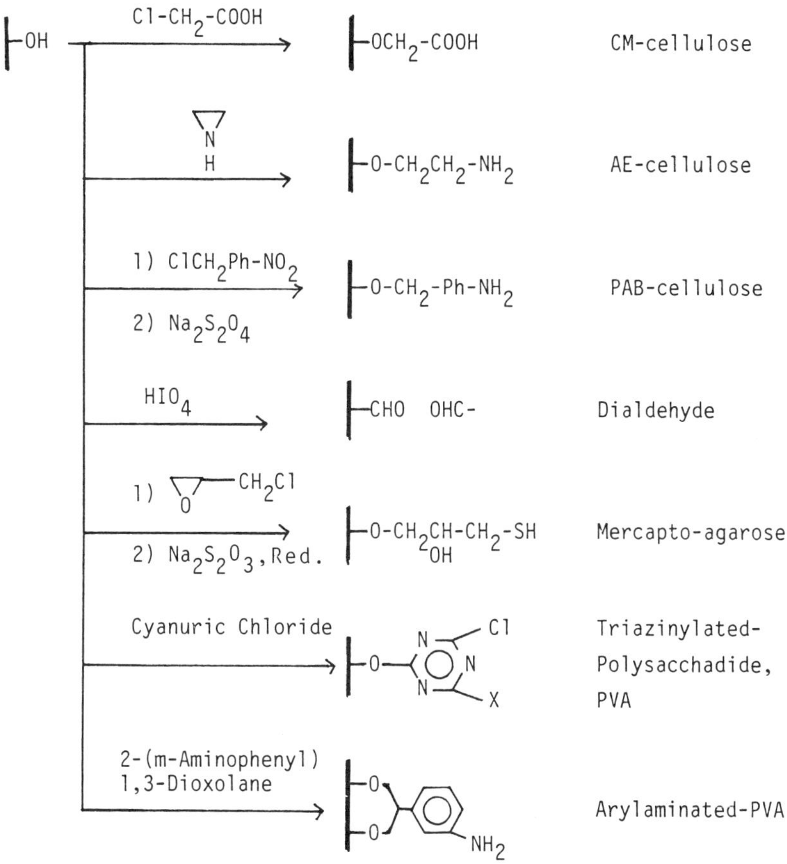

Figure 3

saccharides, poly(vinyl alcohol), poly(hydroxyethyl methacrylate), silicas, and porous glass. These polymers are used as supports not only for direct coupling but also by modification to form other coupling functional groups (Figures 2 and 3).

Naturally occurring polysaccharides have been widely used as support materials. The most important polysaccharide supports are (a) cellulose, (b) dextran, (c) starch, and (d) agar. Cellulose consists of linear chains of 1,4-linked β-D-glucose organized into fibers with high crystallinity. Dextran, a linear water-soluble polysaccharide composed of 1,6-linked α-D-glucose is produced by microorganisms of the genus *Leuconostoc*. The commercially available dextran gels (Sephadex) are prepared by cross-linking the linear dextran with epichlorohydrin. Agarose is one of the components of agar, a complex mixture of polysaccharides, extracted from several species of the Rhodophyceae family of red seawater algae. Agarose is composed of alternating 1,3-linked β-D-galactose and 1,4-linked 3,6-anhydro-α-L-galactose.

Polysaccharides are activated by various reagents and the derivatives obtained are used as support for enzyme immobilization (Figure 3). The carboxymethyl ether of cellulose (CM-cellulose) is prepared by the reaction of cellulose with chloroacetic acid. The hydrazide derivative is prepared from the CM-cellulose by treatment of the methyl ester of CM-cellulose with hydrazide. The immobilization of enzymes onto the carboxyl groups is described in Section 9.4.

Various derivatives of polysaccharides containing alkylamino and arylamino groups have been prepared; aminoethyl cellulose (AE-cellulose), diethylaminoethyl cellulose (DEAE-cellulose), and *p*-aminobenzyl cellulose (PAB-cellulose). These polymers are used for enzyme coupling after diazotization or isothiocyanation (described in Section 9.5).

Several derivatives of polysaccharides, containing alkylating or arylating functional groups, have been reported. For example, bromoacetyl cellulose is prepared by treating cellulose with bromoacetic acid and bromoacetyl chloride in dioxane. Polymer supports containing monochloro- or dichloro-*s*-triazinyl groups have been prepared by treating various polysaccharides with 2,4,6-trichloro-*s*-triazine (cyanuric chloride) or dichloro-*s*-triazine derivatives. These coupling reactions are described in Section 9.8.

Dialdehyde derivatives of polysaccharides, prepared by controlled periodate or dimethyl sulfoxide oxidation of vicinal hydroxyl groups, have been used as supports. Partially oxidized cellulose is used to bind enzymes, presumably through the formation of Schiff's base linkages with the protein amino groups, described in Section 9.6.

One of the most extensively used procedures for immobilization involving polysaccharide supports has been the cyanogen bromide activation method. This method is carried out at low temperatures to yield conjugates containing relatively high amounts of bound protein. This method has been particularly important for the chemical modification of agarose gels. The procedure of cyanogen bromide activation of polysaccharides is described in detail in this section. Another activation method of polysaccharides involves cyclic *trans*-2,3-carbonate derivatives. The *trans*-2,3-carbonate group is analogous to the cyclic *trans*-2,3-imidocarbonate structure, assumed to be the reactive intermediate in cyanogen bromide activation.

Hydroxy-containing vinyl polymers such as poly(vinyl alcohol) and poly(hydroxyethyl methacrylate) are useful as support polymers. Poly(vinyl alcohol) cross-linked with terephthalaldehyde gives a mechanically stable xerogel that is being explored as starting material for the preparation of hydrophilic, nonbiodegradable supports, devoid of charged groups. Cross-linked poly(vinyl alcohol) gels can be activated directly with 2,4,6-trichloro-*s*-triazine or converted to a diazotizable derivative, *p*-aminobenzyl ether, by treatment with 4-nitrobenzylchloride followed by reduction. By transacetalation, a diazotizable polymer is prepared from poly(vinyl alcohol) with 2-(*m*-aminophenyl)-1,3-dioxolane.

Macroporous copolymers containing neutral hydroxyl groups and prepared by suspension copolymerization of hydroxyalkyl acrylate and/or hydroxyalkyl methacrylate monomers are useful polymer support of enzymes. They are prepared in the form of spherical particles, which in the dry state exhibit measurable porosity and rather high specific surface values.

Hydroxyalkyl methacrylate gels are formed by radical suspension copolymerization in an aqueous dispersing medium containing a suitable suspension stabilizer. Most widely used are the copolymers of 2-hydroxyethyl methacrylate and ethylene dimethacrylate, produced under the trade name Spheron.

The hydroxyethyl groups of Spheron gel may be chemically modified by employing the known procedures for polysaccharide modification. In contrast to polysaccharides, the modification and activation reactions can be conducted under much more drastic reaction conditions. The gels can be activated with cyanogen bromide, diisocyanates, dithioisocyanates, anhydrides and dicarboxylic acid chlorides, phosgen, bromoacetyl bromide, and by a number of other procedures.

Inorganic supports offer several advantages, such as high mechanical strength, resistance to solvents and microbial attack, and regenerability; moreover, inorganic supports do not change in structure over wide ranges of pH, pressure, and temperature. Inorganic materials that have been used as supports include alumina, bentonite, glass, nickel oxide, silica, titanium, zirconium, and magnetic iron oxide powders.

The immobilization of enzymes onto controlled-pore glass or ceramics and onto magnetic iron oxide has been investigated extensively. The methods most commonly employed for the introduction of chemically reactive functional groups onto a glass or ceramics surface are shown in Figure 4. The inorganic support material is first treated with a trialkoxysilane derivative containing an organic functional group; γ-aminopropyltriethoxysilane (2). Coupling to the carrier takes place presumably with the surface silanol or oxide groups, polymerization most likely occurs between adjacent silanes. The product of the reaction (3), alkylamine glass can be converted to the arylamino derivative (4) by treatment with *p*-nitrobenzoyl chloride followed by reduction. Alkylamine glass can also be converted to the isothiocyanate

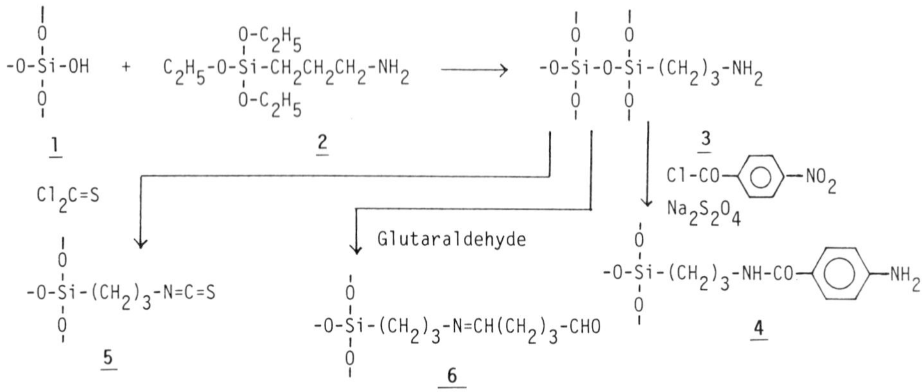

Figure 4

(5) derivative, or activated directly with glutaraldehyde (6). The covalent attachment of enzymes to porous glass directly activated with cyanogen bromide is also possible.

9.3.1 The Cyanogen Bromide Method

The coupling method most extensively used is based on the reaction between cyanogen halides with the hydroxyl-containing polymers [32-52]. The hydroxyl-containing polymer is activated with cyanogen bromide at pH 10-11.5. Compounds containing free amino groups can be attached covalently to the activated polymer at mildly alkaline pH values. The cyanogen bromide activation reaction is considered to proceed via the 2,3-*trans*-imidocarbonate intermediate (7), leading to the formation of three possible types of bonds with protein amino groups: N-substituted imidocarbonate (8), N-substituted carbamates (9), and N-substituted isoureas (10) (Figure 5). The isourea structure (10) is sensitive to attack by nucleophiles as well as being positively charged at physiological pH values. The cyanogen bromide methods are used for immobilization of enzymes on cellulose [36-40], agarose [41-52], dextran [47,48,53,54], and copolymers of hydroxyethyl methacrylate with acrylamide [32-35].

9.3.1.a Immobilization of chymotrypsin on hydroxyethyl methacrylate gels (Spheron) [34]

Cyanogen bromide-activated Spheron (7)

To 100 mL of swollen gel suspended in distilled water (volume ratio 1:1), 100 mL of freshly prepared 10% cyanogen bromide solution is added gradually with constant stirring. The pH of the suspension is kept at 11 by the addition of 6 N, and 4 N, and finally 2 N NaOH solution. The reaction is completed in 15 min. The activated Spheron is separated on a sintered-glass filter and washed quickly with 2 L of precooled 0.1 M $NaHCO_3$ (pH 9.0).

Immobilization of enzyme

The washed activated Spheron (100 mL) is suspended in 100 mL of 0.1 M $NaHCO_3$, pH 9.0 at 4°C and immediately 10 g of solid chymotrypsin are added. The reaction is performed with constant stirring at 4°C for 20 h. After the coupling is completed, the gel-attached chymotrypsin is transfered onto a sintered-glass filter and the unreacted chymotrypsin is filtered off. The chymotrypsin-gel is suspended in 100 mL of 1 M ethanolamine (pH 9) to block the remaining active groups. After reacting for 2 h, the excess uncoupled ethanolamine is washed away with a borate buffer solution (0.1 M, pH 8.5) and an acetate buffer solution (0.1 M, pH 4.1), each containing 1 M NaCl. The gel is then placed on a column and washed alternately with borate and acetate buffers until the effluent shows no absorbance at 280 nm.

9.3.2 Cyclic *trans*-2,3-Carbonate Method

The reaction of cellulose with ethyl chloroformate in anhydrous organic solvents gives cyclic *trans*-2,3-carbonate derivative of cellulose (11). The *trans*-2,3-carbonate group is analogous to the cyclic *trans*-2,3-imidocarbonate structure (7) which is assumed to be the reactive intermediate in cyanogen bromide activation. Nucleophilic attack on

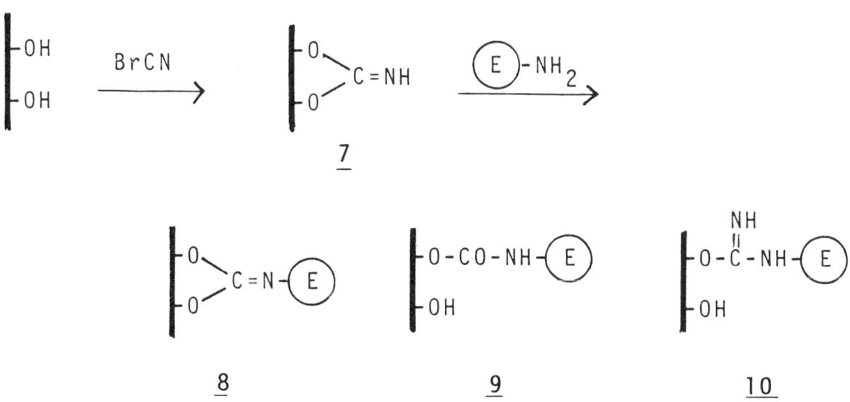

Figure 5

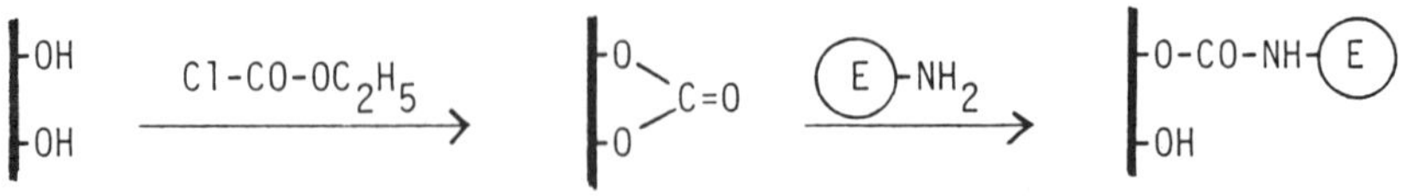

11 12

Figure 6

the cyclic carbonate by an amino group hence leads to ring opening and the formation of N-substituted carbonate (12). This method has been used for immobilization of enzyme on cellulose [55-58] (Figure 6).

9.3.2.a Immobilization of chymotrypsin on cellulose [57]

Macroporous cellulose carbonate (11)

An aqueous suspension of macroporous cellulose (1 g dry weight of cellulose), is collected on a sintered-glass filter. The solid is then stirred in a closed vessel with several changes of dry dimethyl sulfoxide (6 × 25 mL) for 18 h. The cellulose, collected on a sintered-glass filter, contains 9 to 10 mL of dimethyl sulfoxide.

Small batches of this material are suspended in dry dimethyl sulfoxide (10 mL) and water (0 to 1 mL) is added to each. Each suspension is stirred in an ice-salt bath with dry 1,4-dioxane (1.5 mL) and dry triethylamine (8 mL). Ethyl chloroformate (16 mL) is added dropwise over a 30-min period and the mixture is stirred for a further 15 min at this low temperature. Dry dioxane (20 mL) is then added, the suspension is filtered, and the solid is washed with dry dioxane (9 × 20 mL). The product is freed from triethylamine by washing with dry ethanol (10 × 20 mL), then with dry diethyl ether (3 × 20 mL), and dried. The resulting off-white granular material shows IR absorptions at 1750 cm^{-1} (COOEt), 1815 cm^{-1} and 1850 cm^{-1} (cyclic carbonate).

Coupling of chymotrypsin to macroporous cellulose carbonate

Macroporous cellulose carbonate (0.2 g) is swollen by suspension in dimethyl sulfoxide for 6 to 35 days, and is dried. This dry, solvent-free cellulose carbonate is washed with water (5 × 5 mL). The sample is suspended in 0.1 M citric acid-0.2 M sodium phosphate buffer (pH 5.4, 4.5 mL) and stirred with an aqueous solution of chymotrypsin (20 mg/mL in 0.5 mL) for 18 h at 4°C.

Immediately after coupling, the insoluble enzyme sample is washed first with 0.005 M sodium acetate buffer (pH 5.0, 5 × 10 mL) and then stirred in 1 M sodium chloride-sucrose solution (5 × 7.5 mL) for 2 to 3 h.

The sample is finally washed with acetate buffer (2 × 10 mL) and is isolated on a sintered-glass filter.

9.4 POLYMERS CONTAINING CARBOXYL GROUPS

Enzymes can be immobilized onto polymers containing carboxyl groups, by various coupling reactions via amino groups of the protein to form an amido bond. For the coupling reaction, the carboxyl groups are activated as (a) acyl azide, (b) acid anhydride, (c) active ester by condensation reagents, and (d) four-centered condensation reaction (4CC) (Figure 7).

Poly(acrylic acid) and poly(methacrylic acid) are typical polymers with carboxyl groups. These polymers are used as homopolymer or copolymers with acrylamide. Polymers containing acid anhydride groups, such as homopolymer and copolymer of methacrylic anhydride or copolymers of maleic anhydride, are also useful polymer supports.

Various carboxyl containing polymers can also be prepared from hydroxyl-containing polymers, such as polysaccharides, poly(hydroxyethyl methacrylate), and porous glass (see Section 9.3).

9.4.1 The Curtius Azide Method

Immobilization of enzyme onto the carboxyl-containing polymers by the Curtius azide method is shown in Figure 8. The carboxyl group is modified to the methyl ester, which is reacted with hydrazine hydrate to form hydrazide (13). The hydrazide derivative is then reacted with aqueous sodium nitrite to give the azide derivative (14). The azide

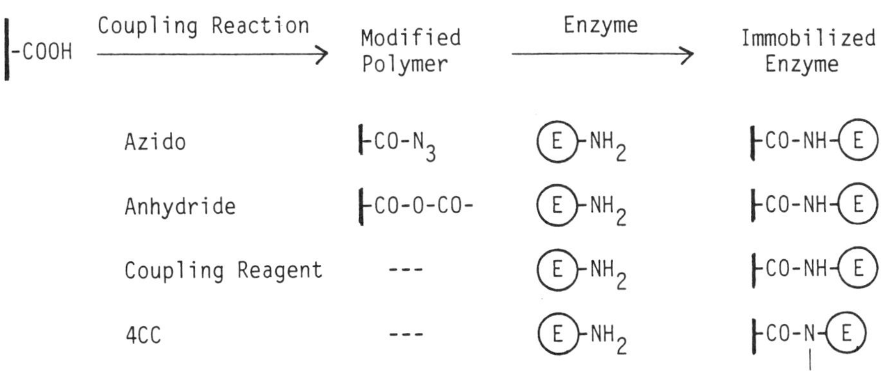

Figure 7

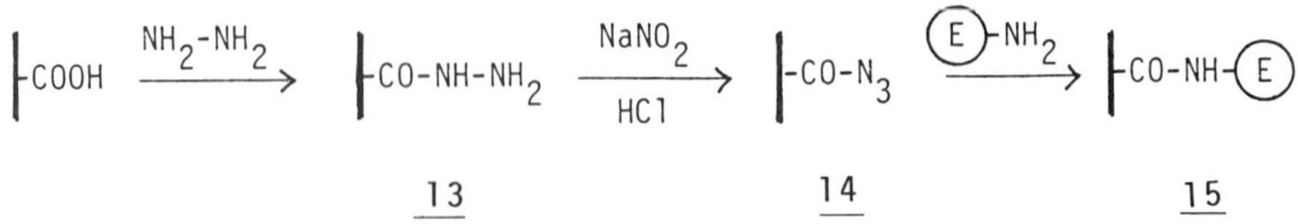

Figure 8

derivative can be reacted with an enzyme amino group under mild alkali conditions to give 15 [59-65].

The copolymer of acrylamide containing acyl hydrazide units (18) is available commercially as Enzacryl AH (Koch-Light Ltd.) [66,67]. Enzacryl AH (18) is prepared according to Figure 9.

9.4.1.a Preparation of CM-cellulose bromelain (CMCB) [63]

CM-cellulose hydrazide (13)

Carboxymethylcellulose, CM-11 (Whatman, exchange capacity 0.6 meq/g) (100 g), is suspended in dry methanol (1.5 L) by stirring the mixture vigorously. Concentrated HCl (10 mL) is added and the mixture is heated carefully until the methanol is just refluxing. The mixture is stirred and heated under reflux for 4 h. The CM-cellulose methyl ester is then collected and washed thoroughly on a Büchner funnel with dry methanol (3 × 500 mL).

The conversion of the CM-cellulose methyl ester to the hydrazide is effected by stirring a mixture of methyl ester (100 g), methanol (1 L), and hydrazine hydrate (100 mL) in a water bath at 40°C for 72 h. The CM-cellulose hydrazide is collected by vacuum filtration and washed successively with 2 × 250 mL of methanol, methanol

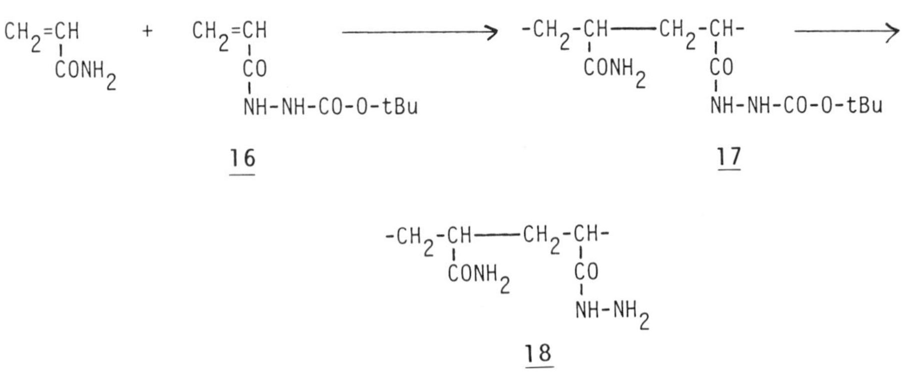

Figure 9

saturated with CO_2, 90% aqueous methanol, and dry methanol. This procedure removes unreacted hydrazine as hydrazine carbonate.

CM-cellulose azide (14)

CM-cellulose hydrazide (5 g) is suspended in 0.6 N HCl (200 mL) cooled to 0°C in an ice bath. Aqueous 5% sodium nitrite (25 mL) is added slowly with vigorous stirring over 20 min. The suspension is stirred at 0°C for a further 20 min and then CM-cellulose azide is collected by vacuum filtration. The azide is washed with ice-cold water (2 × 200 mL), ice-cold 1 M NaCl (200 mL), and again with ice-cold water (2 × 200 mL). The product is collected on a Büchner funnel and air dried.

CM-cellulose bromelain (15)

To a solution of purified bromelain (1 g) in water (100 mL) immersed in an ice bath is added freshly prepared CM-cellulose azide (5 g) all at once with stirring. The pH is rapidly adjusted to 8.7 with a saturated sodium tetraborate solution. After the suspension is stirred with cooling by an ice bath for 20 min, the pH is readjusted to 8.7. The suspension is then stirred for a further 40 min. The solid is collected by vacuum filtration and washed on the filter as follows: H_2O (1 × 200 mL), 1 M NaCl (2 × 200 mL), 0.05 M acetic acid (2 × 200 mL), 1 M NaCl (2 × 200 mL), 0.5 M $NaHCO_3$ (2 × 200 mL), 1 M NaCl (2 × 200 mL), and H_2O (4 × 200 mL). The solid CMCB is then suspended in a known volume of water to give a standard suspension which is stored at 4°C until required. For long periods of storage 1 drop of purified chloroform or sodium azide (0.1% w/v) is added to the suspension to prevent bacterial contamination.

9.4.1.b Immobilization of thermolysin on enzacryl AH [66]

Enzacryl AH (18)

A mixture of acrylamide (1.05 g, 15 mmol), *N*,*N*'-methylenediacrylamide (0.28 g, 1.8 mmol), and *N*-acryloyl-*N*'-*t*-butoxycarbonyl hydrazide (0.34 g, 2.7 mmol) is dissolved in ethylene glycol (15 mL) and irradiated from above with an 80-W ultraviolet lamp at a distance of 8 cm over 15 h. The resulting gel is broken up, washed exhaustively with water, equilibrated with ethanol, collected by filtration, and dried under reduced pressure. Selective hydrolysis of the *t*-butoxycarbonyl protecting groups is effected by stirring with an excess of 2 M HCl at ambient temperature for 2 days.

Thermolysin on Enzacryl AH

A sample of acyl hydrazide derivatized copolyacrylamide (Enzacryl AH) (100 mg) is suspended in 2 M HCl (5 mL) at 0°C, and ice-cold aqueous sodium nitrite solution (2% w/v, 5 mL) is added. Activation is allowed to proceed for 15 min with stirring; the suspension is then centrifuged, and the supernatant is discarded.

The diazotized sample is equilibrated with 25 mM phosphate buffer (pH 8.0) at 0°C, and a 0.25-mL aliquot of a solution of thermolysin (20 mg/mL) in the same buffer is added. The reaction is allowed to proceed at 0°C for 4 h with gentle stirring. The Enzacryl AH-thermolysin is then washed alternately with 25 mM phosphate buffer (pH 8.0, 15 mL) and 1% casein solution (15 mL) in the same buffer for 10 cycles. Each washing is for a 20-min duration at 0°C with vigorous stirring. After washing, the sample is stored as a suspension in 50 mM tris-HCl buffer (pH 7.25, 10 mL) containing 10 mM $CaCl_2$.

9.4.2 Acid Anhydride Method

Polymers containing acid anhydride functional groups have been used to prepare a variety of polyelectrolyte derivatives of enzymes. The most commonly used polymer supports of this type are the 1:1 copolymers of maleic anhydride and ethylene (19). Copolymers with other vinyl comonomers can also be used. As shown in Figure 10, opening of the anhydride ring by $-NH_2$ groups of protein or by spontaneous hydrolysis in the aqueous medium generates free carboxyl groups. The immobilized enzyme is therefore located in an environment of excess negative charge; this fact is reflected in the kinetic behavior and stability properties of the bound enzyme.

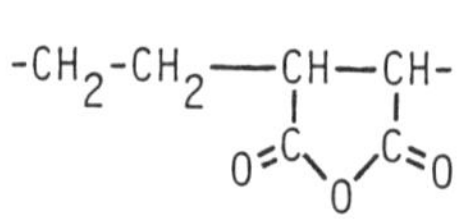

19

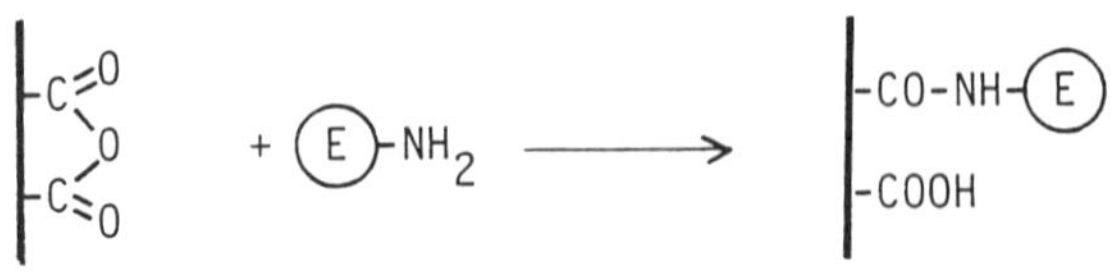

Figure 10

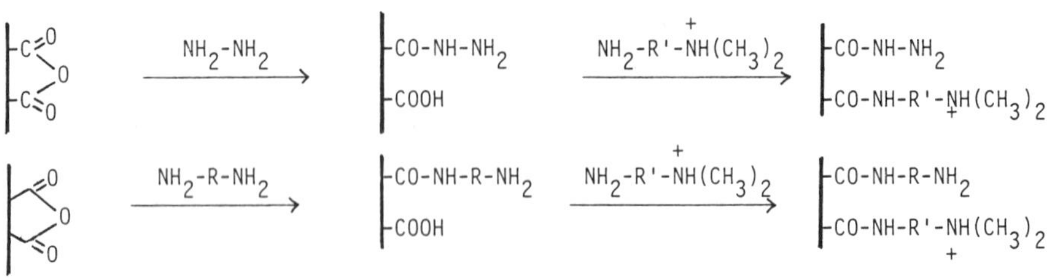

Figure 11

Maleic anhydride copolymers could serve as starting materials for further chemical modifications, such as cross-linking and the preparation of cross-linked polyanionic and polycationic polymers containing arylamino [79], acid hydrazide [79] or alkylamino [80] functional groups (Figure 11). As comonomers, other than ethylene [68-77], butandiol divinylether [65], methyl vinyl ether [76], and styrene [73,78], are used.

9.4.2.a Immobilization of trypsin on ethylene-maleic anhydride (1:1) copolymer (EMA-trypsin) [68,69]

EMA-trypsin

Ethylene-maleic anhydride (1:1) copolymer (EMA) (100 mg) is suspended in ice-cooled, 0.2 M potassium phosphate buffer (pH 7.5, 5 to 10 mL). The suspension is homogenized and the cross-linking agent, hexamethylenediamine (1% aqueous solution, 1 mL), is added with stirring. The appropriate amount of trypsin (10 to 400 mg) dissolved in the same buffer (5 to 30 mL) is added and the reaction mixture left overnight at 4°C with stirring. The precipitate is separated by centrifugation and washed by suspending in 1 M KCl (100 mL) and recentrifuging at 4°C until no residual enzymatic activity can be detected in the supernatant after filtration through a Millipore filter.

9.4.3 Coupling Reagents Method

The methods for the activation of carboxyl groups involve water-soluble carbodiimides and similar coupling reagents that are used for peptides or polynucleotide synthesis (Figures 12 to 14).

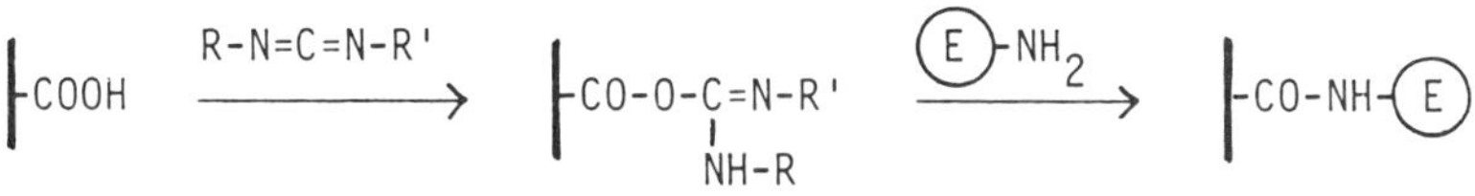

20 21

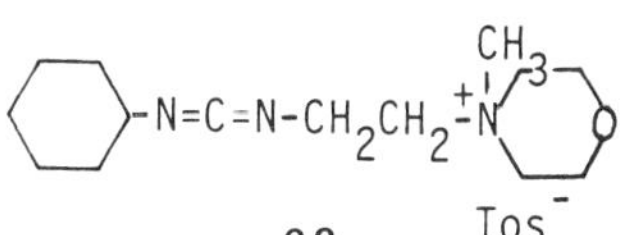

22

$CH_3CH_2-N=C=N-CH_2CH_2CH_2-N(CH_3)_2$

23

Figure 12

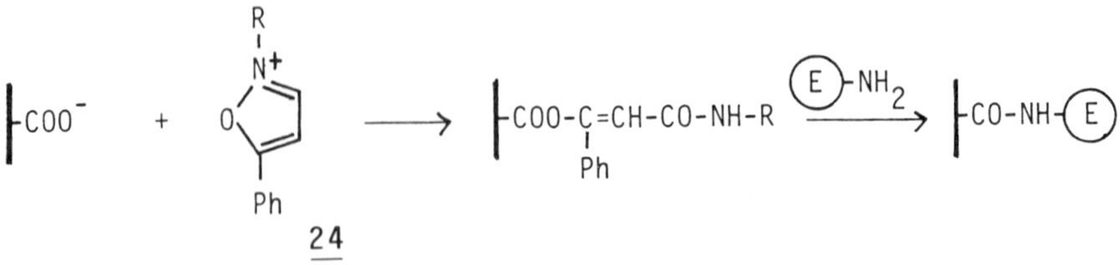

Figure 13

Carbodiimides react with carboxyl groups at slightly acidic pH values (pH 4.75 to 5) to give *O*-acyl isourea derivatives (20). These highly reactive intermediates either condense with amines to yield the corresponding amides or rearrange to an acyl urea. Several water-soluble carbodiimides have been used for the activation of carboxyl groups [33,84-87]. The most widely used water-soluble carbodiimides are 1-cyclohexyl-3,2-(4-*N*-methylmorpholinium)ethyl carbodiimide tosylate (22) and 1-ethyl-3-(3-dimethylaminopropyl) carbodiimide (23), which are commercially available.

Other coupling reagents, *N*-alkyl-5-phenylisoxazolium salts (Woodward K reagent) and *N*-ethoxycarbonyl-2-ethoxy-1,2-dihydroquinoline [85,88], can be used in a similar manner (24 and 25). These reagents react rapidly with carboxyl groups only at pH values below 4.75. The active intermediates are enol esters in the case of Woodward K reagent, and mixed anhydrides in the case of the EEDQ (Figures 13 and 14).

9.4.3.a Immobilization of trypsin onto acrylamide-acrylic acid copolymer using carbodiimide [86]

Acrylamide-acrylic acid-trypsin (21)

Swollen beads of acrylamide-acrylic acid copolymer (50/50, w/w, 100 mg dry weight) are washed in 0.2 M $NaHCO_3$ after which 2 mL of a solution of 1-ethyl-3(3-dimethylaminopropyl) carbodiimide (EDC) · HCl

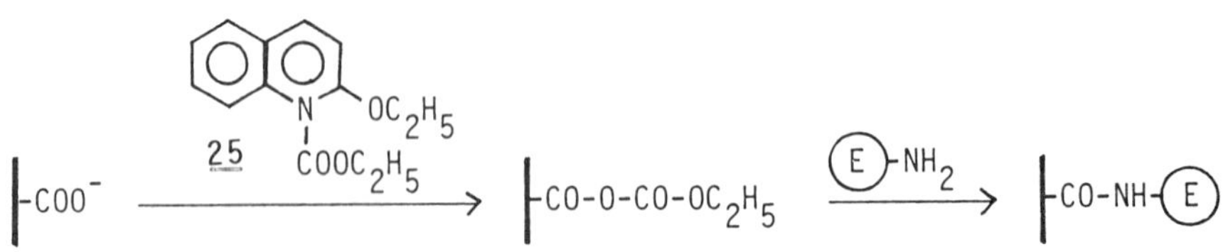

Figure 14

(2 mg of EDC per milliliter of distilled water) and 2 mL of enzyme solution (2 mg of trypsin per milliliter of 0.2 M $NaHCO_3$) are added. (Alternatively, the coupling can be performed at pH 3.5 using diluted HCl.) The coupling proceeds for 2 to 4 h at 4°C. The beads are then washed in 0.1 M $NaHCO_3$, 0.5 M NaCl, distilled water, 1 mM HCl, and buffer. When coupling glycine and using equimolar amounts of EDC, the coupling yield is 86%, resulting in 120 μmol of bound glysine per gram of dry polymer.

9.5 POLYMERS CONTAINING AMINO GROUPS

Functional amino groups containing polymers are used as polymer supports using various coupling method (Figure 15). The amine groups can be also activated by phosgen or thiophosgen to isocyanate or isothiocyanate functional groups, by cyanuric chloride, and by glutaraldehyde. These activated functional groups can react with amine groups of an enzyme. As described in this section, isocyanates and isothiocyanates react with most protein nucleophiles. Only the amino groups result in the formation of stable substituted urea or thiourea products. The activation by cyanuric chloride or glutaraldehyde is described in another section.

Four component condensation (4CC) reactions involving carboxylate, amine, aldehyde, and isocyanide lead to the formation of an N-

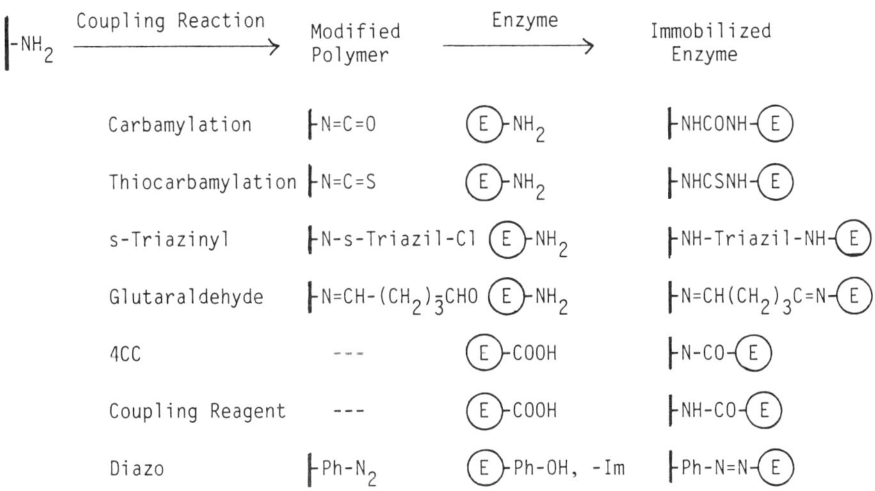

Figure 15

substituted amide (Ugi reaction). In this reaction the carboxyl groups on the enzyme and the amine groups on the polymers combine to form an amide bond, the aldehyde and isocyanide components appearing as the side chain on the amide nitrogen. This experimental procedure is described in this section.

The carboxyl groups of enzymes are coupled with the amino groups of the polymers by direct coupling using carbodiimides [89-91] or similar coupling reagents [92] to form amide bonds. This procedure is the same as the coupling of carboxyl-containing polymer with the amino groups of enzyme, described above.

Various arylamino derivatives can be used, after diazotization, for the coupling of proteins mainly through tyrosyl or histidyl residues. Azo coupling of proteins can be effected by polymers containing aryldiazonium functional groups. Polydiazonium reagents are prepared by treating polymers with pendant arylamino side chains with nitrous acid. The electrophilic aryldiazonium ion attacks mainly activated aromatic rings, such as phenol (tyrosine) or imidazole (histidine), to form the corresponding azo derivatives.

Polymers containing alkylamino groups have been prepared with polysaccharides, porous glass, polyacrylamide, and maleic anhydride copolymers. Chemical modification of polysaccharides by ethyleneimine or *N,N*-diethyl-2-chloroethylamine gave aminoethyl (AE) or diethylaminoethyl (DEAE) derivatives of polysaccharides; AE-cellulose; and DEAE-cellulose, respectively. These polysaccharides, with attached alkylamino groups, are commercially available.

Porous glass, inorganic support material, is treated with trialkoxysilane derivatives containing an organic functional group, γ-aminopropyltriethoxysilane. Coupling to the carrier takes place presumably through the surface silanol or oxide groups; polymerization most likely occurs between adjacent silanes to give alkylamine glass.

The chemical modifications of commercially available cross-linked polyacrylamide beads such as Bio-Gel, or ethylene-maleic anhydride copolymer (EMA) by ethylenediamine, give corresponding aminoethyl derivatives.

Polymers containing arylamino groups are poly(*p*-aminostyrene), poly(acrylamido-co-*p*-aminoacrylanilide) (Enzacryl AA) (26), and poly(leucine-co-*p*-aminophenylalanine) (27). The arylamine-containing polymers are also obtained by chemical modification of polysaccharides, alkylamino glass, and vinyl polymers, such as polyacrylamide, maleic anhydride copolymers, poly(vinyl alcohol), and poly(hydroxyethyl methacrylate).

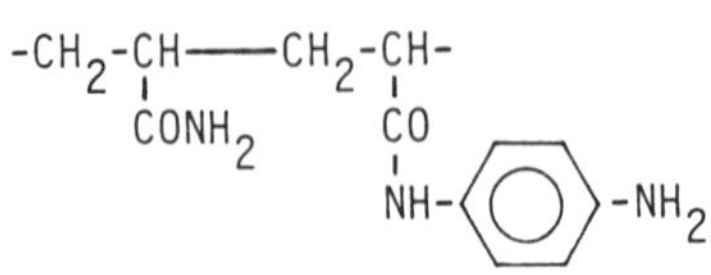

26

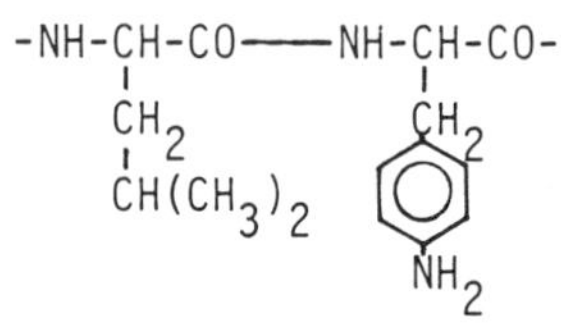

27

The *p*-aminobenzyl ether of cellulose (PAB-cellulose) is commercially available and is prepared by the reaction of cellulose with *p*-nitrobenzyl chloride followed by reduction. Alkylamine glass can be converted to the arylamino derivative by treatment with *p*-nitrobenzoyl chloride followed by reduction. Maleic anhydride copolymers react with 4,4'-methylenedianiline to give the arylamino derivative. Cross-linked poly(vinyl alcohol) gels can be converted to the *p*-aminobenzyl ether by treatment with 4-nitrobenzyl chloride followed by reduction. By transacetalation of poly(vinyl alcohol) with 2-(*m*-aminophenyl)-1,3-dioxolane, *m*-aminobenzacetal derivative can be obtained.

9.5.1 Four-Component Condensation Reaction

Four-component condensation (4CC) reactions combine the carboxyl and amine groups to form an amide bond (Figure 16). Therefore, this reaction can be used both for carboxyl-containing polymers and amine-containing polymers. Four-component condensation reactions can be carried out in an aqueous medium at neutral pH and allow for considerable versatility and high selectivity.

9.5.1.a Immobilization of pepsin on agarose by 4CC reaction [93]

p-Phenylenediamine agarose

Agarose gel as spherical beads (4%, 25 g) is cross-linked with epichlorohydrin. The cross-linked gel is activated with cyanogen bromide at pH 11. The activated gel is allowed to react with *p*-phenylenediamine (0.5 g) in 0.1 M sodium bicarbonate solution (50 mL). Nitrogen is bubbled through the coupling suspension for 15 min and the reaction is allowed to continue in a closed vessel for 16 h in the dark at room temperature. The amino polymer is washed carefully with the following solutions in order: 0.1 M sodium borate buffer (1 M in NaCl, pH 8.5); 0.1 M sodium acetate (1 M in NaCl, pH 4.5); and 0.1 M sodium acetate buffer (pH 4.5). The product is stored in the last buffer with 0.02% NaN_3.

Coupling of pepsin

Phenylenediamine-agarose (10 g) is suspended in distilled water (40 mL) and acetaldehyde (250 μL) is added. The reaction suspension is stirred gently. After 15 min, pepsin (250 mg) is added together with

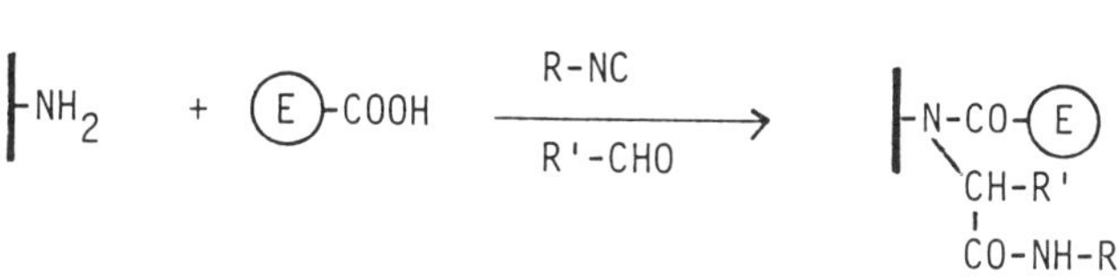

Figure 16

cyclohexyl isocyanide (250 μL). The pH is kept in the range 5.6 to 6.0 for 6 h. The product is washed with water and thereafter with the following solutions in order: 0.1 M sodium citrate buffer (1 M in NaCl, pH 4.0); 0.1 M sodium acetate buffer (1 M in NaCl, pH 5.4); and 0.01 M sodium citrate buffer (pH 4.0). The immobilized pepsin is stored in the last buffer with 0.02% NaN_3 at 4°C. The product contains 150 to 190 mg of enzyme per gram of dried conjugate. The activity of immobilized pepsin toward denatured hemoglobin at pH 4 is 15% of that for native enzyme at the same pH.

9.5.1 Carbamylation and Thiocarbamylation Reactions

Many polymers containing isocyanate or isothiocyanate groups have been used as carbamylating or thiocarbamylating reagents for enzyme immobilization [66, 67, 94-104]. Polymers containing isocyanate or isothiocyanate functional groups are prepared by the reaction of the polymers containing amino groups with phosgen or thiophosgen. Alternatively, poly(isocyanate styrene), poly(isothiocyanate styrene), or copolymers containing these styrene units can be used directly for the carbamylation or thiocarbamylation reactions (Figure 17).

9.5.2.a Immobilization of amylase on enzacryl AA [66]

Aminoaryl-derivatized copolyacrylamide (Enzacryl AA)

A mixture of acrylamide (0.639 g, 9 mmol), *N,N'*-methylenediacrylamide (0.154 g, 1 mmol), and 1-acryloylamino-2-(4-nitrobenzoylamino)ethane (0.263 g, 1 mmol) is dissolved in ethylene glycol (5 mL). Nitrogen is continuously bubbled through the solution while it is irradiated from above with an 80-W ultraviolet lamp at a distance of 8 cm over 15 h. The soft gel produced is broken up, equilibrated with ethanol, collected by filtration, and dried under reduced pressure.

Selective reduction of the aryl nitro groups in this copolymer is effected by an adaptation of the procedure of Kolthoff and Robinson [181]. Sodium citrate solution (20% w/v) is added dropwise with stirring to an acidic solution of titanous chloride (15% w/v) until the pH of the

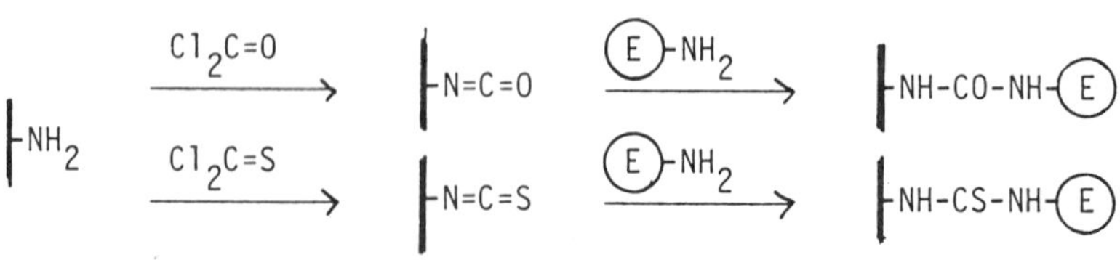

Figure 17

mixture reaches 5.0. The copolymer (2.5 g) is suspended in this solution (250 mL) with stirring and left for 3 days at ambient temperature in the dark. The resulting aminoaryl-derivatized copolyacrylamide is washed with distilled water, equilibrated with ethanol, and dried under reduced pressure.

Isothiocyanate coupling

Enzacryl AA (100 mg) is stirred vigorously with 3.5 M phosphate buffer (pH 6.9, 1 mL) at ambient temperature. An aliquot (0.2 mL) of a 10% (v/v) solution of thiophosgene in carbon tetrachloride is added. Some 20 min later, another aliquot (0.2 mL) of thiophosgene solution is added, and the stirring is continued for another 20 min. The isothiocyanate derivatives of Enzacryl AA are then washed once with acetone (15 mL) and twice with 0.5 M $NaHCO_3$ (15 mL) and equilibrated with the 50 mM borate buffer (pH 8.6) in preparation for enzyme coupling.

Prior to enzyme binding, the activated samples are centrifuged and the excess buffer is discarded. A solution of amylase (5 mg/mL) is prepared and a 0.5 mL aliquot is added to a sample of isothiocyanated Enzacryl AA. Coupling is allowed to proceed for 48 h at 0°C with gentle stirring. The isothiocyanate Enzacryl AA-amylase is washed alternately with 20 mM phosphate buffer (pH 6.9, 15 mL) and a 0.5 M NaCl solution (15 mL) in the same buffer over five cycles. Each washing is of 20 min duration with vigorous stirring. The immobilized enzyme is stored as a suspension in phosphate buffer (10 mL).

9.5.3 Diazotization Reaction [60, 70, 67, 68, 97, 98, 105-127]

Polymers containing arylamino functional groups can be used for coupling of enzymes by an diazotization reaction (Figure 18).

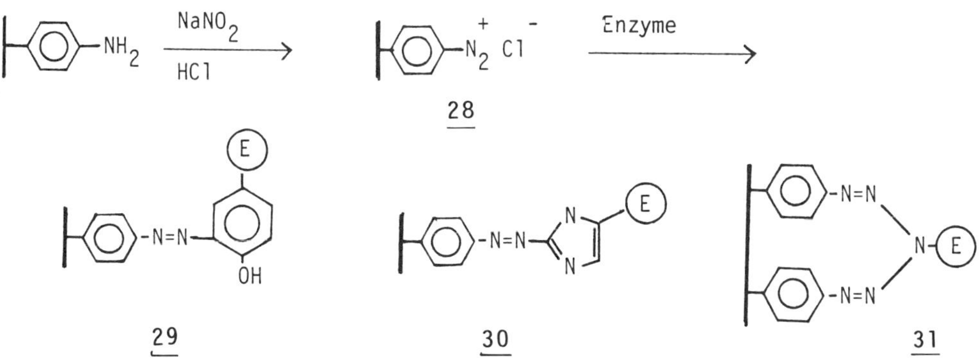

Figure 18

Arylamino-containing polymers are reacted with nitrous acid (acid + $NaNO_2$) to form the polymers containing aryldiazonium functional groups (28). The electrophilic aryldiazonium ion mainly attacks activated aromatic rings, such as phenols of tyrosine or imidazole of histidine to form the corresponding azo derivatives (29 and 30). Tyrosine and histidine react at mildly alkaline pH values (8 to 9) at comparable rates to form monoazo derivatives. Amino groups (α-amines and the ε-amino groups of lysine) react under similar conditions with 2 mol of diazonium salt to give the disubstituted bisazo derivatives known as triazenes (31). The guanido (arginine) and indole (tryptophan) also undergo coupling. Their reactions are slower than those of phenol, imidazole, and amino groups.

9.5.3.a Immobilization of papain on poly(vinyl alcohol) [107]

Cross-linked poly(vinyl alcohol)

Poly(vinyl alcohol) (PVA) (22 g, 0.5 unit mol) is cross-linked with terephthalaldehyde (3.35 g, 25 mmol). The insoluble polymer is ground and sieved.

Arylaminated PVA

Cross-linked PVA (1.0 g) is allowed to swell in water (40 mL) and treated with HCl (d = 1.19, 1 mL). Then 2-(*m*-aminophenyl)-1,3-dioxolane (0.60 mL, 0.72 g, 4.37 mmol) is added dropwise with stirring. The temperature of the reaction mixture is raised to 45 to 50°C in 1 h and stirring is continued for another 15 h at this temperature. The yellowish product is filtered, washed with water until neutral, extracted with acetone (72 h) in a Soxhlet apparatus, dried under reduced pressure, and sieved again. The content of amino groups (elementary analysis) is 1.4 mmol of NH_2 per gram of polymer.

Diazotized PVA

Arylaminated PVA (50 mg) is allowed to swell in water (2 mL) and treated at 4°C with 0.5 M HCl (10 mL) and 1 M $NaNO_2$ solution (1 mL). After stirring for another 1 h at this temperature, the polymer is filtered by suction and washed carefully with ice water until neutralized.

Immobilization of papain

The freshly prepared diazonium product is immediately treated with 20 mL of a 2% papain solution in 1 M $NaHCO_3$ and stirred for 15 h at 4°C. Unreacted and adsorbed enzyme is washed out with vigorous stirring using successively small portions of 1 M $NaHCO_3$ (200 mL), 1 M NaCl (200 mL), and water (80 mL). The product is stored under water or in the freeze-dried state at 4°C.

9.6 POLYMERS CONTAINING ALDEHYDE GROUPS

Several types of polymers containing aldehyde or acetal functional groups can be used as supporting polymers for enzyme immobilization (Figure 19).

9.6.1 Schiff Base Formation Reaction

Aldehydes react with amino groups on the protein to form Schiff base (aminol, azomethine, aldimine) linkages. Although the reaction of polymeric aldehydes with proteins can be carried out under mild conditions, the method has found only limited application because of the uncertainty about the stability of the bonds formed with the amino groups of enzyme. Schiff base formation is reversible and the equilibrium is usually unfavorable for Schiff base formation in aqueous solution, particularly at low pH values.

Dialdehyde derivatives of polysaccharides (32) prepared by controlled periodate or dimethyl sulfoxide oxidation of vicinal hydroxyl groups have been used as supports. Partially oxidized cellulose was used to bind enzymes through the formation of Schiff base linkages with protein amino groups [128,129]. Dialdehyde starch has been used in a similar fashion [130,131].

The carbonyl function of partially oxidized polysaccharides has been utilized to bind enzymes through their amine or carboxyl groups via four-component condensation reactions (4CC) in the presence of water-soluble isocyanide and carboxylate or amine. Dialdehyde derivatives of polysaccharides can be used as starting materials for the preparation of supports containing various functional groups, such as the arylamino group. Dialdehyde starch, the periodate-oxidation product of starch (90% oxidized), can be obtained commercially (Sumstar 190, Miles Chemical Co.).

Vinyl polymers containing aldehydr or acetal functional groups are poly(acryloylaminoacetaldehydedimethylacetal-co-*N*,*N*'-methylene-diacrylamide (34) [135] and poly(allyl alcohol-co-vanillinmethacrylate (33) [132-134] (Figure 20). Polymers containing amino groups can be modified to aldehyde-containing polymers using glutaraldehyde. Polymeric dialdehydes are prepared by periodate or dimethyl sulfoxide oxidation of polysaccharides, such as cellulose and dextran.

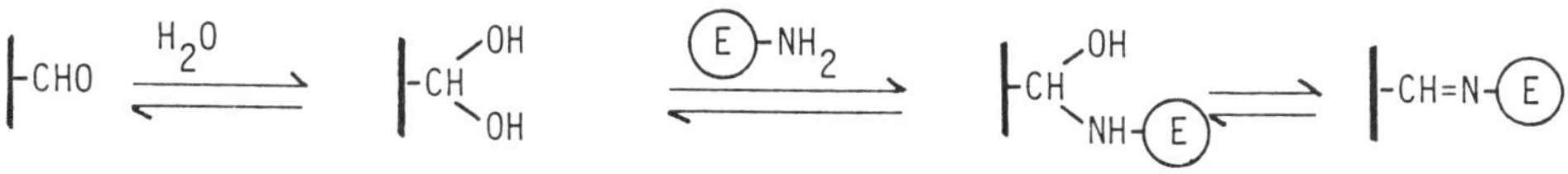

Figure 19

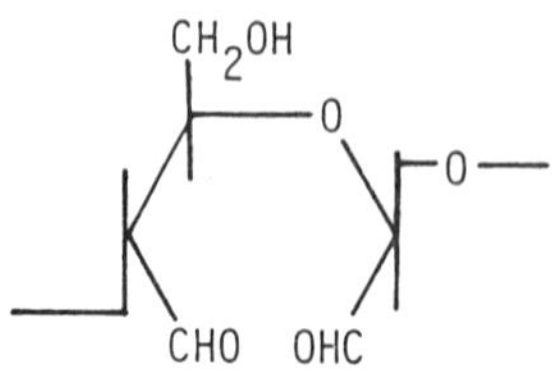

32

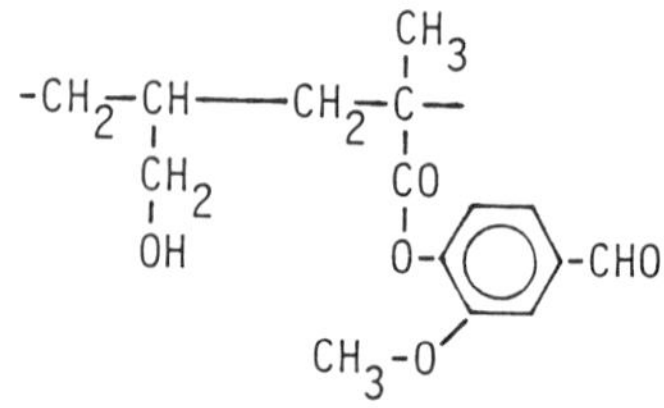

33

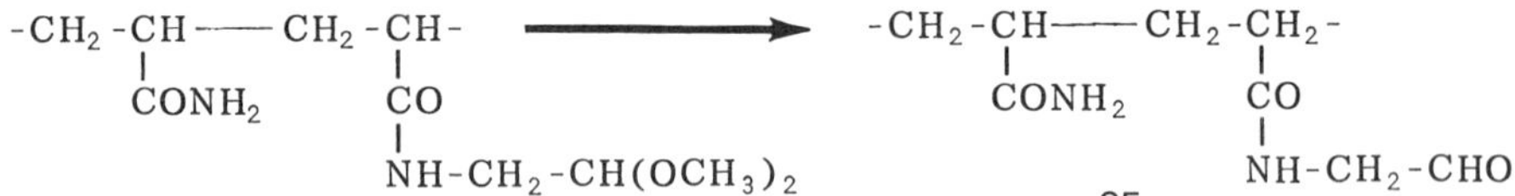

34 35

Figure 20

9.6.1.a Immobilization of urease on enzacryl polyacetal [135]

Enzacryl polyacetal (34)

Enzacryl polyacetal, a cross-linked copolymer of acryloylaminoacetaldehydedimethylacetal and *N*,*N*'-methylenediacrylamide (molar ratio 8:1), is available in aqueous xerogel form (Koch-Light Ltd.).

Enzacryl polyaldehyde (35)

Enzacryl polyacetal is activated for enzyme binding by shaking with aqueous 0.25 M HCl at 30°C for 48 h. The activated xerogel product, Enzacryl polyaldehyde, has an aldehydrol content of 5.4 mmol/g (71% theoretical); determined by reaction of a sample with buffered hydroxylamine hydrochloride followed by titration.

Immobilization of urease

Samples of hydrated Enzacryl polyaldehyde (150 mg dry weight of copolymer) is allowed to equilibrate with 25 mM phosphate buffer (15 mL, pH 7.5) for 1 h. After centrifugation, the excess buffer is discarded, and an aliquot (0.5 mL) of urease solution (20 mg/mL) in 25 mM phosphate buffer (pH 7.5) is added. The suspensions are stirred at 0 to 5°C for 18 h. The resulting urease derivatives are washed six times with 25 mM phosphate buffer (pH 7.5, 20 mL) containing 5 mM EDTA and finally stored in the same buffer at 0 to 5°C.

9.6.2 Glutaraldehyde Method

Glutaraldehyde is an intermolecular cross-linking agent that produces stable and insoluble three-dimensional networks of protein (Figure 21).

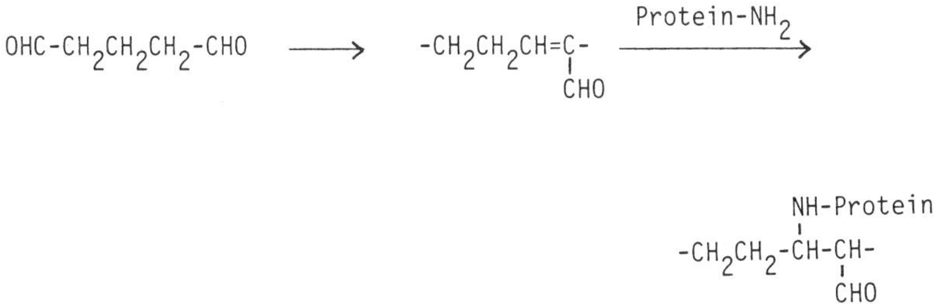

Figure 21

This reagent can be reacted with polymers containing primary amino groups to yield matrices containing the aldehyde function. Enzymes are bound irreversibly to the glutaraldehyde-treated polymer by a reaction presumably analogous to that occurring during cross-linking with glutaraldehyde (Figure 22) [35,136-141,124].

The nature of the reaction of glutaraldehyde with enzymes and amine-containing polymers is not fully understood, and not consistent with the formation of the unstable Schiff base (36), which is normally assigned to the adduct of aldehyde with an amino group.

9.6.2.a Immobilization of enzyme on porous glass [141]

Alkylamine glass (3)

To clean porous glass (1 g) is added distilled water (18 mL) with γ-aminopropyltriethoxysilane (2 mL). After the addition of the silane solution, the pH is adjusted between 3 and 4 with 6 N HCl. The neutralization of the basic silane solution may cause heating. After pH adjustment, the reaction mixture is placed in a 75°C water bath for 2 h. The beads are filtered on a Büchner funnel, washed with 20 mL of distilled water, and dried in a 115°C oven for at least 4 h. The product may be stored at this point for later use.

Glutaraldehyde treatment (6)

Alkylamine glass (1 g) is added to 25 mL of 2.5% glutaraldehyde in 0.05 M Na_2HPO_4 buffer (pH 7.0) solution. The reaction is continued

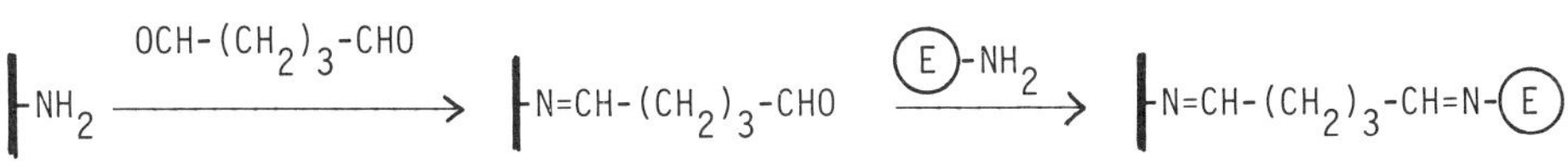

36

Figure 22

for at least 60 min. The treated glass is washed exhaustively with distilled water on a Büchner funnel. It is important to remove all excess glutaraldehyde before adding the enzyme; otherwise, cross-linking will occur.

Immobilization of enzyme (36)

To the glutaraldehyde-activated alkylamine glass is added the enzyme in as small a volume as possible with a minimum concentration of 1% in the same buffer as that used in the activation step. The quantity of enzyme added should be between 50 and 100 mg per gram of carrier. Allow 2 to 4 h for the reaction; then wash with distilled water.

Soaking in 6 M urea has been used successfully for removing adsorbed enzyme without inactivating the coupled enzyme in most cases. However, before subjecting the entire preparation to a urea soak, it is advisable to test a small batch first. Do not dry the sample of immobilized enzyme. Retain as a damp preparation in a sealed container in the refrigerator.

9.7 POLYMERS CONTAINING MERCAPTO GROUPS

Polymers containing mercapto functional groups can be used for specific reversible coupling to protein mercapto groups (cystein) via thiol-disulfide interchange reactions (Figure 23).

9.7.1 Thiol-Disulfide Interchange Reactions

Polymers containing mixed disulfide are prepared by treating a thiolated polymer with 2,2'-dipyridyldisulfide. The polymeric mixed disulfides are in effect polymeric analogs of low-molecular-weight disulfides used for the quantitative estimation of thiol groups in protein: 2,2'-dipyridyldisulfide and 5,5'-dithiobis(2-nitrobenzoic acid), Ellman's reagent (37)

The coupling of an enzyme through a thiol group (cysteine) is accompanied by the liberation of 2-thiopyridone; this offers a means

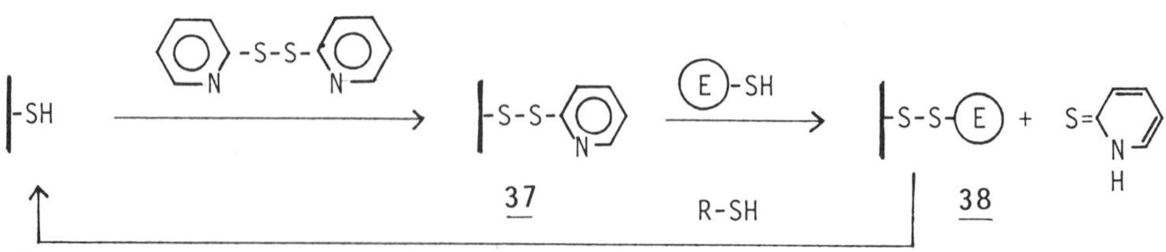

Figure 23

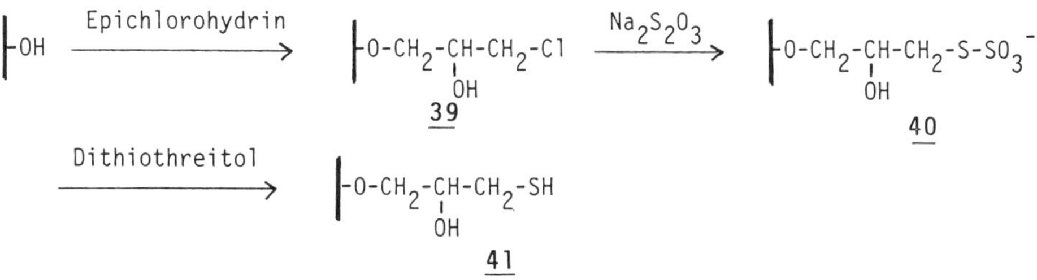

Figure 24

of monitoring the reaction. The disulfide (38) linkages are stable under nonreducing conditions and can be reversed with low-molecular-weight mercapto reagents such as ethanthiol or cysteine.

Polymers containing mercapto groups are glutathione-agarose [142], and thiolated-agarose (41) [144,145]. Agarose is treated with epichlorohydrin (39) followed by $Na_2S_2O_3$ to give 40. The thiolated-agarose is obtained by reduction of 40 with dithiothreitol (Figure 24).

Another type of polymer containing mercapto groups is Enzacryl polythiol (43) [143]. Enzacryl polythiol is prepared by acid hydrolysis of a cross-linked copolymer of acrylamide and *N*-acryloyl-4-carboxymethyl-2,2-dimethylthiazolidine (42) (Figure 25). Enzacryl polythiol is used for the reversible binding of enzymes. Since most enzymes contain either no free thiol groups or a very small number, prior enrichment by treatment with *N*-acetylhomocysteine thiolactone is desirable. Subsequently, oxidative coupling with Enzacryl polythiol

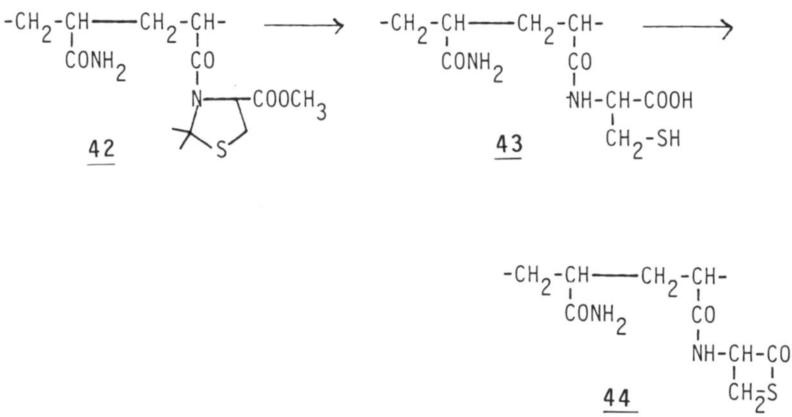

Figure 25

is brought about with potassium ferricyanide. The enzyme may be detached by washing with a solution of ethanthiol or cysteine.

Carbodiimide treatment of Enzacryl polythiol yields Enzacryl polythiolactone (44). Enzacryl polythiolactone is effective in coupling enzymes without further activation.

9.7.1.a Immobilization of Amyloglucosidase on agarose [145].

Mercapto-agarose gel (41) [144]

Agarose gel (6% beads), washing with water and suction drying, is suspended (30 g) in 24 mL of NaOH solution, followed by dropwise addition of epichlorohydrin at room temperature under stirring. To obtain a mercapto gel with a degree of substitution of about 50 μmol of SH groups per gram of dried product, 0.75 mL of epichlorohydrin is added during a 15-min period. The temperature is then raised to 60°C, and stirring continued for 2 h. The product is washed with water and with 0.5 M sodium phosphate buffer of pH 6.25. The gel is suction dried and suspended in 30 mL of the foregoing buffer; then 30 mL of 2 M sodium thiosulfate solution is added, followed by stirring for 6 h at room temperature.

The alkyl thiosulfate ester (40) is washed with water and suspended in 0.1 M $NaHCO_3$ solution (60 mL). The alkyl thiosulfate ester gel is then reduced with dithiothreitol (8 mg/mL) dissolved in 1 mM aqueous EDTA solution. The reduction of agarose gel, with a low degree of substitution, requires 4 mL of dithiothreitol solution, whereas the reduction of agarose gel, having a high degree of substitution (700 μmol -SH per gram), uses 50 mL of dithiothreitol solution. The mercapto gel obtained is washed with 300 mL of 0.1 M $NaHCO_3$ solution with 1 M NaCl and 1 mM EDTA, and finally with 100 mL of 1 mM EDTA solution. The gel is stored in 0.01 M deaerated sodium acetate buffer of pH 4 (1 mM in EDTA).

2-Pyridine disulfide agarose gel (37)

Mercapto-agarose gel (50 mol of -SH groups per gram of dry product) is washed with water and a solution of 50% acetone-water. The gel is suspended in the acetone-water solution. 2-Pyridine disulfide (100 mg) is dissolved in 50% acetone-water solution and added to the gel suspension (reaction time 30 min). The gel is washed with 50% aqueous acetone and 1 mM EDTA solution.

Immobilization of amyloglucosidase (38)

Two milliliters of a solution containing amyloglucosidase (41.4 mg) in 0.1 M $NaHCO_3$ (1 mM in EDTA, 6 mL) is passed into a reductor column containing 17 mL of thiol agarose (about 700 mol of SH per gram of dry gel), which is operated in tandem with another column (3.2 mL volume) containing 2-pyridine disulfide agarose. The enzyme solution is followed by 20 mL of 0.1 M $NaHCO_3$ (1 mM EDTA), 1 M sodium acetate buffer (pH 4.8, 160 mL), 1 M sodium acetate containing 1% starch (30 mL), and finally 0.01 M sodium acetate (pH 4.8, 30 mL). The flow rate is kept at about 10 mL/h. This procedure immobilizes 0.32 enzyme activity units per 1.25 units applied.

9.8 POLYMERS CONTAINING ACTIVE HALOGEN GROUPS

Polymers containing various active halogen groups can be used for coupling via arylation or alkylation with the amino groups of protein. Polymers containing 3-fluoro-4,6-dinitrophenyl group are commonly used arylating polymers. These polymers are poly(methacrylic acid-co-methacrylic acid 3-fluoro-4,6-dinitroanilide) [95,146-148], and poly(methacrylic acid-co-3-fluoro-4,6-dinitrostyrene) (45) [95,149] (Figure 26).

Polymers containing monochloro- or dichloro-s-triazinyl groups are used with polysaccharides. These hydroxyl- or amino group-containing polymers are treated with 2,4,6-trichloro-s-triazine (cyanuric chloride) or with a 4,6-dichloro-s-triazine to give activated polymers (47) (Figure 27). CM-cellulose, CM-Sephadex, DEAE-

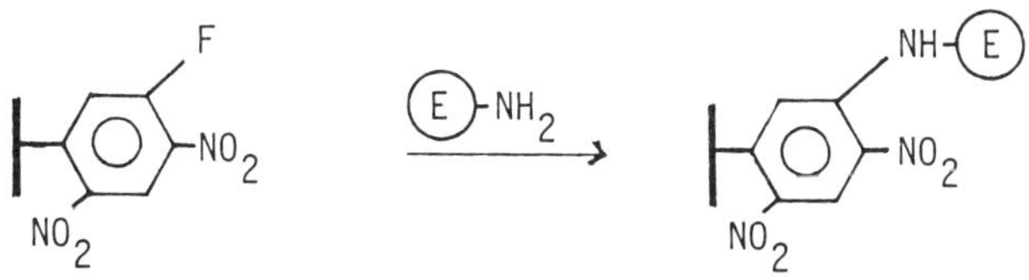

45

Figure 26

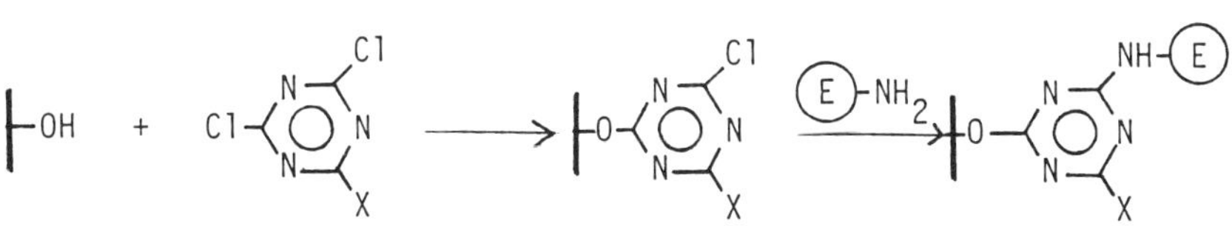

46 47 48

Figure 27

cellulose, DEAE-Sephadex, and cross-linked poly(vinyl alcohol) are used as starting materials.

Polymers having monohaloacetyl groups have been used in a similar manner for the coupling of enzymes by alkylation of amino groups on protein (49). Bromoacetyl cellulose is prepared by treating cellulose with bromoacetic acid and bromoacetyl chloride in dioxane. The chloro- and iodoacetates of cellulose have also been used [161-163, 38]. Polymethacrylates containing ω-iodoalkyl side chains are also the alkylating polymers for coupling with amino groups of enzymes (50) [164-167].

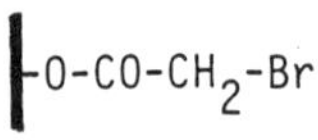

49

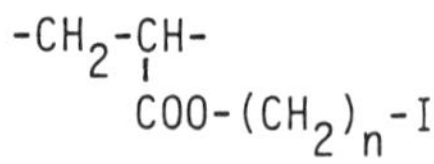

50

9.8.1 3-Fluoro-4,6-dinitrophenyl Derivatives

Polymers containing 3-fluoro-4,6-dinitrophenyl functional groups are prepared by nitration of cross-linked copolymers of methacrylic acid and methacrylic acid 3-fluoroanilide (52) (Figure 28), or cross-linked copolymer of methacrylic acid and fluorostyrene (53).

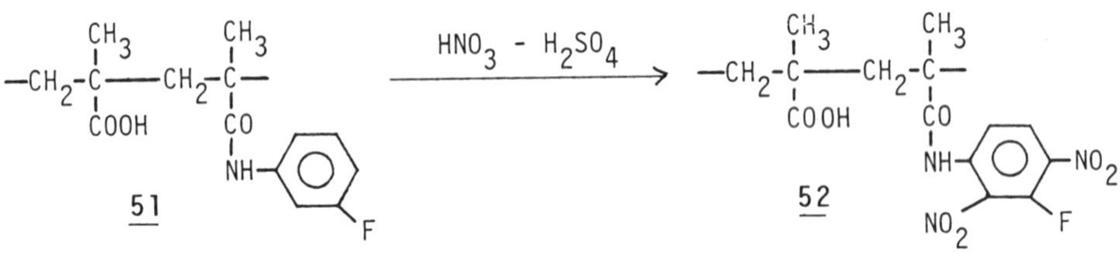

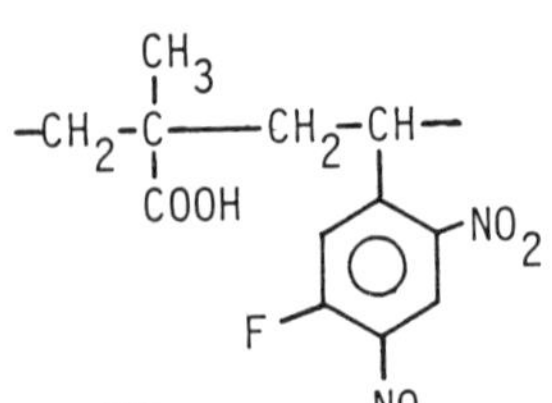

53

Figure 28

9.8.1.a Immobilization of papain on poly(methacrylic acid-co-3-fluoro-4,6-dinitroanilide) [146,147]

Polymerization (51)

Methacrylic acid (0.86 g, 10 mmol), methacrylic acid 3-fluoroanilide (1.79 g, 10 mmol), and divinylbenzene (26.5 mg, 0.20 mmol) are mixed with dibenzoyl peroxide (2.6 mg, 0.1 wt%) in a polymerization flask and heated for 48 h at 75°C and for additional 48 h at 90°C. The polymer is extracted with methanol (8 h). The composition of the copolymer (elementary analysis) is F:COOH = 1:2.

Nitration (52)

The copolymer (2.0 g) is added at -10°C in small portions with stirring to the nitrating acid (75 mL) consisting of 1 part HNO_3 (d = 1.52) and 3 parts H_2SO_4 (d = 1.84). During the reaction the temperature rises to -2°C. The reaction mixture is stirred another 6 h at this temperature and is poured over ice (500 g). The polymer is filtered under vacuum, washed with water until neutral, and exhaustively extracted with acetone. The degree of nitration (elementary analysis) is 1.4 NO_2 groups per fluoroaryl group.

Immobilization of papain (45)

The nitrated copolymer (100 mg) is vigorously stirred in a closed vessel for 24 h with a solution of papain (2%, 20 mL) in 0.1 M $NaHCO_3$. After separation of the immobilized enzyme from supernatant solution by filtration, unreacted adsorbed enzyme is washed out with successively small portions of 0.1 M $NaHCO_3$ (200 mL), 1 M NaCl (200 mL), and water (80 mL). The product can be stored under water or in the freeze-dried state at 4°C.

9.8.2 s-Triazinyl Derivatives

This method has been used mainly in conjunction with polysaccharide supports (e.g., cellulose, agarose, and cross-linked dextran [150-160]. Dichloro-s-triazinyl derivatives of polysaccharides are made with 2,4,6-trichloro-s-triazine (cyanuric chloride). The products react rapidly with protein. The remaining chlorines may be treated with an amine to prevent further reaction, including cross-linking of the polysaccharides. As the reaction of the second chlorine with protein is so fast, it is preferable to replace this chlorine and use the third chlorine for the coupling reaction.

Monochloro-s-triazinyl derivatives of polysaccharides can be made by allowing the polysaccharide to react initially with cyanuric chloride and then with a low-molecular-weight amine under controlled conditions

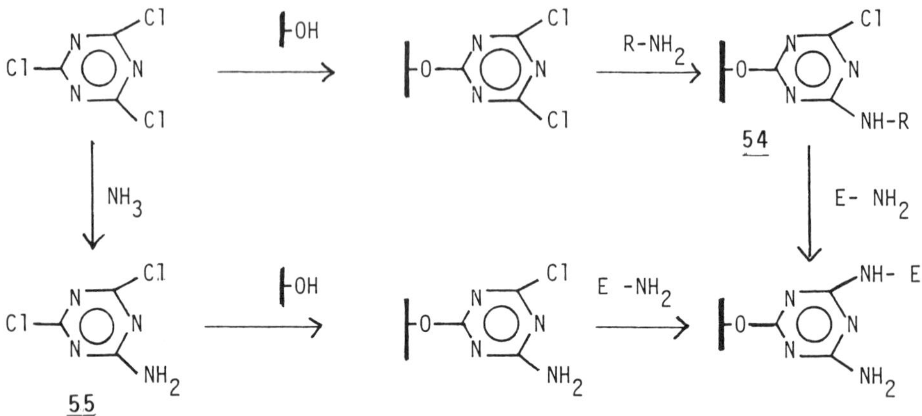

Figure 29

to leave the third chlorine unreacted (54). It is, however, preferable to use 4,6-dichloro-s-triazine derivative, 2-amino-4,6-dichloro-s-triazine, or the so-called Procion M dyes (55) (Figure 29).

9.8.2.a Immobilization of enzyme on *s*-triazinyl cellulose [153]

2-Amino-4,6-dichloro-s-triazine (55)

A stream of gases, generated by blowing N_2 through a warmed solution of ammonia and dried by passage through NaOH pellets, is passed into a cold slurry (5°C) of cyanuric chloride (184 g) in a mixture of dioxane (1 L) and toluene (0.2 L). When the reaction products form a thick suspension, the solid is filtered, washed with dioxane (0.5 L), and discarded. The filtrate and washings are mixed and evaporated to dryness under reduced pressure. The product is purified by dissolving it in a minimum volume of acetone/water (1:1, v/v) followed by removal of the acetone under reduced pressure, and then redissolving the recovered solid in boiling water and cooling to 20°C within 10 min to prevent any hydrolysis. The yield is about 90%.

Aminochloro-s-triazinyl cellulose (47)

Cellulose (10 g dry weight) is added to 50 mL of acetone/water (1:1, v/v) at 50°C containing 2-amino-4,6-dichloro-s-triazine (1 g) and stirred for 5 min. Then an aqueous solution of Na_2CO_3 (15%, 20 mL, containing 0.6 volume of 1 M HCl) is poured into the reaction liquor and the slurry is stirred for a further 5 min at 50°C. Concentrated HCl is added to reduce the pH rapidly to below 7. The product is recovered by filtration, washed three times with acetone/water (1:1, v/v), followed by extensive washing with water and finally 0.05 M phosphate buffer (pH 7), and stored at 2°C.

Immobilization of enzyme (48)

The aminochloro-*s*-triazinyl cellulose suspension is adjusted to pH 8.0 with alkali. A solution of the enzyme in 0.05 M phosphate (pH 8.0) is added, and the suspension is stirred gently for 18 h at 20°C. The product is recovered by filtration and washed thoroughly with phosphate buffer (pH 8.0), the same buffer containing 1 M NaCl, and the phosphate buffer. The immobilized enzyme may be stored in buffer at 2°C.

9.9 POLYMERS CONTAINING AMIDE GROUPS

Polymers containing amide bonds, polyacrylamide, and polyamides such as nylon are useful hydrophilic and electrically neutral supports. As mentioned, copolymers of acrylamide-containing functional groups have been prepared and used as polymer supports. Preformed polyacrylamide gel or beads (e.g., Bio-Gel) can be chemically modified. Polyacrylamide beads are treated with hydrazine or with ethylenediamine to obtain the corresponding acyl hydrazide [168] or aminoethyl derivatives [169]. These functional groups can be used for the immobilization of enzymes by using the reaction described above (Figure 30).

Polymers containing amides in the main chain are polyamides. Synthetic polyamides of ω-aminocarboxylic acid or α,ω-dicarboxylic acid and α,ω-diamine are known as nylons: nylon 6, nylon 6,6, or nylon 6,10, designated according to the number of carbon atoms in the repeating units. These nylons are available commercially in a variety of physical forms, such as membranes, powders, tubes, hollow fibers, and spin-bonded films. They are mechanically strong and nonbiodegradable and therefore are useful polymer as supports.

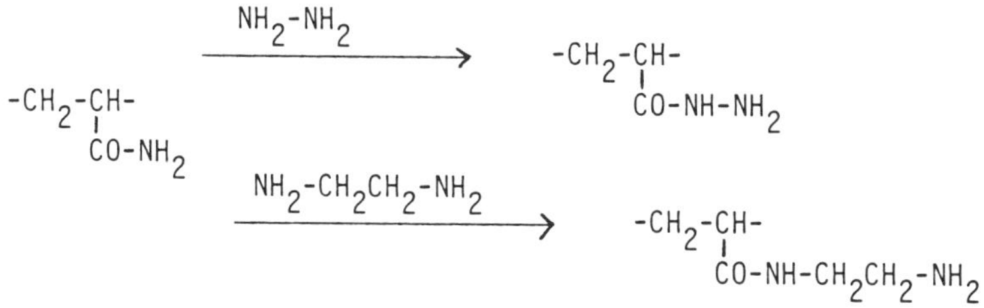

Figure 30

Nylons, however, contain only a terminal carboxyl and an amine group per polymer chain as possible reactive sties for covalent immobilization of enzymes. To increase the binding capacity of nylon, the following approaches have been adopted: (a) controlled cleavage of amide bonds to increase the number of carboxyl and amine groups; (b) *O*-alkylation of the amide bonds to form reactive centers; and (c) *N*-alkylation of the amide bonds to form reactive side chains (Figure 31).

9.9.1 Peptide Bond Cleavage

A small percentage of the amide bonds in the nylon backbone are cleaved by mild acid hydrolysis (56). The carboxyl and amino groups generated are activated by the methods described for the polymers containing carboxyl or amino groups [170-177]. In these methods, however, a residue of ionized carboxyl or amino group is left on the surface of the nylon (57).

The controlled cleavage of the amide bonds in organic solvents has also been accomplished with nucleophiles such as *N*,*N*-dimethylaminopropylamine [176,177]. This method blocks the carboxyl group with a cationic *N*,*N*-dimethylaminopropylamide residue, leaving only the complementary amino group free for further chemical modification (58) (Figure 32).

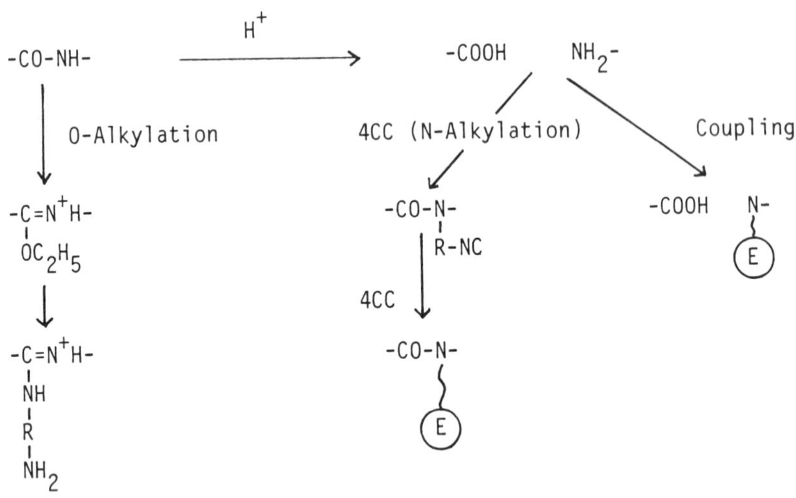

Figure 31

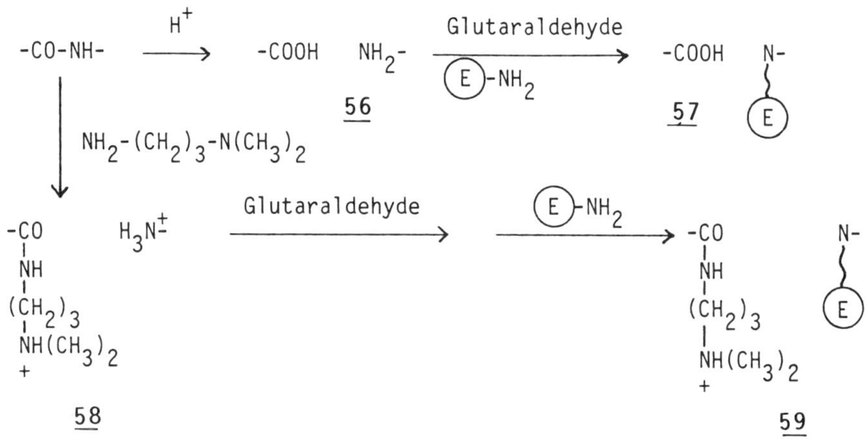

Figure 32

9.9.1.a Immobilization of enzyme on nylon [177]

Controlled cleavage of nylon

A 3-m length of nylon-6 tube, 1 mm bore, is filled with a mixture of 18.6% (w/w) CaCl , 18.6% (w/w) water in methanol and incubated at 50°C for 20 min. This process etches the inside surface of the nylon tube by dissolving the regions of amorphous nylon, thereby increasing both the available surface area and its wettability. The amorphous nylon is then purged from the tube by perfusion with water for 30 min at a flow rate of 5 mL/min.

For nonhydrolytic cleavage of the nylon, the tube is dried first by perfusing it with methanol for 30 min at a flow rate of 2 mL/min, then filled with *N*,*N*-dimethyl-1,3-propanediamine and incubated for 12 h at 70°C. The excess amine is removed by washing the tube with water for 12 h at a flow rate of about 2 mL/min.

For the hydrolytic cleavage of the nylon, the tube is perfused for 40 min with 3.65 M HCl at 45°C at a flow rate of 2 mL/min and then washed through with water for 12 h at a flow rate of about 2 mL/min.

Immobilization of enzyme by glutaraldehyde method

The cleaved nylon is activated with 5% (w/v) glutaraldehyde in 0.2 M borate (pH 8.5) for 15 min at 20°C at a flow rate of 2 ml/min. The tubes are washed free of excess glutaraldehyde by perfusion for 10 min with 0.1 M phosphate buffer (pH 7.8, 50 mL) and immediately filled with a solution of the enzyme (0.5 to 1.0 mg/mL) in the same buffer. After 3 h of incubation at 4°C, excess protein is removed by washing thoroughly with 0.5 M NaCl in 0.1 M phosphate buffer (pH 7.0, 1 L), at a flow rate of 5 mL/min. Nylon tube-immobilized enzymes are stored filled with assay buffer at 4°C.

9.9.2 N-Alkylation of Nylon

The introduction of chemically reactive side chains on nylon by N-alkylation of the secondary amide groups has been carried out by two-step procedures: (a) mild acid hydrolysis of nylon generating -COOH and NH_2- pairs on the surface of nylon, and (b) the newly formed -COOH and NH_2- pairs are resealed by a four-component condensation (4CC) reaction involving the neighboring carboxyl and amino groups on the nylon chain, an aldehyde, and an isocyanide (Figure 33).

In principle the 4CC reactions allow for considerable versatility, since by proper choice of aldehyde and isocyanide various functional groups can be introduced on the N-alkyl side chains of the re-formed amide groups of nylon. The procedure most commonly employed utilizes acetaldehyde or isobutyral as the aldehyde component with a bifunctional isocyanide, 1,6-diisocyanohexane. This reaction leads to nylon derivatives containing isonitrile functional groups (60). Enzymes can be coupled to the isonitrile group of the isonitrile-nylon by a 4CC reaction (61) [178]. The isonitrile functional groups on nylon can be modified to other functional groups, such as arylamino derivative by the reaction with diaminodiphenylmethane [178].

9.9.2.a Immobilization of enzyme on N-alkylated nylon [178]

Controlled hydrolysis of nylon powder (56)

Commercial nylon 6 pellets are suspended in a 20% methanolic solution of $CaCl_2$ (30 g/L) and stirred at room temperature until a homogeneous solution is obtained. The nylon solution is added dropwise with strong

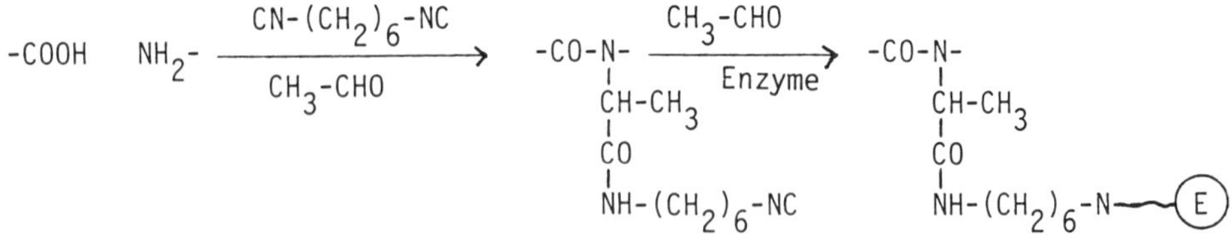

60 61

Figure 33

stirring into a large excess of water; the powder thus obtained is separated on a suction filter, washed with water, resuspended in water, and stirred vigorously. The powder is separated, washed again with water and ethanol, and air dried.

The dry nylon powder (10 g) is suspended in 3 N HCl (300 mL) and stirred at room temperature for 4 h. The powder is separated on a suction filter, washed exhaustively with water, ethanol, and ether, and dried under vacuum over phosphorus pentoxide.

Polyisonitrile-nylon (60)

Partially hydrolyzed nylon powder (2 g) is suspended in isopropanol (80 mL). Acetaldehyde (20 mL, 0.3 mol) or isobutyral (32 mL (0.3 mol) is then added, followed by 1,6-diisocyanohexane (8 mL, 0.06 mol), and the reaction is allowed to proceed in a closed vessel for 24 h with stirring at room temperature. The polyisonitrile-nylon powder is separated on a suction filter, washed with isopropanol (50 mL) and then with ether (200 mL), and air dried and vacuum dried. The polyisonitrile-nylon is stored at -5°C in a dark vial over silica gel. The mean carboxyl content is 20 μmol/g, and the isonitrile content is 40 to 50 μmol/g.

Immobilization of enzyme (61)

Polyisonitrile-nylon (50 mg) is suspended in 2 mL of cold 0.1 M sodium phosphate (0.5 M in sodium acetate, pH 7). A cold aqueous solution of enzyme (5 to 10 mg in 1 mL) is then added, followed by 0.1 mL of acetaldehyde. The reaction mixture is allowed to stir overnight at 4°C. The water-insoluble enzyme derivative is separated on a filter, washed with water, then with 1 M KCl, 0.1 M in $NaHCO_3$, and again with water, resuspended in water (4 mL), and stored at 4°C.

9.9.3 Imidoester Reaction: *O*-Alkylation of Nylon

The secondary amides in nylon can be alkylated by powerful alkylating reagents, triethyloxonium tetrafluoroborate, to yield imidate salts of nylon (62). Imidate salts and their free bases, imido esters, will react with nucleophiles such as amines to yield positively charged substituted amidines. In this way it is possible to introduce reactive centers onto nylon without breaking any of its secondary amide bonds in the process (Figure 34) [180].

The imidate salts of nylon afford several routes for enzyme immobilization. The enzyme can be allowed to react directly through its amino groups with the imidate salt of nylon. The imidate salt of nylon can be allowed to react with a bis-acid hydrozide, such as adipic acid dihydrazide, to give a hydrazide substituted nylon (64). The imidate

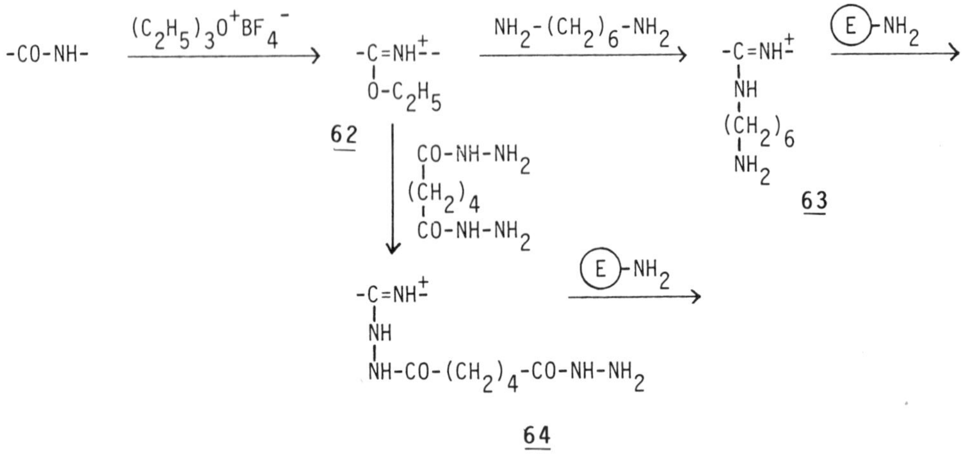

Figure 34

salt of nylon can be allowed to react with a bis-amine, such as 1,6-diaminohexane, to give 63, amine-substituted nylon. These chemically modified nylons can be used for enzyme immobilization by using the procedure described for the polymers containing carboxyl or amino groups.

9.9.3.a Immobilization of enzyme on O-alkylated nylon [179]

O-Alkylation of nylon (62)

A 3-m length of nylon 6 tube is filled with the triethyloxonium tetrafluoroborate solution (10%, in dichloromethane) and incubated at room temperature for 15 min. Excess alkylating agent is removed by suction and the tube is washed for 2 min with dichloromethane and for 2 min with dioxane.

Chemical modification of O-alkylated nylon

The O-alkylated nylon tube is filled immediately with a solution of either amine (10%, w/v, 1,6-diaminohexane in methanol) or acid hydrazide (3%, w/v, adipic acid dihydrazine in formamide) and incubated for 3 h at room temperature. The excess amine or hydrazide is removed by perfusing the tube with water for 12 h.

Immobilization of enzyme

Both the amine- and hydrazide-substituted nylon can be reactivated with glutaraldehyde and coupled to enzyme as described in Section 9.9.1.a.

REFERENCES

1. O. R. Zaborsky, Immobilized Enzymes, CRC Press, Cleveland, Ohio (1973).
2. J. J. Gutcho, Immobilized Enzymes; Preparations and Engineering Techniques, Noyes Data Corporation, Park Ridge, N.J. (1974).
3. M. Salmona, C. Saronia, and S. Garattini, Eds., Insolubilized Enzymes, Raven Press, New York (1974).
4. H. H. Weetall, Ed., Immobilized Enzymes, Antigens, Antibodies, and Peptides; Preparation and Characterization, Marcel Dekker, Inc., New York (1975).
5. K. Mosbach, Ed., Methods in Enzymology, Vol. 44, Immobilized Enzymes, Academic Press, Inc., New York (1976).
6. L. B. Wingard, Jr., E. Katchalski, and L. Goldstein, Eds., Applied Biochemistry and Bioengineering, Vol. 1, Immobilized Enzyme Principles, Academic Press, Inc., New York (1976).
7. T. K. Ghose, A. Fiechter, and N. Blakebrough, Eds., Advances in Biochemical Engineering, Vol. 10, Immobilized Enzymes I, Springer-Verlag, New York (1978).
8. L. B. Wingard, Jr., E. Katchalski, and L. Goldstein, Eds., Applied Biochemistry and Bioengineering, Vol. 2, Enzyme Technology, Academic Press, Inc., New York (1979).
9. T. K. Ghose, A. Fiechter, and N. Blakebrough, Eds., Advances in Biochemical Engineering, Vol. 12, Immobilized Enzymes II, Springer-Verlag, New York (1979).
10. L. B. Wingard, Jr., E. Katchalski, and L. Goldstein, Eds., Applied Biochemistry and Bioengineering, Vol. 3, Analytical Applications of Immobilized Enzymes and Cells, Academic Press, Inc., New York (1981).
11. I. Chibata, S. Fukui, and L. B. Wingard, Jr., Eds., Enzyme Engineering, Vol. 6, Plenum Publishing Corporation, New York (1982).
12. G. Manecke, Pure Appl. Chem., 4, 507 (1962).
13. G. Manecke, Naturwissenschaften, 51, 25 (1964).
14. I. H. Silman and E. Katchalski, Annu. Rev. Biochem., 35, 873 (1966).
15. L. Goldstein and E. Katchalski, Z. Anal. Chem., 243, 375 (1968).
16. A. S. Lindsey, J. Macromol. Sci. Rev. Macromol. Chem., 3, 1 (1969).
17. E. M. Crook, K. Brocklehurst, and C. W. Wharton, in Methods in Enzymology, Vol. 19, Proteolytic Enzymes (G. Perlmann and L. Lorand, Eds.), Academic Press, Inc., New York (1970), p. 963.
18. L. Goldstein, in Methods in Enzymology, Vol. 19, Proteolytic Enzymes (G. Perlmann and L. Lorand, Eds.), Academic Press, Inc., New York (1970), p. 935.

19. E. Katchalski, I. H. Silman, and R. Goldman, Adv. Enzymol., 34, 445 (1971).
20. K. Mosbach, Sci. Am., 224, 26 (1971).
21. G. J. H. Melrose, Rev. Pure Appl. Chem., 21, 83 (1971).
22. R. Goldman, O. Kedem, and E. Katchalski, Biochemistry, 10, 165 (1971).
23. R. Goldman, L. Goldstein, and E. Katchalski, in Biochemical Aspects of Reactions on Solid Supports (G. R. Stark, Ed.), Academic Press, Inc., New York (1971), p. 1.
24. K. L. Smiley and G. W. Strandberg, Adv. Appl. Microbiol., 15, 13 (1972).
25. H. D. Orth and W. Brümmer, Angew. Chem. Int. Ed. Engl., 11, 249 (1972).
26. R. Epton, in Reactions on Polymers (J. A. Moore, Ed.), D. Reidel Publishing Co., Dordrecht, The Netherlands (1973), p. 286.
27. W. R. Vieth and K. Venkatasubramanian, Chem. Technol., 4, 47 (1974).
28. C. Kent, A. Rosevear, and A. R. Thomson, Top. Enzyme Ferment. Biotech., 2, 12 (1978).
29. S. A. Barker and P. J. Somers, Top. Enzyme Ferment. Biotechnol., 2, 120 (1978).
30. K. Mosbach, Adv. Enzymol., 46, 205 (1978).
31. P. Brodelius and K. Mosboch, Adv. Appl. Microbiol., 28, 1 (1982).
32. J. Coupek, M. Kriváková, and S. Pokorny, J. Polymer Sci. Polym. Symp., 42, 185 (1973).
33. K. Mosbach, Acta Chem. Scand., 24, 2084 (1970).
34. J. Turkova, O. Hubalkava, M. Krivakova, and J. Coupek, Biochim. Biophys. Acta, 322, 1 (1973).
35. A. C. Johansson and K. Mosbach, Biochim. Biophys. Acta, 370, 339 (1974).
36. R. Axén and S. Ernback, Eur. J. Biochem., 18, 351 (1971).
37. A. B. Patel, R. Stasiw, H. D. Brown, and C. Ghiron, Biotechnol. Bioeng., 14, 1031 (1972).
38. H. Maeda and H. Suzuki, Agric. Biol. Chem., 36, 1581 (1972).
39. G. J. Bartling, H. D. Brown, L. J. Forrester, M. T. Koes, A. N. Mather, and R. O. Stasiw, Biotechnol. Bioeng., 14, 1039 (1972).
40. M. P. Coughlan and D. B. Johnson, Biochim. Biophys. Acta, 302, 200 (1973).
41. J. Porath, R. Axén, and S. Ernback, Nature, 215, 1491 (1967).
42. J. Porath, K. Aspberg, H. Drevin, and R. Axén, J. Chromatogr., 86, 53 (1973).
43. R. Axén, E. Heilbronn, and A. Winter, Biochim. Biophys. Acta, 191, 478 (1969).
44. P. Cuartrecasas, J. Biol. Chem., 245, 3059 (1970).
45. P. Cuartrecasas, Proc. Nat. Acad. Sci. USA, 69, 1277 (1972).
46. P. Cuartrecasas, Adv. Enzymol., 36, 29 (1972).

47. D. Gabel, P. Vretblad, R. Axén, and J. Porath, Biochim. Biophys. Acta, 214, 561 (1970).
48. D. Gabel, I. Z. Steinberg, and E. Katchalski, Biochemistry, 10, 4661 (1971).
49. S. C. March, I. Parikh, and P. Cuatrecasas, Anal. Biochem., 60, 149 (1974).
50. T. Kristiansen, L. Sundberg, and J. Porath, Biochim. Biophys. Acta, 184, 93 (1969).
51. R. Jost, T. Miron, and M. Wilchek, Biochim. Biophys. Acta, 362, 75 (1974).
52. M. Wilchek, T. Oka, and Y. J. Topper, Proc. Nat. Acad. Sci. USA, 72, 1055 (1975).
53. R. Axén, J. Porath, and S. Ernback, Nature, 214, 1302 (1967).
54. R. Axén, P. Myrin, and J. Janson, Biopolymers, 9, 401 (1970).
55. S. A. Barker, S. H. Doss, C. J. Gray, J. F. Kennedy, M. Stacey, and T. H. Yeo, Carbohydr. Res., 20, 1 (1971).
56. J. F. Kennedy, S. A. Barker, and A. Rosevear, J. Chem. Soc., Perkin Trans. 1, 2568 (1972).
57. J. F. Kennedy, S. A. Barker, and A. Rosevear, J. Chem. Soc., Perkin Trans. 1, 2293 (1973).
58. J. F. Kennedy and A. Zamir, Carbohydr. Res., 29, 497 (1973).
59. F. Michael and J. Ewers, Makromol. Chem., 3, 200 (1949).
60. M. A. Mitz and L. J. Summaria, Nature, 189, 576 (1961).
61. C. J. Epstein and C. B. Anfinsen, J. Biol. Chem., 237, 2175 (1962).
62. W. E. Hornby, M. D. Lilly, and E. M. Crook, Biochem. J., 98, 420 (1966).
63. C. W. Wharton, E. M. Crook, and K. Brocklehurst, Eur. J. Biochem., 6, 565 (1968).
64. C. W. Wharton, E. M. Crook, and K. Brocklehurst, Eur. J. Biochem., 6, 572 (1968).
65. W. Brümmer, N. Hennrich, M. Klockow, H. Lang, and H. D. Orth, Eur. J. Biochem., 25, 129 (1972).
66. S. A. Barker, P. J. Somers, R. Epton, and J. V. McLaren, Carbohydr. Res., 14, 287 (1970).
67. S. A. Barker, P. J. Somers, and R. Epton, Carbohydr. Res., 14, 323 (1970).
68. Y. Levin, M. Pecht, L. Goldstein, and E. Katchalski, Biochemistry, 3, 1905 (1964).
69. L. Goldstein, Y. Levin, and K. Katchalski, Biochemistry, 3, 1913 (1964).
70. L. Goldstein, M. Pecht, S. Blumberg, D. Atlas, and Y. Levin, Biochemistry, 9, 2322 (1970).
71. L. Goldstein, Anal. Biochem., 50, 40 (1972).
72. L. Goldstein, Biochemistry, 11, 4072 (1972).

73. B. Solomon and Y. Levin, Biotechnol. Bioeng., 16, 1161, (1974).
74. E. G. Ong, Y. Tsang, and G. E. Perlmann, J. Biol. Chem., 241, 5661 (1966).
75. T. L. Westman, Biochem. Biophys. Res. Commun., 35, 313 (1969).
76. R. A. Zingaro and M. Uziel, Biochim. Biophys. Acta, 213, 371 (1970).
77. H. H. Weetall and L. S. Hersh, Biochim. Biophys. Acta, 206, 54 (1970).
78. L. Goldstein, A. Lifshitz, and M. Sokolovsky, Int. J. Biochem., 2, 448 (1971).
79. L. Goldstein, Biochim. Biophys. Acta, 315, 1 (1973).
80. L. Goldstein, Biochim. Biophys. Acta, 327, 132 (1973).
81. R. P. Patel and S. Price, Biopolymers, 5, 583 (1967).
82. R. P. Patel, D. V. Lopiekes, S. P. Brown, and S. Price, Biopolymers, 5, 577 (1967).
83. A. B. Patel, S. N. Pennington, and H. D. Brown, Biochim. Biophys. Acta, 178, 626 (1969).
84. N. Weliky, F. S. Brown, and E. C. Dale, Arch. Biochem. Biophys., 131, 1 (1969).
85. G. J. Bartling, H. D. Brown, and S. K. Chattopadhyay, Biotechnol. Bioeng., 16, 361 (1974).
86. K. Martensson and K. Mosbach, Biotechnol. Bioeng., 14, 715 (1972).
87. K. Mosbach and B. Mattiasson, Acta Chem. Scand., 24, 2093 (1970).
88. P. V. Sundaram, Biochem. Biophys. Res. Commun., 61, 667 (1974).
89. W. F. Line, A. Kwong, and H. H. Weetall, Biochim. Biophys. Acta, 242, 194 (1971).
90. I. C. Cho and H. E. Sawisgood, Biochim. Biophys. Acta, 258, 675 (1972).
91. G. Baum, F. B. Ward, and H. H. Weetall, Biochim. Biophys. Acta, 268, 411 (1971).
92. P. Cuartrecasas and I. Parikh, Biochemistry, 11, 2291 (1972).
93. P. Vretblad and R. Axén, FEBS Lett., 18, 254 (1971).
94. R. Axén and J. Porath, Nature, 210, 367 (1966).
95. G. Manecke, G. Günzel, and H. J. Förster, J. Polym. Sci. Polym. Symp., C30, 607 (1970).
96. G. Manecke and G. Günzel, Naturwissenschaften, 54, 531 (1967).
97. H. H. Weetall, Biochim. Biophys. Acta, 212, 1 (1970).
98. M. K. Weibel, W. Dritschhilo, H. J. Bright, and A. E. Humphrey, Anal. Biochem., 52, 402 (1973).
99. W. M. Herring, R. L. Laurence, and J. R. Kittrell, Biotechnol. Bioeng., 14, 975 (1972).
100. A. D. Traher and J. R. Kittrell, Biotechnol. Bioeng., 16, 413 (1974).

101. H. Brandenberger, Angew. Chem., 67, 661 (1955).
102. G. Manecke and S. Singer, Makromol. Chem., 37, 119 (1960).
103. G. Manecke, S. Singer, and K. E. Gillert, Naturwissenschaften, 45, 440 (1958).
104. E. Van Leemputten and M. Horisberger, Biotechnol. Bioeng., 16, 385 (1974).
105. M. D. Lilly, C. Money, W. E. Hornby, and E. M. Crook, Biochem. J., 95, 45P (1965).
106. R. Datta, W. Armiger, and D. F. Ollis, Biotechnol. Bioeng., 15, 993 (1973).
107. G. Manecke and H. G. Vogt, Makromol. Chem., 177, 725 (1976).
108. N. Grubhofer and L. Schleith, Naturwissenschaften, 40, 508 (1953).
109. G. Manecke and K. E. Gillert, Naturwissenschaften, 42, 212 (1955).
110. H. Filippusson and W. E. Hornby, Biochem. J., 120, 215 (1970).
111. W. M. Ledingham and W. E. Hornby, FEBS Lett., 5, 118 (1969).
112. W. M. Ledingham and M. S. S. Ferreira, Carbohydr. Res., 30, 196 (1973).
113. A. Bar-Eli and E. Katchalski, Nature, 188, 856 (1960).
114. A. Bar-Eli and E. Katchalski, J. Biol. Chem., 238, 1960 (1963).
115. J. J. Cebra, D. Givol, H. I. Silman, and E. Katchalski, J. Biol. Chem., 236, 1720 (1961).
116. I. H. Silman, M. Albu-Weissenberg, and E. Katchalski, Biopolymers, 4, 441 (1966).
117. H. H. Weetall, Nature, 233, 959 (1969).
118. H. H. Weetall, Science, 166, 615 (1969).
119. H. H. Weetall and L. S. Hersh, Biochim. Biophys. Acta, 185, 464 (1969).
120. H. H. Weetall and G. Baum, Biotechnol. Bioeng., 12, 399 (1970).
121. G. P. Royer and G. M. Green, Biochem. Biophys. Res. Commun., 44, 426 (1971).
122. M. J. Grove, G. W. Strandberg, and K. L. Smiley, Biotechnol. Bioeng., 13, 709 (1971).
123. R. D. Mason and H. H. Weetall, Biotechnol. Bioeng., 14, 637 (1972).
124. J. E. Dixon, F. E. Stolzenbach, J. A. Berenson, and N. O. Kaplan, Biochem. Biophys. Res. Commun., 52, 905 (1973).
125. H. H. Weetall and R. D. Mason, Biotechnol. Bioeng., 15, 455 (1973).
126. G. P. Royer and R. Uy, J. Biol. Chem., 248, 2627 (1973).
127. G. P. Royer and J. P. Andrews, J. Biol. Chem., 248, 1807 (1973).
128. C. Flemming, A. Gabert, and P. Roth, Acta Biol. Med. Ger., 30, 177 (1973).

129. E. Van Leemputten and M. Horisberger, Biotechnol. Bioeng., 16, 997 (1974).
130. F. B. Weakly and C. L. Mehltretter, Biotechnol. Bioeng., 15, 1189 (1973).
131. C. Flemming, A. Gabert, and P. Roth, Acta Biol. Med. Ger., 31, 365 (1973).
132. E. Brown and A. Racois, Tetrahedron Lett., 5077 (1972).
133. E. Brown and A. Racois, Tetrahedron, 30, 675 (1974).
134. E. Brown and A. Racois, Tetrahedron, 30, 683 (1974).
135. R. Epton, J. V. McLaren, and T. H. Thomas, Carbohydr. Res., 22, 301 (1972).
136. P. D. Weston and S. Avrameas, Biochem. Biophys. Res. Commun., 45, 1574 (1971).
137. T. Ternynck and S. Avrameas, FEBS Lett., 23, 24 (1972).
138. A. C. Johansson and K. Mosbach, Biochim. Biophys. Acta, 370, 348 (1974).
139. G. Baum and M. Lynn, Process Biochem., 10, 14 (1975).
140. P. J. Robinson, P. Dunnill, and M. D. Lilly, Biochim. Biophys. Acta, 242, 659 (1971).
141. H. H. Weetall, in Methods in Enzymology, Vol. 44, Immobilized Enzymes (K. Mosbach, Ed.), Academic Press, Inc., New York (1976), p. 134.
142. J. Carlsson, R. Axén, K. Brocklehurst, and E. M. Crook, Eur. J. Biochem., 44, 189 (1974).
143. S. A. Barker and R. Epton, Process Biochem., 5, 14 (1970).
144. R. Axén, H. Drevin, and J. Carlson, Acta Chem. Scand., B29, 471 (1975).
145. K. Brocklehurt, J. Carlsson, M. P. J. Kierstan, and E. M. Crook, Biochem. J., 133, 573 (1973).
146. G. Manecke and S. Singer, Makromol. Chem., 39, 13 (1960).
147. G. Manecke and G. Günzel, Makromol. Chem., 51, 199 (1962).
148. G. Manecke, S. Singer, and K. E. Gillert, Naturwissenschaften, 47, 63 (1960).
149. G. Manecke and H. J. Förster, Makromol. Chem., 91, 136 (1966).
150. G. Kay and E. M. Crook, Nature, 216, 514 (1967).
151. G. Kay, M. D. Lilly, A. K. Sharp, and R. J. H. Wilson, Nature, 217, 641 (1968).
152. D. Self, G. Kay, M. D. Lilly, and P. Dunnill, Biotechnol. Bioeng., 11, 337 (1969).
153. G. Kay and M. D. Lilly, Biochim. Biophys. Acta, 198, 276 (1970).
154. R. J. H. Wilson, G. Kay, and M. D. Lilly, Biochem. J., 108, 845 (1968).
155. R. J. H. Wilson, G. Kay, and M. D. Lilly, Biochem. J., 109, 137 (1968).
156. J. F. Kennedy, Adv. Carbohydr. Chem. Biochem., 29, 305 (1974).

157. R. O. Stasiw, H. D. Brown, and F. X. Hasselberger, Can. J. Biochem., 48, 1314 (1970).
158. R. O. Stasiw, A. B. Patel, and H. D. Brown, Biotechnol. Bioeng., 14, 629 (1972).
159. J. R. Wykes, P. Dunnill, and M. D. Lilly, Biochim. Biophys. Acta, 250, 522 (1971).
160. P. Monsan and G. Durand, FEBS Lett., 16, 39 (1971).
161. A. T. Jagendorf, A. Patchornik, and M. Sela, Biochim. Biophys. Acta, 78, 516 (1963).
162. S. Shaltiel, S. Mizrahi, Y. Stupp, and M. Sela, Eur. J. Biochem., 14, 509 (1970).
163. T. Sato, T. Mori, T. Tosa, and I. Chibata, Arch. Biochem. Biophys., 147, 788 (1971).
164. E. Brown, A. Racois, and H. Gueniffey, Tetrahedron Lett., 2139 (1970).
165. E. Brown, A. Racois, and H. Gueniffey, Bull. Soc. Chim. Fr., 4341 (1971).
166. E. Brown and A. Racois, Bull. Soc. Chim. Fr., 4357 (1971).
167. E. Brown and A. Racois, Bull. Soc. Chim. Fr., 4351 (1971).
168. J. K. Inman and H. M. Dintzis, Biochemistry, 8, 4074 (1969).
169. D. Zabriskie, D. F. Ollis, and M. M. Burger, Biotechnol. Bioeng., 15, 981 (1973).
170. W. E. Hornby and H. Filippusson, Biochim. Biophys. Acta, 343 (1970).
171. C. Horvath and B. A. Solomon, Biotechnol. Bioeng., 14, 885 (1972).
172. P. V. Sundaram and W. E. Hornby, FEBS Lett., 10, 325 (1970).
173. J. P. Allison, A. Davidson, A. Gutierrenz-Hartman, and G. B. Kitto, Biochem. Biophys. Res. Commun., 47, 66 (1972).
174. H. Filippusson, W. E. Hornby, and A. McDonald, FEBS Lett., 20, 291 (1972).
175. P. S. Bunting and K. J. Laidler, Biotechnol. Bioeng., 16, 119 (1974).
176. W. E. Hornby, D. J. Inman, and A. McDonald, FEBS Lett., 23, 114 (1972).
177. D. J. Inman and W. E. Hornby, Biochem. J., 137, 25 (1974).
178. L. Goldstein, A. Freeman, and M. Sokolovsky, Biochem. J., 143, 497 (1974).
179. J. Campbell, W. E. Hornby, and D. L. Morris, Biochim. Biophys. Acta, 384, 307 (1975).
180. D. L. Morris, J. Campbell, and W. E. Hornby, Biochem. J., 147, 593 (1975).
181. I. M. Kolthoff and C. Robinson, Recl. Trav. Chim. Pays-Bas, 45, 109 (1926).

10

The Synthesis of Enzymelike Polymers

Yoshiaki Inaki

Osaka University, Osaka, Japan

10.1 INTRODUCTION

Enzymes are proteins that composed of polypeptide chains. The polypeptide chains are folded irregularly, but precisely, to form a compact ellipsoid. Most of the polar amino acid residues are located on the surface in contact with the aqueous media and the hydrophobic residues are largely buried in the interior of the molecule. Catalysis takes place at a small region called the "active site." This is usually located in a hydrophobic hole in the enzyme surface. The active site of the enzyme contains the catalytic amino acid residues, as shown in Table 1.

The enzymelike polymer catalysis is an attractive problem to organic and polymer chemists. Studies on the enzymelike synthetic polymers have been conducted by Overberger et al. [1-39], Kunitake et al. [40-63], and others [63-97]. These enzyme model studies were concerned mostly with the hydrolysis of phenyl esters, because the molecular mechanism of the action of hydrolytic enzymes such as α-chymotrypsin was the first to be elucidated in detail (Figure 1).

Table 1. Catalytic Amino Acid Residues

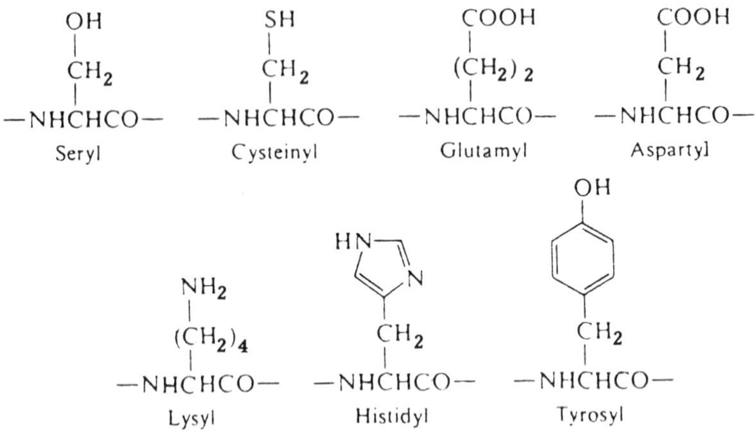

The polymeric nature of an enzyme molecule is essential for the maintenance of the three-dimensional structure and catalytic activity. Therefore, it may be expected that synthetic polymers which incorporate some enzyme features might display interesting catalytic properties. The approach for the preparation of enzymelike polymers is to introduce the type of catalytic groups found in enzymes, such as those listed in Table 1, into polymers. A large number of enzyme model polymers have been prepared by using these functional groups singly or in combinations. The present chapter aims at surveying the preparation of the enzymelike polymers.

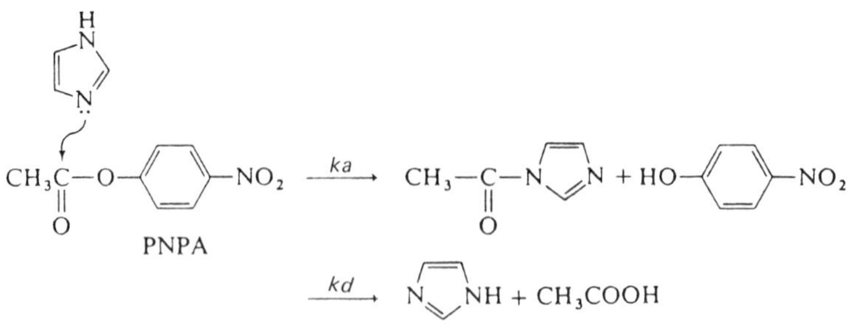

Figure 1

10.2 VINYL POLYMERS CONTAINING IMIDAZOLE

10.2.1 Poly[4(5)-vinylimidazole] (4)

The imidazole group of a histidyl residue is involved in the catalytic action of hydrolytic enzymes. This fact attracted much attention among organic and polymer chemists and imidazole and imidazole compounds became the first organic bases to be established as catalysts for ester hydrolysis. Since the first report on the catalytic behavior of imidazole containing vinyl polymers, a large number of investigations have been reported by Overberger and his group [1-39], Kunitake and his group [40-63], and others [63-97]. Overberger synthesized 4(5)-vinylimidazole for the first time and polymerized it by using a free-radical initiator [2,3].

4(5)-Vinylimidazole (3) was synthesized by the route shown in Scheme 1, starting with histidine [2,3,37]. Histidine is diazotized in concentrated hydrochloric acid, the resulting α-chloro acid (1) is dehydrochlorinated with trimethyamine in a steel bomb. The resultant urocanic acid (2) is decarboxylated in vacuo at melting-point temperature.

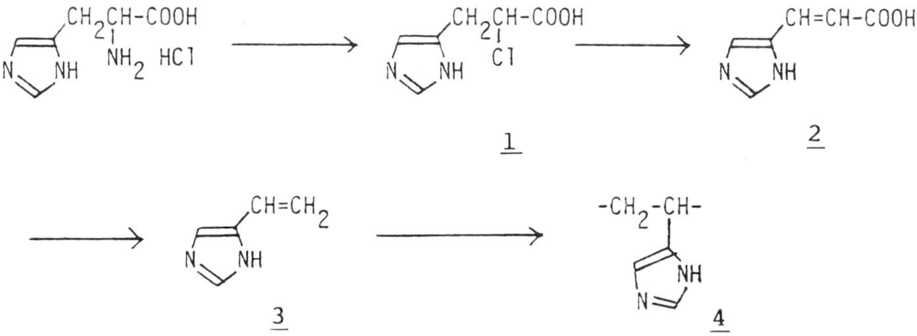

Scheme 1

Homopolymerization of 4(5)-vinylimidazole (3) proceeds readily with free-radical initiation in benzene solution to give a white powder.

2-Chloro-3(4-imidazoyl)-propionic acid (1) [37]

A dispersion of fine L-histidine monohydrochloride (150 g, 0.78 mol) in concentrated hydrochloric acid is cooled in an ice-water bath, followed by the dropwise addition of a solution of sodium nitrate (150 g) in water (300 mL) with stirring. The addition is controlled to keep the reaction temperature below 10°C. After the addition of the sodium nitrate solution is complete, the reaction mixture is stirred for an additional 3 h without adding more ice. The reaction mixture is filtered to remove sodium chloride, and the filtrate is evaporated in vacuo. Evaporation and filtration are repeated until a syrup free of sodium chloride is obtained. The syrup is then dissolved in water (500 mL) and stirred in an ice-water bath with 2 N ammonium hydroxide (500 mL). The solution is evaporated to dryness, leaving a precipitate of 2-chloro-3-(4-imidazoyl)propionic acid in 35% yield (40 to 50 g).

Urocanic acid (2) [37]

A solution of 2-chloro-3(4-imidazoyl)propionic acid (1) (90 g, 0.51 mol) in 25% trimethylamine (500 mL) is placed in a bomb, which is heated at 65°C for at least 2 days. The reaction mixture is refluxed until the odor of trimethylamine is no longer detected. The solution is cooled in a ice-water bath and crude urocanic acid crystallizes immediately as long needles. Recrystallization from water gives urocanic acid dihydrate. The dihydrate crystals are dehydrated by heating overnight at 100°C in a vacuum oven to yield 28 g (40%), mp 228 to 230°C (dec.).

4(5)-Vinylimidazole (3) [37]

A 10-mL distillation apparatus containing anhydrous urocanic acid (2) (2.5 g, 0.018 mol) is placed in vacuo (~0.1 torr) and an air bath preheated to 230°C is applied. Decarboxylation takes place immediately and crude 4(5)-vinylimidazole distills slowly as a viscous liquid. The product (60% yield) slowly crystallizes in the receiver and is purified by sublimation at 50 to 55°C under high vacuum (10^{-4} torr), mp 83.2 to 84.5°C, bp 118°C (0.3 mm).

Poly[4(5)-vinylimidazole] (4) [37]

4(5)-Vinylimidazole (3) (0.5 g, 5.3 mmol) is polymerized in benzene (10 mL) using azobisisobutyronitrile (0.9 mg, 0.1 mol %) as an initiator at 65°C for 24 h in a vacuum-sealed tube. The polymer is obtained as a precipitate from benzene in 80% yield (0.4 g) and is purified by reprecipitation from methanol in benzene.

10.2.2 Copolymers Containing Imidazole

The introduction of additional functional groups onto imidazole containing polymers produces two major results. One is the change in the microenvironment around the catalytic imidazole group, such as to give enhanced catalytic efficiency and/or enhanced substrate selectivity. The incorporation of hydrophobic or electrostatic groups augment the substrate-binding capacity of the polymer catalysts. Another result is the formation of cooperative-catalytic sites involving imidazole and other functional groups.

10.2.2.a Cooperative catalyst

4-Vinylimidazole-vinyl alcohol copolymer (6)

The combination of imidazole with a hydroxy group is a particularly attractive candidate for cooperative catalysis, since a cooperative interaction of these two functional groups is found in the serine esterase-catalyzed hydrolysis. A copolymer of 4(5)-vinylimidazole with vinyl alcohol was prepared (Scheme 2) and its catalytic properties examined [9]. It was found that 4(5)-vinylimidazole is a much more reactive monomer than vinyl acetate in free-radical polymerization; therefore, the polymers always contained larger amounts of imidazole.

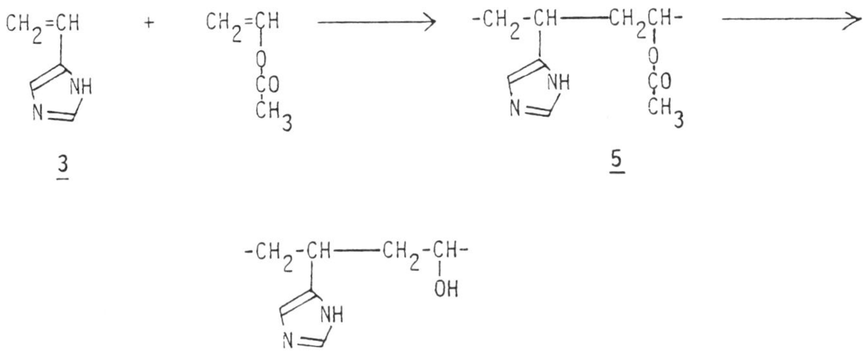

3 5 6

Scheme 2

The composition of the copolymers could only be varied within a limited range. The copolymers are water soluble and exhibit association at high concentrations as shown by gel formation.

The acetate groups are successfully hydrolyzed in dilute acid solution; base hydrolysis generally leads to degradation. Some copolymer hydrolysis occurs by mere dissolution in water. It is assumed that this ready water hydrolysis occurs by participation of neighboring imidazole groups on the polymer chain.

Copolymers of 4(5)-vinylimidazole and vinyl alcohol have solvolytic rates somewhat greater than those of the vinylimidazole homopolymer. It is difficult, however, to assume that the vinyl alcohol groups plays any significant role.

Poly[4(5)-vinylimidazole-co-vinyl acetate] (5) [9]. Vinyl acetate (4.00 g, 47 mmol), 4(5)-vinylimidazole (0.52 g, 5.5 mmol) and azobisisobutyronitrile (8 mg) in benzene (14 mL) are placed in a polymerization tube, degassed, and sealed under reduced pressure. The tube is heated in a refluxing hexane bath for 30 h. The precipitated polymer

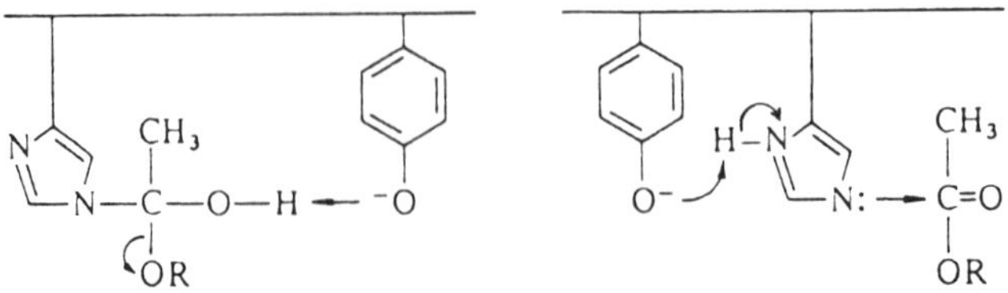

Figure 2

is filtered and washed with benzene to yield an off-white solid (0.62 g, 13.7% conversion). This polymer contains 0.83 mol of imidazole and 0.26 mol of acetate per 100 g of polymer.

Poly[4(5)-vinylimidazole-co-vinyl alcohol] (6) [9]. A solution of the copolymer (5) (0.30 g) in water (100 mL) is heated for 20 h at reflux. The solution is cooled, filtered, and lyophilized. The product is an off-white solid (0.23 g). The infrared spectrum indicates that no acetyl functions are present.

4(5)-Vinylimidazole-4-vinylphenol copolymer (7)

More interesting catalytic properties are obtained with the copolymer of 4(5)-vinylimidazole and 4-vinylphenol (7) [9,20]. It is a superior catalyst at high pH values toward neutral, anionic, and cationic substrates compared with any of its monomeric or polymeric analogs. For example, the solvolytic rate of cationic substrate with the copolymer is 63 times greater than that of imidazole. The rate enhancement can be induced by bifunctional attack involving imidazole and phenolate ion. The imidazole group may act as a nucleophile with the phenolate anion acting as a general base; alternatively, the phenolate anion can assist the nucleophilic attack of a neutral imidazole unit (Figure 2).

p-Vinylphenol was prepared by pyrolysis of *p*-hydroxycinnamic acid in a sublimator at 220°C (0.01 mm) in 58% yield [9,64]. The monomer undergoes free-radical copolymerization with 4(5)-vinylimidazole in solution. It also polymerizes in the solid state on standing at room temperature to give a polymer of DP ca. 40 (Scheme 3).

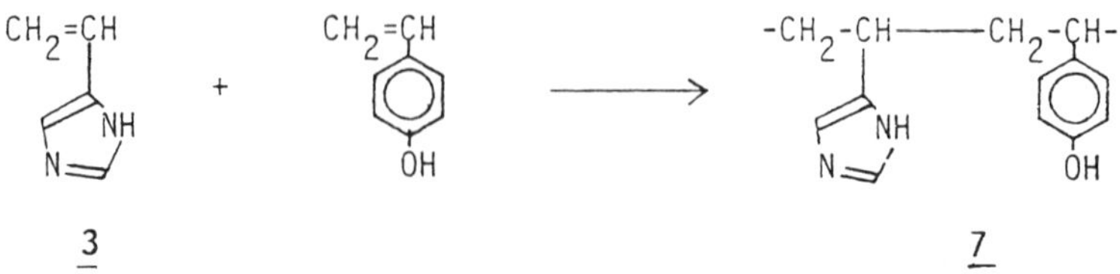

Scheme 3

Poly[4(5)-vinylimidazole-co-p-vinylphenol] (7) [9]. A solution of *p*-vinylphenol [64] (0.97 g, 7 mmol), 4(5)-vinylimidazole (3) (0.25 g, 2.7 mmol), and azobisisobutyronitrile (11 mg) in benzene (10 mL) in a polymerization tube is degassed, and sealed under reduced pressure. This tube is heated for 24 h in refluxing hexane, during which time the polymer precipitates. After filtration and washing with benzene, the off-white polymer is dissolved in methanol and precipitated into ethyl acetate twice. The polymer contains 0.30 mol of imidazole and 0.59 mol of phenol per 100 g of polymer.

4(5)-Vinylimidazole-vinyl mercaptan copolymer (11)

Papain, a proteolytic enzyme, is known to achieve catalysis by the cooperative action of histidyl and cysteinyl residues. As synthetic models for the proteolytic enzyme, ficin and papain, copolymers containing imidazole and mercaptan groups were investigated [30]. Three different types of thiol-protecting groups were screened to prepare the polymers containing free mercaptan functions: a base-sensitive group (vinyl thiolacetate), an acid-sensitive group (methoxymethyl vinyl sulfide), and a heat-sensitive group (*S*-vinyl-*O*-*t*-butyl thiocarbonate).

Vinyl thiolacetate (9) is prepared by pyrolysis of 2-acetoxyethyl thiolacetate (8) at 508 to 514°C in 35% yield [1]. Copolymerization of 4(5)-vinylimidazole (3) and vinyl thiolacetate (9) is carried out in bulk or in benzene solution with a free-radical initiator. Removal of the acetyl group occurs in water at neutral pH, presumably due to some intramolecular imidazole catalysis (Scheme 4).

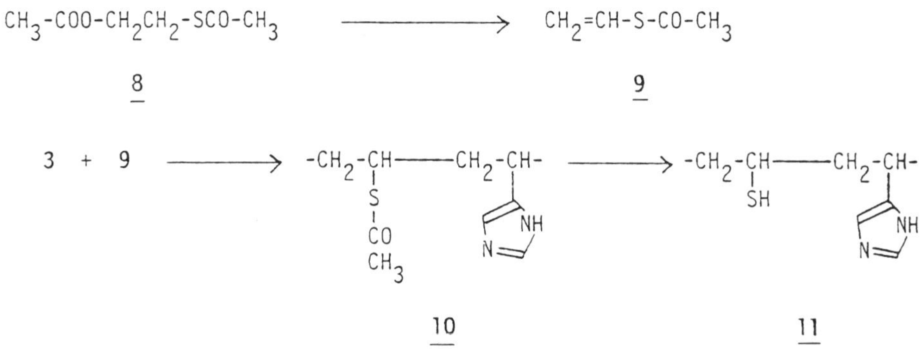

Scheme 4

From hydrolysis studies it was concluded that the catalytic activity of these polymers were due mostly to the additive hydrolytic effect of the imidazole groups and mercaptan groups and that there is only minimal kinetic evidence for some cooperative effects in the copolymers containing imidazole groups and mercaptan groups. However, cooperative effects seem to be operative in the hydrolysis of the anionic substrate in the presence of poly(vinyl mercaptan). The importance of the contribution of nonpolar binding on the hydrolysis rate is demonstrated by employing long-chain anionic esters [30].

Vinyl thiolacetate (9) [1]. To 2-mercaptoethanol (150 g, 1.92 mol) in pyridine (303.6 g, 3.84 mol) is added acetyl chloride (304.5 g, 3.84 mol), dropwise, over 6 h, while the temperature is kept below 40°C. After 15 h, 1500 mL of water is added cautiously to the mass, the mixture is stirred for 15 min, and the yellow oily layer is separated, washed with two 200-mL portions of water, dissolved in 250 mL of ether, and dried over anhydrous magnesium sulfate. From the ether solution is obtained 2-acetoxyethyl thiolacetate (8) (179 g, 58% yield), bp 92 to 93°C/9 mm.

2-Acetoxyethyl thiolacetate (179 g, 1.10 mol) is pyrolyzed at 508 to 514°C in a 1.4 × 30 cm tube packed with cut glass tubing through which nitrogen is passed at the rate of 60 g/17 min. The condensate is flash distilled at atmospheric pressure and the fraction bp 98 to 183°C is separated. The residue is again pyrolyzed by the same procedure. The combined fractions, impure vinyl thiolacetate, are dissolved in 150 mL of ether and shaken cautiously with a 10% sodium carbonate solution until neutral. This is followed by successive washing with 10% sodium chloride solution and two portions of water. After drying with magnesium sulfate the ether solution is fractionated at atmospheric pressure. A second distillation gives 21.3 g (21% based on starting material), bp 115°C/760 mm.

Poly[4(5)-vinylimidazole-co-vinyl thiolate] (10) [30]. A solution of 4(5)-vinylimidazole (3) (0.8 g, 9 mmol), vinyl thiolacetate (9) (0.4 g, 4 mmol), and azobisisobutyronitrile (20 mg) in benzene (20 mL) is degassed by the freeze-thaw technique in a polymerization tube. The tube is sealed in vacuo and heated at 75°C for 3 days. The precipitated polymer is filtered from the solvent and washed with dry benzene, yielding 44% (0.52 g) conversion of a light yellow solid. The polymer contains 0.77 mol fraction of imidazole and 0.23 mol fraction of thiolacetate.

Poly[4(5)-vinylimidazole-co-vinyl mercaptan] (11) [30]. *Alkaline hydrolysis.* To 7% aqueous sodium hydroxide (60 mL) is added copolymer (1:3.5 ratio of imidazole to thiolacetate). Refluxing under nitrogen for 15 min results in complete dissolution of the copolymer. After cooling, the solution is acidified by the addition of concentrated hydrochloric acid. A colorless precipitate, the hydrochloride salt of the hydrolyzed copolymer is formed and isolated by filtration. The precipitate is redissolved in methanol and reprecipitated into ethyl acetate. When the molar amount of imidazole in the copolymer is equivalent to or higher than that of the thiolester, refluxing in aqueous alkali under nitrogen always results in precipitation of a cross-linked material.

Water hydrolysis. Copolymer (1:0.3 ratio of imidazole to thiolacetate) is suspended in distilled-deionized water and refluxed for 1 h under nitrogen. The copolymer becomes soluble during this time. After cooling, the reaction mixture is freeze dried. The amount of free mercaptan in the copolymer is determined by the reaction of mercaptan groups with a known concentration of 1,1-diphenylpicrylhydrazyl (DPPH); 0.79 mol of imidazole, 0.21 mol of mercaptan per 100 g of copolymer.

10.2.2.b Anionic polymer catalyst

4(5)-Vinylimidazole-acrylic acid copolymer (12)

The pH profiles of the poly[4(5)-vinylimidazole] indicates that electrostatic attraction increases the catalytic efficiency when anionic phenyl esters are cleaved by partly protonated poly[4(5)-vinylimidazole]. The electrostatic attraction would be used more effectively by employing carboxylic acid groups which completely dissociate under most solvolytic conditions. Copolymers of 4(5)-vinylimidazole and acrylic acid have been prepared [5]. This polymer catalyst shows a substrate selectivity toward cationic substrate, because of the electrostatic attraction. Higher catalytic efficiency was observed for the copolymers containing 42 to 50 mol % imidazole. It was concluded from this result that the most catalytically active species is the carboxylate-imidazole-carboxylate triad [16,17,65] (Scheme 5).

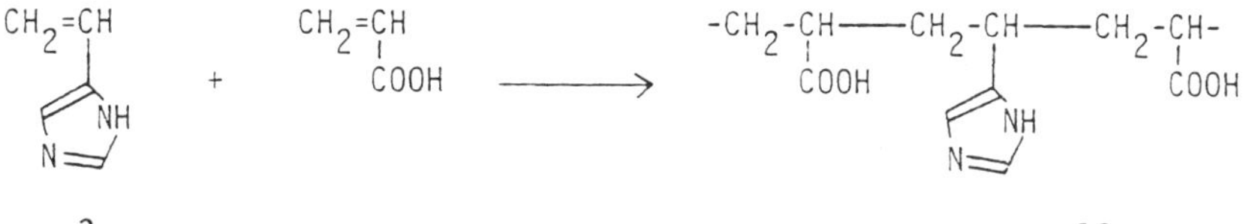

3 12

Scheme 5

Poly[4(5)-vinylimidazole-co-acrylic acid] (12) [16]. In polymerization tubes various molar ratios of 4(5)-vinylimidazole and acrylic acid are dissolved in methanol with azobisisobutyronitrile. After flushing with nitrogen, the tubes are sealed and heated to 70°C. The copolymers precipitate from solution during polymerizations. These suspensions are poured into acetone and the precipitated polymers are collected by filtration. The monomer reactivity ratios are determined to be r_1 = 0.06, r_2 = 0.11, where M_1 is acrylic acid and M_2 is 4(5)-vinylimidazole.

Copolymer of 4(5)-vinylimidazole and acrylic acid is also obtained by the copolymerization of 4(5)-vinylimidazole with methyl acrylate, followed by hydrolysis [2].

4(5)-Vinylimidazole-vinylsulfonic acid copolymer (15)

Studies of the esterolytic activities of 4(5)-vinylimidazole and vinylsulfonic acid (15), as compared with those of 4(5)-vinylimidazole and acrylic acid were undertaken to elucidate the involvement of pendant carboxylate groups in the latter copolymers. It was found that the copolymers that are rich in vinylsulfonic acid are catalytically inactive. These results support the cooperative interaction of pendant imidazole and carboxylate functions in the imidazole-acrylic acid copolymer catalyzed reactions [16,17].

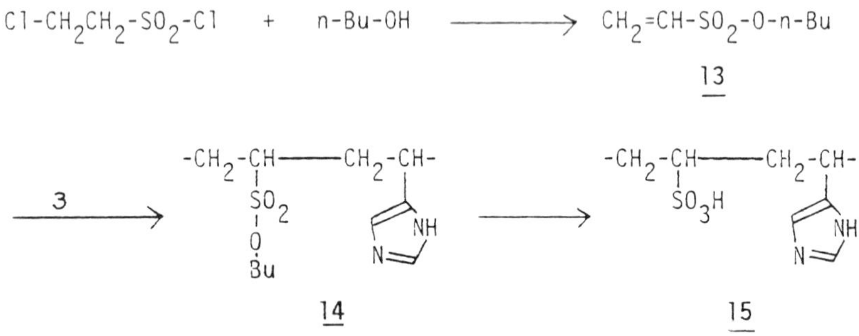

Scheme 6

n-Butyl vinylsulfonate (13) [66]. A solution of *n*-butanol (0.1 mol) and 2-chloroethanesulfonyl chloride (0.11 mol), in methylene dichloride (75 mL), is cooled to 0°C. Dry pyridine (0.22 mol) is added slowly with efficient stirring. The temperature is kept below 10°C during the addition period of 2 to 3 h. The reaction mixture is then allowed to come slowly to room temperature. Dilute hydrochloric acid (100 mL) is added and the lower layer, after washing with water, is drawn off and dried over sodium sulfate. The ester is recovered by evaporation of the solvent and vacuum distilled; yield 61%, bp 115 to 117°C (15 mm).

Poly[4(5)-vinylimidazole-co-n-butyl vinylsulfonate] (14) [17]. In polymerization tubes 4(5)-vinylimidazole (3), freshly distilled *n*-butyl vinylsulfonate (13) (total amount of these monomers, 1.0 g), and azobisisobutyronitrile (0.02 g) are dissolved in *n*-butyl alcohol (20 mL). After flushing with dry nitrogen, the tubes are sealed and then shaken in a bath at 60°C for 5 h. The polymers precipitate during polymerization. The dispersions are added to 500 mL of acetone and left overnight. The precipitated polymers are collected by filtration, washed with acetone, and dried in a vacuum oven overnight. A copolymer containing 52.5 mol % imidazole is obtained from monomers containing 53.0 mol % 4(5)-vinylimidazole in 59.5% conversion.

Poly[4(5)-vinylimidazole-co-vinylsulfonic acid] (15) [17]. The copolymers above are placed in a 100-mL round-bottomed flask with 50 mL of deionized water. The copolymers of compositions 38.6 and 52.6 mol % in imidazole gradually dissolved in water, but the other copolymer (83.9 mol % in imidazole) only swelled. After refluxing overnight, the first copolymer remained in solution, while the others form thin films on the walls of the flasks. After evaporating the solutions to a few milliliters, the copolymers are recovered by freeze drying. These copolymers are hygroscopic and contain some water even after drying at 100°C under vacuum.

4(5)-Vinylimidazole-γ-vinyl-γ-butyrolactone copolymer (17)

The active site of the α-chymotripsin consists of histidine, serine, and aspartic acid. These amino acids contain imidazole, hydroxy, and carboxylic acid moieties, respectively. Those three moieties act as a solvolytic catalyst with cooperation. To obtain a polymer containing these three functional groups, the copolymerization of 4(5)-vinylimidazole and γ-vinyl-γ-butyrolactone (16) was carried out [6,7]. The copolymer obtained was hydrolyzed to give a polymer with 3-carboxy-1-hydroxy-propyl groups (17). The terfunctional polymer obtained is more than twice as active in the solvolysis of *p*-nitrophenyl ester than that of a random copolymer containing imidazole, carboxylic acid, and hydroxy groups (18) (Scheme 7).

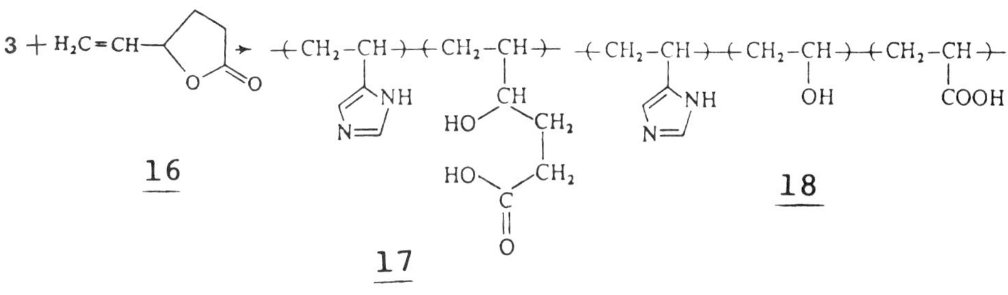

Scheme 7

10.2.2.c Optically active catalyst

4(5)-Vinylimidazole-2,5-(S)-dimethyl-1-heptene-3-one copolymer (20)

Most of the naturally occurring enzymes exhibit a characteristic stereospecificity as well as structural specificity in their catalytic actions toward their various substrates. To investigate the stereospecificity of enzyme-catalyzed reactions, an optically active copolymer of 4(5)-vinylimidazole and 2,5-(*S*)-dimethyl-1-heptene-3-one was synthesized, and its effects on the solvolytic rate of the *p*-nitrophenyl ester of optically active carboxylic acid investigated [11].

The attempted free-radical homopolymerization of the optically active monomer (19) failed. However, it was found that the monomer (19) copolymerized readily with 4(5)-vinylimidazole under free-radical polymerization conditions. A copolymerization monomer feed of ratio of the monomer (19) to 4(5)-vinyl-imidazole of 1:3.5 yielded a copolymer with a composition of 1:0.96 (20). The copolymer was soluble in alcohol and insoluble in water and in ether (Scheme 8).

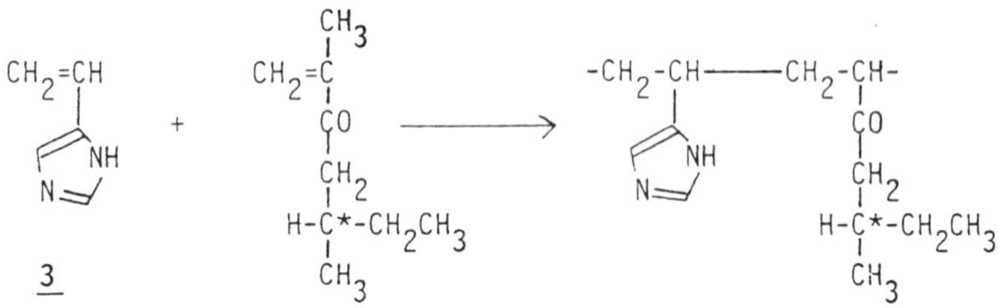

3 19 20

Scheme 8

For the optically active copolymer-catalyzed solvolysis of the optically active esters, performed at pH 6 to 8, no significant differences between the solvolytic rates of (R) and (S) isomers of substrates was observed.

10.2.3 Polymers Containing Imidazole Derivatives

10.2.3.a Poly[4(5)-vinylimidazole] derivatives

Poly[N-acyl-4(5)-vinylimidazole] (22)

Poly[4(5)-vinylimidazole] (4) catalyzes the hydrolysis of *p*-nitrophenyl acetate several times more efficiently than imidazole. However, the catalytic behavior of the polymer is completely different from that of imidazole when a long-chain substrate is used [14,18,19,24]. For the catalytic hydrolysis of 3-nitro-4-dodecanoyloxybenzoate by poly[4(5)-vinylimidazole], the catalytic efficiency relative to that of imidazole increases from 4 in 60% aqueous ethanol to 1600 in 20% aqueous ethanol. A viscosity study showed that the unusually large rate enhancement for the long-chain substrate is related to the shrink-

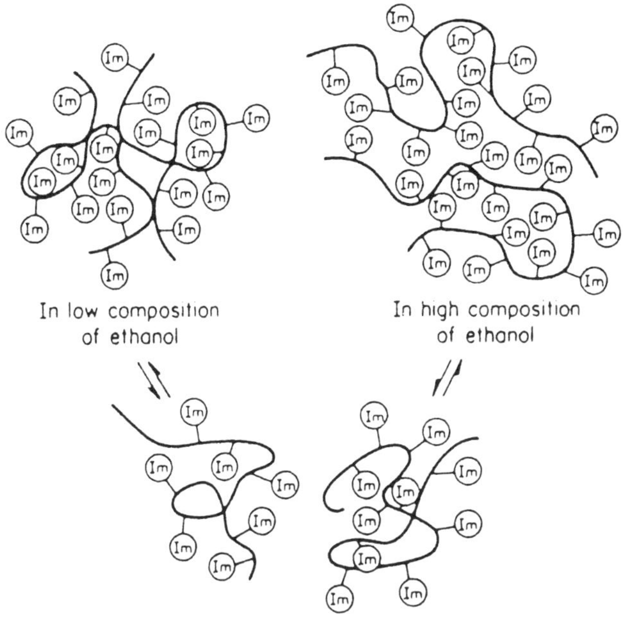

Figure 3

age of the polymer chain. The shrinkage of the polymer chain at low and high ethanol compositions is attributed to increased hydrophobic and hydrogen-bonding interaction, respectively (Figure 3).

This interpretation was confirmed by studying the acylation and deacylation processes separately using acylated polymer catalyst. The acylated poly[4(5)-vinylimidazole] (22) is prepared by free-radical polymerization of the corresponding monomer (21), which is obtained by the reaction of 4(5)-vinylimidazole (3) with acyl chloride [25] (Scheme 9).

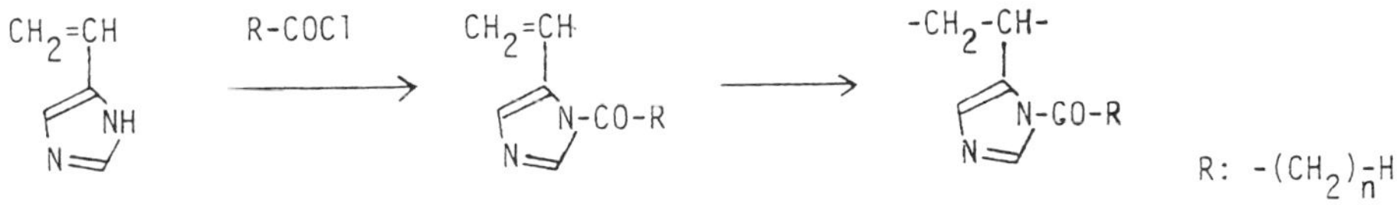

3 21 22

Scheme 9

N-Heptanoyl-4-vinylimidazole (21) [25]. To 4(5)-vinylimidazole (3) (0.94 g, 0.01 mol) is added triethylamine (1.10 g, 0.01 mol). To this well-stirred solution is slowly added heptanoyl chloride (1.48 g, 0.01 mol) and stirring continued for 30 min at room temperature. The mixture is filtered and the filtrate evaporated under reduced pressure to leave a white solid in 95% yield. It is purified by recrystallization from hexane or by sublimation; mp 69°C.

Poly(N-heptanoyl-4-vinylimidazole) (22) [25]. *N*-Heptanoyl-4-vinylimidazole (21) (1.34 g, 6.5 mmol) is polymerized in bulk at 85 to 90°C for 200 h with benzoyl peroxide (0.25 mL of 2.6×10^{-2} M). After the polymerization, the unreacted monomer is recovered by sublimation. The polymer is purified by reprecipitation from chloroform into hexane, 18% yield.

Poly(1-alkyl-4-vinylimidazole) and poly(1-alkyl-5-vinylimidazole) (23 and 24)

As mentioned, poly [4(5)-vinylimidazole] is a more efficient catalyst than monomeric imidazole toward the hydrolysis of the neutral ester, *p*-nitrophenyl acetate. This rate enhancement is attributed to cooperative interactions between pendant imidazole residues.

Poly(1-alkyl-4-vinylimidazole) (23) and poly(1-alkyl-5-vinylimidazole) (24) are similar to poly[4(5)-vinylimidazole] in many respects; however, there are some notable differences [33, 34, 35, 38, 39]. With respect to cooperativity, the absence of the amine hydrogen precludes general base catalysis of attack by an N-alkylated residue. In addition, nucleophillic attack by an N-alkylated residue must occur with the intermediate formation of a positively charged acyl intermediate. This fact has important kinetic ramifications, particularly with respect to the hydrolysis of long-chain nitrophenyl ester.

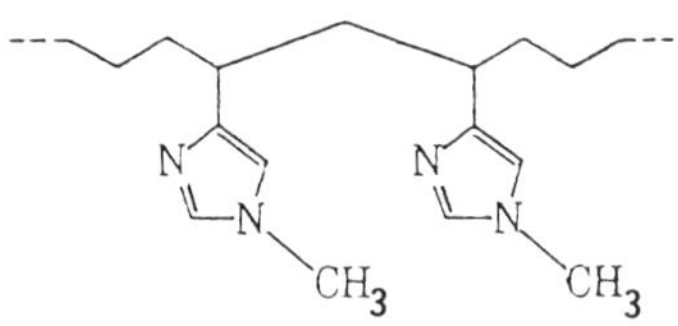

Poly(1-Me-**4**-VIm).

23

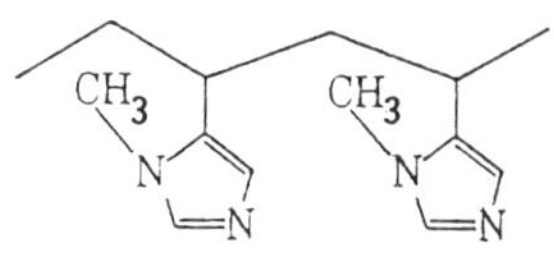

Poly(1-Me-**5**-VIm).

24

1-Methyl-4-vinylimidazole (25) and 1-methyl-5-vinylimidazole (26) are synthesized simply by alkylation of 4(5)-vinylimidazole. The mixed isomers are thus obtained in 86% yield, the isomer ratio being 60:40 as determined by NMR spectroscopy [33] (Scheme 10). Separation of the isomers is achieved by GLC on a 25% QF-1 on Chromosorb W column, or by preparative thin-layer chromatography.

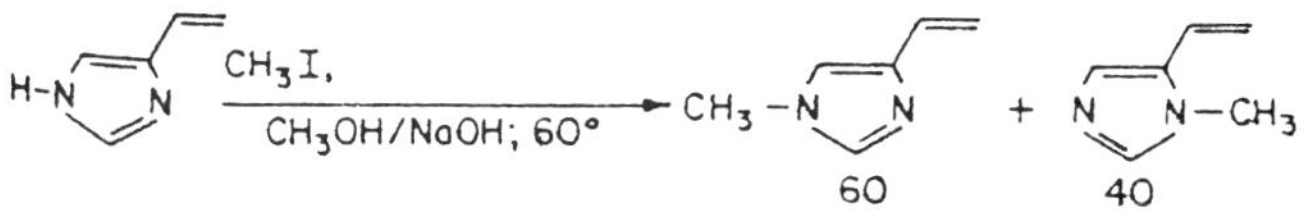

25 26

Scheme 10

Regiospecificity is improved when sodium hydride is used in the preparation of 1-methyl-4-vinylimidazole (25) [38] (Scheme 11). The reaction affords an 81% yield of alkylated imidazoles, 90% of which is 1-methyl-4-vinylimidazole. For the preparation of 1-methyl-5-vinylimidazole (26), methyl iodide quaternization of 1-trimethylsilyl-4-vinylimidazole is used (Scheme 11). The product from the reaction is pure 1-methyl-5-vinylimidazole when the reaction is carried out under a dry nitrogen atmosphere (Scheme 11). These monomers are homopolymerized or copolymerized with 4(5)-vinylimidazole (3) by a free-radical initiator [38].

25 90 : 10 26

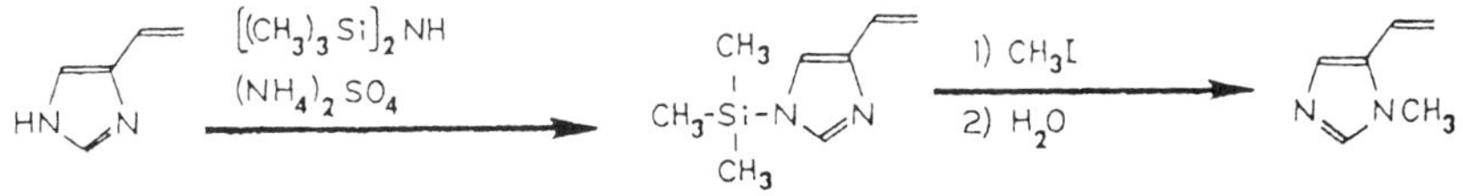

26

27

Scheme 11

From the study of the ester hydrolysis, it should be recognized that there is indeed no possibility for general-base cooperative interactions of the type postulated in catalysis by poly[4(5)-vinylimidazole]. However, hydrophobic interactions play an important role in the hydrolysis of negatively charged long alkyl chain ester substrate in the presence of poly(1-methyl-4-vinylimidazole) or poly(1-methyl-5-vinylimidazole) [39].

1-Methyl-4-vinylimidazole (25) [38]. To a suspension of sodium hydride (57% in oil, 4.4 g, 0.105 mol) washed with hexane in dimethylformamide (100 mL), 4(5)-vinylimidazole (3) (9.4 g, 0.1 mol) is added portionwise under nitrogen at room temperature. The mixture is stirred for 2 h at room temperature. Methyl iodide (14.2 g, 0.1 mol) is added slowly with cooling in an ice bath during a 2 h period. The reaction mixture is then warmed to room temperature, stirred an additional 4 h and quenched with water. The reaction mixture is saturated with potassium carbonate and extracted with chloroform. The solvent is removed under vacuum to give an oil which is passed through a Florisil column with chloroform to remove unreacted 4(5)-vinylimidazole. The column eluent is concentrated and distilled under vacuum to give crude product in 81% (8.75 g) yield, which contains 90% 1-methyl-4-vinylimidazole. The mixture is separated by column chromatography on active, basic aluminum oxide and acrylonitrile as the eluting solvent. Pure 1-methyl-4-vinylimidazole is obtained in 64.5% (7.0 g) yield; bp 64°C (0.20 mm).

1-Methyl-5-vinylimidazole (26) [38]. 4(5)-Vinylimidazole (3) (9.4 g, 0.1 mol) and hexamethyldisilazane (16 g, 0.1 mol) are heated to reflux for 15 h in benzene (20 mL) with a trace of ammonium sulfate as a catalyst. The reaction mixture is distilled under vacuum to give a colorless viscous liquid of 1-trimethylsilyl-4-vinylimidazole (27) in 98% (16.3 g) yield; bp 54°C (0.2 mm).

A pressure bottle is flushed with nitrogen and is charged with 1-trimethylsilyl-4-vinylimidazole (27) (16.6 g, 0.1 mol) and methyl iodide (21.3 g, 0.15 mol). The reaction mixture is maintained at 20 to 25°C for 1 h. The bottle is then immersed into an oil bath at 50°C for 15 h, cooled to room temperature, and quenched with water (50 mL). The reaction mixture is saturated with potassium carbonate and extracted with chloroform. The chloroform is evaporated and the resulting oil is distilled under vacuum to give pure 1-methyl-5-vinylimidazole (26) in 76% (8.2 g) yield; bp 60°C (0.2 mm).

Homopolymerization and copolymerization [38]. Polymerizations are carried out in sealed ampuls under degassed conditions in bulk at 65°C with azobisisobutyronitrile as the initiator. The polymers are precipitated from methanol with acetone. Monomer reactivity ratios of M_1, 4(5)-vinylimidazole, and M_2, 1-methyl-4-vinylimidazole, are $r_1 = 1.04$ and $r_2 = 0.94$, respectively.

Poly(N-vinylimidazole) (29)

The increased catalytic efficiency of poly[4(5)-vinylimidazole] that is observed at high pH values may be explained on the basis of microenvironmental effect. In the case of poly(*N*-vinylimidazole) (29), which lacks an association capacity, an enhanced catalytic activity at high pH media [7,68,69] is not observed. Copolymers of *N*-vinylimidazole and *N*-vinylpyrrolidone (30) were prepared in an attempt to combine the catalytic function of the imidazole group and the hydrophobic binding function of the pyrrolidone unit [42,43,70].

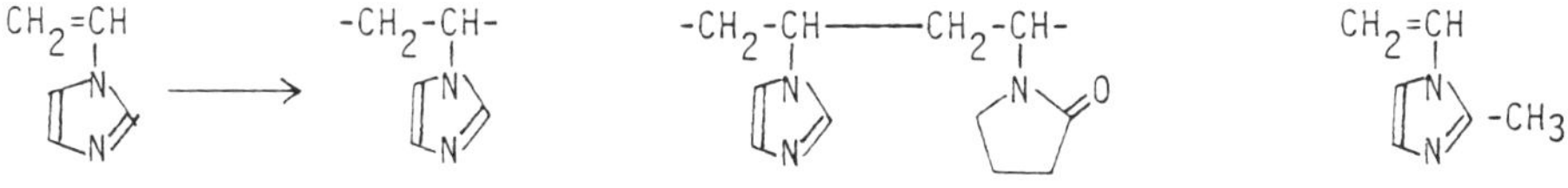

28 29 30 31

Quaternized poly(*N*-vinylimidazole) (33) was used [49] as the positively charged polymer catalyst. Copolymers of 4(5)-vinylimidazole and quaternized *N*-vinylimidazole (34) were used as polymer catalyst for the solvolysis of neutral and anionic phenyl ester. The reaction follows second-order kinetics and the catalytic efficiency toward anionic substrates was generally inferior to that of poly[4(5)-vinylimidazole].

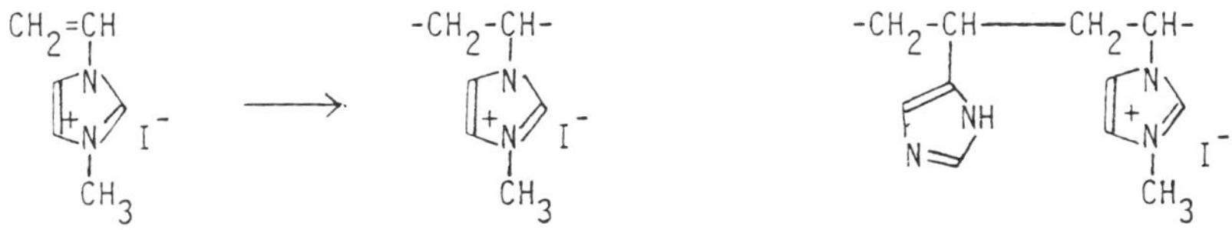

32 33 34

N-Vinylimidazole (28) and 2-alkyl-*N*-vinylimidazole (31) are available commercially. Polymerization of these monomers was carried out by free-radical initiation. Quaternized *N*-vinylimidazoles were prepared by the reaction of *N*-vinylimidazole with corresponding alkyl iodide [29,71,72]. Polymerizations were performed in aqueous solution or in methanol solution using a free-radical initiator. The short-side-chain polyions (methyl and *n*-propyl) are water soluble, whereas the more hydrophobic intermediate-side-chain polyions (*n*-hexyl and *n*-heptyl) are water insoluble. The long-side-chain polyions (*n*-dodecyl and *n*-hexadecyl), which are highly hydrophobic, appear to form polysoaps and remained in aqueous solution.

3-Methyl-1-vinylimidazolium iodide (32) [72]. To freshly distilled 1-vinylimidazole (9.41 g, 0.10 mol) is added methyl iodide (85.20 g, 0.60 mol) in dry ethyl acetate (100 mL) at -10°C. The reaction mixture is stirred for 1 h and then brought to room temperature overnight, during which time off-white crystals form. The mixture is cooled, filtered, and the monomeric salt washed three times with anhydrous diethyl ether, followed by recrystallization from *n*-propanol to give off-white needles (56% yield); mp 77.5 to 78.0°C.

Poly(3-methyl-1-vinylimidazolium iodide) (33) [72]. To a polymerization tube is added aqueous solution of 3-methyl-1-vinylimidazolium iodide (10 mL of 0.45 M) which contains 1 mol % of 4,4'-azobiscyanovaleric acid. After removing oxygen and sealing, polymerization is carried out at 60°C for 20 h. The polymeric salt solution is exhaustively dialyzed and the product recovered by lyophilization.

10.2.4 Polymers Containing Benzimidazole

Poly[5(6)-vinylbenzimidazole] (37)

Poly[5(6)-vinylbenzimidazole] is also an active catalyst for esterolic hydrolysis [6]. Poly[5(6)-vinylbenzimidazole] (37) was obtained by the dehydrochlorination of 5(6)-(β-chloroethyl)benzimidazole monomer in poor yield [4]. Therefore, another procedure for the synthesis of 5(6)-vinylbenzimidazole was developed. The six-step sequence terminates with the decarboxylation of β-[5(6)-benzimidazole]acrylic acid, which gives reproducible yields of 52% [31] (Scheme 12).

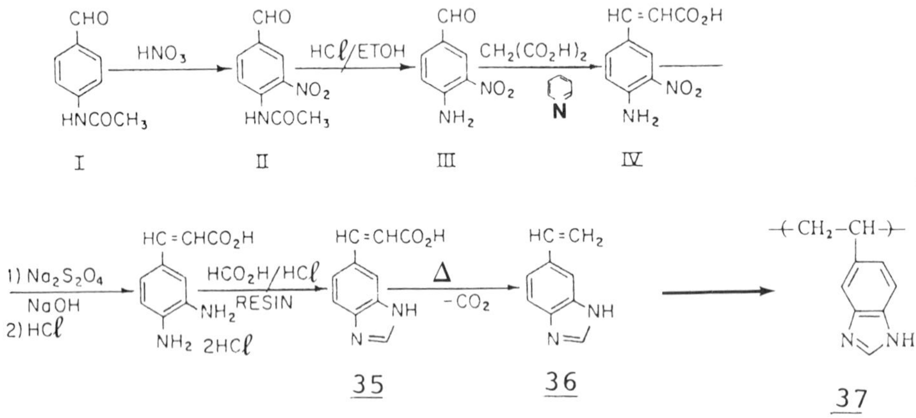

Scheme 12

The monomer was subjected to a free-radical solution polymerization and to bulk thermal polymerization. Poly[5(6)-vinylbenzimidazole], prepared by solution-free-radical polymerization, is a linear addition polymer with pendant benzimidazole groups, whereas the thermally prepared polymer has a different structure. The polymer (37) is insoluble in pure alcohol and pure water, but soluble in the alcohol-water solvent system.

Copolymers of 5(6)-vinylbenzimidazole with acrylic acid were prepared at 75°C in absolute methanol using azobisisobutyronitrile as an initiator. The copolymers formed are insoluble in methanol and precipitated from the solvent. Solutions containing 54.7 mol % 5(6)-

vinylbenzimidazole gave copolymers with 51.4 mol % benzimidazole units in 49.9% conversion [32].

β-[5(6)-Benzimidazole]acrylic acid (35) [31]. Dihydrochloride salt of 3,4-diaminocinnamic acid (5 g, 0.02 mol) which is prepared from 3-nitro-4-aminocinnamic acid by the procedure of Manecke and Rotter [73] with modification, is suspended in water (150 mL) with formic acid (88%, 25 mL). The suspension is placed in a round-bottomed flask together with Dowex-50W-X8 resin (25 g) and then mechanically shaken for 3 days at room temperature. The reaction mixture is filtered and the resin is washed with water. The resin is then extracted with 3 N sodium hydroxide (150 mL). The extract is neutralized with hydrochloric acid and allowed to attain room temperature. The solution is adjusted to a pH of 6.5 with hydrochloric acid where precipitation occurs. Recrystallization of the precipitate from water yields golden brown crystals; 3.1 g (82.3% yield), mp 273 to 274°C.

5(6)-Vinylbenzimidazole (36) [31]. A small sublimator is charged with β-[5(6)-benzimidazole]acrylic acid (35) (1 g, 5.3 mmol) and the acid is covered with a small pad of glass wool. The sublimator is evacuated and heated to 290°C. The solid begins to decarboxylate and a viscous liquid is collected on the probe along with some sublimed acid. The material on the probe is extracted with diethyl ether. The ether is removed and the viscous oil which remains is left to solidify 0.398 g, 52% yield; mp 110 to 112°C.

10.2.5 Polyacrylamide and Polymethacrylamide Derivatives Containing Imidazole Groups

Polyacrylamide and polymethacrylamide derivatives containing imidazole, benzimidazole, histamine, and histidine were prepared and the catalytic behavior of these polymers were studied. Acrylamide and methacrylamide monomers were prepared by the reaction of acryloyl or methacryloyl chloride with the corresponding amines.

Poly[N-(p-(4(5)-imidazolyl)benzyl)acrylamide] and poly[N-(5-benzimidazoyl)acrylamide]

N-[*p*-(4(5)-Imidazolyl)benzyl]acrylamide (38) (Scheme 13) and *N*-(5-benzimidazolyl)acrylamide (39) were prepared from the corresponding amines with acryloyl chloride [41,47]. These monomers were copolymerized with acrylamide by free-radical initiator in methanol [48]. Terpolymer of these monomers with acrylamide and *N*-(*p*-hydroxyphenyl)acrylamide were also prepared. These copolymers containing small amounts of imidazole group (1 to 3 unit mol %) are soluble in water, whereas the copolymers containing more imidazole units are insoluble in water.

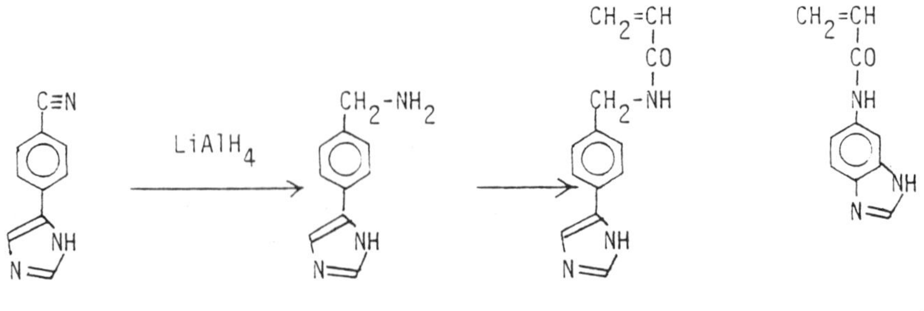

40 38 39

Scheme 13

N-[p-(4(5)-Imidazolyl)benzyl]acrylamide (38) [47]. 4(5)-(*p*-Cyanophenyl)imidazole hydrochloride (40) (5.1 g, 0.03 mol) is reduced by $LiAlH_4$ (4.5 g, 0.15 mol) in tetrahydrofuran (THF) (150 mL) for 2 h under reflux. The reaction is quenched with water (50 mL) and is poured into 10% aqueous NaOH (100 mL). The pale yellow organic layer is separated, and the solvent evaporated in vacuo, leaving pale yellow needles, *p*-[4(5)-imidazolyl]benzylamine, in 78% (4.1 g) yield.

Since the solid amine product is unstable, the organic layer is employed as such for the next step. The THF solution is cooled in an ice bath and added dropwise with acryloyl chloride (3 g, 0.036 mol) and 4 N aqueous NaOH (8 mL) simultaneously. The reaction mixture is stirred during the addition (20 min) and for subsequent 40 min. Then the reaction mixture is adjusted to pH 4 to 5, concentrated, and kept in a freezer. Fine flakes crystallized from water-ethanol (3:1) in 45% yield, mp 187 to 190°C.

N-(5-Benzimidazolyl)acrylamide (39) [46]. 5-Aminobenzimidazole (2 g, 0.013 mol) is dissolved in a mixture of concentrated hydrochloric acid (2 mL) and water (3 mL), and then added to 30% aqueous alkaline (2.5 mL) in dioxane (50 mL). The solution is cooled in an ice bath and 4 N NaOH (4.7 mL) and acryloyl chloride (1.6 g, 0.018 mol). The reaction mixture is stirred for 1 h at pH 8 to 9 at 0 to 5°C and then neutralized to pH 7. Sodium chloride is removed and solvent is evaporated. The residual solid is dissolved in methanol with added concentrated hydrochloric acid and the solvent evaporated. The residue is recrystallized from water-acetic acid (1:1) in 62 to 72% yield; mp 154 to 156°C.

Copolymer with acrylamide [47]. Copolymerization with acrylamide is conducted in methanol at 70° with azobisisobutyronitrile in sealed ampuls under a nitrogen atmosphere. Monomer mixture containing 9% imidazole units gives copolymer containing 8.9 mol % imidazole units in 20% conversion after 80 min).

Poly(N-acrylhistamine)

The acrylamide derivative of histamine (41) was prepared and copolymerized with acrylamide (42). This copolymer does not catalyze the solvolysis of the polymer substrate (43) with *p*-nitrophenyl ester units in water, but is catalytically active in 40% aqueous methanol. It appears that conformations corresponding to extensive contacts between chain segments are significantly favored in poor solvent media such as methanol.

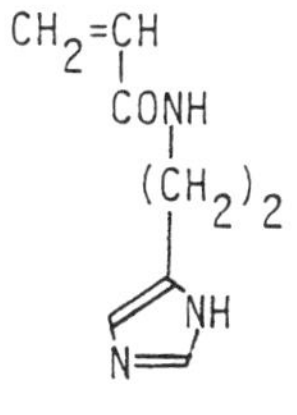

41

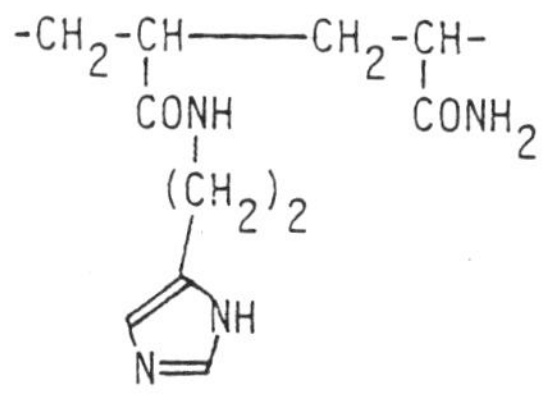

42

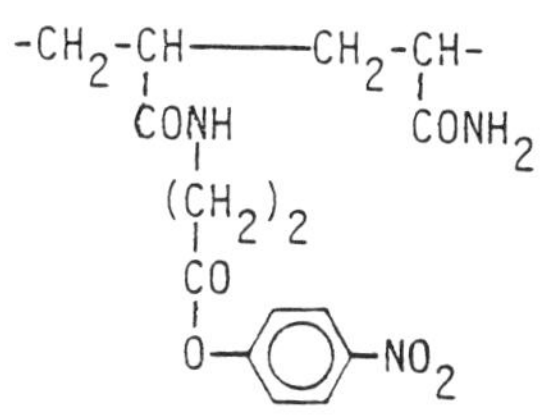

43

N-Acrylhistamine (41) [74]. Histamine hydrochloride (8.4 g) is dissolved in 4 N NaOH (15 mL), 10% sodium carbonate (20 mL), and dioxane (28 mL). To the solution cooled in an ice-salt mixture, acryloyl chloride (5.3 g) and 4 N sodium hydroxide (30 mL) are added dropwise with stirring. The solution is neutralized with HCl to pH 5 and is concentrated. The residue is dissolved in ethanol, filtered, and concentrated again. The material is dissolved in methanol (60 mL) and treated for 1 h with 1 N NaOH (6 mL) at room temperature to hydrolyze the acylimidazole group. The solution is then neutralized to pH 7, concentrated under vacuum, taken up in ethanol, and chromatographed on activated alumina with ethanol as eluent. The first fraction is collected, evaporated, and recrystallized from ethanol-chloroform; mp 142 to 144°C.

Copolymerization with acrylamide is carried out in degassed methanol solution at 50°C. Comonomers have similar reactivity.

Poly(N-acryloyl-L-histidine) and poly(N-methacryloyl-L-histidine)

The acrylamide derivative of L-histidine (44) was prepared and copolymerized with *N*-vinylpyrrolidone using a free-radical initiator (Scheme 14). The resultant copolymer (45) was tested as a catalyst for the hydrolysis of *p*-nitrophenyl esters of carboxylic acids in 20% aqueous dioxane [75].

Poly(*N*-methacryloyl-L-histidine) was also prepared by the free-radical polymerization of the corresponding monomer (46). This polymer was used as a catalyst for the hydrolysis of *p*-nitrophenyl esters of various L-amino acids. However, these polymeric catalysts failed to exhibit specificity toward (R) and (S) substrates [76].

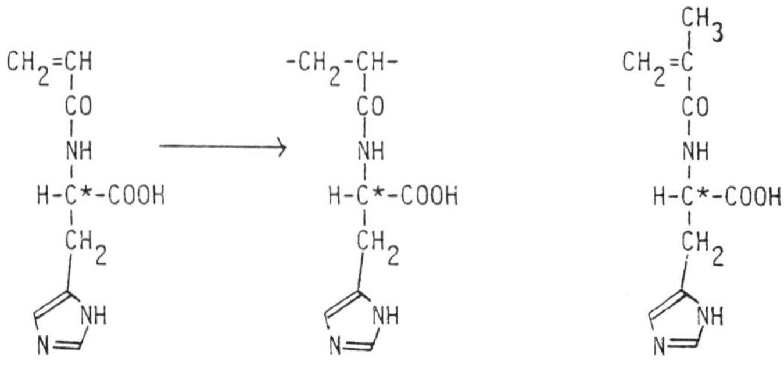

44 45 46

Scheme 14

N-Acryloyl-L-histidine (44) [75]. To an ice-cooled aqueous solution containing L-histidine and 2.2 equivalents of NaOH, a dioxane solution containing 1.2 equivalents of acryloyl chloride is added with vigorous agitation. The solution is extracted with ether and neutralized or weakly acidified with hydrochloric acid. After concentrating and filtering, the filtrate is subjected to a strongly acidic ion-exchange resin (Amberlite IR-120) and subsequently treated with ammonia in water. The solution is evaporated to dryness and the residue is recrystallized from water-methanol-acetone.

Poly(N-acryloyl-L-histidine) (45) [75]. Polymerization of the monomer (44) is carried out in aqueous solution containing 0.3% H_2O_2 and 0.3% NH_3 at 50°C. The precipitated polymer is purified by reprecipitation from water in acetone.

N-Methacryloyl-L-histidine (46) [76]. To a solution of L-histidine in 2 N NaOH, methacryloyl chloride in dichloromethane is added at 0 to 30°C with vigorous stirring. After the reaction the solution is adjusted to pH 2 and methacrylic acid is extracted with diethyl ether. The aqueous solution is adjusted to pH 5.1 and the water is evaporated. The residue is extracted with ethanol and poured into acetone to precipitate the product. The precipitate is reprecipitated from ethanol in ether; 39% yield. Polymerization of the monomer is carried out in ethanol at 65°C using azobisisobutyronitrile.

10.3 POLYMERS CONTAINING NUCLEOPHILE OTHER THAN IMIDAZOLE

10.3.1 Polyvinylpyridine Derivatives

The pyridine group shows nucleophilic catalysis activity similar to that of imidazole. The use of partly protonated polyvinylpyridine as a catalyst for the solvolysis of anionic phenyl ester led to rate enhancements that are due to the coulombic attraction [77,78]. When polyvinylpyridine is alkylated by chloromethylimidazole, the resulting polymer shows solvolytic reactivity caused by the pendant imidazole group (47).

Polyvinylpyridine is obtained by free-radical polymerization of commercially available vinylpyridine monomer. The polymer is alkylated with various alkylchlorides in nitromethane-methanol at 60 to 70°C for 30 to 100 h (48) [79].

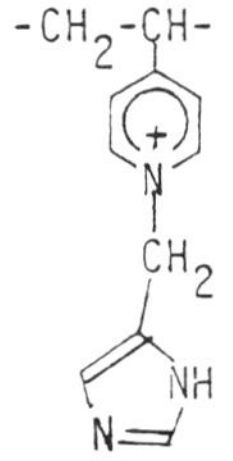

47

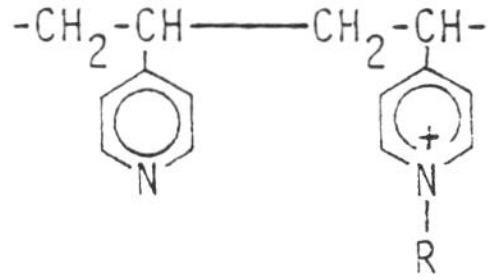

48

Poly(3-vinyl-1,2,4-triazole) (49)

The catalytic effect of some triazoles and poly(3-vinyl-1,2,4-triazole) was investigated in the solvolysis of neutral, negatively, and positively charged phenyl esters [15]. For all systems, the catalytic effect exhibited a linear dependence on the concentration of the respective anionic forms. The polymer exhibited a lower reactivity compared to a small molecule analog (49)

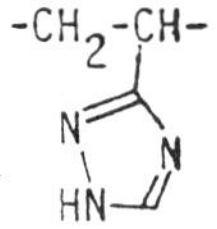

49

Preparation of poly(3-vinyl-1,2,4-triazole) (49) [15]. 3-(β-Acetoxyethyl)-1,2,4-triazole (1.0 g), which is prepared from 3-(β-hydroxyethyl)-1,2,4-triazole with acetic anhydride in ethyl acetate under reflux, is pyrolyzed in a distilling apparatus under reduced pressure at 180 to 200°C. A mixture of monomer and starting material begins to distill at 165°C. The monomer starts to polymerize in the collection flask. The mixture is extracted with ethanol and the insoluble material is precipitated from water to acetone. The polymer is dissolved in water and freeze dried, 30% yield.

10.3.2 Polymers Containing Hydroxamate Groups

Poly(N-phenylacrylohydroxamate) (53)

The hydroxamate anion is known to be highly nucleophilic toward phenyl esters which is attributed to the α effect. Attempts to use this functional group as a catalyst for the hydrolysis of phenyl esters was first carried out by Bender in small-molecule systems [80,81]. Water-soluble copolymers containing the hydroxamate group were prepared and used for the hydrolysis of phenyl esters [50]. In the hydrolysis of the esters by the hydroxamate containing copolymer (53), acylation is fast and deacylation is slow. Therefore, the acyl intermediate accumulates under ordinary reaction conditions. This behavior is in contrast with that of imidazole; the deacylation process is efficient and acylation is usually rate limiting.

Water-soluble copolymers containing protected hydroxamate groups (52) were prepared by free-radical copolymerization of acetyl-*N*-phenylacrylohydroxamate (51) (Scheme 15). The hydroxamic acid group is unmasked by treating it with hydroxylamine to give polymer 53.

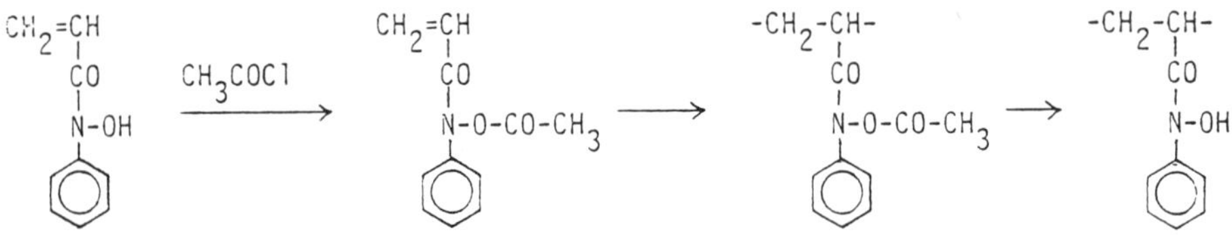

51 52 53

Scheme 15

Acetyl N-phenylacrylohydroxamate (51) [50]. *N*-Phenylacrylohydroxamic acid (50) (38 g, 0.23 mol) (prepared from *N*-phenylhydroxylamine and acryloyl chloride) and triethylamine (24 g, 0.24 mol) are dissolved in a mixture of ether (700 mL) and chloroform (100 mL). To this solution is added dropwise acetyl chloride (17.7 g, 0.23 mol) in ether with stirring at -10°C over 30 min, and stirring is continued for an additional 1 h at room temperature. The reaction mixture, after washing with saturated aqueous NaCl and with dilute Na_2CO_3, is dried over Na_2SO_4. The solvent is evaporated and the tan-brown residue is distilled in vacuo. The distillate is obtained from petroleum ether in a dry ice-methanol bath, as a pale yellow oil, yield 52 to 65%, bp 100 to 110°C (0.08 mm).

Poly(N-phenylacrylohydroxamate-co-acrylamide) (53) [50]. The copolymerization of acetyl *N*-phenylacrylohydroxamate (0.60 M) and acrylamide (3.40 M) is performed with azobisisobutyronitrile (0.02 M) in acetonitrile. After the polymerization for 7 min at 70°C, water is added to the mixture and the polymer is recovered by precipitation in methanol (18% conversion). The copolymer contains 11.0 mol % hydroxamate units.

The copolymer is dissolved in a small amount of water containing a large excess of hydroxylamine and is allowed to stand for 24 h. The product polymer is recovered by precipitation in excess methanol, washed with methanol and dried in vacuo. Complete deacylation is confirmed by the disappearance of an IR peak at 1800 cm^{-1} due to the acetyl hydroxamate group.

In order to increase the deacylation rate of the hydroxamate-containing polymer, a copolymer containing the hydroxamate group and imidazole group was prepared [54]. The terpolymer of *N*-phenylacrylohydroxamate, 2-methyl-*N*-vinylimidazole, and acrylamide (54) was also prepared. The hydrolysis of the acetyl hydroxamate intermediate was greatly accelerated by the assistance of the intrapolymeric imidazole group [52]. The deacylation rate was enhanced further by using the terpolymer of *N*-phenylacrylohydroxamate, 4(5)-vinylimidazole, and acrylamide [58] (55).

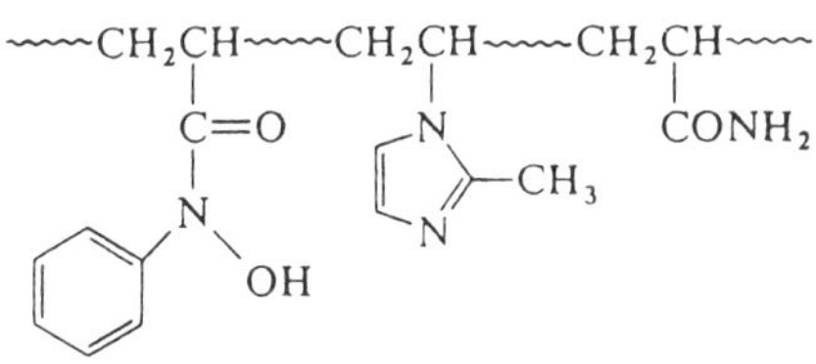

54

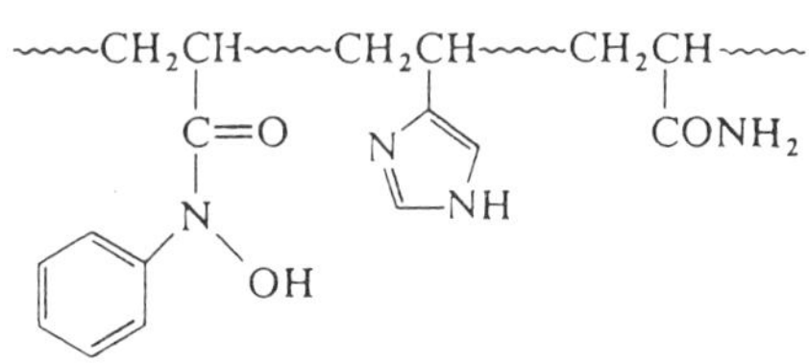

55

Poly[N-phenylacrylohydroxamate-co-4(5)-vinylimidazole-co-acrylamide] (55) [58]. Copolymerization of acetyl *N*-phenylacrylohydroxamate (51) (0.07 M), 4(5)-vinylimidazole (3) (0.40 M), and acrylamide (0.53 M) is carried out using azobisisibutyronitrile (0.25 mol % of total monomer), in benzene at 80°C for 90 min. The precipitate is filtered, washed with acetone and reprecipitated from water in acetone.

The copolymer is dissolved in water, treated with excess hydrazine for 24 h, and poured into acetone (90 to 97% recovery). The composition of the terpolymer is determined by the potentiometric titration and NMR spectroscopy; hydroxamate unit, 5 mol %; imidazole unit, 35 mol %; acrylamide, 60 mol %.

Poly(N-methylacrylohydroxamate)

The system above may be compared to the charge relay system of serine proteases in that the oxygen nucleophile and the imidazole group contribute to the catalysis. However, the concerted bifunctional action characteristic of the enzyme system was not observed. Concerted bifunctional catalysis was observed when a binary copolymer containing *N*-methylhydroxamate units and imidazole units (61) was used for the hydrolysis of phenyl esters [56]. A bifunctional polymer catalyst was prepared by free-radical polymerization of benzyl-*N*-methylacrylohydroxamate (59) and 4(5)-vinylimidazole (3), followed by removal of the benzyl group (Scheme 16).

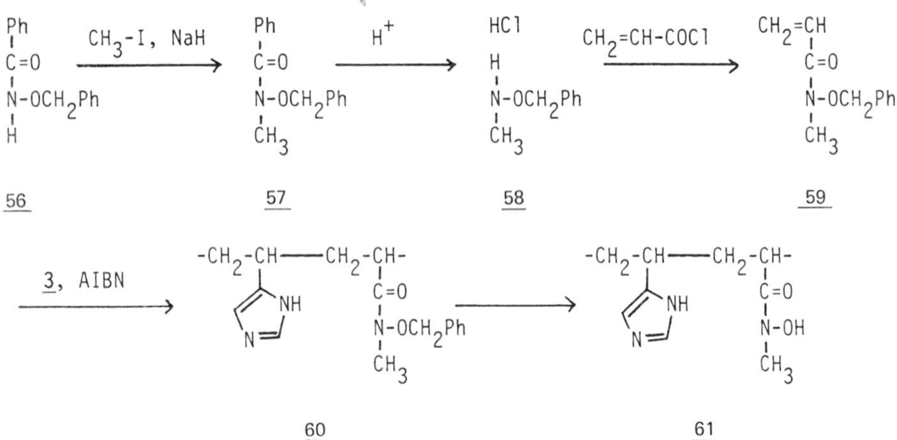

Scheme 16

Benzyl N-methylacrylohydroxamate (59) [56]. To NaH (7.20 g, 0.15 mol, 50% dispersed in oil) in dry tetrahydrofuran (50 mL), benzyl benzohydroxamate (56) (36.0 g, 0.16 mol) in dry tetrahydrofuran (200 mL) is added dropwise over a 30-min period with vigorous stirring at room temperature. After the evolution of hydrogen is almost complete (60 min), methyl iodide (28.4 g, 0.20 mol) in dry tetrahydrofuran (50 mL) is added over a 30-min period at room temperature. The reaction mixture is stirred for an additional 4 h at 60 to 70°C. After cooling to room temperature, a small amount of methanol is added to decompose unreacted NaH, and the mixture is filtered. The filtrate is evaporated in vacuo. The clear oily residue (57) (45 g) is dissolved in 300 mL of ethanol and heated to reflux for 25 min with concentrated hydrochloric acid (400 mL). The clear solution is diluted with water (500 mL), and while still warm, benzoic acid and other by-products are separated by extraction with benzene (200 mL × 3). The aqueous layer is evaporated to dryness under reduced pressure to afford crude *O*-benzyl-*N*-methylhydroxylamine hydrochloride (58), which is recrystallized from ethanol; yield 24 g (85%), mp 95°C.

O-Benzyl-*N*-methylhydroxylamine hydrochloride (21 g, 0.12 mol) and triethylamine (35.0 g, 0.33 mol) are dissolved in chloroform (300 mL). To this solution is added dropwise acryloyl chloride (12.7 g, 0.14 mol) in chloroform (30 mL) with stirring at -5°C over 40 min, and stirring is continued for an additional 1 h at room temperature. The reaction mixture is washed with water (100 mL) and with three 50-mL portions of a dilute Na_2CO_3 solution and dried over Na_2SO_4. The solvent is evaporated and the brown oily residue is distilled in vacuo; yield 60%, bp 105 to 108°C (0.15 mm).

Poly(N-methylacrylohydroxamate-co-4(5)-vinylimidazole) (61) [56]. The copolymerization of benzyl *N*-methylacrylohydroxamate (59) (0.37 M) and 4(5)-vinylimidazole (3) (1.50 M) is conducted in a degassed ampul in methanol with azobisisobutyronitrile (0.5 mol % of total monomer) for 190 min at 80°C. The polymerization mixture (60) is then poured into excess ether. The polymer is dissolved in 30% HBr in acetic acid, the solution stirred for 100 h at room temperature, and the solvent removed in vacuo. The residual (debenzylated) polymer is reprecipitated from methanol and ether, subjected to extensive dialysis in water, concentrated, and freeze dried (34% conversion).

The composition of the copolymer is determined by NMR and UV spectroscopy and potentiometric titration; methylhydroxamate unit, 25 mol %; imidazole unit, 75 mol %.

Hydroxamate-grafted polyvinylpyridine

Partial (ca. 10 mol %) quaternization of poly(4-vinylpyridine) and poly(1-vinyl-2-ethylimidazole) with benzyl *N*-benzylchloroacetohydroxamate and subsequent removal of the benzyl group produces water-soluble polymers containing zwitterionic hydroxamate groups and free pyridine or imidazole groups (62 and 63). The catalytic efficiency of these bifunctional polymers is much better than monofunctional polymers which contain either hydroxamate or imidazole [61].

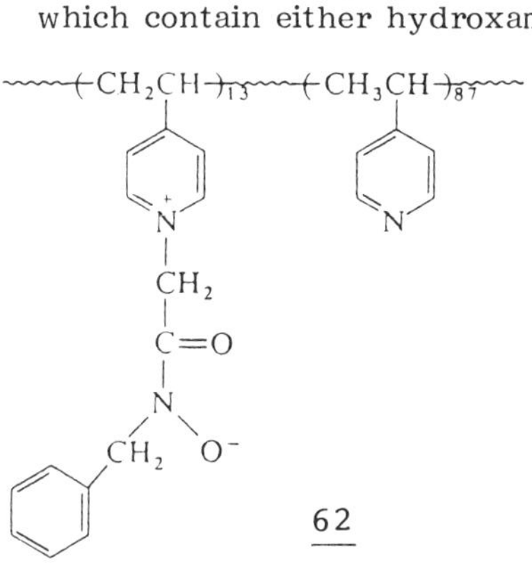

62

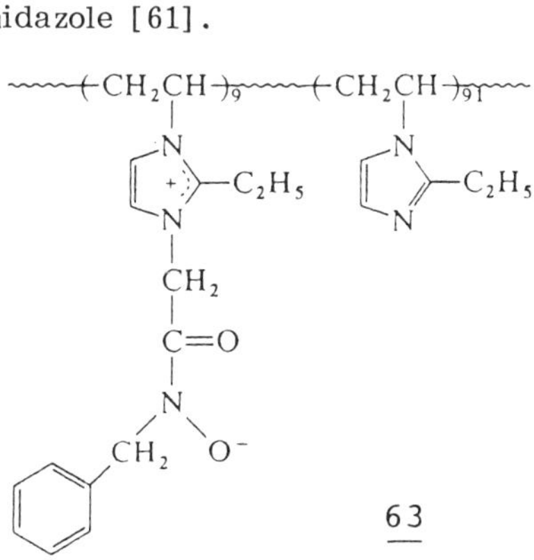

63

As well as hydroxamate, oxime-bound poly(4-vinylpyridine) was prepared by quaternization (64) [83].

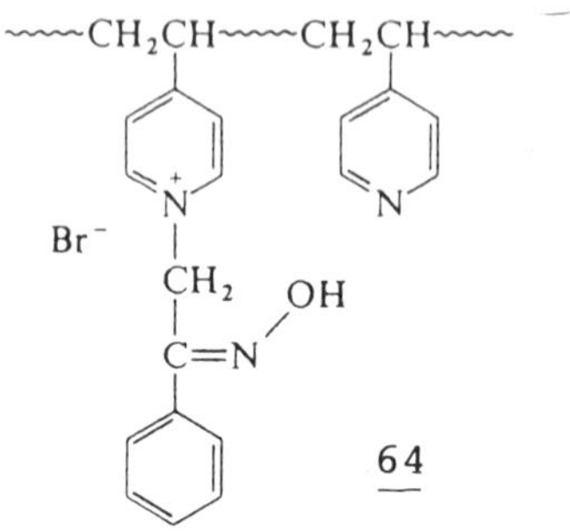

64

Hydroxamate-grafted polyvinylpyridine (62) [61]. To a solution of *N,O*-dibenzylhydroxylamine hydrochloride [82] (25 g, 0.1 mol) and triethylamine (22 g, 0.22 mol) in chloroform (200 mL), chloroacetyl chloride (11.3 g, 0.1 mol) in benzene (30 mL) is added dropwise at 0°C with stirring over 30 min, and for 1 h at room temperature. The reaction mixture is washed with dilute alkali and water. Removal of solvent and distillation at reduced pressure give a pale yellow oil in 68% yield; bp 150 to 155°C (0.03 mm).

Quaternization is conducted in dimethylformamide for 50 to 100 h at room temperature, and the product is recovered by precipitation in ether. *O*-Benzyl group is removed by 33% HBr in acetic acid for 40 h at room temperature. Solvent is removed, dialyzed, and freeze dried.

10.3.3 Polyethyleneimine Derivatives

Polyethyleneimine and its modified polymers were applied to the solvolysis of phenyl esters [84-87]. The reaction involves acyl transfer from the phenyl esters to NH groups in the polymer. The acylated amine group is resistant to hydrolytic cleavage, and therefore the polymers act as acyl-transfer agents but not as true catalysts.

Polyethyleneimine is reacted with lauryl iodide, and the remaining primary amines are reacted with chloromethylimidazole (Scheme 17). This polymer (66) possesses better catalytic activity for the hydrolysis of *p*-nitrophenyl ester and other reactions [88-96].

Polyethyleneimine-containing benzyl groups are also active catalysts for the hydrolysis of *p*-nitrophenyl esters. This polymer catalyst was prepared by grafting of benzyl chloride on polyethyleneimine [97]. A copolymer of L-histidine grafted onto polyethyleneimine by amide bonds is an efficient catalyst for the hydrolysis of phenyl esters in

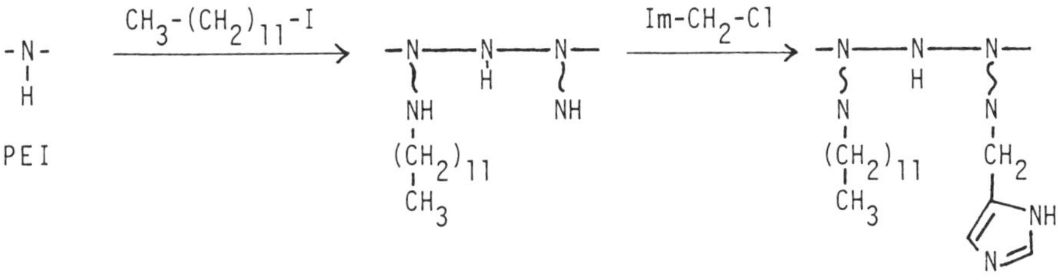

65 66

Scheme 17

aqueous solution [36]. However, stereoselective hydrolysis of enantiomeric esters is not observed.

Commercial polyethyleneimine, which is prepared by ring-opening polymerization of ethyleneimine, is a highly branched, relatively compact, water-soluble polymer with approximately 25% primary, 50% secondary, and 25% tertiary amine groups. Linear polyethyleneimine is prepared by polymerization of 2-oxazoline followed by hydrolysis [98].

Lauryl-polyethyleneimine (65) [93]

Lauryl iodide is added to polyethyleneimine in ethanol. The solution is stirred for 24 h at 50°C. The modified polymer is isolated by passage of the reaction mixture through a mixed bed resin (AG 501 × 8, Bio-Rad Co.) with 80% ethanol. The polymer containing solution is dialyzed against 40% ethanol-water and then water. The aqueous solution is lyophilized.

Lauryl-imidazole-polyethyleneimine (66) [93]

Imidazolemethylenechloride monohydrochloride is added to a solution of lauryl-polyethyleneimine (65) in ethanol. Potassium iodide is used as a catalyst. The solution is stirred for 24 h at 50°C. The polymer is purified by the same procedure as that described for lauryl-polyethyleneimine (65). Contents of the substituents are determined by NMR spectroscopy.

REFERENCES

1. C. G. Overberger, H. Biletch, and R. G. Nickerson, J. Polym. Sci., 27, 381 (1958).
2. C. G. Overberger and N. Vorchheimer, J. Am. Chem. Soc., 85, 951 (1963).

3. C. G. Overberger, T. St. Pierre, N. Vorchheimer, and S. Yaroslavsky, J. Am. Chem. Soc., 85, 3513 (1963).
4. C. G. Overberger, T. St. Pierre, N. Vorchheimer, J. Lee, and S. Yaroslavsky, J. Am. Chem. Soc., 87, 296 (1965).
5. C. G. Overberger, R. Sitaramaiah, T. St. Pierre, and S. Yaroslavsky, J. Am. Chem. Soc., 87, 3270 (1965).
6. C. G. Overberger, T. St. Pierre, and S. Yaroslavsky, J. Am. Chem. Soc., 87, 4310 (1965).
7. C. G. Overberger, T. St. Pierre, C. Yaroslavsky, and S. Yaroslavsky, J. Am. Chem. Soc., 88, 1184 (1966).
8. C. G. Overberger, J. C. Salamone, and S. Yaroslavsky, Pure Appl. Chem., 15, 453 (1967).
9. C. G. Overberger, J. C. Salamone, and S. Yaroslavsky, J. Am. Chem. Soc., 89, 6231 (1967).
10. H. Morawetz, C. G. Overberger, J. C. Salamone, and S. Yaroslavsky, J. Am. Chem. Soc., 90, 651 (1968).
11. C. G. Overberger and I. Cho, J. Polym. Sci., A1(6), 2741 (1968).
12. C. G. Overberger, R. Corett, J. C. Salamone, and S. Yaroslavsky, Macromolecules, 1, 331 (1968).
13. C. G. Overberger and J. C. Salamone, Acc. Chem. Res., 2, 217 (1969).
14. C. G. Overberger, M. Morimoto, I. Cho, and J. C. Salamone, Macromolecules, 2, 553 (1969).
15. C. G. Overberger and P. S. Yuen, J. Am. Chem. Soc., 92, 1667 (1970).
16. C. G. Overberger and H. Maki, Macromolecules, 3, 214 (1970).
17. C. G. Overberger and H. Maki, Macromolecules, 3, 220 (1970).
18. C. G. Overberger and M. Morimoto, J. Am. Chem. Soc., 93, 3222 (1971).
19. C. G. Overberger, M. Morimoto, I. Cho, and J. C. Salamone, J. Am. Chem. Soc., 93, 3228 (1971).
20. C. G. Overberger and C. M. Shen, J. Am. Chem. Soc., 93, 6992 (1971).
21. C. G. Overberger and C. M. Shen, Bioorg. Chem., 1, 1 (1971).
22. C. G. Overberger and Y. Okamoto, Macromolecules, 5, 363 (1972).
23. Y. Okamoto and C. G. Overberger, J. Polym. Sci. Polym. Chem. Ed., 10, 3387 (1972).
24. C. G. Overberger, R. C. Glowaky, and P. H. Vandewyer, J. Am. Chem. Soc., 95, 6008 (1973).
25. C. G. Overberger and R. C. Glowaky, J. Am. Chem. Soc., 95, 6014 (1973).
26. C. G. Overberger and C. J. Podsiadly, Bioorg. Chem., 3, 35 (1973).
27. C. G. Overberger and T. J. Pacansky, J. Polym. Sci. Polym. Symp., 45, 39 (1973).

28. C. G. Overberger and T. W. Smith, Reactions on Polymers (J. A. Moor, Ed.), D. Reidel Publishing Co., Dordrecht, The Netherlands (1973).
29. C. G. Overberger, and T. J. Pacansky, J. Polym. Sci. Polym. Symp., 45, 39 (1974).
30. C. G. Overberger, T. J. Pacansky, J. Lee, T. St. Pierre, and S. Yaroslavsky, J. Polym. Sci. Polym. Symp., 46, 209 (1974).
31. C. G. Overberger and C. J. Podsiadly, Bioorg. Chem., 3, 16 (1974).
32. C. G. Overberger and C. J. Podsiadly, Bioorg. Chem., 3, 35 (1974).
33. C. G. Overberger and T. W. Smith, Macromolecules, 8, 401 (1975).
34. C. G. Overberger and T. W. Smith, Macromolecules, 8, 407 (1975).
35. C. G. Overberger and T. W. Smith, Macromolecules, 8, 416 (1975).
36. C. G. Overberger and K. W. Dixon, J. Polym. Sci. Polym. Chem. Ed., 15, 1863 (1977).
37. C. G. Overberger, R. C. Glowaky, T. J. Pacansky, and K. N. Sannes, Macromol. Synth., Col. Vol. 1 (J. A. Moore, Ed.), 619 (1977).
38. C. G. Overberger and Y. Kawakami, J. Polym. Sci. Polym. Chem. Ed., 16, 1237 (1978).
39. C. G. Overberger and Y. Kawakami, J. Polym. Sci. Polym. Chem. Ed., 16, 1249 (1978).
40. C. Aso, T. Kunitake, and S. Shinkai, Chem. Commun., 1483 (1968).
41. C. Aso, T. Kunitake, and F. Shimada, J. Polym. Sci., B6, 467 (1968).
42. T. Kunitake, F. Shimada, and C. Aso, J. Am. Chem. Soc., 91, 2716 (1969).
43. T. Kunitake, F. Shimada, and C. Aso, Makromol. Chem., 126, 276 (1969).
44. T. Kunitake, S. Shinkai, and C. Aso, Bull. Chem. Soc. Jpn. 43, 1109 (1970).
45. T. Kunitake, S. Shinkai, and C. Aso, Bull. Chem. Soc. Jpn., 43, 2581 (1970).
46. T. Kunitake and S. Shinkai, J. Am. Chem. Soc., 93, 4247 (1971).
47. T. Kunitake and S. Shinkai, J. Am. Chem. Soc., 93, 4256 (1971).
48. T. Kunitake and S. Shinkai, Makromol. Chem., 151, 127 (1972).
49. S. Shinkai and T. Kunitake, Polym. J., 4, 253 (1973).
50. T. Kunitake, Y. Okahata, and R. Ando, Bull. Chem. Soc. Jpn., 47, 1509 (1974).
51. T. Kunitake and Y. Okahata, Chem. Lett., 1057 (1974).
52. T. Kunitake, Y. Okahata, and R. Ando, Macromolecules, 7, 140 (1974).
53. S. Shinkai and T. Kunitake, J. Chem. Soc. Perkin, Trans. 2, 980 (1974).

54. T. Kunitake and Y. Okahata, Bioorg. Chem., 4, 136 (1975).
55. T. Kunitake, Y. Okahata, and T. Sakamoto, Chem. Lett., 459 (1975).
56. T. Kunitake and Y. Okahata, J. Am. Chem. Soc., 98, 7793 (1976).
57. T. Kunitake and Y. Okahata, Macromolecules, 9, 15 (1976).
58. T. Kunitake, S. Shinkai, and Y. Okahata, Bull. Chem. Soc. Jpn., 49, 540 (1976).
59. T. Kunitake, Y. Okahata, R. Ando, and S. Hirotsu, Bull. Chem. Soc. Jpn., 49, 2547 (1976).
60. T. Kunitake and Y. Okahata, Adv. Polym. Sci., 20, 159 (1976).
61. Y. Okahata and T. Kunitake, J. Polym. Sci. Polym. Chem. Ed., 15, 2571 (1977).
62. Y. Okahata and T. Kunitake, J. Polymer Sci. Polymer Chem. Ed., 16, 1865 (1978).
63. T. Kunitake, in Polymer-Supported Reactions in Organic Synthesis (P. Hodge and D. C. Sherrington, Eds.), John Wiley & Sons, Inc., New York (1980), p. 195.
64. R. C. Sovish, J. Org. Chem., 24, 1345 (1959).
65. T. Shimizu, A. Furuta, and Y. Naka, Macromolecules, 7, 160 (1974).
66. W. F. Whitmore and E. Landan, J. Am. Chem. Soc., 68, 1797 (1946).
67. T. Shimizu, A. Furuta, T. Watanabe, and S. Kato, Makromol. Chem., 175, 119 (1974).
68. R. L. Letsinger and I. S. Klaus, J. Am. Chem. Soc., 86, 3884 (1964).
69. R. L. Letsinger and I. S. Klaus, J. Am. Chem. Soc., 87, 3380 (1965).
70. K. D. Kopple, Biopolymer, 6, 1417 (1968).
71. J. C. Salamone, S. C. Israel, P. Taylar, and B. Snider, J. Polym. Sci. Polym. Symp., 45, 65 (1974).
72. J. C. Salamone, S. C. Israel, P. Taylar, and B. Snider, Polymer, 14, 639 (1973).
73. G. Manecke and U. Rotter, Makromol. Chem., 127, 264 (1969).
74. H. Morawetz and W. R. Song, J. Am. Chem. Soc., 88, 5714 (1966).
75. Y. Imanishi, Y. Amimoto, T. Sugihara, and T. Higashimura, Makromol. Chem., 177, 1401 (1976).
76. Y. Okamoto, Nippon Kagaku Kaishi, 870 (1978).
77. R. L. Letsinger and T. J. Savereide, J. Am. Chem. Soc., 84, 114 (1962).
78. R. Letsinger and T. J. Savereide, J. Am. Chem. Soc., 114, 3122 (1962).
79. S. G. Starodubtzev, Y. E. Kirch, and V. A. Kabanov, Eur. Polym. J., 10, 739 (1974).
80. W. B. Gruhn and M. L. Bender, J. Am. Chem. Soc., 91, 5883 (1969).

81. R. Hershfield and M. L. Bender, J. Am. Chem. Soc., 94, 1376 (1972).
82. R. Behrend and K. Leuchs, Ann. Chem., 257, 203 (1890).
83. Y. E. Kirsh, A. A. Rahnanskaya, G. M. Lukovkin, and V. A. Kabanov, Eur. Polym. J., 10, 393 (1974).
84. I. M. Klotz and V. H. Stryker, J. Am. Chem. Soc., 90 2717 (1968).
85. I. M. Klotz, G. V. Roger, and A. R. Sloniewsky, Biochemistry, 8, 4752 (1969).
86. G. P. Royer and I. M. Klotz, J. Am. Chem. Soc., 91, 5885 (1969).
87. Y. Birk and I. M. Klotz, Bioorg. Chem., 1, 275 (1971).
88. I. M. Klotz, G. P. Roger, and I. S. Scarpa, Proc. Nat. Acad. Sci. USA, 68, 263 (1971).
89. H. C. Kiefer, W. I. Congdon, I. S. Scarpa, and I. M. Klotz, Proc. Nat. Acad. Sci., USA, 69, 2155 (1972).
90. T. W. Johnson and I. M. Klotz, Macromolecules, 6, 788 (1973).
91. J. Suh and I. M. Klotz, Bioorg. Chem., 6, 165 (1977).
92. W. J. Spetnagel and I. M. Klotz, J. Polym. Sci. Polym. Chem. Ed., 15, 621 (1977).
93. M. Nango and I. M. Klotz, J. Polym. Sci. Polym. Chem. Ed., 16, 1265 (1978).
94. J. Suh and I. M. Klotz, J. Polym. Sci. Polym. Chem. Ed., 16, 1943 (1978).
95. J. Suh, I. S. Scarpa, and I. M. Klotz, J. Am. Chem. Soc., 98, 7060 (1976).
96. W. J. Spetnagel and I. M. Klotz, J. Am. Chem. Soc., 98, 8199 (1976).
97. V. S. Pshezhetskii, G. A. Murtazaeva, and V. A. Kabanov, Eur. Polym. J., 10, 571 (1974).
98. T. Saegusa, H. Ikeda, and H. Fujii, Macromolecules, 5, 108 (1972).

Index